Farm Management

Farm Management

Edited by **Carlos Hassey**

SYRAWOOD
PUBLISHING HOUSE
New York

Published by Syrawood Publishing House,
750 Third Avenue, 9th Floor,
New York, NY 10017, USA
www.syrawoodpublishinghouse.com

Farm Management
Edited by Carlos Hassey

International Standard Book Number: 978-1-68286-368-8 (Hardback)

Printed in China.

Contents

	Preface	**IX**
Chapter 1	**Economics of farm safety: The Nigerian scenario** Olowogbon S. T., Fakayode S. B., Jolaiya A. J. and Adenrele A. Y.	**1**
Chapter 2	**Producer characteristics and determinants of technical efficiency of tomato based production systems in Ghana** B. O. Asante, M. K. Osei, A. A. Dankyi, J. N. Berchie, M. B. Mochiah, J. N. L. Lamptey, J. Haleegoah, K. Osei and G. Bolfrey-Arku	**6**
Chapter 3	**Economic welfare effects of the increase in labor costs on U.S. tobacco farms** Shiferaw T. Feleke, Jane H. Starnes, Harwood D. Schaffer and Kelly J. Tiller	**18**
Chapter 4	**Factors affecting household access to enough food in the Eastern Cape Province of South Africa** L. Musemwa, L. Zhou, S. Ndhleve and F. Aghdasi	**26**
Chapter 5	**Estimating technical efficiency of small scale beef cattle fattening in the lake zone in Tanzania** S. N. Mlote, N. S. Y. Mdoe, A. C. Isinika and L. A. Mtenga	**34**
Chapter 6	**The dilemma of biodiversity and specialization in Agricultural development** Price Edwin, Kibriya Shahriar and Zhang Yu	**45**
Chapter 7	**Farm household level impacts of information communication technology (ICT)-based agricultural market information in Ghana** Ramatu M. Al-Hassan, Irene S. Egyir and James Abakah	**51**
Chapter 8	**Production efficiency and risks in limited resource farming: The case of Bulgarian peanut industry** Carel Ligeon, Curtis Jolly, Nelly Bencheva, Stanko Delikostadinov and Naveen Puppala	**58**
Chapter 9	**Economic analysis of selected food crops in Ile-Ife of Osun State, Nigeria** Adisa O. D. and Sofoluwe N. A.	**69**
Chapter 10	**Effects of international fund for agricultural development (IFAD) credit supply on rural farmers in Rivers State, Nigeria** Tasie C. M.	**76**

Chapter 11 **Determinants of income diversification among rural households: The case of smallholder farmers in Fedis district, Eastern hararghe zone, Ethiopia** 82
Amare Demissie and Belaineh Legesse

Chapter 12 **On-farm evaluation of maize varieties in the transitional and savannah zones of Ghana: Determinants of farmer preferences** 91
Prince Maxwell Etwire, Tahirou Abdoulaye, Kwadwo Obeng-Antwi, Samuel S. J. Buah, Roger A. L. Kanton, Henry Asumadu, Mashark S. Abdulai, Alidu Haruna and John C. Etwire

Chapter 13 **The performance of potato markets in South Western Uganda** 99
J. Bonabana-Wabbi, S. Ayo, B. Mugonola, D. B. Taylor, J. Kirinya and M. Tenywa

Chapter 14 **Regional differences of technology modernization in agriculture of Pará State** 110
Mario Miguel Amin Garcia Herreros, Heriberto Wagner Amanajás Pena, Allan Klynger da Silva Lobato and Raimundo Thiago Lima da Silva

Chapter 15 **Analysis of economic capacity utilization in the Nigerian sugar industry (1970 - 2010)** 120
Sunday B. Akpan, Inimfon V. Patrick, Daniel E. John and Samuel J. Udoka

Chapter 16 **Pomegranate marketing margins in the Neyriz, Iran** 133
Mahdi Babaei, Elahe Asemani, Saeid Gazmeh, Soheyla Rezazadeh Shamkhal and Ebrahim Shirmohammadi

Chapter 17 **Effects of planting depth on agronomic performance of two potato varieties grown in the Sahel region of Tunisia** 137
Sayed Chehaibi, Wissem Hamdi and Khaoula Abrougui

Chapter 18 **The implications of refuge requirements for Bt cotton in India on world cotton markets** 142
Rohit Singla, Phillip Johnson and Sukant Misra

Chapter 19 **Participation in land market and technical efficiency in Southern Ethiopia: A case study after 2005 land proclamation of Ethiopia** 154
Tewodros Tefera and D. V. Subaro

Chapter 20 **Viability of tobacco production under smallholder farming sector in Mount Darwin District, Zimbabwe** 164
James Masvongo, Jackqeline Mutambara and Augustine Zvinavashe

Chapter 21 **The role of small scale irrigation in poverty reduction** 171
Adugna Eneyew, Ermias Alemu, Mekonnen Ayana and Mihret Dananto

Chapter 22 **Eradicating extreme poverty among the rural poor in Uganda through poultry and cattle improvement programmes - A Review** 181
J. Ekou

Chapter 23 **Economic assessment of yam production in Kabba-Bunu Local Government Area of Kogi State, Nigeria** 187
Ibitoye, Stephen Jimoh and Onimisi J. A.

Chapter 24 **Understanding the performance of food production in sub-Saharan Africa and its implications for food security** 193
M. Demeke, F. Di Marcantonio and C. Morales-Opazo

Chapter 25 **Tenure security and soil conservation investment decisions: Empirical evidence from East Gojam, Ethiopia** 212
Akalu Teshome

Chapter 26 **Oil pollution and technical efficiency of food crop farmers in the Niger Delta region of Nigeria** 223
F. O. Idumah and F. Y. Okunmadewa

Chapter 27 **Determinants of smallholder sweet potato farmers' participation in different market options: The case of Vihiga County, Kenya** 231
Benjamin K. Mutai, Elvi N. Agunda, Augustus S. Muluvi, Lawrence K. Kibet and Mary C. Maina

Chapter 28 **Impact of the adoption of semi-mechanized technologies of shea processing on rural women's income in Northern Benin (West Africa)** 238
Jacob Afouda Yabi, Patrice Igue Adegbola, Dansi Silvere Tovignan, Morenike Cendrine Ahouandjinou and Adeyemi Suleiman Adekambi

Chapter 29 **Synergies between urban agriculture and urban household food security in Gweru City, Zimbabwe** 246
Admire Jongwe

Chapter 30 **Consumer demand system of agri-food in Tanzania** 254
Joel Chongela, Vasudev Nandala and Suhasini Korabandi

Chapter 31 **Analysis of determinants of farm-level adaptation measures to climate change in Southern Africa** 261
Charles Nhemachena, Rashid Hassan and James Chakwizira

Permissions

List of Contributors

Preface

Every book is a source of knowledge and this one is no exception. The idea that led to the conceptualization of this book was the fact that the world is advancing rapidly; which makes it crucial to document the progress in every field. I am aware that a lot of data is already available, yet, there is a lot more to learn. Hence, I accepted the responsibility of editing this book and contributing my knowledge to the community.

Farm management is concerned with the management of farm operations. It involves both maximization of crop yield as well as financial management of farm resources. The topics included in this book such as agricultural engineering, crop harvesting, commercial management and large scale operations, erosion control systems, risk management, and management of work force, are of utmost significance and bound to provide incredible insights to readers. It picks up individual concepts and explains their need and contribution in the context of a growing economy. This book, with its detailed analyses and data, will prove immensely beneficial to professionals and students involved in this area at various levels.

While editing this book, I had multiple visions for it. Then I finally narrowed down to make every chapter a sole standing text explaining a particular topic, so that they can be used independently. However, the umbrella subject sinews them into a common theme. This makes the book a unique platform of knowledge.

I would like to give the major credit of this book to the experts from every corner of the world, who took the time to share their expertise with us. Also, I owe the completion of this book to the never-ending support of my family, who supported me throughout the project.

Editor

1

Economics of farm safety: The Nigerian scenario

Olowogbon S. T.[1]*, Fakayode S. B.[2], Jolaiya A. J.[2] and Adenrele A. Y.[1]

[1]Occupational Health and Safety Division of Vertext Media Ltd., Ibadan, Oyo State, Nigeria.
[2]Department of Agricultural Economics, University of Ilorin, Ilorin, Kwara State, Nigeria.

Investing in farm safety is a win-win game. Apart from improving the financial bottom line of a farm business, it increases and improves the labour productivity of farm workers, which in turn improve agricultural sector. Consequently, agricultural accidents place a great burden on the economy resulting in reduced return on investment (ROI) in agriculture and the burden of injuries/illness (BOI) is increased. This impedes efficiency of the workers, decreased agricultural output and profitability. The dearth of information on the economic relationship of farm safety and ROI, especially in agriculture with focus on a developing country like Nigeria stimulated this research by assessing the economic implication of farm safety practices on agriculture in Nigeria. 100 respondents with agriculture as primary occupation were randomly selected from farming households in Ibadan, Nigeria. Health and work performance questionnaire (HPQ), work limitation questionnaire and interview schedule were adopted for data collection. The research revealed that about 80 and 75% of the respondents had severe wrist/hand injury and lower back pain, respectively which are work related. 63% of the respondents on average have been prevented from doing their work as a result of occupational injuries and illness which leads to lost time, restricted work cases and fatalities leading to production loss. Also, 80% of the respondents do not use any form of personal protective equipment (PPE). Inadequate information and awareness coupled with cost implication were ranked highest on the farm safety adoption constraint list. Thus, information dissemination on farm safety, a subsidized well implemented farm safety programme and farm accident record keeping are recommended.

Key words: Nigeria, farm safety, economics.

INTRODUCTION

The realization of the African Green Revolution and its contribution to food security and economic growth in Sub-Saharan Africa is threatened by many factors (Ngigi, 2009). The direct effect of these factors on agricultural production and food security will be exacerbated by greater exposure to occupational diseases and illnesses that reduce labour productivity. Sub-Saharan Africa countries have more than 54,000 fatal occupational accidents annually; approximately, 42 million work-related accidents took place that caused at least three days' absence from work; the fatality rate of the region is 21 per 100,000 workers; the accident rate is 16,000 per 100,000 workers; and the fatal accident rate in agriculture is 22.5 per 100,000 (Hämäläinen, 2005). On a global scale, agricultural accidents place a great burden on the economy, resulting in reduced return on investment (ROI) in agriculture; moreover, the burden of injuries/illness (BOI) is on the increase. This impedes workers' efficiency, decreases agricultural output and weakens productivity.

Farm safety is therefore a focal issue for improved agricultural productivity; with the expansion of agricultural technology, there is a growing health concern that agricultural workers will face, in addition to traditional health risks with the new occupational health and safety hazards. Even in countries where the primary health care is well developed, occupational health care in agriculture is often non-existence. Nigerian farmers have little or no knowledge of health and safety as it relates to agriculture. Agriculture as practiced today in some countries is crude

*Corresponding author. E-mail: olowogbonsam@yahoo.com.

and unsafe, exposing millions of workers to hazards with many sustaining injury and probably death (Mostafa, 2003). The issue of health and safety in agriculture is inconsequential in a developing nation like Nigeria or if at all considered, it is still at a very low ebb or non-existence despite the fact that agriculture is the mainstay of her economy. Agriculture was the most important sector of the economy before the decline in the 1970s due to oil boom. It created jobs and generated employment, and engaged about 70% of the labor force (Kwanasahi et al., 1998). A greater fraction of the country's population of over 167 million depends on agriculture for their livelihood today either directly or indirectly. Unfortunately, this sector is deep seated culture of unwise risk taking and lack appreciation of the role of good health and safety management (Chris, 2008). Farm workers are exposed to hazard ranging from chemical, physical, ergonomics, biological and psychosocial hazards thereby leading to disorders such as musculoskeletal disorder and diseases like carcinogenicity (cancer causing), mutagenicity (induce mutations), tetragenicity (affect the fetus), psychiatric disorder and delayed neuropathy, and the dusts have been known to cause diseases ranging from byssinosis, occupational asthma, pneumonitis and non-specific chronic obstructive pulmonary disease (COPD) (Mostafa, 2003). After construction, agricultural sector has the worst record for work related fatalities and one of the worst for occupational ill health and injuries (Chris, 2008).

A healthy workforce is definitely one of the most important economic assets to a nation. An unhealthy workforce on the other hand is an economic loss to a nation which in turn affects productivity (Karen et al., 2005). At the individual level, good health can directly increase general output (e.g. through reduced absenteeism, injury and illnesses as well as improved morale and general work-well being. Consequently, economists and historians have focused much attention on better understanding of productivity determinants. There is an increasing awareness that human capital is a key factor. Recently, however, increasing attention has been given to health as a form of human capital (Emile, 2002). Poor health state erodes human capital and reduces resilience to shock and reduces productivity as well as impaired physical and mental capacity (Soekirman, 2010).

A need for farm safety programme

According to International Labour Organization (ILO) report in 2000, as cited by Oluwagbemi (2007), revealed that 170,000 (50.7%) agricultural workers are killed each year from the total of 335,000 in fatal workplace accidents worldwide. Therefore, special attention is needed for the health and safety of agricultural workers, who constitute three quarters of the Nigerian working population. Health and safety has been identified to have a strong influence and invaluable to any sector of the economy. Effective management of health and safety can help to deliver improved productivity and efficiency. It has been tied to positive performance indicators (PPIs) (Smallman and John, 2001). "Agriculture" covers not only farming, but also many other associated activities such as crop processing and packaging, irrigation, pest management, grain storage, animal husbandry, construction and agricultural work also refer to as domestic tasks (carrying water or fuel-wood, etc). The distinguishing characteristics of agricultural work is the fact that it is carried out in a rural environment where there is no clear-cut distinction between working and living conditions. As agricultural work is carried out in the countryside, it is subject to the health hazards of a rural environment as well as those inherent in the specific work processes involved (ILO, 1996). Due to the nature of agricultural work, workers in this sector are prone to many hazards; hence, there is a need for farm safety programme for workers. Recent innovation in agriculture has also added drastically to the dangers or hazards faced by farm workers. Effective health and safety programme should be in place to mitigate the effect of such hazards.

Nigeria has little or no data on workers health and safety record in agriculture probably because the workers are peasant and rural dwellers coupled with the fact that they are insensitivity to the role safety can play in the development and productivity of the sector. The dearth of information on the economic relationship of farm safety and ROI especially in agriculture with focus on a developing country like Nigeria stimulated this research. This research intends to bridge this gap by assessing the economic implication of farm safety practices on agriculture in Nigeria.

MATERIALS AND METHODS

This study was carried out in Ibadan metropolis in Oyo State, Nigeria. The state has thirty-three local government areas. Its population is 1,338,659 according to census results for 2006, with coordinates 7°23′47″N, 3°55′0″E and Area 1,189.2 (3,080 km^2). Ibadan has a tropical wet and dry climate, with a lengthy wet season and relatively constant temperatures throughout the course of the year. Ibadan wet season runs from March through October, though August sees to an extent a break in precipitation. This lull nearly divides the wet season into two different wet seasons. The remaining months form the city's dry season. Like a good portion of West Africa, Ibadan experiences harmattan between the months of November and February. Agriculture is the main occupation of the people of Oyo State. The climate in the state favours the cultivation of crops like maize, yam, cassava, millet, rice, plantains, cocoa, palm produce, cashew, etc (Ibadan Wikipedia, 2011). Data were obtained using a health and work performance questionnaire coupled with structured interview schedule, administered to 100 respondents with agriculture as the primary occupation. These respondents were randomly selected from the faming population in the area; this was supplemented with secondary data from reports, publications and internet. Simple statistics such as mean, mode, and Likert scaling were used for the data analysis. BOI are the cost

Table 1. Result.

Demographic characteristics	Frequency	%
Age (years)		
20-30	30	30
31-40	40	40
>40	30	30
Sex		
Males	80	80
Females	20	20
Marital status		
Single	20	20
Married	80	80
Religion		
Christianity	40	40
Islam	40	40
Traditional	20	20
Educational background		
No formal education	30	30
Primary	45	45
Secondary	15	15
Tertiary	10	10
Hazard awareness		
Yes	40	40
No	60	60
Personal protective equipment (PPE) usage		
Yes	20	20
No	80	80

of occupational injuries and ROI is the returns on investment.

RESULTS AND DISCUSSION

Table 1 reveals that 80% of the respondents were male, signifying a male dominated occupation, with a modal age group of 31 to 40 which denote age range of vigour and strength. 10% of the respondents had tertiary education, and 45% had primary education as the highest educational level attained, which implies that they have little or no formal education which has be proven to have a strong tie to adoption of innovation and modern practices. This study revealed that 60% of the respondents are not aware of any inherent hazard in agriculture, this connote the fact that health and safety in agriculture as it relates to occupational health and safety is at a lower ebb in the country; hence, there is a need for awareness campaign. Also, 80% of the respondents do not use any form of personal protective equipment which still reflect the poor awareness level.

Health and work performance questionnaire (HPQ) questions response

Statistics of respondents that have experienced one form of discomfort in the past 12 months related to their occupation

Figure 1 reveals that 80% of the respondents had wrist/hand injury, 75% had lower back pain and 55% knee injury in the last 12 months resulting from their job which shows that occupational injury in agriculture is prominent, which is probably due to the poor awareness level of stakeholders in the sector in identifying the roles that health and safety can play in maximizing productivity.

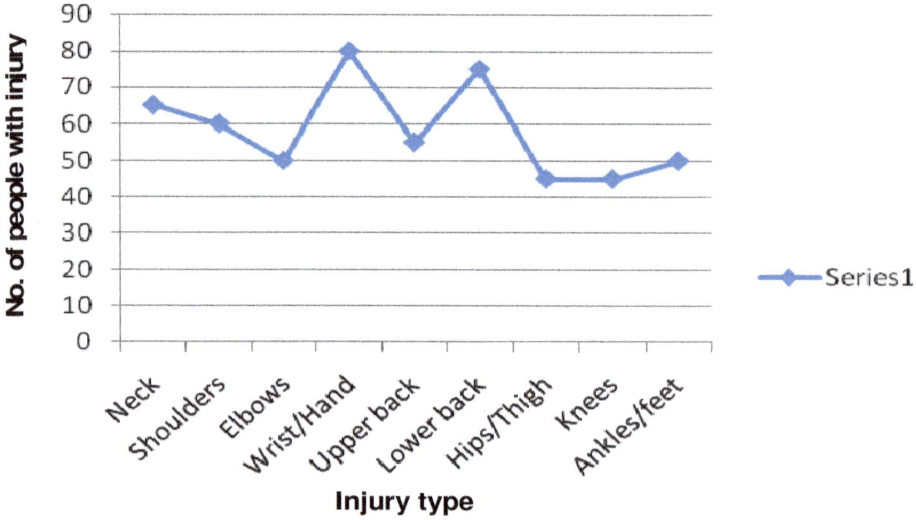

Figure 1. The distribution chart of injury and type

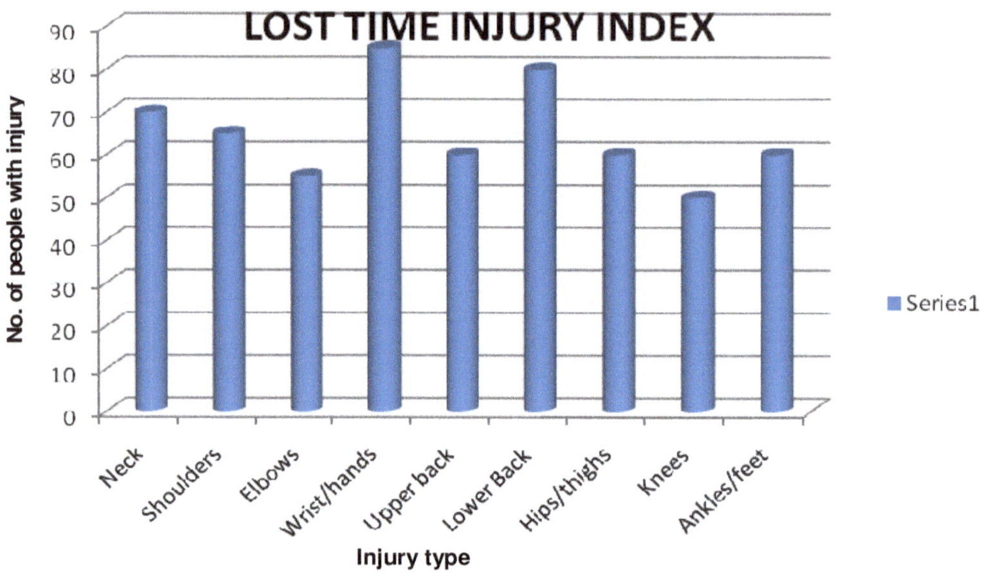

Figure 2. Lost time injury index chart

Statistics of respondents that have at any time during the last 12 months been prevented from doing their work because of discomfort in their body part

Figure 2 reveals that 80% of the respondents have been prevented from doing their job as a result of wrist/hand injury and 75% have also been prevented as a result of lower back injury. These occupational injuries or work related injuries have resulted in lost time or lost production and restricted work cases which has reduced productivity. The study also revealed that poor information and awareness (27.6%) and cost implication (26.9%) rank first and second in the adoption constraints list for farm safety programme, respectively. This fact could be linked to poverty and unavailability of farm accidents record and other information on farm safety in Nigeria.

RECOMMENDATIONS

Based on the findings of this study, the following are recommended:

(1) Formulation and implementation of farm safety policies in Nigeria.

(2) Adequate information dissemination on farm safety.
(3) Workers in agriculture should be educated/trained on agricultural hazards and means of prevention.
(4) Massive poverty reduction programme, since it is a bottleneck to adoption.
(5) Awareness on usage and the provision of personal protective equipment (PPE) for workers ranging from hang cloves, coveralls, footwear, ear muffs at a subsidized rate.
(6) Extension agents, cooperative societies, and trade unions should be equipped with accurate information so that they can enlighten farmers on the hazards.
(7) Proper record keeping on agricultural health and safety should be encouraged for performance review.

Conclusively, the importance of farm safety to the development of the sector cannot be overemphasized. Human capital such as health has been described as a determinant for productivity, hence the wealth, financial and holistic growth of the sector is a function of the health and well being of workers in the sector. Investing in farm safety programme is therefore a win-win game, because it is better and cheaper to invest in safety rather than bear the cost of working in an unsafe environment.

REFERENCES

Chris I (2008). *Pastoral care,* Safety Health Practitioner Magazine, United Media.

Emile T (2002). The Impact of Health on Productivity :Empirical Evidence and Policy Implications. The review of economic Performance and Social Progress.

Hämäläinen P, Takala J, Saarela KL (2005), Global estimates of occupational accidents International Labour Office (ILO), SafeWork, Geneva Switzerland.

Ibadan Wikipedia, (2011) http://www.en.wikipedia.or/wiki/Ibadan---- retrieved available on line 30:06-11.

Karen D. Collins,SR Michelle MD, Alice H, Alyssa LH (2005) www.commonwealthfund.org/usr doc/856 Davis hlt productivity US workers available on line. 18:03-11.

Kwanasahi M, Ajilima .I, Garba .A (1998). Nigerian economy: response of Agriculture to adjustment policy. www.aerfrica/documents/RD78 .PDF available online. 26:04-10.

Mostafa AB (2003) *Health of Agricultural Workers in Agriculture:* World Health Regional Publication, Eastern Mediterranean Series 25, Egypt.

ILO (1996): International Labour Organization, Yearbook of Labour Statistics, Geneva)

Ngigi NS (2009). Climate Change Adaptation Strategies: Water Resources Management Options for Smallholder Farming Systems in Sub-Saharan Africa. The MDG Centre for East and Southern Africa of the Earth Institute at Columbia University, New York,USA.

Oluwagbemi BF (2007). *Basic Occupational Health and Safety:* published by Vertext Media Limited, Ibadan.

Smallman C, John G (2001). 'British directors perspectives on the impact of health and safety on corporate performance', *Safety Science*, 38: 227-239

Soekirman (2010). Food Security and Nutrition Nexus: Food security seminar at Manila July, 2010.

2

Producer characteristics and determinants of technical efficiency of tomato based production systems in Ghana

B. O. Asante, M. K. Osei*, A. A. Dankyi, J. N. Berchie, M. B. Mochiah, J. N. L. Lamptey, J. Haleegoah, K. Osei and G. Bolfrey-Arku

Crops Research Institute, Box 3785, Kumasi, Ghana.

This paper uses cross-sectional data collected from 126 small scale tomato farmers in the forest, forest-savannah transition and the guinea savannah agro-ecological zones in 2011 to examine the characteristics of tomato based producing households in Ghana. It involved a description of the demographic characteristics of sampled farmers and an investigation of the production practices. This study further investigated the determinants of technical efficiency in tomato production among small scale farmers in Ghana. Descriptive statistics was used to present the characteristics of tomato producing households and the stochastic frontier analysis was used to estimate the determinants of technical efficiency and the inefficiency effect models. Results indicated that majority of the sampled tomato farmers were males. Almost half of the farmers had formal education at least up to the basic level. Almost half of the farmers cultivated tomato in both the major and minor seasons. Most of the varieties cultivated are exotic. Pests and diseases were predominant on the farms of majority of the sampled households. Consequently the use of pesticide to control pests and diseases is very common among sampled household. Our analysis further suggests average technical efficiency of 85.4%. In addition, factors such as extension services, land, frequency of weeding and fertilizer positively influenced technical efficiency of tomato farmers. Conversely, factors such as pesticide, labour and the frequency of pesticide application had negative effects on technical efficiency. The findings reveal the need for an effective and sustainable integrated approach to pest and weed management to enhance technical efficiency and productivity in tomato production.

Key words: Tomato, production practice, technical efficiency, stochastic frontier production function, pests, diseases, Ghana.

INTRODUCTION

Tomato, *Solanum lycopersicum* is one of the popular vegetables commonly cultivated by small scale farmers in Ghana (Osei et al., 2010). It is also one of the most important income-generating vegetables produced in Ghana. Tomato cultivation has been a significant economic activity in Ghana especially in the Upper East region. The crop is cultivated continuously throughout the year because apart from the rain-fed system that normally spans between June and November in the southern part of the country, there is the dry-season system between October and April mainly in the north (especially in the Upper East). It is consumed daily by most households as a source of lycopene, vitamins A and C. Tomato production has intensified over the years however, yields continue to remain low due to several production constraints which include biotic and abiotic factors. The abiotic factors are erratic rainfall, high temperature, and poor soils, among others.

Biotic constraints of significant economic importance in tomato production in Ghana include diseases such as tomato yellow leaf curl virus, bacterial wilt, bacterial spot, early blight, and tomato mosaic viruses. Spraying with

*Corresponding author: E-mail: oranigh@hotmail.com.

copper fungicides and other pesticides can help control some of these pests and diseases (personal communication). Dependency on chemical products for vegetable production is high compared to other food crops among small scale farmers. Also in Ghana, indiscriminate use of chemical pesticides has been reported (Osei et al., 2008) and this tends to pose potential health risk for farmers, consumers, and other stakeholders besides the attendant environmental effects. One of the main reasons adduced for low productivity in agriculture is the inability of farmers to fully exploit the available technologies (Sreenivasa Murthy et al., 2009). This fact which has been emphasized in many studies, particularly on cereals and pulses (Bagi, 1982; Battese, 1992; Battese and Coelli, 1988, 1992; Battese and Broca, 1997) and pertains to small scale tomato farmers in Ghana. For instance, the area cropped to tomato alone accounted for 69.45% of the total vegetable area cropped in the country in 2010. However, the average yield of tomato was about 7.5 mt which is far below the attainable of 15.0 mt hence leaving farmers with a yield gap of about 7.5 mt (MOFA, 2010). It is also believed that the use of improved seeds and the recommended rate of fertilizer and other key inputs in tomato production will help farmers to reach the maximum attainable yield level (MOFA, 2010). In view of this, efforts has been made by government, non governmental organizations like the international water management institute (IWMI), and other institutions like research and extension to improve tomato production through the development of improved varieties that are disease resistant, high yielding among other desirable attributes. Studies on tomato production in Ghana has mostly focused other aspects rather than investigating producer characteristics and technical efficiency in among tomato based producing households (Osei et al., 2008; Asare-Bediako et al., 2007; Donkoh et al., 2012; Tambo and Gbemu, 2010). Consequently, in addition to characterising tomato farmers in Ghana, it is therefore imperative to investigate the determinants of technical efficiency and productivity in tomato production among small scale farmers. This will unearth the magnitude of the effect to which these inputs and other factors account for these variations in yield. Such information will be handy for farmers, research and policy. This paper seeks to examine the characteristics of tomato based producing households in Ghana. It involved a description of the demographic characteristics of sampled farmers and an investigation of the production practices. This study further investigates the determinants of technical efficiency in tomato production among small scale farmers in Ghana.

METHODOLOGY

Study area

This study was conducted in the forest, forest-savannah transition and the guinea savannah agro-ecological zones in 2011. It involved tomato farmers in the Brong Ahafo, Ashanti and Upper East regions. These regions are the major tomato growing areas in the country. The forest, forest-savannah transition zone is characterized with a bi-modal rainfall regime of 1300 to 2200 mm per annum and therefore two distinct growing seasons (major and minor seasons). The guinea savannah has a uni-modal rainfall pattern (May to August) with an average of 1100 mm per annum. It also has a long dry season (November to March). Major tomato growing districts and communities within these agro-ecological zones were selected for the survey. Farming is the major economic activity and tomato cultivation is the main cash crop in these communities. The districts and communities included in this study are presented in Table 1.

Sampling technique

Multi-stage sampling was used for this study. Purposive sampling was first employed to select the five (5) districts in the major tomato growing regions of the country. Six communities were further selected purposively from the districts. A total of 126 tomato farmers regardless of acreage were then selected randomly from the communities with at least 19 farmers per district. The distribution of the respondents by the districts is presented in the Table 2.

Data collection and analysis

Both formal and informal approaches were employed to collect data for the survey. Participatory Rural Appraisal (PRA) was used to generate information on tomatoes from small scale farmers at the community level. Each group consisted of at least 40 tomato farmers with almost equal proportions of males and females. In addition, triangulation was applied to the groups to test the robustness of the responses from the groups. This was followed by a formal survey with the use of structured questionnaires. This enabled individual farmers to express their own views without any community influence. During the interview, useful demonstrations and drawings were frequently used to illustrate difficult points. Data collected included demographic characteristics, farm characteristics and farming practices including weed and pest control practices. Descriptive statistics such as graphs, charts, tables and diagrams were used to summarise the data and present a description of the characteristics of sampled tomato producing households and other summaries of production practices. The technical efficiency of tomato producers was estimated using the stochastic frontier production function, details of which have been discussed in the next section.

Methodological framework

Measurement of efficiency draws on the seminal work of Farell (1957) in which Farrell suggested that the efficiency of a firm consists of two components: technical and allocative efficiency. Technical efficiency is a measure of the ability of a firm to obtain maximum output from a bundle of inputs given the best available technology. In estimating the determinants of technical efficiency, two broad approaches are generally been proposed: parametric and non-parametric methods (more details of these methods are in Lovell, 1993; Coelli et al., 1998; Zhu, 2003; Ray, 2004). For the purpose of this study, the parametric approach is preferred to the nonparametric one because the production environment in which our respondents operate is prone to exogenous shocks.

The parametric approach specifies some functional form to represent the relationship between output and inputs. A preferred

Table 1. Districts and communities included in the study.

District	Communities
Bongo	Goo, Gori, Gor Kuakua and Gori Tengre
Navrongo	Nagal Rinia, Tono and Yigwama
Offinso North	Akumadan
Talensi Nabdam	Pawlugu, Yindowli and Moori
Techiman	Tuobodom, Tanoso, Epishya. Asuobrofo and Pinkyem

Table 2. Distribution of respondents by district.

District	Frequency
Bongo	22
Navrongo	19
Offinso North	22
Talensi-Nabdam	21
Techiman	42
Total	126

functional form exhibits the properties identified by Coelli et al. (2005) as flexibility, linearity in parameters, regularity and parsimony. Both the A Cobb-Douglas and the transcendental logarithmic (translog) function developed by Christensen et al. (1973) satisfy these properties and it is widely used in econometric estimation. A transcendental logarithmic (translog) functional form was considered to represent the production model. However, a hypothesis test result suggests that it is not an adequate representation of the data given the assumptions of the translog stochastic frontier model and was therefore not employed in this study.

Following Battese and Coelli (1995), Cobb-Douglas production function can be specified as follows:

$$\ln Y_i = \beta_0 + \sum_{j=1}^{2} \beta_{0j} D_{ji} + \sum_{j=1}^{8} \beta_j \ln X_{ji} + V_i - U_i \quad (1)$$

, i=1,2,3,.....,N

Where the subscript i indicate the i-th farmer in the sample: Y represents the quantity of tomatoes harvested for the sampled farmer (in kilograms). D_1 is the dummy variable which has a value of 1 if the farmer planted an exotic seed and 0 if otherwise (hereafter, Seed dummy). D_2 is the dummy variable which has a value of 1 if the farmer has access to extension services, and 0 if otherwise (hereafter, Extension dummy). X_1 is the total area (in Acres) planted to tomato (hereafter, Land). X_2 is the total labour (in man days) used in tomato cultivation (hereafter, Labour). X_3 is the quantity of fertilizer (in kilograms) used in tomato cultivation (hereafter, Fertilizer). X_4 is the quantity of pesticides (in kilograms) used in tomato cultivation (hereafter, Pesticides). X_5 is the quantity of seeds (in grams) used in tomato cultivation (hereafter, Seed).
X_6 is the frequency of fertilizer application per season (hereafter, Fertilizer frequency). X_7 is the frequency of weeding per season (hereafter, Weeding frequency). X_8 is the frequency of pesticide application (hereafter, Pesticide frequency).

The V_i^s are the random errors, assumed to be independent and identically distributed as $N(0,\sigma^2)$.

The U_i^s are the non-negative technical inefficiency effects assumed to be independently distributed among themselves and between the V_i^s, such that U_i^s is defined by the truncation of the $N(\mu_i,\sigma^2)$ distribution, where μ_i is defined by:

$$\mu_i = \delta_0 + \delta_{01}^{*} D_1 + \sum_{j=1}^{3} \delta_0 Z_{ji} \quad (2)$$

Where: D_1 is the dummy variable which has a value of 1 if the farmer is a male and 0 if otherwise (hereafter, Gender dummy). Z_1 represents age (in years) of the farmer (hereafter, Age); Z_2 represents the level of formal education (in years) of the farmer (hereafter, Education); Z_3 represents experience in tomato cultivation (in years) (hereafter, Experience); N denote the number of sampled rice farmers involved

The variation in output levels largely depends on the quantity of inputs used in production while differences in technical efficiencies are explained by productivity-enhancing factors. The dependent variable is the total quantity of tomato fruits harvested. The explanatory variables included in the frontier production function comprise area or land, quantity of seed, quantity of pesticides, access to extension, quantity of labour, frequency of fertilizer application, frequency of pesticide application, frequency of weeding and the type of seed planted. These variables are important physical inputs used in tomato production. The variables for extension and type of seed planted were included in the model to assess their effects on technical efficiency.

Besides describing the relationship between inputs and rice output, this paper is also concerned with those factors that influence farmers' technical inefficiency in their decisions making. The model for the technical inefficiency effects contains variables related to human capital, such as age, years of schooling, gender and experience in tomato production. These variables have been used in the models for the technical inefficiency effects in several previous studies, Oduol et al. (2011), Villano and Fleming (2006), Sriboonchitta and Wiboonpongse (2004a, b), Battese et al. (1996, 1997), Battese and Broca (1997), and all the variables have been clearly described and explained in the model.

Descriptive statistics of variables

The models presented in the previous section are applied to the cross sectional of tomato farmers in Ghana. Basic summary statistics of the key variables used in the stochastic frontier models are presented in Table 3. The average production of tomato was approximately 3975.03 kg per household, which translates to a mean yield of approximately 1967.84 kg ha^{-1}. Tomato output was highly variable, ranging from 260 kg to a maximum of 17940.0 kg per household. Average fertilizer use was 69.5 kg ha^{-1}. The average labour use was approximately 51 person-days per hectare. The

Table 3. Descriptive statistics of variables in the stochastic frontier production models and inefficiency models.

Variable	Mean	SD	Minimum	Maximum
Output (kg)	3975.03	2526.01	260.0	17940.0
Land (Acres)	2.02	1.28	0.2	6.0
Labour (man-days)	1190.98	1801.07	3.0	8662.0
Fertilizer (kg)	69.05	23.96	25.0	100.0
Pesticides (l)	1.88	1.08	0.0	6.5
Seed (g)	9.27	1.19	5.0	12.0
Fertilizer frequency (number)	2.71	0.54	1.0	4.0
Weeding frequency (number)	2.67	0.80	1.0	4.0
Pesticide frequency (number)	4.18	1.30	0.0	5.0
Age (years)	41.59	10.42	20.0	70.0
Education (years)	6.87	4.43	0.0	16.0
Experience (years)	12.50	7.96	1.0	40.0

coefficient of education is expected to have a negative sign because a higher level of educational attainment would result in lower inefficiency. The educational attainment of the farmer is a proxy for human capital. The sign on the coefficient of the age of the farmer could be negative or positive. If older farmers were not willing to adopt better practices whereas younger farmers were more motivated to embrace better agricultural production practices that reduce technical inefficiency effects, then the coefficient would be positive (greater technical inefficiency). However, if older farmers have more experience and knowledge of the production activities and are more reliable in performing production tasks, then the coefficient would be negative. The technical efficiency of the i-th farmer given the specification if given by:

$$TE_{it} = \exp(-U_{it}) \tag{3}$$

Where $-U_{it}$ is defined by the specification of the inefficiency model in Equation (2).

The prediction of the technical efficiencies is based on its conditional expectation, given the observable value of $(V_{it} - U_{it})$ (Jondrow et al., 1982; Battese and Coelli, 1988). The technical efficiency index is equal to 1 if the farm has an inefficiency effect equal to zero and it is less than 1 if otherwise.

Estimation procedure

To estimate the determinants of technical efficiency in tomato production, the stochastic frontier production functions, defined by equation (1), and the technical inefficiency model, defined by equation (2), are jointly estimated by the maximum likelihood method using Frontier 4.1 (Coelli, 1996). Various tests of null hypotheses for the parameters in the frontier production functions and in the inefficiency models are performed using the generalized likelihood ratio test statistic defined by:

$$\lambda = -2\{\ln[L(H_0) / \ln[L(H_1)]\}, \tag{4}$$

Where L (H_0) and L (H_1) denote the values of the likelihood function under the null (H_0) and alternative (H_1) hypotheses, respectively. If the null hypothesis is true, the test statistic has approximately a chi-squared or a mixed chi-squared distribution with degrees of freedom equal to the difference between the parameters involved in the null and alternative hypotheses. If the inefficiency effects are absent from the model, as specified by the null hypothesis, $H_0: \gamma = \delta_0 = \delta_{01} = \delta_{02} = \delta_3 ..... = \delta_8 = 0$, then λ is approximately distributed according to a mixed $\chi 2$ distribution with at least 4 degrees of freedom. In this case, critical values for the generalized likelihood ratio test are obtained from table 1 of Kodde and Palm (1986).

In estimating the elasticities of the model, since the variables of the Cobb-Douglas model were mean-corrected to zero, the first-order coefficients are the estimates of elasticities at the mean input levels. The elasticity of mean rice output with respect to the *j*-th input variable is defined by the following expression (Battese and Broca 1997):

$$\frac{\partial \ln E(Y_i)}{\partial \ln X_{ji}} = \{\beta_j + \beta_{jk} \ln X_{ki}\} - C_i \left(\frac{\partial \mu_i}{\partial \ln X_{ji}} \right) \tag{5}$$

where μ_i is defined in (2); C_i is defined by

$$C = 1 - \frac{1}{\sigma} \left\{ \frac{\phi\left(\frac{\mu_i}{\sigma} - \sigma\right)}{\Phi\left(\frac{\mu_i}{\sigma} - \sigma\right)} - \frac{\phi\left(\frac{\mu_i}{\sigma}\right)}{\Phi\left(\frac{\mu_i}{\sigma}\right)} \right\}$$

And ϕ and Φ represent the density and distribution functions of the standard normal random variable, respectively.

RESULTS AND DISCUSSION

Characteristics of sampled farmers

Majority of the respondents were males. Almost half of the farmers were educated up to the basic level (middle/JHS) and majority were married (Table 4). Further analysis on education suggests male dominance at both extremes of no formal education (70%) as well as at the tertiary level (75%) (Figure 1). Age distribution by gender of farmers demonstrates that most of the age catogories were dominated by males. The maximum

Table 4. Demographic characteristics of the respondents.

Variable	Percent
Sex	
Male	81.0
Female	18.3
Educational level	
No formal education	21.4
Primary	14.3
JHS/Middle	43.7
Secondary	14.3
Tertiary	3.2
Marital status	
Single	9.5
Married	84.1
Divorced	4.8
Widowed	1.6

proportion of females is 35% which is in the age group 61 to 70 (Figure 2).

Farm level characteristics

Half of the farmers were operating on farm sizes ranging between one and two acres, 22.2% were farming on less than one acre and the remaining 27% had farms greater than two acres (Table 5).

Over 50% of the farmers cultivated only exotic tomato varieties whiles 16.7% cultivated only local varieties. The remaining 23.8% cultivated both local and exotic varieties (Figure 3). Most of the exotic varieties were cultivated by female farmers (73.9%). However, there is no association between the varieties cultivated by males and females. This as illustrated by the chi square was not significant (Table 6).

About 45.2% of the local varieties were cultivated by farmers in the forest savannah transition agro-ecological zone. With over 91.9% of exotic varieties cultivated in the Guinea Savannah, there was no local variety cultivated in this zone. In the forest zone however, a combination of both varieties were cultivated (Table 7). There is a strong relationship between the Agro-ecological zone and type of varieties cultivated (Chi-Square sig. 0.000)

Figure 4 shows how farmers acquired tomato seeds for cultivation. Sixty-seven percent of the farmers obtained their tomato seeds from agro- input shops, 26% of them use seeds from their previous harvest whiles 7% obtained seeds from other farmers. More than half of females were cultivating farm sizes below one (1) acre whiles the males were cultivating farm sizes of between 1 and 2 acres. There is a strong relationship between farm size and gender of the farmers (Table 8).

Most of the farmers in all three zones cultivate between 1 to 2 acres of land. However, about 36% of farmers in the forest zone are cultivating more than two acres. The correlation between farm sizes of households and agro-ecological zones is significant (Table 9)

In general, more of each of the categories was allocated to the exotic variety. However, most (82%) of this variety was planted in farm sizes less than one acre. The local variety was planted more on 1 to 2 acres of land. The relationship between farm size and planted variety is strong (Table 10).

Tomato production practices

About 53.2% of the respondents applied the slash and burn method of field preparation, while 43.7% of them carried out ploughing and harrowing (mechanized) and 2.4% employed the zero tillage (use of herbicides) to prepare their field (Table 11). About 96.8% of the farmers applied fertilizer on their tomato field with over 70% applying three times per season. About 23.8% of the farmers applied fertilizer twice per season while 3.2% of them only once. Almost half of the farmers (46.8%) irrigated their farms once during the rainy season planting, 31.0% did not irrigate at all, 16.7% irrigated three times and 5.6% irrigated twice. However, during the dry season, 97.6% of the farmers irrigated more than thrice and the rest only thrice. About 50.0% of the farmers utilized the furrow irrigation system, 46.0% used manual irrigation, while 2.4% employed the drip irrigation system. About 44.4% of the farmers weeded their farms twice before harvesting, 34.9% weeded three times whereas 17.5% weeded four times. Weed control was predominantly manual (99.2%) with less than 1% herbicide use. Almost all the farmers (93%) reported of effective weed control practices.

Pests and diseases management

Over 99% of the farmers encountered pests and diseases in their farms and the use of pesticides was widespread in all agro-ecological zones. More than 70% of the farmers sprayed more than four times during the growing season whilst 0.8% sprayed only once (Table 6), 13.5% sprayed thrice per planting period and 15.1% sprayed four times (Table 6). Almost all the farmers (96.8%) encountered diseases in their farms.

Significantly, the use of chemicals to control diseases was noticeable, involving 108 farmers representing over 80% (Table 12). The term pesticide is comprehensive for all chemicals used to control pests such whiteflies, thrips etc (weeds inclusive) and diseases such as tomato yellow leaf curl virus, bacterial wilt, nematodes etc. Generally farmers sprayed four times or more per planting period. However, the majority of farmers (85.7%)

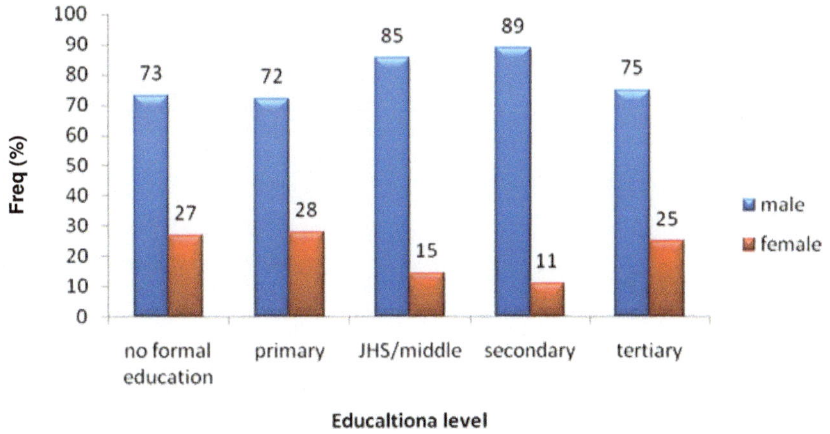

Figure 1. Education level of households by gender.

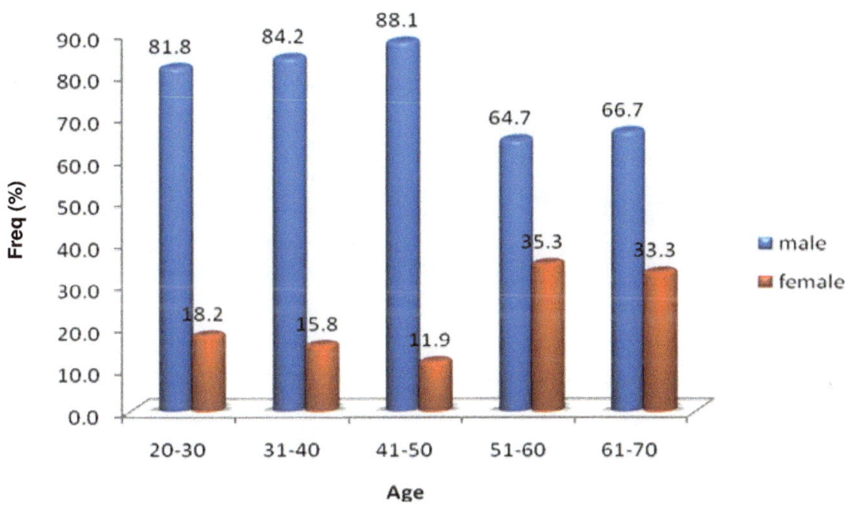

Figure 2. Age of households by gender.

Table 5. Distribution of Respondents by farm size.

Size (acre)	Frequency	Percent
<1	28	22.2
1-2	64	50.8
>2	34	27.0
Total	126	100.0

Table 6. Cross tabulation of gender and type of variety cultivated.

Gender	Type of variety (%)		
	Local	Exotic	Both
Male	19.6	55.9	24.5
Female	4.3	73.9	21.7

Chi-Square: 3.692, Asymp. Sig. (2-sided): 0.158.

Table 7. Cross tabulation of agro-ecological zone and type of variety cultivated.

Agro-ecological zone	Type of variety (%)		
	Local	Exotic	Both
Forest	9.1	27.3	63.6
Forest savannah transition	45.2	28.6	26.2
Guinea savannah	0.0	91.9	8.1

Chi-Square: 74.421, Asymp. Sig. (2-ided): 0.000.

Table 8. Distribution of farm size by gender.

Gender	Farm size (%)		
	<1 acre	1-2 acres	>2 acres
Male	14.7	55.9	29.4
Female	56.5	26.1	17.4

Chi-Square: 18.953, Asymp. Sig. (2-sided): 0.000.

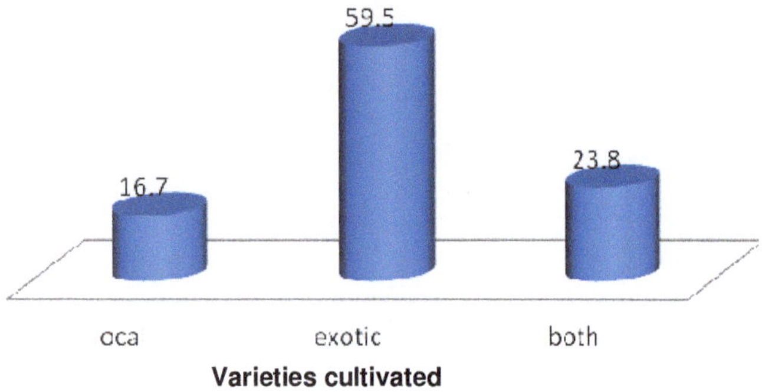

Figure 3. Varieties cultivated by respondents.

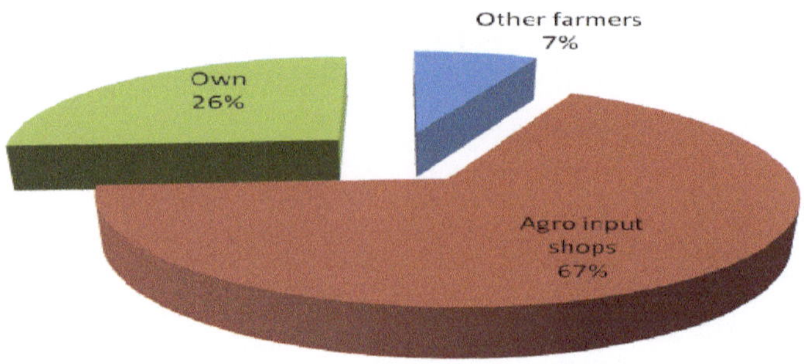

Figure 4. Source of seeds.

Table 9. Farm sizes of households in different agro-ecological zones.

Agro-ecological zone	Farm size (%)		
	<1acre	1-2acres	>2acres
Forest	9.1	54.5	36.4
Forest savannah transition	4.8	64.3	31.0
Guinea savannah	38.7	40.3	21.0

Chi-Square: 19.752, Asymp. Sig. (2-sided): 0.001.

Table 10. Distribution of farm size and type of variety cultivated by households.

Farm size	Type of variety (%)		
	Local	Exotic	Both
<1acre	7.1	82.1	10.7
1-2acres	26.6	46.9	26.6
>2acres	5.9	64.7	29.4

Chi-Square: 14.606, Asymp. Sig. (2-sided): .006.

in the Forest Savannah Transition sprayed more than four times.

Table 11. Tomato production practices employed by farmers in the study area.

Frequency of fertilizer application per season	Percent
Once	3.2
Twice	23.8
Thrice	72.2
More than thrice	0.8
Frequency of irrigation (rainy season)	
Once	46.8
Twice	5.6
Thrice	16.7
None	31.0
Frequency of irrigation in (dry season)	
Thrice	2.4
More than thrice	97.6
Type of irrigation system applied	
Manual	46.0
Drip	2.4
Furrow	50.0
Frequency of weeding	
Once	3.2
Twice	44.4
Thrice	34.9
Four times	17.5
Weed control methods	
Manual(weeding)	99.2
Chemical (use of weedicides)	0.8
Effectiveness of existing weed control methods	92.9
Type of field preparation	
Zero tillage(use of herbicides)	2.4
Ploughing and harrowing(mechanized)	43.7
Slashing and burning(manual)	53.2
Fertilizer application	96.8
Production seasons	
Wet season	2
Dry season	45
Both wet and dry seasons	53

Empirical results of determinants of technical efficiency of small scale tomato farmers

The maximum likelihood estimates of the parameters of the Cobb-Douglas stochastic frontier production function are presented in Table 14. The maximum likelihood estimates of the parameters of the inefficiency models for the rice producers are also presented in Table 15. The values of the explanatory variables in the Cobb-Douglas stochastic frontier model were mean-corrected by subtracting the means of the variables so that their averages were zero. This approach dictates that the first-order parameters are estimates of output elasticities for the individual inputs at the mean values.

Table 12. Pest and disease prevalence.

Status	Percent
Incidence of pests on farm	99.2
Use of pesticides	99.2
Frequency of spraying per season	
Once	0.8
Thrice	13.5
Four times	15.1
More than four	70.6
Incidence of diseases on farm	96.8
Use of pesticides for control	85.7

All estimated first-order coefficients in the model fall between zero and one, satisfying the monotonicity condition that all marginal products are positive and diminishing at the mean of inputs. The results indicate that land, labour, fertilizer and pesticides are essential factors for tomato production and the impacts of these variables on the mean tomato outputs are significant. Labour, quantity of pesticides, frequency of pesticide application and the use of exotic seeds had a negative effect on tomato output. Conversely, a positive relationship was found between tomato output and land, fertilizer frequency of weeding and access to extension services.

The positive effect of the land on tomato output indicated that increasing the land results in lower technical inefficiencies in tomato production. This is so because increases in area cultivated to tomato has been a major factor to increase in tomato output over the years. In addition, the extension dummy has positive effect on technical efficiency and was highly significant. This indicates that increased access to extension services is likely to increase technical efficiency. Farmers with access to extension services are more likely to be abreast with current good practices in tomato production and technical advice from extension officers and are likely to be more technically efficient in their production. The results of the inefficiency effects model are presented in Table 15. These results further suggest the age variable has a negative association with technical inefficiency, indicating that older farmers tend to be more inefficient in tomato production. This is not surprising due to the labour and capital intensive nature of tomato cultivation. In addition, experience in tomato cultivation also negatively influenced technical inefficiency, indicating that more experienced farmers are likely to be more efficient that less experienced one. Gender was found to have a negative effect on technical inefficiency, this indicates that male farmers are more likely to have low levels of inefficiency that their female counterparts.

Socio-economic variables and other covariates included in both the technical efficiency and the inefficiency models such as formal education, age, gender, household size, extension contacts etc. have widely been employed in the literature (Oduol et al., 2011; Chaovanapoonphol et al., 2009; Villano and Fleming, 2006; Sriboonchitta and Wiboonpongse 2004a, b; Battese and Ceolli, 1995).

This result is supported by the second hypothesis test in which the null hypothesis, $H_0: \gamma = \delta_0 = \delta_{01} = \delta_{02} = \delta_3 ... = \delta_8 = 0$, indicates that the inefficiency effects in the frontier model are not present. If γ=0 and all the δ-coefficients are zero, and that the stochastic frontier production function is the same as the mean production function that does not account for the inefficiency effects.

From Table 16, it can be seen that this null hypothesis is rejected at the 5% level of significance. This rejection indicates that the traditional production function is not an adequate representation of the data given the assumptions of the stochastic frontier model.

The first null hypothesis, $H_0: \beta_{ij} = 0$ for all $i \leq j = 1, 2, ..., 8$, states that the second-order coefficients in the translog production function have zero values and so, if this hypothesis is true, then the Cobb-Douglas production function applies. This null hypothesis is not rejected, even if the size of the test is as small as 5% level of significance. This implies that the Cobb-Douglas production function is an adequate representation of the data given the assumptions of the stochastic frontier model.

Finally the third null hypothesis that is considered is, $H_0: \delta_{01} = \delta_1 = \delta_2 = \delta_3 = 0$, which indicates that all the coefficients of the explanatory variables in the inefficiency model are equal to zero (technical inefficiency effects have half-normal distribution). If this hypothesis is true, then the explanatory variables in the inefficiency model do not influence the technical inefficiencies of tomato production. This second null hypothesis is also rejected. This means that the technical inefficiency effects have a truncated normal distribution indicating that the coefficients of the explanatory variables in the inefficiency model are not equal to zero and thus influence the level of inefficiency in tomato production.

Elasticities and returns to scale

The estimates of the elasticities of output with respect to inputs of production are presented in Table 17. For the Cobb-Douglas model, the elasticities of mean tomato output with respect to the different inputs depend on several parameters and values of the inputs.

The empirical results show that from the estimates of the Cobb-Douglas production function model, the estimated elasticities of mean tomato output with respect to land, labour, fertilizer, pesticide and seed at mean input values, are 0.130, -0.052 , 0.124, -0.001 and -0.376, respectively, at the mean input values.

This indicates that if the amount of land and fertilizer were to be individually increased by 1%, then the mean production of tomato is estimated to increase by 13.0 and 12.4%, respectively. However, the estimated labour, seed and pesticides elasticities were found to be elasticity is found to be negative. This suggests that if the amount of seeds planted with the required quantities of pesticide application and labour were to be individually increased by 1%, then the mean production of tomato is estimated to decrease by 37.6, 0.1 and 5.2%, respectively.

The returns to scale estimates, evaluated at the mean input values is -0.175 as presented in the bottom of Table 17. This value is significantly less than one, which indicates decreasing returns to scale in tomato production in Ghana.

Technical efficiency indexes

Table 18 presents the distribution of the predicted technical efficiencies of the sampled tomato producers. The mean technical efficiency was estimated to be 0.78, with the maximum of 0.95 and the minimum of 0.40. This implies that, on the average, the tomato farmers were producing tomato about 79% of the potential (stochastic) frontier production levels, given the technology currently being used. Thus, in the short run, there is capacity for increasing tomato production by 22% by adopting and using techniques used by the best practice tomato farms in the country. In the present study, in general, more than half of tomato producers had a mean technical efficiency in the range of 0.81 to 0.99. The remaining proportion of the tomato producers had mean technical efficiency ranging from 0.51 to 0.80. This means that generally, most of the sample tomato producers on the average are technically efficient in the allocation and use of inputs.

Characteristics of tomato producers and production system

Over 99% of the farmers encounter pests in their farms and almost all the farmers use pesticides to control pest in their farms hence the importance of pests in tomato cultivation in Ghana. Over 70% of the farmers spray their farms more than four times per planting period. Recognisance survey conducted revealed that most of the farmers even spray their farms just before harvesting because buyers believe the fruits are fresh when they see traces of the pesticide on it. The high incidence of pest and disease in the forest savannah transition accounts for high intensity of spraying than the other agro-ecological zones. In most cases, farmers do not adhere to recommended rates and practice during spraying. Abuse of agrochemicals; the use of unrecommended and heavy doses of even recommended pesticides endanger the lives of consumers. This might account for the non-significance of these two variables. Likewise, frequency of irrigation was not significant at 10%. Farm size was identified as insignificant at 10%. The adaptability and performance of the local varieties in the forest savannah transition as well as the exotic ones in the guinea savannah accounts for their dominance in these agro-ecological zones. However, the smaller acreages allocated to the production of the exotic varieties could be because it requires much care with its attendant high capital requirement which farmers may not be able to meet on larger acreages.

Factors influencing technical efficiency and productivity in tomato production

The coefficient of the education variable in the inefficiency model has a negative sign, which implies that more educational training acquired by farm operators is associated with higher technical efficiency of tomato production. Further, education improves farmers' ability to read and understand basic instructions on rates of application of agrochemicals, fertilizer, seeds and other inputs, since there instructions in most cases are written in the English language. This means tomato production requires at least some level of basic education for farmers to be efficient producers. Apart from experience, the ability to process information about an innovation is enhanced by the educational status of a person. In addition, the educated understand more the need for information and are better motivated to look out for innovations which can improve their tomato production. Certainly, the design of agricultural interventions should consider promotion of education as a major component in the short term. However, in the long term, the country needs to develop procedures to ensure the citizens have at least basic education. Our findings are consistent with earlier studies by Weir and Knight (2000) and Solis et al. (2009), which found that education enhances the ability of the farmers to acquire and make judicious use of information about production inputs, thus improving efficient use of the inputs.

The inefficiency effects model indicated a negative effect of gender on technical inefficiency in tomato production. The intensive nature of tomato production accounts for why males are dominating in tomato production. This situation is however not different across ago ecological zones (Table 13). However, preliminary investigation shows that, the females provide support to their male counterparts in most tomato production activities. The observed male dominance in the tomato production, Ghana is in consonance with what Wiredu et al. (2011) argues as an obvious and unique characteristic of the agricultural based production systems in the country.

This research was conducted among small scale tomato farmers who own just small piece of land therefore

Table 13. Distribution of spraying frequency by agro-ecological zone.

Agro-ecological zone	Type of variety (%)			
	Once	3 times	4 times	More than 4
Forest	0.0	9.1	22.7	68.2
Forest savannah transition	0.0	2.4	11.9	85.7
Guinea Savannah	1.6	22.6	14.5	61.3

Chi-Square: 12.271 Asymp. Sig. (2-sided): 0.056

Table 14. Maximum-likelihood estimates for parameters of the Cobb-Douglas stochastic frontier production model for small scale tomato farmers in Ghana.

Variable	Parameter	Coefficient	SE
Constant	β_0	2.294	0.510^{ns}
Seed Dummy	β_{01}	-0.032	0.049^{a}
Extension dummy	β_{02}	0.019	0.072^{a}
Land	β_1	0.130	0.206^{a}
Labour	β_2	-0.052	0.037^{a}
Fertilizer	β_3	0.124	0.120^{a}
Pesticides	β_4	-0.001	0.017^{a}
Seed	β_5	-0.376	0.014^{ns}
Fertilizer freq	β_6	0.899	0.111^{ns}
Weeding freq.	β_7	0.213	0.132^{b}
Pesticide freq.	β_8	-0.018	0.026^{a}
Gamma	γ	0.07	0.006
Variance parameters	σ^2	0.309	0.051
Log likelihood function		7.35	

[a], [b], and [c] denote significance at the 1, 5 and 10% level, respectively, and [ns] denote not significant.

Table 15. Maximum likelihood estimates for parameters of the inefficiency effects model of the Cobb-Douglas stochastic frontier production function for tomato farmers in Ghana.

Variable	Parameter	Coefficient	SE
Constant	δ_0	-1.699	0.340
Gender dummy	δ_{01}	-1.129^{b}	0.128
Age	δ_1	-0.001^{a}	0.009
Education	δ_2	-0.034^{b}	0.011
Experience	δ_3	-0.006^{a}	0.016

[a], and [b] denote significance at the 1 and 5 level, respectively.

increasing farm sizes is not of prime interest to the farmers. In spite of this, farm size is positively related to technical efficiency implying that farmers with larger farm sizes tend to be more technically efficient in tomato production. Exploiting economies of scale in cultivating large outputs could account for this result.

Further, the results suggest a positive effect of fertilizer application and its frequency on technical efficiency. Continuous cropping and its resultant depletion of soil fertility resulting in lower yields without fertilizers application could account for this finding. In addition, most of these farmers have been cultivating on the same piece of land for a long time for various reasons such as unavailability and high cost of land rent, proximity of a piece of land to water source etc, hence fertilizer application is one of the major ways through which farmers are able to increase yield. In addition, the frequency of weeding also positively affected technical efficiency of tomato production. These findings compare with that of Chaovanapoonphol et al. (2009), and Villano and Fleming (2006), who found a positive effect of fertilizer on technical efficiency in their various studies.Both the quantity of pesticide and the frequency of application were found to be negatively related with technical efficiency. This indicates that farmers are not applying pesticides efficiently in tomato production. This findings is consistent with Osei et al. (2008) who reported on the indiscriminate of use of pesticides.

Conclusion

Prevalence of pests and diseases was common on most farms and the major means of control is the use of agrochemicals. In addition, regardless of the fact that most farmers spray with pesticides and copper fungicides, dependency on these chemical products is very high. Key factors influencing technical efficiency of tomato production include farm size or land, labour pesticide and the frequency of pesticide application, fertilizer and the frequency of fertilizer application, frequency of weeding, and extension services. A positive effect was found between extension services, land, frequency of weeding and fertilizer on technical efficiency. However, other factors such as pesticide labour and the frequency of pesticide application had negative effects on technical efficiency. Efforts towards

Table 16. Tests of null hypotheses for parameters in the stochastic frontier production function and the inefficiency effect models for tomato farmers in Ghana.

Null hypothesis	λ	Critical value[a]	Decision
$H_0: \beta_{ij} = 0$ for all $i \leq j = 1, 2, ..., 8$	21.23	24.4	Do not reject H_0
$H_0: \gamma = \delta_0 = \delta_{01} = \delta_1 = \delta_2 = \delta_3 = 0$	32.63	16.3	Reject H_0
$H_0: \delta_{01} = \delta_1 = \delta_2 = \delta_3 = 0$	15.42	5.14	Reject H_0

Table 17. Output elasticity estimates for inputs in the stochastic frontier production functions.

Input	Estimated output elasticity
Land	0.13 (0.206)
Labour	-0.052 (0.037)
Fertilizer	0.124 (0.120)
Pesticides	-0.001 (0.017)
Seed	-0.376 (0.014)
Returns to scale	-0.175 (0.394)

Standard errors are in parenthesis.

Table 18. Percentages of technical efficiencies of tomato farmers in Ghana within decile ranges.

Interval	Percentage
<0.5	4.6
0.51-0.60	3.1
0.61-0.70	10.3
0.71-0.80	19.6
0.81-0.90	52.6
0.91-1.00	9.8
Total	194
Mean	0.78
SD	0.10
Maximum	0.95
Minimum	0.40

increasing tomato production should be targeted at these factors. These must not come alone but with appropriate training in good tomato management practices through research, extension and NGOs and existing farmer based organisations. An integrated approach to pest and disease management is recommended to manage tomato farmers' persistent dependence on pesticides and other agro chemicals in their production activities.

ACKNOWLEDGEMENTS

The authors express profound gratitude to the USAID, IPM CRSP (Cooperative Agreement Number EPP-A-00-04-00016-00) for financial support of this study. Many thanks go to the Agricultural extension officers in the study areas for the enormous role played in organizing farmers for the survey. Mr. A.Y. Kwarteng, a chief technical officer of the CSIR-Crops Research Institute is appreciated for his assistance in obtaining weed information.

REFERENCES

Asare-Bediako E, Showemimo FA, Buah JN, Ushawu Y (2007). Tomato production constraints at Bolgatanga irrigation project in the Northern Region of Ghana. J. Appl. Sci. 7:459-461.

Bagi FS (1982). Economic efficiency of sharecropping: Reply and some further results. Malaysia Econ. Rev. 27:86-95.

Battese G (1992). Frontier production functions and technical efficiency: A survey of empirical applications in agricultural economics. Agric. Econ. 7:185-208.

Battese G, Broca S (1997). Functional forms of stochastic frontier production functions and models for technical inefficiency effects: A comparative study for wheat farmers in Pakistan. J. Product Anal. 8:395-414

Battese G, Coelli T (1992). Frontier production functions, technical efficiency and panel data: With application to paddy farmers in India. J. Product. Anal. 3:153-169.

Battese GE, Coelli TJ (1995). A model for technical inefficiency effect in stochastic frontier production for panel data. Empirical Econ. 20:325–345

Battese G, Coelli T (1988). Prediction of firm-level technical efficiencies with a generalized frontier production function and panel data. J. Economet. 28:387-399.

Battese GE, Malik SJ, Gill MA (1996). An investigation of technical inefficiencies of production of wheat farmers in four districts of Pakistan. J. Agric. Econ. 47:37–49.

Battese GE, Rambaldi AN, Wan GH (1997). A stochastic frontier production functions with flexible risk properties. J. Productivity Anal. 8:269–80

Chaovanapoonphol Y, Battese GE, Chang HS (2009). The impact of agricultural loans on the technical efficiency of rice farmers in the Upper North of Thailand, In: Lee J-D, Heshmati A (Eds.), Productivity, Efficiency, and Economic Growth in the Asia-Pacific Region, Berlin, Heidelberg: Springer-Verlag. pp. 279–295.

Christensen LR, Jorgenson DW, Lau LJ (1973). Transcendental logarithmic production frontiers. Rev. Econ. Statist. 55(1):28–45.

Coelli TJ (1996). A guide to FRONTIER version 4.1: A computer program for stochastic frontier production and cost function estimation. CEPA Working Papers, No. 7/96. Department of Econometrics, University of New England, Armidale.

Coelli T, Rao DSP, O'Donnell CJ, Battese GE (2005). An Introduction to Efficiency and Productivity Analysis, New York, USA: Springer.

Coelli TJ, Rao DSP, Battese GE (1998). An Introduction to Efficiency and Productivity Analysis. Kluwer Academic Publishers, Boston.

Donkoh SA, Tachega M, Amowine N (2012). Estimating technical efficiency of tomato production in Northern Ghana. Am. J. Exp. Agric. 3(1):56-75.

Farell MJ (1957). The measurement of productive efficiency. J. Royal Statistics Assoc. Series A (general) 21:253-281.

Jondrow J, Lovell CAK, Materov IS, Schmidt P (1982). On the estimation of technical inefficiency in the stochastic frontier production function model. J. Economet. 19:233–238.

Kodde DA, Palm FC (1986). Wald criteria for jointly testing equality and inequality restrictions. Economet 54:1243–1246.

Lovell CAK (1993). Production frontiers and productive efficiency', in Fried HO, Lovell CAK, Schmindt SS (eds), Measurement of productive efficiency: Techniques and applications, Oxford University Press, New York. pp. 3-67.

Ministry of Food and Agriculture (2010). Agriculture in Ghana: Facts and Figures, Statistics, Res. Inform. Directorate, Accra. pp. 12–15,

Oduol JBA, Binam AJN, Olarinde L, Diagne A, Adekunle A (2011). Impact of adoption of soil and water conservation technologies on technical efficiency: Insight from smallholder farmers in Sub-Saharan Africa. J. Develop. Agric. Econ. 3(14):655-669 DOI: 10.5897/JDAE11.091.

Osei MK, Akromah R, Shilh SL, Lee LM Green SK (2008). First Report and Molecular Characterization of DNA A of Three Distinct Begomoviruses Associated with Tomato Leaf Curl Disease in Ghana: APS-Plant Dis. 92(11):1585

Osei MK, Akromah R, Shilh SL, Green SK (2010). Evaluation of Some tomato Germplasm for resistance to Tomato Yellow Leaf curl Virus disease (TYLCV) in Ghana. Aspects Appl. Biol. 96:315-323.

Ray SC (2004). Data Envelopment Analysis: Theory and Techniques for Economics and Operation Research. Cambridge University Press, New York.

Solis D, Bravo-Ureta B, Quiroga ER (2009). 'Technical efficiency among peasant farmers participating in Natural Resource Management Programmes in Central America'. J. Agric. Econ. 60:202-219.

Sreenivasa MD, Sudha M, Hegde MR, Dakshinamoorthy V (2009). Technical efficiency and Its determinants in tomato production in Karnataka, India: Data Envelopment Analysis (Dea) Approach Division of Extension and Training, Indian institute of horticultural research. Agric. Econ. Res. Rev. 22:215-224.

Sriboonchitta S, Wiboonpongse A (2004a). On estimation of stochastic production-frontiers with self-selectivity: Jasmine and non-Jasmine rice in Thailand. Asia-Pacific Productivity Conference, University of Queensland, Brisbane, 14-16 July.

Sriboonchitta S, Wiboonpongse A (2004b). The effect of production inputs, technical efficiency and other factors on Jasmine and non-Jasmine rice yields in production year 1999/2000 in Thailand. Asia-Pacific Productivity Conference, University of Queensland, Brisbane, 14-16 July.

Tambo AJ, Gbemu T (2010). Resource-Use Efficiency in Tomato Production In The Dangme West District, Ghana. Faculty of Life Sciences, University Of Copenhagen. AGRINATURA.

Villano RA, Fleming EM (2006).Technical inefficiency and production risk in rice farming: Evidence from Central Luzon, Philippines, Asian. Econ. J. 20(1):29–46

Weir S, Knight J (2000). Education externalities in rural Ethiopia. evidence from average and stochastic frontier production functions, Working paper CSAEWPS/2000-4, Centre for the Study of African Economies, University of Oxford.

Wiredu AN, Mensah-Bonsu A, Andah EK, Fosu KY (2011)..Hybrid Cocoa and Land Productivity of Cocoa Farmers in Ashanti Region of Ghana. World J. Agric. Sci. 7(2):172-178.

Zhu J (2003). Quantitative models for performance evaluation and benchmarking: data envelopment analysis with spreadsheets and excel solver. Kluwer Academic Publishers, Boston.

Economic welfare effects of the increase in labor costs on U.S. tobacco farms

Shiferaw T. Feleke , Jane H. Starnes, Harwood D. Schaffer and Kelly J. Tiller

Agricultural and Resource Economics, University of Tennessee, 2621 Morgan Circle Drive, Knoxville, TN 37996, United States.

The United States Department of Labor (USDOL) amended elements of the 2008 H-2A Final Rule and adopted the 2010 Final Rule, which came into effect on March 15, 2010. Compared to the 2008 H-2A Final Rule, the 2010 H-2A Final Rule is arguably more expensive for farm employers. The H-2A Final Rule provides regulations which govern the H-2A program, a guest worker program allowing U.S. farm employers to bring foreign workers into the country to perform seasonal and temporary agricultural work. The most influential cost-increasing element of the 2010 H-2A Final Rule is a change in the methodology by which the Adverse Effect Wage Rate (AEWR) is calculated. The AEWR is the minimum wage rate that must be offered and paid to H-2A workers and similarly employed U.S. farm workers. This paper estimates the economic welfare effects of the additional labor costs on tobacco farms using the Equilibrium Displacement Model. Considering the prospect of U.S. immigration reform and revision of the current H-2A program, results provide an insight into the potential effects of changes in policy and labor regulations affecting wages and admission of foreign guest farm workers.

Key words: Tobacco, adverse effect wage rate, H-2A Final Rule, equilibrium displacement model, economic welfare effects.

INTRODUCTION

The United States Department of Labor (USDOL) amended elements of the 2008 H-2A Final Rule and adopted the 2010 Final Rule that came into effect on March 15, 2010. Compared to the 2008 Final Rule, the 2010 H-2A Final Rule is arguably more expensive for farm employers. The most influential cost-increasing element of the 2010 H-2A Final is a change in the methodology in which the Adverse Effect Wage Rate (AEWR) is calculated. The AEWR is the minimum wage rate that the USDOL has determined "must be offered and paid to U.S. and alien workers by agricultural employers of nonimmigrant H-2A agricultural workers" (Federal Register, 2010). In the 2008 Final rule, the AEWR was estimated based on the Occupational Employment Statistics (OES) wage survey that is under-taken by the Bureau of Labor Statistics (BLS) of the USDOL. However, in the 2010 Final Rule, the USDOL amended the methodology such that the AEWR would be set based on the Farm Labor Survey (FLS) of the United States Department of Agriculture (USDA). Since OES pand FLS surveys cover different farm-related employers and provide different levels of detail by occupation and geographic area, they provide different estimates of AEWR. The FLS-based AEWR is an annual weighted average hourly wage rate for field and livestock workers (combined) by region while wages from the OES survey are established by levels of skill and experience. The change in methodology from the OES wage survey to the FLS survey raised the national average of the AEWR by $1.02 per hour (Federal Register, 2010).

Considering the importance of labor in tobacco production, accounting for 30% to 50% of the cost of production depending on the tobacco type (Foreman and McBride, 2011) and the importance of the H-2A program among tobacco farms, accounting for nearly 25% of the H-2A labor petitions (U.S.DOL, 2009), the amended AEWR-setting rule could have a significant effect on the cost of tobacco production and hence the welfare of tobacco growers. In 2008, 75% of burley tobacco farms and 95% of flue-cured tobacco farms used hired and contract labor, which in turn provided 61% of all labor hours needed for burley production and 84% of the hours needed for flue-cured tobacco production (Foreman and McBride, 2011). Burley and flue-cured are the major tobacco types used in cigarettes.

This paper assesses the economic welfare effects of the additional increase in AEWR due to the 2010 Final Rule on tobacco farms using the Equilibrium Displacement Model (EDM) in the short-run when there is (1) no adjustment in the production process and output market, and in the long-run when there is (2) adjustment in the tobacco production process as well as in the output market.

Background

The H-2A program is a guest worker program allowing U.S. farm employers who anticipate a shortage of domestic farm labor to bring foreign workers into the country to perform seasonal and temporary agricultural work. It was authorized for agricultural work as part of the Immigration Reform and Control Act (IRCA) of 1986 pursuant to the amendment of the H–2 temporary guest worker program that was earlier established as part of the Immigration and Nationality Act of 1952.

In order to participate in the H-2A program, farm employers must comply with a set of requirements, one of which is filing an application with the DOL stating that there are not sufficient able, willing and qualified U.S. workers available to perform the intended job(s) and that the employment of foreign workers will not have an adverse effect on the wages and working conditions of similarly employed U.S. workers. To do so, they must offer the job at the highest of AEWR or the prevailing hourly, or the agreed upon collective bargaining rate, or the Federal or State minimum wage rate. In addition, they must provide for free housing, pay for inbound and outbound transportation, and abide by the 50% rule and 75% guarantee. The 50% rule requires employers of H-2A workers to provide employment to any able, willing, qualified and available U.S. worker who applies for the job until 50% of the period of the work contract has elapsed. The 75% rule requires employers of H-2A workers to guarantee to offer to each covered worker employment for a total number of hours equal to at least 75% of the workdays in the contract period. Further information on requirements is given in Vol. 75, No. 29 of the Federal register (2010).

The H–2A program is managed by three federal agencies (DOL, Department of HomeLand Security, and State Department). The DOL is responsible for labor certifications and oversight of compliance with labor laws. The Department of Home Land Security is responsible for adjudication of petitions and approval of admission in H–2A classification while the State Department is responsible for approval of visas for entry into the country.

The way the H–2A program works is that a farm employer who anticipates a shortage of farm labor during a certain period of the year needs to initiate pre-filing 75 to 60 days before their first date of need by submitting a job order to a State Workforce Agency (SWA) serving the area of intended employment. After receiving clearance with the SWA by way of an SWA–approved job order, interested farm employers apply for certification from the Secretary of Labor by filing an Application for Temporary Employment Certification with the National Processing Center located in Chicago. If they receive approval of labor certification, they then proceed with filing petitions requesting H–2A workers with the Citizenship and Immigration Services of the Department of Homeland Security. If the petitions are approved, then the prospective temporary workers need to apply to a U.S. embassy for an entry visa, which allows them to travel to the U.S. port of entry (airport, seaport or land border crossing) at which point they apply with the Customs and Border Protection of the Department of Homeland Security for admission to the country under H–2A classification. If the Customs and Border Protection approves the admission, the guest workers are issued an admission ticket known as an I–94 card (also known as the Arrival/Departure Record) stamped with a date indicating the length of time they are allowed to stay in the country as per the approved petition. They must leave the country before the expiration of the I-94 card unless an extension is granted.

LITERATURE REVIEW

A number of studies have looked into the employment, price, and profit effects of the increase in minimum wages in non-agricultural sectors (Katz and Krueger, 1992; Card and Krueger, 1994; Neumark et al., 2004). Detailed studies on the effects of the rise in minimum wages is given in Neumark and Wascher (2006) who provided a comprehensive review of more than 100 minimum wage studies published since the 1990s. Most of these studies are focused on the employment effects while the number of studies on the price and profit effects is limited (Aaronson, 1998; Card and Krueger, 1994). This is probably because the price effects of minimum wage increases in non-agricultural sectors are too small to draw significant research attention. In the case of tobacco, however, it cannot be ignored because of the labor-intensive nature of tobacco production and the relatively larger labor cost share. Comparing tobacco with grain crops in terms of the intensity of labor (that is, the number of labor hours required per acre), tobacco is much more labor intensive than field crops. Tobacco's labor requirement ranges from 72 h per acre for flue-cured production to 151 h per acre for burley production. In contrast, the labor requirement for field crops is less than 3 h per acre (Foreman and McBride, 2011). As a result of the labor-intensive nature of tobacco production, the cost share of labor in tobacco production is high. Labor accounts for 30% of the cost of production per acre for flue-cured tobacco and 50% for burley tobacco production (Foreman and McBride, 2011). In contrast, labor cost in grain production accounts for only 5% of the variable costs of production. The labor cost share of tobacco is comparable to the most labor-intensive agricultural enterprises: vegetables, fruits and nuts, greenhouse and nursery and dairy.

MODEL

The effect of the increase in AEWR on tobacco farms due to the 2010 H-2A Final Rule is assessed using the EDM. Starting with the input side, the labor cost function of a given tobacco farm can be given by

$$C = WL, \tag{1}$$

where C is labor cost; W is AEWR, and L is total employment. Totally differentiating Equation (1) we found that

$$dC = WdL + LdW, \tag{2}$$

where dC is the change in cost of tobacco production; dL is the change in total employment due to the 2010 H-2A,and dW is the change in AEWR due to the 2010 H-2A Final Rule.
Rewriting Equation (2) in log form, we found

$$CdLnC = (WL)dLnL + (LW)dLnW, \tag{2.1}$$

where $dLnC$ is the percentage change in cost of tobacco production; $dLnL$ is the percentage change in total employment due to the 2010 H-2A Final Rule, and $dLnW$ is the percentage change in AEWR due to the 2010 H-2A Final Rule.
Rearranging the terms in Equation (2.1), we found:

$$dLnC = \gamma_{dt} dLnW + \gamma_{dt} dLnL, \tag{2.2}$$

where $\gamma_{dt} = WL/C$ is labor cost share of tobacco production.
Decomposing $dLnL$ (the percentage change in total employment) into domestic and H-2A employment, we have:

$$dLnL = \alpha dLnL_d + (1-\alpha) dLnL_{H-2A}, \tag{2.3}$$

where α is the share of domestic employment; $(1-\alpha)$ is the share of H-2A employment; $dLnL_d$ is the percentage change in domestic labor employment, and $dLnL_{H-2A}$ is the percentage change in H-2A labor employment.
Substituting Equation (2.3) into Equation (2.2) we found:

$$dLnC = \gamma_{dt} dLnW + \gamma_{dt}\left[\alpha dLnL_d + (1-\alpha) dLnL_{H-2A}\right]. \tag{2.4}$$

The demand for domestic labor can be given as

$$dLnL_d = \omega dLnW, \tag{3}$$

where ω is the own price elasticity of the demand for domestic labor. The import demand for H-2A labor can be given as

$$dLnL_{H-2A} = \omega_{H-2A} dLnW, \tag{4}$$

where ω_{H-2A} is the own price elasticity of the import demand for H-2A labor. Substituting Equations (3) and (4) into Equation (2.4), the percentage change in cost of production $(dLnC)$ due to the increase in AEWR can be given by

$$dLnC = \theta\gamma_{dt} dLnW, \tag{5}$$

where $\theta = 1 + A$ is the coefficient of adjustment in the input market and production process, and $A = \alpha\omega + (1-\alpha)\omega_{H-2A}$ is the price elasticity of the derived demand for labor.
Now considering the output side, the tobacco supply function is given by:

$$dLnQ = \varepsilon(dLnP - dLnC), \tag{6}$$

where $dLnQ$ is the percentage change in tobacco production; ε is the price elasticity of the supply of tobacco; $dLnP$ is the percentage change in tobacco price; $dLnC$ is the percentage change in cost of production due to the increase in the AEWR representing the percentage change in the initial equilibrium price.
The percentage change in domestic demand for tobacco is given as

$$dLnQ_d = \eta dLnP, \tag{7}$$

where $dLnQ_d$ is the percentage change in the quantity of tobacco for domestic use; η is the price elasticity of the domestic demand for tobacco, and $dLnP$ is the percentage change in tobacco price.
The percentage change in export demand for tobacco is given as:

$$dLnQ_e = \eta_e dLnP_e,$$

(8)where $dLnQ_e$ is the percentage change in the quantity of tobacco for export, and η_e is the price elasticity of the export demand for tobacco. The market–clearing identity for tobacco can thus be given as

Table 1. Parameters estimates of the price elasticity of demand for tobacco and labor from published sources.

Parameter	Definition	Flue-cured	Burley	Sources
η	Price elasticity of the demand for tobacco	-1.46	-1.46	Rezitis et al. (1998)
η_e	Price elasticity of export tobacco	-3.0	-3.0	Brown and Martin (1996)
ε	Price elasticity of tobacco supply	7	7	Fulginiti and Perrin (1993)
β	Proportion of domestic sales of tobacco	38%	40%	Predicted from data in NASS/USDA (2008)
α	Proportion of domestic hired labor	77%	71%	Foreman and McBride (2011)
$1-\alpha$	Proportion of H-2A labor	23%	29%	Foreman and McBride (2011)
ω, ω_e	Price elasticity of labor	-0.40	-0.40	Moss et al. (2010)
γ_{dt}	Labor cost share in the tobacco price	30%	50%	Foreman and McBride (2011)

$$dLnQ = \beta dLnQ_d + (1-\beta) dLnQ_e, \tag{9}$$

where β is the proportion of domestic tobacco sales and $(1-\beta)$ is the proportion of export tobacco sales.
Substituting Equations (7) and (8) into Equation (9) and then substituting the resulting Equation into Equation (6), we find

$$\varepsilon(dLnP - dLnC) = \beta\eta dLnP + (1-\beta)\eta_e dLnP_e \tag{10}$$

Now multiplying out and collecting similar terms, the percentage change in tobacco price $(dLnP)$ due to the change in cost of production can be given as

$$dLnP = \lambda dLnC, \tag{11}$$

where $\lambda = \frac{\varepsilon}{\varepsilon - B}$ is the coefficient of adjustment in the output market, and $B = \beta\eta + (1-\beta)\eta_e$.

Then, substituting Equation (5) into Equation (11), the $dLnP$ (that is, the percentage change in tobacco price due to the increase in AEWR) can be given as

$$d\ln P = \kappa\gamma_{dt} dLnW, \tag{12}$$

where $\kappa = \lambda\theta$ is the pass-through elasticity. In the absence of adjustment, $\kappa = 1$ so that the increase in cost of production due to the increase in AEWR is fully passed on totobacco leaf purchasers.The adjustment in input and output markets mitigates the effect of the AEWR increase on tobacco price. The adjustment in the input market occurs in the production process as growers respond to the increase in the AEWR by making input substitution and reducing the number of labor hours or reducing the number of workers. This adjustment is represented by θ. The adjustment in the output market represented by λ occurs as tobacco leaf purchasers respond to the increase in tobacco leaf price. The product of the adjustment coefficients in the input market (θ) and output market (λ) provides the pass-through elasticity (κ).

Finally, with the proportional change in tobacco price and quantity supplied thus computed, the change in producer surplus can be calculated as

$$\Delta PS = (dLnP_i - dLnC_i) P_{i0} \times Q_{i0} (1 + 0.5 dLnQ_i) \tag{13}$$

where ΔPS is the change in producer surplus; P_{io} is the initial equilibrium tobacco price; Q_{io} is the initial equilibrium tobacco output and the other terms have been defined previously.

Data

Data used in this study came from the Center for Tobacco Grower Research (CTGR) at the University of Tennessee, the USDOL and the National Agricultural Statistics Service of the U.S. Department of Agriculture (NASS/USDA). Table 1 provides published parameter estimates (e.g., the price elasticity of the demand for labor, the price elasticity of the demand and supply of tobacco, the proportion of domestic and H-2A labor, the proportion of tobacco for domestic and export sales) and used in the simulation for burley and flue-cured tobacco.

The share of domestic labor versus H-2A labor, the labor cost share, the share of total tobacco production used for domestic consumption and exports varies by type of tobacco. Comparing burley and flue-cured tobacco, domestic labor (migrant hired and contract labor plus operator and family labor) account for 77 and 71% of the total number of labor hours needed in flue-cured production and burley production, respectively (Foreman and McBride, 2011).USDA defines migrant farm workers as those who cross county or state lines and stay away from a U.S. home overnight to do farm work for wages.

The remaining 23% flue-cured production and 29% in burley production are provided by non-immigrant hired and contract labor such as H-2A labor. Table 2 provides the base value of the variables used in the simulation. With regard to the wage rate, the 2008 AEWR in Tennessee and Kentucky where burley is grown was $9.13 per hour while the 2011 AEWR in the same states in 2011 is $9.48 per hour, indicating that the wage cost of burley tobacco production would increase by nearly 4%.With regard to flue-cured tobacco, the 2008 AEWR in North Carolina and Virginia where flue-cured is grown was $8.85 per hour while the 2011 AEWR in the same states in 2011 was $9.30 per hour, indicating that the wage cost of flue-cured tobacco production would increase just over 5%.

Table 2. Wages, production and price variables by tobacco type and sources.

Variable	Flue-cured	Burley	Source
Acres per farm	97.0	15.0	CTGR
Yield (lb. per acre)	2382	2122	CTGR
Total production in 2008 ('000,000)	499.2	201.5	NASS/USDA
Labor hours per acre	72 h	151 h	Foreman and McBride (2011)
AEWR ($/h in 2008)	8.85	9.13	DOL
AEWR ($/h in 2011)	9.30	9.48	DOL
Tobacco price ($/lb. in 2008)	1.76	1.67	NASS/USDA

CTGR, Center for Tobacco Grower Research; NASS/USDA: National Agricultural Statistics Service/ U.S. Department of Agriculture; DOL: Department of Labor.

RESULTS AND DISCUSSION

The increase in cost of tobacco production due to the increase in AEWR would affect the welfare of tobacco growers depending on whether and how much the additional cost is translated into higher price. In a perfectly competitive market with constant returns to scale, average cost equals marginal cost, which in turn equals the output price. Thus, any increase in labor costs will add to the tobacco farm's average and marginal costs, and therefore to the output price by the amount of the labor share of operating cost. The higher labor cost is then passed on to tobacco leaf purchasers. The question is how much of the increase in costs due to the increase in the AEWR is passed on to tobacco leaf purchasers. As tobacco growers are faced with a higher labor cost due to the increase in AEWR, they have three options. The first one is simply absorbing the added labor cost through reduced profits, but that is unlikely to be sustainable. Under this option, the added labor cost is fully borne by tobacco growers. The second option is a case where growers will continue employing the same number of workers as before and keep production constant as AEWR increases, and simultaneously tobacco leaf purchasers will keep their tobacco purchases constant with increasing tobacco prices. Under this option, the added labor cost is fully passed on to tobacco leaf purchasers. The third option is the case where both tobacco growers and tobacco leaf purchasers make adjustments in their production and purchase plans, respectively. Tobacco growers make adjustments to pay for the added costs through reduced employment of workers or cutting employee work hours. Tobacco leaf purchasers make the adjustments through reduced purchases. Under this option, the added labor cost is shared between tobacco growers and tobacco leaf purchasers.

Option I: No adjustment in the input and output markets but growers absorb the added labor cost

Under this option, there is no employment effect but there is a profit effect because employing the same number of workers at the new higher wage rate increases the cost of production. Since growers are assumed to bear the full cost of the increase in AEWR under this option, and employment remains the same, there is no change in output supply and consequently no change in price. As a result, tobacco leaf purchasers are expected to buy the same quantity of tobacco, leaving total revenue unchanged. However, growers would be worse off with a loss in producer surplus. Results indicate that in the aggregate flue-cured growers will be worse off with a loss of $13.4 m while burley growers will be worse off with a loss of $6.4 m in producer surplus. This option is unlikely to be sustainable because most tobacco growers cannot afford to absorb the extra costs and keep the price steady. Therefore, the most *sustainable* options for most tobacco growers are allowing the additional labor costs to be fully or partially passed on to tobacco leaf purchasers.

Option II: No adjustment in both the input and output markets but tobacco growers passes the added labor cost on to tobacco leaf purchasers

Under this option, there is no adjustment in both the input and output markets in the sense that neither tobacco production nor tobacco purchase is affected as higher AEWR increases the cost of production. Growers continue employing the same number of workers and keep the input proportion constant, and tobacco leaf purchasers do not respond when the added cost is passed on to them in the form of higher tobacco price.

Consequently, the full effect of the increase in cost induced by the higher AEWR is passed on to tobacco leaf purchasers in the form of a higher tobacco leaf price. The increase in tobacco leaf prices due to the increase in cost induced by the increase in the AEWR depends on the importance of labor in the production of tobacco. Table 3 shows that with no change in employment, the 5.1% increase in the AEWR in flue-cured growing states such as North Carolina and Virginia between 2008 and 2011 would result in cost of flue-cured production increasing 1.53% in view of a cost share of 30% (0.3×5.1). In contrast, Table 4 shows that the 3.8% increase in AEWR in burley growing states such as Kentucky and Tennessee between 2008 and 2011 would result in the cost of burley tobacco production increasing 1.90% in

Table 3. Effects (%) of the AEWR increase on flue-cured tobacco farms under two likely options.

Variable	Option II	Option III
Tobacco price	1.53	0.68
Domestic sales decrease	0.00	-1.00
Export sales decrease	0.00	-2.05
Total production	0.00	-1.65
Marginal cost	1.53	0.92
Revenue	1.53	-0.97

Option II applies to the short-run case where the increase in cost due to the AEWR is passed onto tobacco leaf purchasers; Option III is the case where the increase in cost due to the AEWR is shared between tobacco growers and tobacco leaf purchasers due to the adjustment in production process and purchase plan; The percentage change in tobacco price $(dLnP)$ due to the 5.1% increase in AEWR was determined using Equation (12).

Table 4. Effects (%) of the AEWR increase on burley tobacco farms under two likely options.

Variable	Option II	Option III
Tobacco price	1.90	0.85
Domestic sales decrease	0.00	-1.24
Export sales decrease	0.00	-2.55
Total production	0.00	-2.03
Marginal cost	1.90	1.14
Revenue	1.90	-1.18

Option II applies to the short-run case where the increase in cost due to the AEWR is passed on tobacco leaf purchasers; Option III is the case where the increase in cost due to the AEWR is shared between tobacco growers and tobacco leaf purchasers due to the adjustment in production process and purchase plan; The percentage change in tobacco price $(dLnP)$ due to the 3.8 percent increase in AEWR was determined using Equation (12).

view of a cost share of 50% (0.5 ×3.8). These increases in cost of production are passed onto tobacco leaf purchasers as proportional tobacco price increases. Flue-cured tobacco purchasers would experience a 1.53% price increase while burley tobacco purchasers would experience a 1.90% price increase. Since tobacco leaf purchasers are not expected to respond to the increase in tobacco prices in the short- run, tobacco revenue would increase. However, despite the increase in revenue, in the aggregate, growers are neither better off nor worse off. In other words, the price increase offsets the same percentage increase in production costs, thus creating a welfare-neutral situation.

Option III: Adjustments in the input and output markets

As the methodology that led to the increase in the AEWR was set by regulation, the wage increase is permanent in which case the tobacco leaf price will eventually reflect the increase in cost of production. Obviously, the increase in AEWR leads to an increase in cost of production, which will in turn result in higher tobacco leaf price. However, growers mitigate the higher cost by making some adjustment in input substitution, employing fewer workers or reducing the number of labor hours in their operations. As a result, the increase in cost will get smaller and put less upward pressure on the tobacco leaf price. Also, by reducing their tobacco purchases, tobacco leaf purchasers further mitigate the effect of higher AEWR on tobacco prices.

The increase in AEWR will result in a 0.90% cut in total employment in flue-cured tobacco farms compared to a 0.68% cut in burley tobacco farms. This adjustment in employment will lead to a reduction in the cost of production. The coefficient of adjustment in the input market (θ) is 60% from Equation (5). Therefore, as a result of the adjustment in the production process, the cost of flue-cured production is reduced from 1.53% under option II to 0.92% (0.6 ×1.53) under option III (Table 3). In short, as a result of the input adjustment, the cost of flue-cured production would only increase 0.92% as opposed to 1.53%. Similarly, the adjustment in burley production process would reduce the rate of increase in costs from 1.90% under option II to 1.14% (0.6 ×1.90) under option III (Table 4). In short, the cost of burley tobacco production would only increase 1.14% as opposed to 1.90%.

Therefore, in the absence of any further adjustment in the output market, flue-cured tobacco leaf purchasers would have to experience a 0.92% increase in price while burley tobacco purchasers would have to experience a 1.14% increase in price. However, since tobacco leaf purchasers had enough time now, they would respond to the price increase and mitigate the increase through reduced purchases. The coefficient of adjustment in the output market (λ) is 74% from Equation (11). The adjustment in flue-cured tobacco leaf purchases would cause the price of flue-cured tobacco to increase by only 0.68% (0.74 ×0.92) (Table 3). Similarly, the adjustment in burley tobacco leaf purchases would cause the price of burley tobacco to increase by only 0.85% (0.74 ×1.14) (Table 4). These prices increases are higher than the case in other industries. Aaronson (1998) shows that a 10% hike in the minimum wage increased restaurant prices on the whole by 0.68% and prices at limited service establishments by 1.6%.

Now putting together the effects of the two adjustments (that is, input adjustment representing 60% reduction and output adjustment representing 74% reduction), the pass-through elasticity is 44% (0.60 ×0.74) from Equation (12). This means that a one-dollar increase in cost of production would result in the price of tobacco increasing $0.44. The 1.53% increase in price of flue-cured tobacco under option II where tobacco purchasers were assumed to absorb the increase in costs in the form of higher price

Table 5. Change in tobacco revenue and producer surplus due to the increase in AEWR.

Variable	Flue-cured				Burley			
	Aggregate ('000,000)		Individual farms		Aggregate ('000,000)		Individual farms	
	Option II	Option III	Option II	Option III	Option II	Option III	Option II	Option III
Actual supply (lb.)	0.00	-8.23	0.00	-3.808	0.00	-4.09	0.00	-645
Actual revenue ($)	13.44	-8.48	6.222	-3.927	6.39	-3.96	1.010	-626
Producer surplus ($)	0.00	-2.05	0.00	-950	0.00	-0.96	0.00	-152

is now reduced to 0.68% (0.44 × 1.53) under option III. Similarly, the 1.90% increase in price of burley tobacco under option II is now reduced to 0.85% (0.44 × 1.90) under option III. Because of these increases, sales of both flue-cured and

burley tobacco would decline. The domestic and export sales of flue-cured would decline by 1.0 and 2.0%, respectively (Table 3). Accounting for the proportion of domestic sales (38%) and export sales (62%), the total production of flue-cured tobacco would decrease by 1.65%. Therefore, given a 0.68% price increase and a 1.65% production decrease, the total flue-cured tobacco revenue would decrease by 0.97% (0.68-1.65).

Similarly, the domestic and export sales of burley tobacco would decline by 1.24 and 2.55%, respectively (Table 4).

Accounting for the proportion of domestic sales (40%) and export sales (60%) given in Table 1, the total production of burley tobacco would decrease by 2.03%. Therefore, given a 0.85% price increase and a 2.03% production decrease, the total burley tobacco revenue would decrease by 1.18% (0.85-2.03).

Change in tobacco revenue and producer surplus

With 499.1 million lbs. of flue-cured tobacco leaf produced in the country in 2008 (Table 1), the cut in flue-cured tobacco leaf supply induced by the 5.1% increase in the AEWR would be 8.23 million lbs. (1.65% × 499.1 million lbs.) (Table 5). Given the flue-cured price of $1.76 per lb., total flue-cured revenue would decline by $8.48 m from its level in 2008.

Further, in the aggregate, flue-cured growers would lose just over $2.0 m in producer surplus. In terms of individual farms, on average each flue-cured farm would cut supply by 3,808 lbs. of tobacco leaf, or reduce tobacco acreage by about 1.5 acre. As a result, each flue-cured tobacco farm would lose about $ 3,927 in gross revenue and about $950 in producer surplus.

In contrast, on average each individual burley tobacco farm would cut supply by 645 lbs. of tobacco leaf, or reduce tobacco acreage by about one-third of an acre. As a result, each burley tobacco farm would lose about $ 626 in gross revenue and about $152 in producer surplus. Other results on burley tobacco are given in Table 5.

CONCLUSION AND IMPLICATIONS

The paper has examined the major cost-increasing element of the 2010 H-2A Final Rule (that is, a change in wage calculation methodology that led to the increase in AEWR) that went into effect on March 15, 2010 and provided estimates of the effect of the additional labor costs on tobacco farms. Results are presented for both short-term and long-term. The short-term effects are presented under two options (I and II) while the long-term effects are presented under option III.

Results presented under option I suggest that in the short-run when neither tobacco growers nor tobacco leaf purchasers are able to make adjustment in their production process and purchase plans, respectively, the increase in cost of production due to the increase in AEWR would be fully absorbed by tobacco growers, creating a profit effect that makes them worse off. In contrast, results of option II suggest that the increase in cost of production due to the increase in AEWR would be fully passed on to tobacco leaf purchasers, creating a welfare-neutral situation.

In the long-run (option III), given the fact that the change in methodologythat led to the increase in the AEWR set by regulation, the increase in wage remains permanent and therefore the additional costs associated with the AEWR prompt both farm employers and tobacco leaf purchasers to make some adjustments in their production process and purchases, respectively. The adjustment made by both tobacco growers and purchasers mitigates the effects of the increase in the AEWR on tobacco price. Results presented under option III indicated that the coefficient of adjustment in the production process is 60% and that in the output market is 74%. Over all, the pass-through elasticity is 44% meaning that a one dollar increase in cost of tobacco production would result in the price of tobacco increasing $0.44. In terms of welfare, flue-cured growers, in the aggregate, would be worse off with a loss of over $2m in producer surplus while burley growers would be worse off with a loss in producer surplus of nearly $1m. On individual basis, flue-cured growers would be much worse off than burley tobacco growers.

Considering the increased prospect of U.S. immigration reform and revision of the current H-2A program, these results provide an insight into the potential effects of changes in policy and labor regulations affecting wages. Given the importance of labor in tobacco production, unfavorable changes in immigration policy and labor regulations affecting wages could have significant impact on tobacco farms.

REFERENCES

Aaronson D (1998). Price pass-through and the minimum wage:The impact of the minimum wage on prices of food away from home in the U.S. and Canada. Federal Reserve Bank of Chicago.

Brown AB, Martin LL (1996).Price versus quota reductions: U.S. flue-cured tobacco policy. J. Agric. Appl. Econ. 28(2):445–452.

Card D, Krueger AB (1994).Minimum wages and employment: A case study of the fast food industry. Am. Econ. Rev. 84(4):772-793.

Federal Register (2010).Rules and regulations 75:29, Friday, February 12, 2010.

Foreman L, McBride W (2011). Policy reform in the tobacco industry: Producers adapt to a changing market.Economic Research Service/U.S. Department of Agriculture, Economic Information Bulletin 77, May 2011.

Fulginiti LE, Perrin RK (1993).The theory and measurement of producer response under quotas. Rev. Econ. Stat. 75:97-106.

Katz L, Krueger A (1992). The effect of the minimum wage on the fast food Industry. Ind. Lab. Rel. Rev. 46(1):6-21.

Moss C, Livanis G, Schmitz A (2010). The effect of increased energy prices on agriculture: A differential supply approach. J. Agric. Appl. Econ. 42(4):711–718.

NASS/USDA (National Agricultural Statistics Service/ U.S. Department of Agriculture (2008). available at: http://www.nass.usda.gov/Statistics_by_Subject/index.php?sector=CROPS

Neumark D, Schweitzer M, Wascher W (2004). Minimum wage effects throughout the wage distribution. J. Hum. Res. 39(2):425-450.

Neumark D, Wascher W (2006). Minimum wages and employment: A review of evidence from the new minimum wage research. National Bureau of Economic Research, Working Paper no. 12663.

Rezitis AN, Brown BA, Foster WE (1998). Adjustment costs and dynamic factor demand for U.S. cigarette manufacturing. Agric. Econ.18:217-231.

USDOL (United States Department of Labor) (2009). Foreign Labor Certification Report: 2009 Data, Trends and Highlights across Programs and States. Available at: http://www.foreignlaborcert.doleta.gov/pdf/2009_Annual_Report.pdf.

Factors affecting household access to enough food in the Eastern Cape Province of South Africa

L. Musemwa[1]*, L. Zhou[2], S. Ndhleve[1] and F. Aghdasi[3]

[1]Department of Agricultural Economics and Extension, University Of Fort Hare, P. Bag X1314, Alice 5700, RSA.
[2]Risks and Vulnerability Centre, University of Fort Hare, P. Bag X1314, Alice 5700, RSA.
[3]Faculty of Science and Agriculture, University of Fort Hare, P. Bag X1314, Alice 5700, RSA.

South Africa is experiencing rapid increase in food prices while household income is increasing at a slower rate. This antagonistic state has positioned ordinary South Africans, at present struggling to meet their basic household requirements, more vulnerable to food security. The objectives of the study were to determine factors that affect household access to enough food in Eastern Cape Province of South Africa and to determine the role played by agriculture and fisheries in enhancing household access to enough food. Data were collected from 159 households using a questionnaire as the main instrument. Most households interviewed were male headed (50.3%). Unemployment level was high with 73.6% of the interviewed household heads being unemployed. On average, households had five members and range from 1 to 13. The average age of household heads was 59 years. Most of the household heads owned home gardens and had access to arable land though the majority did not cultivate their land (83%). Poultry and livestock production were practiced by 61.6 and 52.8% of the households, respectively. Government grants were the main source of income for the majority of households. The mean household income for all households was ZAR2 987. From the 159 respondents only 29 had access to enough food. Access to enough food was affected by gender of head of household, household size, education level of household head, agricultural training, poultry production and monthly total income. Most of the interviewed households depend mostly on food purchases rather than own production. This therefore mean that the people in the area to do not explore agriculture up to its potential.

Key words: Food security, garden, household, income, land access, livestock.

INTRODUCTION

Despite the political and economic advances made since 1994, South Africa continues to experience major challenges of poverty, unemployment and, more recently, steep increases in food and fuel prices, and energy tariffs. These adverse conditions have placed ordinary South Africans, already struggling to meet their basic household needs, in an ever more vulnerable situation (Labadarios et al., 2009). In most of the developing countries, including South Africa, agriculture is an essential sector of the proportion of the economic activities that take place in rural areas. However, agricultural activities are vulnerable to climate change and this put the lives of the poor in developing countries at risk (Fraser et al., 2003). In the Eastern Cape province of South Africa, climate change is seemingly increasing the vulnerability of the households to income losses; poverty and food insecurity and this is becoming increasingly visible in most of the rural communities in the province (Bank and Minkley, 2010).

Programmes implemented to solve the problem of poverty and food security in the Eastern Cape are continuously failing to produce the desired results, the hectarage of abandoned farmlands is increasing every year and the abandonment of agricultural projects is

*Corresponding author. E-mail: lmusemwa@yahoo.co.uk.

common across the province (Hebinck and Lent, 2007; Hall and Aliber, 2010; Bank and Minkley, 2010). In the province, extreme weather events like droughts and floods, gradual increases in temperatures and increased variability in annual rainfall appear to be common. These changes are seemingly having a damaging effect on the rural poor (Hall and Aliber, 2010). Official statistics suggest that conditions of the poor have worsened, and that poverty levels in the Eastern Cape have deepened five years after the inauguration of the first democratic government, especially in rural areas, where 65% of the province's 6.3 million people live (ECSECC, 2000).

This paper seeks to provide socio-economic factors that are affecting household access to enough food[1]. Household sources of food are also explored in this paper. This will enable us to assess the causes of households not having access to food and this will in turn help policy makers in designing policies that could enhance household access to food.

MATERIALS AND METHODS

The study area

The study was conducted in Ngqushwa Local Municipality, one of the eight municipalities that fall under the Amatole District Municipality located in the Eastern Cape Province of South Africa. The municipality is composed of one hundred and fourteen rural villages and two towns namely, Hamburg and Peddie. With its natural beauty and character (especially in the coastal areas), Ngqushwa is a wonderful tourist attraction that prides itself in its rich history and heritage. It is bounded on the East by the Fish River and on the South by the Indian Ocean and has 118 villages under its jurisdiction and a population of 84,234 made up of 20,757 households. The municipality has a climate which varies with the elevation from cool humid sub-topical at the coast to hot and sub-arid inland. The climate is characterized by variable moderate to low rainfall ranging between an annual average of 700 mm at the coast and 400 mm in the inland with about 60% of rainfall occurring in summer and peaks being in October and February. The natural vegetation has been heavily transformed by grazing and other land use practices. Even though certain parts of the vegetation have been degraded especially with the presence of alien plants, a greater portion of the region is still favorable for livestock production.

Sampling procedure

Ngqushwa Local Municipality was randomly selected from the eight municipalities that fall under the Amatole District Municipality. The municipality has a diverse range of economic activities which include agriculture, tourism and fisheries. During this study, 159 randomly selected households in Ngqushwa were interviewed at their homesteads by trained enumerators under the supervision of the researcher in January 2012 using a questionnaire as the main household heads, any adult member of the household was interviewed. The questionnaire consisted of both open-ended and closed questions, in order to improve the quality of data collected. Open-ended questions gave the respondents greater freedom of expression as they offered respondents an opportunity to qualify their answers thus reducing bias due to unlimited response ranges. The fear of researcher/interviewee bias from using only open ended questions, the questionnaire was balanced with closed ended questions. Data on household socio-economic characteristics, agricultural systems, household income, household expenditures and access to food and clean water were comprehensively collected.

Data analysis and description of variables used in the analysis

Descriptive statistics was applied to the basic characteristics of the sampled households. This employed both frequency and means to describe the data which included data related to gender, marital status, age, education level, occupation of head of household, monthly average household income and access to food, water and land.

Household from the area of study differ on how they have access to enough food. Some households have access to enough food while others do not have access to enough food. Therefore, this implies that the problem that needs to be analysed needed a method that was able to explain a binary endogenous variable (yes /no) by a set of covariates that determine the outcome of the decision. A typical method used to analyse such dichotomous variables is the logistic regression (Hosmer and Lemeshow, 2000). According to Kleinbaum (1994), there are two main reasons for using logistic regression in economics research. Firstly, the logistic function is flexible and easily applicable, and secondly the interpretation of the results is straight forward and meaningful. The logistic model also imposes for threshold and interaction effects and allows for examination of social interaction (Musemwa et al., 2010).

Following Gujarat (2003), the cumulative logistic distribution function for factors affecting access to enough food was specified as

$$P = \frac{1}{1+e^{-Z}} \quad (1)$$

Where P is the probability of a household to have access to enough food, e is the error term and Z is a function of explanatory variables (X) and was expressed as

$$Z = B_0 + B_1X_1 + B_2X_2 + \ldots\ldots\ldots\ldots\ldots B_mX_m \quad (2)$$

The probability of not having access to enough food was given by

$$1 - P = \frac{1}{1-e^{Z}} \quad (3)$$

The conditional probability of the outcome variable follows a binomial distribution with probability given by the conditional means $P_{(i)}$. The logistic model in terms of logs is

$$\log\left(\frac{p}{1-p}\right) = \beta_0 + \beta_1X_1 + \beta_2X_2 + \cdots + \beta_kX_k \quad (4)$$

Where

$$\log\frac{p}{1-p} = Z$$

[1]Household's ability to provide future physical and economic access to sufficient, safe, and nutritious food that fulfills the dietary needs and food preferences for living an active and healthy lifestyle.

Table 1. Characteristics of household heads.

Characteristics	Frequency (%)
Gender	
Male	50.3
Female	49.7
Marital status	
Married	58.5
Singe	25.7
Widowed	14.5
Divorced	1.3
Education level	
None	3.1
Primary	5.1
Secondary	72.3
Tertiary	19.5
Employment	
Employed	26.4
Unemployed	73.6

The log of odds ratio is not only linear in X but also linear in the Bi variable and as a result, OLS is used. Taking the stochastic term μ into account, the logit econometric model to be used will be

$$Z = B_0 + B_1X_1 + B_2X_2 + \ldots\ldots\ldots\ldots\ldots B_mX_m + \mu \quad (5)$$

This econometric model was used and treated against the potential variables, which are assumed to affect household access to enough food. Potential variables which may influence access to enough food were obtained from Alaimo et al. (2001), Fraser et al. (2003). Variables included in the model are shown in Table 2.

RESULTS AND DISCUSSION

Farmers' socioeconomic profile

The domination of males in the rural areas of South Africa is still common (Montshwe, 2006). Most households interviewed were male headed (50.3%) and 49.7% were female headed. This indicates that the effect of rural-urban migration, where the males go to urban areas in search for employment is minimal in the study area. The number of household heads that were not employed was 73.6% as shown in Table 1. Those (26.4%) who were employed work in low income jobs. On average, households have five members but sizes range from 1 to 13. A larger family size means that the required labour for agricultural production as well as fishing is provided, however pressure is set on consumption. The average age of household heads was 59 years. The youngest household head was 21 years old while the oldest was 98 years old. The majority of the household heads were married (58.5%) and 25.7% were single household heads. Widowed household heads constituted 14.5% and divorcees constituted the least proportion of household heads (1.3%).

The majority of the interviewed household heads had secondary education (72.3%) whereas 19.5% had qualifications from tertiary institutions.

The problem of household heads having never attended school is diminishing quite significantly over the years as access to education is improving significantly in rural areas (Nkhori, 2004) this is evidenced by only 3.1 and 5.1% of household heads having no education and primary education respectively. Many of the existing household heads are elderly and today's youths will have had considerably more basic education by the time they become household heads as the youths have better access to education nowadays. However the problem may be that most of the youths may be employed in the formal sector and other informal sectors in urban areas where there are bright lights as most of them view agriculture as a dirty business and backward (Musemwa et al., 2007).

Land ownership

An important area of focus when addressing the problem of food insecurity in remote areas where employment opportunities are minimal is how to improve household's access to food production on land. Figure 1 shows land ownership patterns for households of Ngqushwa Local Municipality.

The majority of households interviewed did not have access to arable land (69.8%) while only 30.2% had access to arable land. Most of the interviewed household heads had home gardens (59.1%) while only 40.9% did not own gardens.

Agriculture and land use

Agriculture play vital roles of enhancing food security of household located in remote areas as it is the backbone of the economies of these areas. It plays a direct role through providing food to the household and an indirect role through providing households with income from sales of agricultural produce. Poultry and livestock production was practiced by 61.6 and 52.8% of the households interviewed, respectively. Livestock and poultry meet multiple objectives that are desired by resource-poor households (Musemwa et al., 2007; Coetzee et al., 2004; Chimonyo et al., 1999). People keep livestock and poultry for provision of draught power, manure, cash sales, among other socio-economic functions. Most of the poor resource households keep livestock and poultry because their land is largely marginal and not suitable for cropping. The erratic rainfall and high incidence of droughts in the Eastern Cape, therefore, makes a large majority of the

Table 2. Factors influencing access to enough food.

Analysis of maximum likelihood estimates				
Parameter	**Estimate**	**Error**	**Chi-Square**	**P-value**
Intercept	1.796	4.564	0.155	0.694
Gender	0.728	0.547	1.772	0.038*
Age	0.011	0.016	0.450	0.502
Marital status	-0.020	0.400	0.002	0.961
Household size	-0.158	0.113	1.962	0.031*
Employment	-0.235	0.596	0.155	0.694
Education	0.758	0.482	2.470	0.016*
Own residential site	-0.111	1.044	0.011	0.915
Garden ownership	0.679	0.744	0.833	0.361
Use garden	0.629	0.740	0.721	0.396
Access to arable land	-0.820	0.725	1.279	0.258
Cultivate arable land	0.060	0.573	0.011	0.917
Agricultural training	0.856	0.962	0.792	0.047*
Grow fruit trees	0.049	0.612	0.006	0.936
Preserve vegetables/fruits	-0.850	0.554	2.356	0.125
Poultry ownership	1.802	0.655	7.563	0.006*
Livestock ownership	0.631	0.547	1.332	0.248
Monthly total income	0.458	0.184	6.166	0.013*
Main source of income	0.000	0.000	0.972	0.324
Food from fishing	0.884	0.629	1.976	0.160
Food shortage times	-0.183	0.641	0.081	0.776
No. of meals per day	0.176	0.397	0.197	0.657
Water supply	0.217	0.316	0.470	0.493

*Significant at 5%.

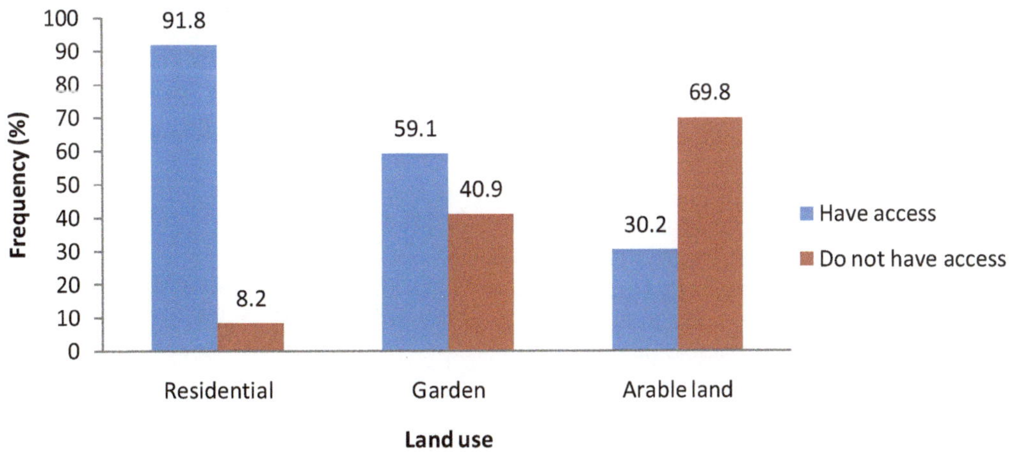

Figure 1. Households' access to land in Hamburg community.

population depend on livestock for their livelihoods.

Crop and vegetable production also play a role in supplementing household food requirements. Most of the interviewed households grow vegetables which included mostly cabbages in their gardens (51.6%). Due to bad climatic conditions in the Ngqushwa Local Municipality which hinders crop production, the majority of the households (83%) interviewed in the area did not cultivate their land. Only 11.5% cultivate all their arable land while 5.7% cultivate part of their arable land. Maize is the main

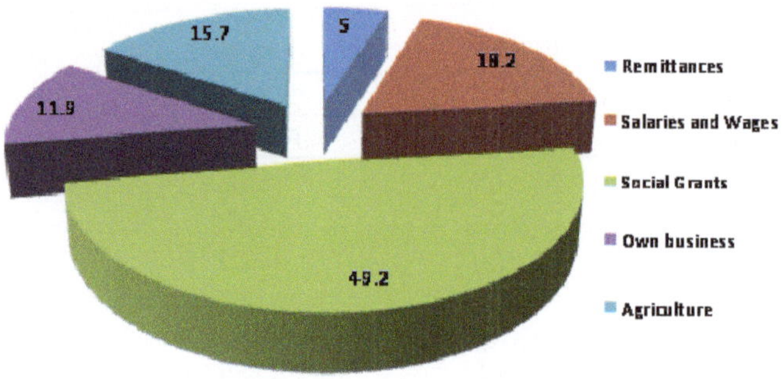

Figure 2. Household main source of income for the Hamburg community.

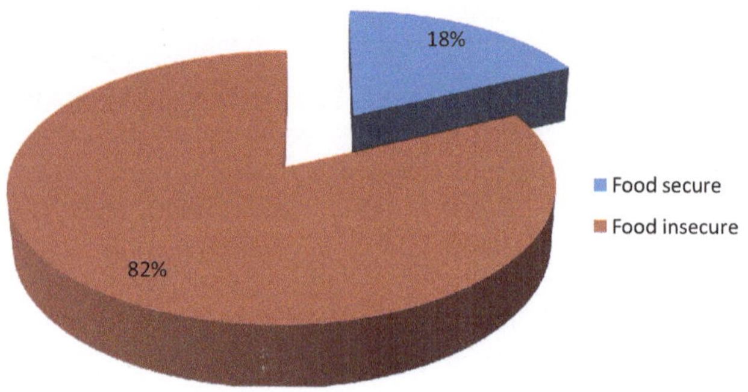

Figure 3. Households' access to enough food.

crop that the households grow. The majority of the households do not have fruit trees on their homesteads (74.8%) while only 25.2% had fruits trees. Fruits play an important role in the dietary requirements of human beings. They provide micro nutrients which are essential in human beings. Fifty five percent of the households preserve vegetables and fruits through freezing and drying while 44.7% do not preserve fruits and vegetables. Preservation helps in improving household access to food throughout the year. Agricultural products are mostly seasonal, therefore preservation helps households in keeping vegetables and fruits in seasons were there are available in abundance in the market. It is at this time that vegetables and fruits would be cheaper. During off season, vegetables and fruits would not be available in some extreme cases or expensive if available due to poor supply.

Fisheries

Ngqushwa Local Municipality is bounded on the East by the Fish River, on the South by the Indian Ocean and other numerous dams; rivers and streams provide plenty of opportunities for fishing. However most of the interviewed household did not use this opportunity. Only 15.7% of the households obtained some of their food from fishing. Fish are an important resource worldwide, especially as food. Fish has been an important source of protein for humans throughout recorded history. Research over the past few decades has shown that the nutrients and minerals in fish, and particularly the omega 3 fatty acids found in pelagic fishes, are heart-friendly and can make improvements in brain development and reproduction. This has highlighted the role for fish in the functionality of the human body.

Household income

In rural areas of South Africa where in most cases employment opportunities are minimal, the main sources of income of households are government grants, remittances and agriculture. Agriculture especially livestock production, is the main economic activities that take place due high incidents of droughts and poor soils which make crop production impossible. The government grants include child support, disability and old age pension grants. Figure 2 illustrates the distribution of households according to main source of income. In the area studied, 49.2, 18.2, 15.7, 11.9 and 5.0% household main source of income were grants, salaries and wages, agriculture, own business and remittances, respectively. Access to food is primarily determined by income since the majority of the households in rural areas of South Africa do not practice agriculture and depend on purchasing food instead of production. Most of the food consumed by the all the interview households were purchased from shops. Household income for January 2012 ranged from 0 to ZAR26 280 with the mean household income of ZAR2 987 considering that the mean household size was 5 members per household, meaning that on average each household member had an average of ZAR600 which is below the poverty datum line and this value is US$2 per day per person.

Access to food

From the 159 respondents, only 29 households had access to enough food while the majority did not have access to enough food. This distribution of access to enough food is shown in Figure 3.

Escalating food prices, particularly of maize and wheat which are the staple diet of the poor in South Africa, pose serious problems for the urban and rural poor as most are net buyers of food. Recent information from the Food and Agricultural Organization (FAO, 2009) supported by independent sources (Heady and Fan, 2008) suggest that food prices will increase steadily over the next decade even if there are some fluctuations and occasional drop in prices (Evans, 2009). Domestic electricity supply constraints and rising oil prices are examples of important factors driving the cost of food. The price of electricity in

South Africa almost doubled between 2008 and 2011. Petrol and diesel prices could spike to highs of around R16 per litre in South Africa under a worst case scenario as tensions between the United States of America and Iran rapidly become a key concern for global markets. The majority of the households had had 3 meals per day (67.3%). Only 1.3% of the household had only 1 meal per day while 11.9 and 19.5% had 2 and varying number of meals per day, respectively. Food shortages were experienced mainly towards the period when households receive grants, remittances and salaries (93.7%). Only 1.9% of the households experienced food shortages all the times while 4.4% of the households experienced food shortages irregularly.

Supply of clean water was very reliable to 74.8% of the households and unreliable to 25.2% of the households. In October 2010, the UN Human Rights Council affirmed that the right to safe, clean drinking water and sanitation is contained in existing human rights treaties and that states have a primary responsibility to ensure the full realisation of this and all other basic human rights. Food and water security are complex sustainable development issues, linked to health through malnutrition, but also to sustainable economic development environment and trade. Food and water security are an aspiration for all and their absence is an affront to human dignity. A major contributor to lack of access to clean water is obviously climate change and although it is impossible to halt its relentless progress, by becoming more environmentally aware at least we can help to slow it down.

Factors influencing access to enough food

There is a positive and significant relationship between the probability of a household accessing enough food and gender of head of household. In this study, descriptive statistics also confirm these findings. The majority of the food secure households were male headed whereas the majority of the food insecure households were female headed. This result is in line with the priori expectations. According to Musemwa et al (2010), females are normally involved in many household activities and most of them in the rural areas are not employed hence they do not have any other source of income. This therefore limits their access to enough food as most of the households in South Africa depend on food purchases. Past cultural practices also hindered women in going to schools. As a result, most women are not qualified to do professional jobs that generate high incomes hence low purchasing power and thus have limited access adequate food (Quisumbing 1996). In many developing countries, according to Udry et al. (1995) land is predominantly owned by men and transferred intergenerationally to males. Therefore women may lack access to land, water rights and livestock. In addition, even when women are able to access land, lack of ownership creates a disincentive to invest time and resources into sustainable farming practices, which in turn lowers production and results in less income and limited access to enough food for the household.

Household size, according to Montshwe (2006), is a useful unit of analysis given the assumptions that within the household resources are pooled, income is shared, and decisions are made jointly by responsible household members. Household requirements are many and one person in most case cannot handle them alone. In this study, there was a significant and negative relationship between household size and access to food. This is in line with the findings of Ankomah (2001). He found that the size of a household also influences the amount spent on food. Household food expenditure peaks at a household size of 4 and decline with an increase in household size beyond 4. In addition, large households have the lowest incomes in society. These groups of households are therefore more vulnerable to food insecurity and malnutrition.

Another important factor to consider is the level of education of the head of household since they are the decision makers in matters concerning the expenditure on food. According to Nkhori (2004), education increases the ability of households to use their resources efficiently and the locative effect of education enhances households` ability to obtain, analyse and interpret information. Education significantly affected access to enough food positively. The more the head of household is educated the more the household is likely to access enough food. Educated household heads were more likely to be absorbed in the labour market and would get remunerations that would improve their access to enough food.

Household Income is a measure of the combined incomes of all people sharing a particular household or place of residence. It includes every form of income, such as salaries and wages, retirement income, near cash government transfers like food stamps, and investment gains. Average household income can be used as an indicator for the monetary well-being of a country's citizens. Mean or median net household income, after taxes and mandatory contributions, are good indicators of standard of living, because they include only disposable income and acknowledge people sharing accommodation benefit from pooling at least some of their living costs. According to the results from the study, monthly household income significantly affected household access to food positively. Household that had more monthly income had more access to enough food. This is in line with earlier reports (Alaimo et al., 2001; Fraser et al., 2003)

Poultry ownership significantly increased household access to enough food. According to Ahuja et al. (2008), increasing the productivity of and returns to poultry birds does not represent a pathway out of poverty for a typical rural household, unless flock size is significantly increased. According to the descriptive statistics on average, each household from the study area kept 14 non-descript low yielding local birds in the backyard. Even though the ownership of few poultry birds does not contribute substantially to food access, it provides a mechanism to improve nutrition (particularly important in children) and alleviate credit constraints faced by the majority of the rural poor. On the one hand, when rural food markets are imperfect, and the availability and prices of grains and animal proteins at marketplaces are unpredictable, poultry farming serves as an inexpensive device for households to generate highly nutritious food items at minimal cost, because of the low input requirements and the low opportunity cost of family labour allocated to poultry care (Ali, 2007). On the other hand, when financial markets are imperfect, which is often the case in rural areas, the sale of birds helps cover recurrent minor expenditures, such as school fees, and to deal with unexpected shocks, such as medical fees (Permin et al., 2001). Investment in backyard poultry farming could thus enhance household access to enough food and reduce the vulnerability of landless and marginal households.

The none or poor provision of agricultural training in rural areas is a key factor that has greatly limited agricultural development in developing countries and contribution of agriculture to food security (Bailey et al., 1999). Access to food was significantly affected by agricultural training positively. Household that had access to agricultural training had better access to enough food. Agricultural training has a direct impact on agricultural productivity and on the performance of ancillary businesses and trade. It also stimulates implementation of knowledge driven economic growth strategies and poverty reduction.

Strategies for enhancing access to enough food

Policy makers in agriculture should partner with the education department such that they design and implement a curriculum that has got a larger bias towards improving land use and productivity starting at primary level such that even if a person drops out before getting to tertiary school, the person would be having the basic needed agricultural knowledge. The curriculum should ensure that the agriculture is sustainable in terms of both productivity and intergenerational continuity. Most farmers in developing countries as also observed in the current study only have access to secondary Education. This, therefore, implies that agricultural education in the curricula should be of high quality, simple and clear as well as to make careers in farming and related branches of agriculture more striking. This can only be made possible by fine-tuning the way agriculture is presented to students at primary level.

Promotion of small-scale agricultural production remains central to food security in most poor countries, to provide food and income for those at risk. Sustainable technology improvements in agriculture can increase the productivity of agriculture thereby improving household access to enough food. Central elements of this policy include research and extension linkages, drawing upon indigenous knowledge, and improved input supply.

Income generating projects including livestock and non-farm activities will allow rural families to use time previously spent on low productivity work to switch to jobs with higher returns. Non-farm work generates incomes not closely connected to farm income, thus helping to stabilize household incomes. Income generation is important in South Africa because families are usually more reliant on purchased foods.

Conclusion

The majority of the household in Ngqushwa Local Municipality in the Eastern Cape Province of South Africa have limited access to enough food and do not utilise the opportunity that they are located in the coast. Coastal location could enhance household access to food through venturing into activities such as fishing and also handcraft since the area is a tourist destination. In addition, all the household depend mostly on food purchases rather than own production. This therefore mean that the people in the area to do not explore agriculture up to its potential. To increase household access to food, income generating projects and the promotion of small-scale agricultural especially poultry production should be implemented. Educating rural households using informal methods is of paramount importance and should be included and prioritized as most of them are elderly and illiterate.

REFERENCES

Ahuja V, Dhawan M, Punjabi M, Maarse L (2008). *Economics of village poultry*. Summary. Mimeo. NDDB-FAO South Asia Pro-Poor Livestock Policy Programme, New Delhi.

Alaimo K, Olson C, Frongillo E, Briefel R (2001). Food insufficiency, family income and health in US preschool and school-aged children. Am. J. Public Health. 91 (5):781-786.

Ali J (2007). Livestock sector development and implications for rural poverty alleviation in India. *Livestock Research for* Rural Development, 19(2). Retrieved January 25, 2012, from http://www.lrrd.org/lrrd19/2/ali19027.htm.

Ankomah SB (2001). Addressing food insecurity in South Africa. The National Institute for Economic Policy. Paper presented at the SARPN conference on Land Reform and Poverty Alleviation in Southern Africa Pretoria 4th and 5th June 2001

Bailey D, Barrett CB, Little PD, Chabari F (1999). Livestock markets and risk management among East African pastoralists: A review and research agenda. Utah University, USA.

Bank L, Minkley G (2010). Going Nowhere Slowly? Land, Livelihoods and Rural Development in the Eastern Cape. Fort Hare Institute for Social and Economic Research, Centre for African Studies at the

University of Cambridge.
Chimonyo M, Kusina NT, Hamudikuwanda H, Nyoni O (1999). A survey on land use and usage of cattle for draught in a smallholder farming area of Zimbabwe. J. Appl. Sci. Southern Afr. 5(2):111-121.
Coetzee L, Montshwe BD, Jooste A (2004). The Marketing of Livestock on communal lands in the Eastern Cape Province: Constraints, Challenges and Implications for the Extension Services. *South* Afr. J. Agric. Exten. 34(1):81-103.
Eastern Cape Socio-Economic Consultative Council (ECSECC) (2000). Rural Development Framework. Draft document.
Evans A (2009). The Feeding of the Nine Billion: Global Food Security for the 21st Century. A Chatham House Report. London: The Royal Institute for International Affairs.
Food and Agricultural Organization (FAO) (2009). The State of Agricultural Commodity Markets. Rome, Italy: Food and Agricultural Organization of the United Nations
Fraser G, Monde N, Van Averbeke W (2003). Food Security on South Africa: A case study of rural livelihoods in the Eastern Cape Province. In: Lieb Nieuwoudt and Jan Groenewald (eds), The Challenge of Change: Agriculture, Land and the South African Economy University of Natal Press, Pietermaritzburg. pp. 171-183.
Gujarat DN (2003). Basic Econometrics (4th Ed). McGraw Hill. New York.
Hall R, Aliber M (2010). The Case for Re-Strategising Spending Priorities to Support Small Scale Farmers in South Africa. Institute for Poverty, Land and Agrarian Studies (PLAAS), University of the Western Cape, Working Paper. p. 17.
Heady D, Fan S (2008). Anatomy of a Crisis: The Causes and Consequences of Surging Food Prices. IFPRI Discussion Paper 00831 (December). International Food Policy Institute: Washington, DC.
Hebinck P, Lent PC (2007). Livelihoods and landscapes: The people of Guquka and Koloni and their resources. Leiden, Boston.
Hosmer DW, Lemeshow S (2000). Applied Logistic Regression (2nd Ed). Wiley-Interscience Publication, New York.
Kleinbaum DG (1994). Logistic regression: A self learning text. Springer-Verlag, New York, USA.
Labadarios D, Davids YD, Mchiza Z, Weir-Smith G (2009). The Assessment of Food Insecurity In South Africa. Unpublished paper, Centre for Poverty, Employment and Growth, Human Sciences Research Council.
Montshwe DB (2006). Factors affecting participation in mainstream cattle markets by small-scale cattle farmers in South Africa. MSc Thesis, University of Free State, RSA.
Musemwa L, Chagwiza C, Sikuka W, Fraser G, Chimonyo M, Mzileni N (2007). Analysis of Cattle Marketing Channels Used by Small Scale Farmers in the Eastern Cape Province, South Africa. Livestock Research for Rural Development 19(9) Article #131.
Musemwa L, Mushunje A, Chimonyo M, Mapiye C (2010). Low cattle market off-take rate in the communal areas of Eastern Cape Province of South Africa: Causes and mitigation strategies. J. Sustainable Develop. 12 (5):209-226.
Nkhori PA (2004). The impact of transaction costs on the choice of cattle markets in Mahalapye district, Botswana. MSc Dissertation, University of Pretoria, RSA.
Permin A, Pedersen G, Riise JC (2001). Poultry as a Tool for Poverty Alleviation: Opportunities and Problems Related to Poultry Production at Village Level. In R.G. Alders, P.B. Spradbrow (eds.) SADC Planning Workshop on Newcastle Disease Control in Village Chickens. Proceedings of an International Workshop. Maputo, Mozambique, ACIAR Proceedings. p. 103.
Quisumbing A (1996). Male-Female Differences in Agricultural Productivity. World Development 24:1579-1595
Udry C, Hoddinott J, Alderman H, Haddad L (1995). Gender Differentials in Farm Productivity: Implications for Household Efficiency and Agricultural Policy. Food Policy 20:407-423.

Estimating technical efficiency of small scale beef cattle fattening in the lake zone in Tanzania

S. N. Mlote[1], N. S. Y. Mdoe[2], A. C. Isinika[3] and L. A. Mtenga[4]

[1]Ministry of Livestock and Fisheries Development, P. O. Box 9152, Dar es Salaam, Tanzania.
[2]Department of Agricultural Economics and Agribusiness, Sokoine University of Agriculture, P. O. Box 3007, Morogoro, Tanzania.
[3]Institute of Continuing Education, Sokoine University of Agriculture, P. O. Box 3044, Morogoro, Tanzania.
[4]Department of Animal Science and Production, Sokoine University of Agriculture, P. O. Box 3004, Morogoro, Tanzania.

This study examines the technical efficiency of small-scale beef cattle fattening in eight districts of Mwanza and Shinyanga regions, the lake zone in Tanzania. Data were collected using a structured questionnaire which was administered to 90 randomly selected cattle fatteners from the eight districts. Data were analyzed using descriptive methods and stochastic frontier production function approach. The estimated stochastic frontier production function showed that the herd size was the critical variable that affected weight gain in beef cattle fattening and hence beef output. The efficiency analysis results show that the estimated farm level technical efficiency ranged between 48 to 98% with a mean of 91%, indicating that the majority of the beef cattle fattening operators in the study area are efficient though not at 100%. The socio-economic determinants of the respondents' technical efficiency were age, education, experience, extension services and ethnicity. These findings can be used by the Ministry of Livestock and Fisheries Development extension agents to promote beef cattle fattening in areas where beef cattle fattening is not practiced in the country.

Key words: Average daily weight gain (ADG), cattle fattening operators, enterprises, technical efficiency, Tanzania shorthorn zebu (TSZ).

INTRODUCTION

The beef sub sector in Tanzania is dominated by indigenous cattle breeds which account for about 94 percent of the national cattle herd kept traditionally under pastoral and agro-pastoral production systems (Ministry of Livestock Development (MLD), 2006). The remaining 6%, fall under improved dairy and beef cattle management systems. Indigenous cattle are made of Tanzania shorthorn zebu (TSZ) and boran and ankole [Ministry of Livestock and Fisheries Development (MLFD), 2011]. The Tanzania short horn zebu (TSZ) and ankole are the dominant indigenous breeds in the country.

These types are well known for their low genetic potential in terms of beef production (MLD, 2006). As a result, the contribution of livestock to the national economy as well as to the household's well-being is low. In 2010, for example the livestock sector contributed only 16 and 3.8% to the agricultural gross domestic product (AGDP) and national gross domestic products (NGDP), respectively and the sector grew by only 3.4% compared to 4.4% for the crop sector (United Republic of Tanzania-URT, 2011). There are two major factors contributing to this situation. First, low average meat production per head especially in the traditional beef production system where average cattle productivity is estimated to be 107 kg per head compared to the average of 166 kg per head for other developing countries and 198 kg per head for the world (FAOSTAT, 2005).

Second, the low rate of adoption of improved livestock production technologies such as use of artificial insemination to improve the size of animals and limited use of fattening technologies is another contributing factor (URT, 2011). Low productivity of beef cattle in addition with other factors such as low purchasing power for the most of the population in the country has prevented Tanzania from meeting the World Health Organization (WHO) recommendation of a minimum per capital consumption of meat of 50 kg per year. The per capita consumption of meat in Tanzania is estimated to be 12 kg, compared to the world average of 46.6 kg/year (FAOSTAT, 2005; MLFD, 2012).

In order to improve the contribution of livestock to the national economy and well-being at the household level, the government has formulated the Livestock policy and strategy geared towards improving livestock productivity and commercialization of the livestock in the traditional sector (MLD, 2006, 2010, 2011). In the course of implementing the strategy, efforts have been made to encourage nomadic pastoralists with large cattle herds to reduce their herds and settle in areas where they are allocated land for cattle grazing. Besides encouraging pastoralists to settle, beef cattle fattening has been earmarked as one of the means to improve beef cattle productivity. Beef cattle fattening is a recent undertaking in the country, but it is increasingly gaining importance in some traditional livestock keeping regions of Tanzania. According to MLFD (2012), the number of fattened animals in the country is estimated to be 132,229 heads of cattle. Out of these, the lake zone (Mwanza, Shinyanga and Kagera regions) which is the study area account for 46.6%. Beef cattle fattening has the potential to improve the Tanzanian economy, if well harnessed. However, this depends, to a large extent, on the efficiency of the production system, hence livestock productivity. Efficiency is concerned with relative performance of the processes used in transforming given input into output (Abu and Asember, 2011; Otieno et al., 2012). Efficiency measurement has received significant attention from researchers in different fields of study (Abu and Asember, 2011; Facayode et al., 2011; Dawing et al., 2011). Measuring efficiency is vital because it can guide resource utilization and may lead to considerable resource savings, which have important implication for both policy formulation and farm management (Bravo-Ureta and Rieger, 1991). However, to date there is no study carried out to investigate the technical efficiency of beef cattle fattening enterprises undertaken in the country.

This study investigates the technical efficiency of beef cattle fattening enterprises practiced in Shinyanga and Mwanza regions of Tanzania. The specific objectives of the study are to: (i) estimate the technical efficiency of beef cattle fattening in the study area and (ii) examine the relationship between technical efficiency and socio-characteristics of cattle fattening operators. Subsequent parts of the paper are organized into five main sections which include: the theoretical framework, methodological section, results and discussion, conclusion and recommendations.

Theoretical framework

Economic theory identifies three important efficiency measures (Boris et al., 1997; Effiong and Onyenweaku, 2006). These include; the allocative, economic and technical efficiency. The allocative efficiency (AE) reflects the ability of the farm to use the inputs in optimum proportions given their respective prices and the production technology. Economic efficiency (EE) is defined as the capacity of a firm to produce a predetermined quantity of output at minimum cost for a given level of technology. Technical efficiency (TE) is the measure of the farms success in producing maximum output from a given set of inputs. Alternatively, it is the ability to operate on the production frontier or the isoquant frontier (Effiong and Onyenweaku, 2006). This study examines the TE for beef cattle fattening operators in the study area.

METHODOLOGY

Sampling and data collection

Data for this study were collected from beef cattle fattening enterprises in Mwanza and Shinyanga regions in the lake zone in Tanzania. Geographically, Mwanza and Shinyanga regions lie in the Northern part of the country.

The sampling frame included all beef cattle operators who are engaged in beef cattle fattening in selected districts of Magu, Nyamagana and Sengerema in Mwanza region and Kahama, Kishapu, Meatu, Bariadi and Maswa districts in Shinyanga region as shown in Figure 1. The list of beef cattle fattening entrepreneurs was collected from the respective local government authority offices in the districts, from which 90 respondents were randomly selected. Global positioning system (GPS) devices were used to geo-reference all points where the interview were held.

Data were collected through face to face interviews conducted by the principal researcher during the dry season between July and September, 2011, with the assistance of local experienced interviewers who were adequately trained prior to the surveys. The data were collected during dry season because this is when fattening is practiced. During the wet season animals are able to get enough forage from the rangelands and are normally in good condition hence hardly any fattening was done. Data gathered include information on socio-economic variables of the entrepreneurs such as age, education, household size, location, experience, access to credit and extension services. Information was also collected on the inputs as well as output in terms of weight gain for fattened beef cattle. The inputs data collected included; feeds used, amount of minerals used, and treatment/drugs used while output data collected was the weight of beef cattle after the fattening period.

It was observed during the field survey that beef cattle fattening operators did not weigh their animals before and after fattening.

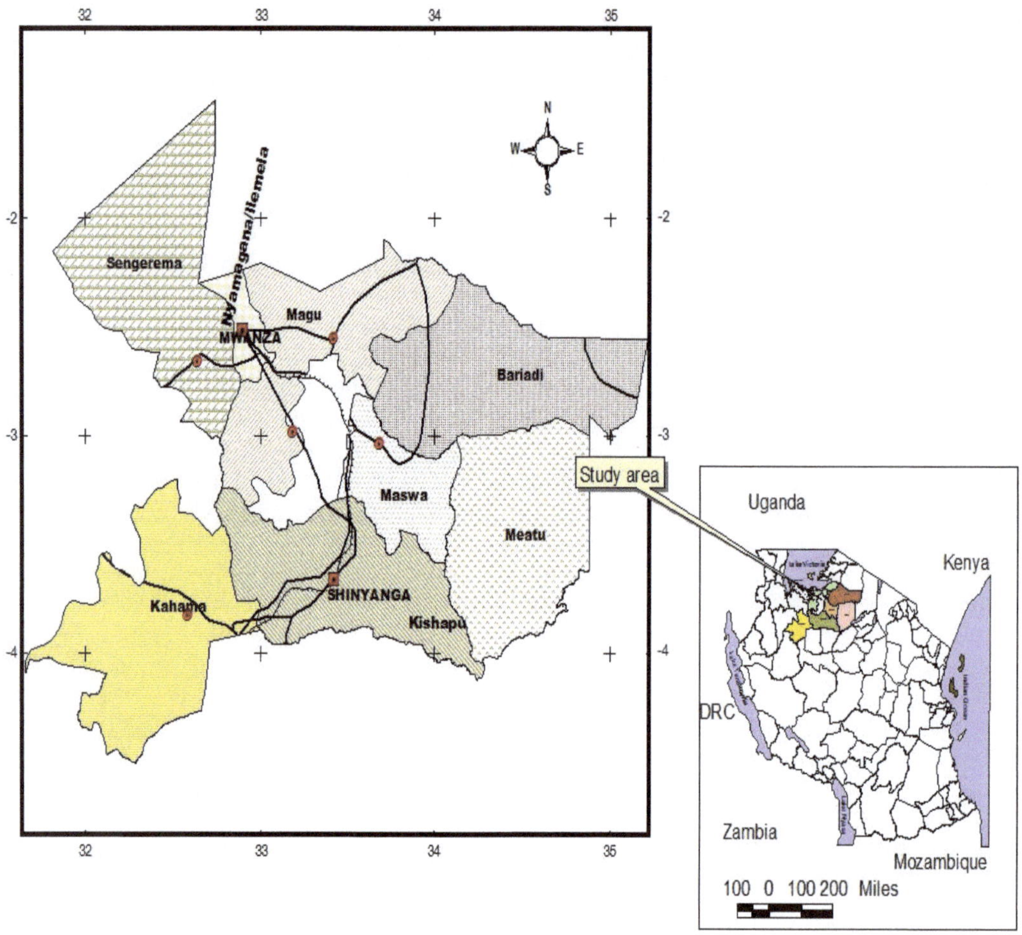

Figure 1. Location of the study area.

Also, they did not keep records of their business. In this regard, a control group of 105 animals were selected from Kahama district and used to determine the Average Daily Gain (ADG) along with 5675 animals (belonging to the respondents) whose weights were tracked during the study period covering 12 to 16 weeks (3 to 4 months). The animals under fattening were of the same breed that is Tanzanian short horn zebu (TSHZ) and same fodder conditions originating within the lake zone. Most of them were mature bulls and few culled cows purchased by the respondents from the livestock markets in the lake zone. A weight band tape, which is designed to take the live weight of animals such as; cattle, pigs, sheep and goats was used to measure the live weight of the study animals from each respondent (Figure 2a and b). The tape measure was used in the field because cattle weighing instruments are costly and heavy to transport and it is a practical field techniques, particularly in less developed countries. The animals' heart girth measurements were taken to predict animal's weights at entry, every two weeks thereafter and at exit.

Analytical framework

Technical efficiency can be estimated using computer software such as LIMDEP econometrics packages (Green, 1993) or the Frontier 4.1 program (Coelli, 1996) to find the maximum likelihood estimates for the parameters of the stochastic frontier production function. The stochastic frontier production function model is popular because of its flexibility and ability to closely marry economic concepts with modeling the reality (Dawang et al., 2011). Battese and Corra, 1997 applied the technique to the pastoral zone of the Eastern Australia.

Recently, many studies have been done in Africa using the technique which include; Facayode et al. (2011), for Technical efficiency analysis on small scale rabbit production in South to West Nigeria, Dawang et al., 2011 used the technique to estimate the profitability and technical efficiency of artisanal fishermen also in Nigeria, while Oleke and Isinika (2008) used the technique to assess the technical efficiency of commercial egg production in Tanzania.

Stochastic frontier production function analysis (SFA)

Farmers in agricultural production aim at maximizing production, minimizing costs and maximizing profits. However, though every producer may attempt to optimize, not all of them may succeed in their efforts. Given the same inputs and technology, some farmers will produce more efficient than others. Econometrics estimation techniques allow for the fact that deviations of observed choices from optimal ones are due to either failure to optimize (inefficient) or due to random shocks.

The SFA produces efficiency estimates or scores of individual producers, thus identifying those who need interventions and corrective measures. The variations of efficiency scores can be

Figure 2. (a) Weight band tape. (b) Estimating live weight using weight band tape.

related to producer's characteristics like age, size, ownership, location, education etc. Thus one can identify source of inefficiency. The SFA also provides a powerful tool for examining effects of interventions. A stochastic production frontier model using the FRONTIER 4.1 statistical software developed by Coelli (1996) was used to estimate the technical efficiency of beef cattle fattening in the study area. This analysis has the advantage of separating the impact of weather and other factors that contribute to variations in technical efficiency.

A frontier model with output-oriented technical efficiency according to Battese and Coelli (1995) is specified in Equation 1.

$$Y_i = X_i \beta + (\varepsilon = v_i - \mu \quad (1)$$

Where; Y_{ij} = is the (logarithm of) output in terms of weight gain obtained by the i-th respondents from the jth animal (for i = 1, 2,3 N; and j = 1,2,3, ...m), X_{ik} = is the corresponding matrix of K inputs, β = is a kx1 vector of unknown parameter to be estimated, ε_i is the disturbance term comprised of two independent components, V_i and U_i, V_i = are random variables which are assumed to be i.i.d. N (0, δ^2_v), and independent of the U_i 's and U_{is} = are non negative random variables which are assumed to account for technical inefficiency in production and are assumed to be independently distributed as truncations at zero of the N (m_i, δ^2_u).

The maximum likelihood estimation of Equation 1 provides the estimators for β_j and variance estimators and other relationships denoted as;

$$\delta^2 = \delta_v^2 + \delta_\mu^2. \quad (2)$$

$$\gamma = \delta_v^2 / \delta^2 \quad (3)$$

$$\lambda = \delta_v^2 / \delta_\mu^2. \quad (4)$$

Where, δ^2, δ^2_v, δ^2_u are the overall variance of the model, variance of the random error, and variance of the technical inefficiencies respectively. According to Battese and Corra (1997), gamma (γ) is the total output made on the frontier function which is attributed to technical efficiency. The parameter gamma (γ) has a value between zero and one. Hence (1- γ) measures the technical inefficiency of the farms. The parameter lambda (λ) is expected to be greater than one. This condition according to Facayode et al. (2011) indicates a good fit for the model and correctly specified error term (V_i and U_i). From Equation 1, the polynomial of degree 2 technical efficiency equation for beef cattle fattening in the study area was specified as follows;

$$\begin{aligned} InY_i &= \beta_o + \beta_1 InX_1 + \beta_2 InX_2 + \beta_3 InX_3 + \beta_4 InX_4 + \beta_{12} InX_1 InX_2 + \beta_{13} InX_1 InX_3 \\ &+ \beta_{14} InX_1 InX_4 + \beta_{23} InX_2 InX_3 + \beta_{24} InX_2 InX_4 + \beta_{34} InX_3 InX_4 \\ &+ \tfrac{1}{2}(\beta_{11} InX_1 + \beta_{22} InX_2 + \beta_{33} InX_3 + \beta_{44} InX_4)^2 + \beta_5 X_5 + \beta_6 X_6 \\ &+ (v_i - \mu_i) \end{aligned} \quad (5)$$

Where; In = denotes the natural logarithm (base e), Y_i = is the total weight gain over the study period for the jth animal of the ith respondent for I = 1, 2, 3...n and for j = 1, 2, 3...k, X_1 = Herd size in numbers denoted as HERDSIZE, X_2 = Supplementary feeds in kg denoted as FEED, X_3 = Treatment index denoted as TREAT, X_4 = Minerals index denoted as MINERAL, X_5 = Location of farmers denoted as LOC (Dummy), X_6 = Fattening potential according to farmers perception denoted as FATPOT (Dummy) and β_{ij} = A vector of parameters to be estimated

Technical inefficiency model

According to Battese and Coelli (1995) in the efficiency model, the one sided error term is specified as;

$$\mu_i = \delta_0 + \sum \delta_m Z_{mi} + \omega_{it} .. \quad (6)$$

Where; Zs are social economic variables used to explain efficiency differences among beef cattle fatteners, where the parameters δ_m are unknown, to be estimated and ω_{it} are iid random variable with zero mean and variance defined by the truncation of the normal distribution. The specific Z variables in the above model can be specified as follows:

$$U_i = \delta_0 + \sum\sum \delta_{ij} Z_{ij} \quad (7)$$

Where: U_i = the technical inefficiency of the ith respondent for i = 1, 2, 3, ..., N; Z_{ij} = the jth socio-economic variable for the ith respondent for i = 1, 2, 3, ..., N and j = 1, 2, 3, ..., k δ_0 = the intercept, and δ_j = the coefficient for the Kth variable.

Table 1. Average daily weight gain per animal.

Variable	Mean	Min	Max.	Std	Mode
Duration of fattening (days)	92	60	240	21.9	90
Average weight gain per animal over the study period (Kg)	58.6	37.7	151.1	13.6	56.6
Average daily weight gain per animal (Kg)	0.646	0.236	1.259	0.116	0.630

Source: Field survey data, 2011/2012.

The socio-economic variables represented in this study include; 1 = age, 2 = educational level in years, 3 = experience, 4 = housing type, 5 = Ethnic group, 6 = access to credit and 7 = extension services. The technical inefficiency in Equation 7 can be estimated if the technical inefficiency effects, U_i are stochastic and have particular distributional properties (Battese and Corra, 1997). Under the null hypothesis gamma (γ) = 0, the stochastic model reduces to a traditional average response function, thus no technical inefficiency effects exist. The null hypothesis was tested using log likelihood ratio test as follows;

$$\lambda = -2 \ln \{ L(H_0) \} - \ln \{ L(H_i) \}. \quad (8)$$

Where; $L(H_o)$ and $L(H_i)$ are values of likelihood function under the null hypothesis and the alternative (H_i) hypothesis, respectively.

RESULTS AND DISCUSSION

Determination of average daily weight gain (ADG)

The ADG was calculated by subtracting the initial weight from the final weight and dividing by the number of fattening days. The computed ADG is as shown in Table 1.

The ADG per animal was 0.646 kg and the mean fattening duration was 92 days. This finding compares well with similar finding by Mwilawa (2012) which revealed that ADG for Boran beef cattle kept under station conditions in Tanzania was 0.889 kg which implies that traditional beef cattle respond to fattening.

Results of technical efficiency analysis

Descriptive statistics of variables used for the production function frontier estimation are presented in Table 3. The mean total animal weight gain per farmer (obtained by multiplying the average daily weight gain per animal (0.646 kg/day) with the number of animals per enterprise and the duration of fattening) was 3607.75 kg with a standard deviation of 4292.2 kg. The large value of standard deviation implies that the beef cattle fattening operators were raising different numbers of animals with a minimum of 4 animals and a maximum of 330 animals.

The mean quantity of supplementary feed used per farmer was 1053.8 kg with a minimum of 0 and maximum of 7800 kg. There was variability in feed usage by cattle fatteners measured by the minimum and maximum amounts, which was confirmed by the large value of the standard deviation. This may be due to differences in feed ranging from exclusive standing hay to different additional quantities of supplementary feeds. Feed scarcity and the high cost of feeds as it was observed during the survey, making it difficult for most operators to provide adequate feed for speedy weight gain of the animals. This implies that improvement in feed availability in the study area may lead to improvements in animals' weight gain and reduce the duration of beef cattle fattening.

The mean mineral index (calculated as a ratio of feeds) was 0.004 while the mean for the treatment index was 0.32. The location of the respondent (urban = 1 or peri-urban = 0) was included in the model as dummy variable. The fattening potential according to farmers perception of the animals (un-castrated bulls = 1 or castrated bulls and cows = 0) was also included as a dummy variable to test farmers' preferences to fatten un-castrated beef cattle bulls over castrated bulls and cows as observed during the survey. A summary of all the descriptive statistics is presented in Table 2.

Most of the entrepreneurs who were engaged in beef cattle fattening were middle aged being 39.5 years on average age with a standard deviation of 9.2 years. This shows that most farmers were middle aged people. The average year of schooling was 6.8 years with a minimum of 0, a maximum of 16 and a standard deviation of 3 years, hence most of the farmers attended at least primary school. The average experience in beef fattening was 5.4 years with a standard deviation of 4.6 years implying that most of the farmers are relatively new in the beef fattening enterprises, but the most experienced operator had been 26 years in the business. Other variables such as the average housing type, credit provision, extension services and ethnic group were included in the model as dummy values.

Results of stochastic frontier production analysis

Equation 8 was used to test the assumption of the trans-log model specification for the beef cattle fattening. The maximum likelihood estimates of the stochastic frontier production function are presented in Table 3. The findings show that, the value of lambda (λ) was greater than one (17.36) implying goodness fit for the estimated model and also correct specification of the error term

Table 2. Descriptive statistics of variables used in the stochastic frontier analysis.

Variable	Notation	Mean	SD	Min	Max
Total animal weight gain per farmer (kg) for the study period	WGAIN	3607.8	4292	151.1	18699
Number of animals per farmer (N)	HERDSIZE	63.6	76	4	330
Feeds usage (kg)	FEED	1053.8	1214	0	7800
Treatment (Index)	TREAT	0.32	1.8	0	16.7
Mineral Intake (Index)	MINERAL	0.004	0.006	0	0.343
Location	LOC	0.45	0.5	0	1
Fattening Potential	FATPOT	0.4	0.5	0	1
Age (years)	AGE	39.5	9.2	24	69
Education (years)	EDUYRS	6.8	3	0	16
Experience (years)	EXP	5.4	4.6	1	26
Housing type	HOUSE	0.8	0.4	0	1
Credit provision	CREDIT	0.06	0.2	0	1
Extension services	EXT	0.6	0.5	0	1
Ethnic group	ETHNIC	0.8	0.4	0	1

Table 3. Maximum Likelihood estimates of trans-log production function.

Variable	Par.	Expected sign	Coeff.	SE	t-ratio
Constant	β_0		4.874	0.13	37.451
LNHERDSIZE	β_1	-	6.036*	3.232	1.868
LNFEED	β_2	+	1.031	1.214	0.85
LNMINERAL	β_3	-	3.276*	1.895	-1.729
LNTREAT	β_4	+/-	1.758*	1.06	-1.659
LNHERDSIZE^2	β_5	+/-	2.581*	1.617	-1.596
LNFEED^2	β_6	-	-0.613	0.608	-1.009
LNMINERAL^2	β_7	+	1.637*	0.949	1.724
LNTREAT^2	β_8	+	0.915*	0.535	1.71
LN HERDSIZE X FEED	β_9	+	0.093**	0.041	2.284
LN HERDSIZE X MINERAL	β_{10}	+	0.005	0.049	0.097
LN HERDSIZE X TREAT	β_{11}	-	-0.043	0.046	-0.942
LN FEED X MINERAL	β_{12}	+/-	0.025	0.033	0.749
LN FEED X TREAT	β_{13}	+/-	-0.006	0.037	-0.168
LN MINERAL X TREAT	β_{14}	+/-	-0.03	0.053	-0.565
LOCATION	β_{15}	+/-	0.015	0.043	0.344
FATPOT	β_{16}	+/-	0.033	0.038	0.867
Variance parameter					
Sigma-squared	σ^2		0.082***	0.026	3.154
Gamma	γ		0.853***	0.066	13.02
log likelihood (LLF)			46.18		
Lambda	λ		17.36		
Inefficiency effect					
AGE	δ_1	-	-0.044**	0.021	-2.1
AGE^2	δ_2	+/-	0*	0	1.97
EDUC	δ_3	-	0.314***	0.09	3.485
EDUC^2	δ_4	-	-0.02***	0.006	-3.542
EXP	δ_5	-	-0.073*	0.04	-1.814
EXP^2	δ_6	+/-	0.004*	0.002	1.83

Table 3. Contd.

HOUSING	δ_7	+/-	0.085	0.155	0.549
CREDIT	δ_8	-	-0.104	0.14	-0.746
EXTENSION	δ_9	-	-0.366**	0.153	-2.387
ETHNIC	δ_{10}	-	0.531***	0.173	-3.068

Level of Significance; *** α = 0.01; ** α = 0.05; * α = 0.1

distribution. This was also evident from the estimated Gamma (Υ) value of 0.853 which measures the level of inefficiency of the production system, this was significant at α = 0.05 and was above the 50 percent of a unity. These results imply that within the study area, about 85.3% of the variation in beef cattle fattening is explained by inefficiency, whereas the remaining 14.7% (that is 1-Υ = 0.147) is due to measurement error and specifications bias as well as other factors that are not incorporated in the stochastic frontier and inefficiency effects models. This further confirms that the technical inefficiency effects are significant in the estimated model. The variance parameter (σ^2) was also positive (0.082) and significantly greater than zero at α = 0.05.

The inefficiency model estimates

The inefficiency analysis of the stochastic frontier indicated the coefficients for age, education^2, experience; extension contact and ethnic group were all negative and significantly different from zero, suggesting that increasing these variables will increase technical efficiency. However, ageing (age^2) does not contribute to technical efficiency such that the coefficient for this variable is zero. The estimated coefficient on education is positive and significantly greater than zero indicating increase in production inefficiency. However, as the education level increases (Education^2) the effect on technical efficiency is negative and statistically significant indicating reduction in production inefficiency. In other words, education (years of schooling) had a positive correlation with technical efficiency. These results are consistent with Ogunniyi (2011) and Dawing et al. (2011) who established that for every extra year of schooling that a farmer/fisher got, there was an increase in the chance or probability of being technically efficient.

The estimated coefficient on experience is negative and statistically significantly different from zero, contributing to decrease in technical inefficiency by 7.3% for every year of increase in experience. Credit access showed a negative relationship with technical inefficiency as expected though it was not significant implying that only few farmers had access to credit. Having access to credit reduced the level of inefficiency of respondents by 1.04%. This means policies that will make micro-credit from government and non-governmental agencies accessible to these farmers will go a long way to address their resource use inefficiency problems. The coefficient for extension service contact was negative and significantly different from zero indicating that if a respondent had contact with extension services they were likely to experience increase in technical efficiency equivalent to about 3.66% average daily weight gain of the fattening animals. The Ethnic group was negative and significant (α < 0.05) indicating that Sukuma group (denoted as 1) were more involved in beef cattle fattening compared to other ethnic groups (denoted by zero).

Technical efficiency score analysis

The frequency distributions of technical efficiency (TE) scores are presented in Table 4. The findings show the beef fattening enterprises achieved on average 91% level of efficiency, ranging from 48 to 98% with a wide range of efficiency variation among the operators (Appendix 1). About 81.6% of beef cattle fattening operators operate at 91% TE or more while 13.8% were able to achieve 81% up to 90% TE levels. The results also suggest that about 98.7% of the respondents were able to achieve above 50% of the technical efficiency. Only 1.1% of the enterprises operated at 50% or less (Appendix 1). These results suggest that the majority of the respondents operate efficiently though not at 100%, there is still room for improving technical efficiency of few beef cattle fattening enterprises in the study area.

CONCLUSION AND RECOMMENDATIONS

This study applied the stochastic frontier model to estimate technical efficiency of beef cattle fattening. Results give efficiency scores that range from 48 to 98%, suggesting that there is room for improving the low performing cattle fattening enterprises. Further, it was found that education, experience, access to credit, extension services and ethnicity of cattle fattening operator contribute negatively to inefficiency that is, positively to efficiency improvement in cattle fattening. On the other hand, location and housing type did not have a positive influence on efficiency in cattle fattening, probably indicating that these variables did not render any technological differences among the cattle fattening entrepreneurs in the study area. All these findings have important implications on improving beef fattening

Table 4. Technical Efficiency scores for the sampled beef cattle operators.

Technical efficiency level (%)	No. of farmers	Percentage
<50	1	1.1
51-60	1	1.1
61-70	1	1.1
71-80	1	1.1
81-90	12	13.8
>91	71	81.6
Total	87	100
Mean	0.91	
Standard Deviation	0.08	
Minimum	0.48	
Maximum	0.98	

efficiency in Tanzania. The positive relationship of education and extension on technical efficiency imply providing appropriate knowledge and skills would facilitate better use of the available cattle fattening technologies by the operators who are currently operating inefficiently. Necessary interventions would include improving fattening operators' knowledge on feeding and disease control methods.

The negative relationship between access to credit and inefficiency implies that policies that will improve access to credit from government and non government agencies would increase the cattle fattening operators' ability to acquire better technologies (better cattle breeds and feeds through acquisition of mini to medium ranches where pastures could be improved and herd size expanded) with the potential of increasing productivity for domestic and export market. The efforts by the government to establish an agricultural bank is likely going to improve access to credit by farmers in Tanzania including cattle fattening operators if the bank conditions for credit will be favorable. These findings can be used by the Ministry of Livestock and Fisheries Development extension agents to promote beef cattle fattening in areas where beef cattle fattening is not practiced in the country.

ACKNOWLEDGEMENT

The authors are grateful to the livestock farmers in the study areas of Mwanza and Shinyanga regions for sparing their time during survey and questionnaire administration and to the permanent secretary, Ministry of Livestock and Fisheries Development for financial support through ASDP programme.

REFERENCES

Abu O, Asember DJ (2011). Opportunities for smallholder spinach farmers in Nigeria; A profit efficiency analysis. J. Econ. 2(2):75-79.
Battese GE, Corra GS (1997). Estimation of a Production Function Model with Applied to the Pastoral zone of Eastern Australia. Aust. J. Agric. Econ. 21:169-179.
Battese GE, Coelli TJ (1995). A model for technical inefficiency effect in Stochastic Frontier production for panel data. Empirical Econ. 20:325-345.
Boris E, Bravo-Ureta and Antonio EP (1997). Technical, Economic, and Allocative efficiency in peasant farming. Evidence from the Dominican Republic. Dev. Econ. 35(1):48-67.
Bravo-Ureta BE, Rieger L (1991). Dairy farm efficiency measurement using stochastic frontier and neo-clssical duality. Am. J. Agric. Econ. 73:421-428.
Coelli TJ, (1996). A guide to FRONTIER Version 4.1. A computer program for Stochastic Frontier Production and Cost Function Estimation. Centre for Efficiency and Productivity Analysis (CEPA) Working Papers. Dept. of Econometrics University of New England, Armidale, NSW 2351, Australia. http://www.une.edu.au/econometrics/cepawp.htm.
Dawang NC, Ayuba D, Matawal OM (2011). Estimates of profitability and Technical Efficiency of Artisanal Fisheries. A case study of Natural Lakes from Plateau State Nigeria. Asian J. Agric. Sci. 3(6):516-523.
Effiong EO, Onyenweaku CE (2006). Profit efficiency in Bloiler production in Akwa Ibom State, Nigeria. Department of Agricultural economics, Michael Okpara University of Agriculture, Umudike, Umuakwa, Abia State, Nigeria. Int. J. Agric. Rural Dev. 7(1):72-79.
Facayode SB, Rahji MAY, Fayeye TR Yinusa SO (2011). Technical Efficiency of Small Scale Rabit Production in South-West, Nigeria. J. Agric. Res. Dev. 10(1).
FAO (Food and Agriculture Organization of the United Nations) (2005). The State of food and Agriculture, Livestock Balance. Produced by the Electronic Publishing Policy and Support Branch. Communication Division.
Green WH (1993). Limdep Version 8.0 Econometric Modeling Guide. Econometric Software, Inc, Plainview, New York. P. 201.
MLD (Ministry of Livestock Development) (2006). National Livestock Policy (NLP). http://www.mifugo.go.tz/
MLFD (Ministry of Livestock and Fisheries Development) (2010). Livestock Sector Development Strategy. Government document.
MLFD (Ministry of Livestock and Fisheries Development) (2011). Livestock Sector Development Program. Government document.
MLFD (Ministry of Livestock and Fisheries Development) (2012). Budget speech 2012/13. http://www.mifugo.go.tz/
Mwilawa AJ (2012). Effects of Different Diets on Weight Gain, Carcass and Meat Quality Characteristics of two Indigenous Cattle Breeds of Tanzania. A Thesis Submitted in Partial Fulfilment of the Degree of Doctor of Philosophy of Sokoine University of Agriculture. P. 245.
Ogunniyi LT (2011). Profit Efficiency among maize producers in Oyo State, Nigeria. ARPN J. Agric. Biol. Sci. pp. 11-17.
Oleke JM, Isinika AC (2008). Assessing the technical efficiency of Commercial egg production in Tanzania for improved livelihoods. J.

Dev. Agric. Econ. 3(8):343-352.
Otieno DJ, Hubbard L, Ruto E (2012). Determinants of Technical Efficincy in beef cattle production in Kenya. Selected paper prepared for presentation at the International Association of Agricultural Economist (IAAE) Triennial Conference, Foz do Iguacu, Brazil.
URT (United Republic of Tanzania) (2011). The Economic Survey 2010. Government document produced by the Ministry of Finance. Dar es Salaam, Tanzania.
URT (United Republic of Tanzania) (2011). Agriculture and Food Security Investment Plan (TAFSP) 2011-12 to 2020-21. Main Document.

Appendix 1. beef cattle fattening operators individual efficiency score

Beef Cattle fattening operators No.	Efficiency score	Percentage
1	0.96214561	96.21
2	0.94622229	94.62
3	0.90990675	90.99
4	0.90836731	90.84
5	0.91718459	91.72
6	0.94651977	94.65
7	0.92162874	92.16
8	0.91154839	91.15
9	0.92729179	92.73
10	0.94453186	94.45
11	0.9178525	91.79
12	0.95143245	95.14
13	0.95393396	95.39
14	0.95476767	95.48
15	0.91568868	91.57
16	0.90804903	90.8
17	0.93534417	93.53
18	0.9089111	90.89
19	0.97801878	97.8
20	0.93575987	93.58
21	0.9321321	93.21
22	0.91532701	91.53
23	0.9009158	90.09
24	0.90640892	90.64
25	0.93834282	93.83
26	0.95582983	95.58
27	0.94903479	94.9
28	0.92827808	92.83
29	0.84726318	84.73
30	0.97484628	97.48
31	0.96947349	96.95
32	0.90510863	90.51
33	0.93306386	93.31
34	0.63422246	63.42
35	0.96445711	96.45
36	0.91327269	91.33
37	0.83504685	83.5
38	0.95367473	95.37
39	0.91304118	91.3
40	0.9195629	91.96
41	0.96544705	96.54
42	0.92534497	92.53
43	0.96375558	96.38
44	0.92175427	92.18
45	0.95109311	95.11
46	0.93799274	93.8
47	0.96986503	96.99
48	0.89269371	89.27
49	0.90672046	90.67
50	0.94312793	94.31
51	0.95021185	95.02

Appendix 1

52	0.91480716	91.48
53	0.87264073	87.26
54	0.9002228	90.02
55	0.96420636	96.42
56	0.94762449	94.76
57	0.96078686	96.08
58	0.95389044	95.39
59	0.9462963	94.63
60	0.91920711	91.92
61	0.93534727	93.53
62	0.94473073	94.47
63	0.92517926	92.52
64	0.9157181	91.57
65	0.89907558	89.91
66	0.95832715	95.83
67	0.90969569	90.97
68	0.90579944	90.58
69	0.92821406	92.82
70	0.85060658	85.06
71	0.95608835	95.61
72	0.94071958	94.07
73	0.88981	88.98
74	0.94438271	94.44
75	0.86631765	86.63
76	0.91320091	91.32
77	0.97630442	97.63
78	0.52971188	52.97
79	0.963581	96.36
80	0.47609183	47.61**
81	0.93853943	93.85
82	0.93368394	93.37
83	0.74611235	74.61
84	0.92489804	92.49
85	0.91071263	91.07
86	0.9428072	94.28
87	0.82394518	82.39

**Below 50%.

6

The dilemma of biodiversity and specialization in agricultural development

Price Edwin, Kibriya Shahriar and Zhang Yu

578 John Kimbrough Blvd, AGSV Building, Room 240, College Station, TX, USA, 77843.

The impact and usefulness of specialization and diversification strategies of agricultural research in developing countries is reviewed. The specialization strategy is based on principles of optimization on resources allocation in the spatial context, and it typically prevails over diversification strategy that considers inter-temporal resources allocation. However, diversification strategies offer biodiversity, self-sufficiency, technical competency and risk neutralization. The benefits of the diversification strategy support farmers' independence and production increase, as well as long run conservation of biological resources. Research and development of new technologies for diversification can play an important role in the agricultural development of fragile developing nations.

Key words: Specialization, diversification, biodiversity, risk-taking.

INTRODUCTION

Neo-classical economics has evolved as a study of resource allocation, especially biological and natural resources. There are two ways to consider resource allocation: spatial and temporal allocation. Ricardo (1817) and Torrens (1815) established the principle that has come to be known as the law of comparative advantage. With subsequent neo-classical and modern refinement, the theory has stood for over 175 years. Contemporary economists (Jones, 1961; Shiozawa, 2007) have formulated and generalized the theory for the modern economic context. The law of comparative advantage suggests that welfare among nations is optimized when each specializes in the productive activities to which its resources are best suited. Specialization enables spatially dispersed economic units to ameliorate the problem of immobile natural resources. The theory is not only regarded as the basis for explaining international trade, but also popularly applied to guide production and trade among smaller economic units. With the development of globalization, economic geography is more and more influenced by international and inter-regional specialization (Davis and Weinstein, 1999). We call this “Specialization Strategy” in human productive activities.

The second way to extend the resources allocation study is to address the efficient distribution of resources over time, which is mainly studied by environmental and ecological economists. It states that a safe minimum standard (SMS) for biodiversity is needed to guarantee an economy's sustainability. A decrease in biodiversity threatens economic sustainability because some biological resources extinct now may become important resources in the future (Randall, 1991; Bishop, 1993). The preservation of biological diversity, as a consequence of the optimal inter-temporal utilization of natural resources, places the capability for bearing the risk of declining technologies with future generations, rather than through the production and savings activities of present generations. However, there is not as yet an established body of theory equivalent to the law of comparative advantage from which to derive the precepts of diversity. These two research approaches are parallel: both methods are designed to maximize the utility, or welfare of human beings, but depend on different dimensions. Nowadays, it is obvious that the former one dominates the latter in practice. Here we intend to argue whether specialization strategy is essentially better than the diversity approach in the developing countries.

IMPACTS OF SPECIALIZATION STRATEGY

The pervasive ramifications of economic behavior according to the theory of comparative advantage have characterized industrial and organizational development for the last century and a half. The persuasion has influenced not only industrialization at a national level, but each industry's approach to its own enterprise. Specialization in the utilization of labor on production lines, indeed the evolution of production lines, was the eventual expression at the level of the firm, of the macroeconomic precept of specialization. Emerging forms of enterprise according to comparative advantage are evident today. The admonition is often heard that "one cannot be all things to all people". Organizations are compelled by internal policies or by external evaluation, to focus on the few activities the organization does best, regardless of diverse attractive opportunities. Moreover, the emphasis upon "focus" in business and public institutions is an entrenched precept of organizational dogma (Deming, 1994). For instance, due to the improvement of transport facilities and the widespread use of chemical fertilizer, agricultural production is becoming regionally more specialized (Gregson, 1996; Lyons, 1988). Chisholm (1962) supported the use of agricultural specialization asserting that it is best suited for climatic and soil conditions in different areas, and thus obtains increased production. Among the varied applications of specialization strategy, at least three essential precepts have evolved as follows:

1. Reduction of the unit cost: Specialization can increase total production from a given set of resources. The Ford automobile manufacturing company assembly line established in Dearborn, Michigan, is the textbook example of specialization in production.
2. Instability: Specialized organizations, tools or tasks have a finite lifespan; a decline in their usefulness is expected such that income streams based upon specialized processes are inherently unstable (Day, 1981). Lines of unemployed persons whose specialized skills are no longer needed are as vivid an image of modern industry as the assembly line itself.
3. Internal risk-taking capacity: Since specialized processes will end, each should be operated to the fullest, and part of the gains saved for the time when the enterprise declines. Risk taking capacity is internal to the economic or organizational unit, or nation, and compensated from current income. "Making hay while the sun shines," aptly describes the situation today in many natural resource based industries, from agriculture to fishing, forestry, mining and petroleum.

To sum up, specialization strategy is an accepted economic theory and has popular derivations, especially in agriculture. The principle has an advantage to minimize the production cost, while specialization can also result in the instability of income streams, for which entrepreneurs prepare by exploiting technologies to their fullest, generating savings that can be used when current technologies fail.

Traditional views on biodiversity

The call to preserve biodiversity is on the ground that it must be preserved for future generations to enjoy and use (Tietenberg, 2009). The key point is that the present generation should not presume its own ability to accurately project technological needs in the future. Therefore, we should leave as many possible of the natural building blocks for future technologies in place, so that future generations can make their own adjustments to the evolving milieu. It is argued, in effect, that the capacity to bear the risk of a possible decline in the effectiveness of present technologies should be external to the current economy. Let future generations have access to the full range of biological resources that we have today, for the purpose of creating the means of their own livelihood.

Biodiversity is, on one hand, endangered by production activities of human beings (Gowdy and McDaniel, 1995). Theoretical analyses have been developed for the optimal inter-temporal utilization of fixed and renewable natural resources (Fisher et al., 1972). They studied the trade-off of natural resources between preservation and utilization, showing that it will be optimal to refrain from utilizing environments and leaving them in its natural state. When it comes to non-renewable resources, the increasing number of humans and the more intensively specialized human activities are pushing biological limits in terms of energy exhausting and species displacement or even species extinction. As for the renewable resources (for example, fishery), it has been argued that they may be exploited till the existence of "reproductive surplus", which is a balance between births, deaths, and somatic growth (Hilborn et al., 1995). However, the "invisible hand" of the market and technologies advances usually leads to the over-exploitation of resources (Regev et al., 1998).

On the other hand, biodiversity is influenced by violent conflicts (McNeely, 2003). This point of view is often neglected because international conflicts are frequently conceived in narrow military terms. Violent conflict often leads to environmental degradation (such as deforestation, pollution of land and water), massive migrations of refugees, and reduction of natural conservation funds. For fragile conflict-prone developing nations, violence and conflicts can be major factors in the destruction of biodiversity (McNeely, 2003; Unruh et al., 2008). For stable developing countries with agrarian economies, promoting biological diversity in agriculture has become a common cause among pressure groups within and outside of agriculture.

It would be very difficult if not impossible to measure the value of biodiversity. In some extreme cases, researchers equal the value of biodiversity with the summed value of all the GNPs of all countries from now until the end of the world (Norton, 1988). When it comes to assessing the value of biodiversity the view of ecologists differs from the view of economists (Solow, 1993). Economic analysis would consider biodiversity as a commodity subjected to trade-offs and substitution at the margin, just as any other market commodities. Moreover, economists also recognize market failure and that market price may not reflect the true value of biodiversity. Gowdy (1997) proposes a more moderate point that the value of biodiversity must consist of market value, non-market value to humans, and the value of biodiversity to ecosystems. He admits that the market exchange value of environment service may justify importance of biodiversity, but it constitutes a small portion of total biodiversity value. To estimate the non-market value of biodiversity, several methods have been developed, such as contingent valuation, travel cost method, and hedonic method (Champ et al., 2003). The limitations of these methods are that they are highly vulnerable to sampling and methodology errors and so costly in time and money that it is impossible to make an estimated valuation of biodiversity on earth

Comparing specialization strategy to traditional views on preserving biodiversity

Law of comparative advantage, the basis for the specialization strategy, describes rational economic behavior. Given the law's state of nature: locales are endowed with different qualities and quantities of immobile resources; specialization in production and trade among the locales improves spatial equity in welfare caused by resource immobility. Biodiversity preservation thought is based upon a similar proposition: At different times the earth is endowed different quantities and qualities of depletable resources; human welfare over time is best achieved by using resources in such a diversified manner that resources will be least depleted. The source of inequity in the former case is the immobility of resources; in the latter case, the depletability of resources.

The above parallel approaches are both reasonable. However, why is the specialization strategy so prevalent around the world? The answer may be that while the specialization strategy for achieving spatial equilibrium is supported by mathematical treatments (Chipman, 1965), the employment of diversified resource use strategy for optimizing welfare over time has not been rigorously treated. Then, the following question arises: why is it so difficult for the inter-temporal approaches to be structured in mathematical models? The answer lies in the nature of the approach: preservation of biodiversity requires dampening the economic instinct to employ present technologies to their fullest, even if such activity would ultimately eliminate its own resource base. The prevailing rationale is to acknowledge that the future is uncertain, to protect against that day when current technologies no longer provide a means of livelihood; one should fully exploit resources and save a portion of the gains for the future.

The temporal approach suggests restraining in the use of present resources, and leaving for the future a full range of biologically diverse factors of production. Future generations would be provided not with the saved capital of their forbearers and a reduced set of natural resources, but with a robust natural environment and restricted capital. The difference is in exchanging present capital which we know how to extract, for future capital whose means of extraction is uncertain. Here, uncertainty means we are not only aware of the impact of present technologies on environment, but also have very weak ability to predict future technologies and thus, future valuable resources (Sull and Escobari, 2004). Uncertainty of the temporal approach makes it a risk-externalizing strategy, that is, the responsibility to bear the risk of a possible decline in the effectiveness of present technologies should be external to the current economy. The contrast of this strategy with the specialization strategy is the shift in the locus of risk-taking. The risk-externalizing strategy is severely impaired by the weakness of predictive models. Faced with uncertainty about how future wealth can be insured, decision makers are likely to continue to insist upon internalized risk models, with risk compensated from current income (Costanza, 1991).

COMBINING THE RISK-INTERNALIZING MODEL WITH THE DIVERSIFICATION STRATEGY

The above discussion shows that one of the most important advantages of the specialization strategy is that, it has the internal risk-taking capacity. The inter-temporal biodiversity preservation strategy is less prevalent because of its external risk-taking feature. It will be useful if risk internalizing model can be introduced into the biodiversity preservation strategy, denoted as the "diversification strategy" as compared to "specialization strategy". In fact, much empirical evidence is given regarding the manner in which diversity serves the enterprise of nations, firms, or even households to internalize risks. For instance, three arguments are provided

Self-sufficiency argument

Political units and some communal organizations maintain multiple enterprises in order to be as independent as possible of others. The need to support political autonomy with economic independence has led many nations to produce strategic commodities, irrespective of unfavorable domestic resource costs, and the economic advantage that would be gained by purchasing such goods abroad. The locus of risk-taking according to the self-sufficiency strategy is internal to the economy. The systems bear features that produce both stability and instability over time. It is posited to provide for stability of benefits with respect to international disruptions of commerce. Domestic enterprise is also protected from international competition, such that income streams, however inefficiently they may produce, are stable. A low level equilibrium can be achieved over time. The system, however, is exposed to disruptions in the supply of needed goods and services resulting from natural disasters or other such events internal to the economy. Multiple international sources of needed goods and services are not easily available to supplement local shortfalls.

Portfolio argument

The motivation here is to assure that within a selected set of enterprises, or securities, whatever the environmental condition, one or more income streams will strengthen and offset losses from enterprises that are unsuited to current conditions.

The portfolio strategy can be found in impoverished economies located in highly variable climatic environments. The Gunung Kidil area near Jogjakarta, Indonesia, is a case in point. In this region, each year farmers plant a combination of staple crops, maize, groundnut, pigeon pea and cassava. In any given year, one of the crops will do well, while the others fail. Heavy or light rainfall, early or late in the crop season, distinguishes the four possible outcomes, and determines which of the four crops thrives best. Since the investment in failed crops is lost, it is a high-cost system of production, limiting farmers to a low-level equilibrium income. The portfolio strategy is also found at level of cellular biology among plant breeders. Multiline pest resistance is argued to be the most effective strategy for breeding plants because when one source of resistance breaks down, another gene for resistance can take over. Breeders argue that such an approach makes it possible for a single cultivar to continue to be productive for a much longer period than a plant that has only a single line for resistance.

Advanced economies of national or household scale may follow such a strategy. Diversified portfolios of securities are advised for the risk-averse investor. Business and industrial diversification is sometimes pursued with the expectation that one product line or service area will remain prosperous when others lag. Like the Gunung Kidil farmers, the concern is for current income, with or without trade, and with no specific expectation of technical interaction among activities.

Technical complementary argument

Diversification is also undertaken to increase overall productivity and/or economic efficiency among related enterprises. During the late 1960s and early 1970s, industrial and business firms aggressively pursued the gains expected from combining related enterprises. Mergers and takeovers were said to produce synergies, or what economists term "complementarities" in production.

This strategy has also been applied to agricultural production. Mixed cropping systems or animal and crop systems are often followed in agriculture because of productivity increasing relationships among different crops grown consecutively or simultaneously in mixed patterns (Birgitta et al., 2002). Also, among different crop, animal and household enterprises there may be opportunities for products or by-products of one enterprise to be used as inputs in another (Thomas et al., 2002).

Diversification may be undertaken also to more fully utilize fixed capital such as land, buildings or family labor. Such within-firm opportunities usually affect favorable output/input relationships compared to those experienced when the separate enterprises are operated independently.

In summary, these three diversification strategies can be identified, as follows:

(1) Diversification for self-sufficiency
(2) Offsetting losses and gains
(3) Increased productivity

However, we should note that the argument for linking the state of nature - the natural diversity of biological resources - to diversification in economic activity for the purpose of cross temporal equilibrium, is not as rigorously derivable as Ricardo' argument for specialization for the purpose of cross-national (or spatial) equilibrium.

MORE APPLICATIONS OF DIVERSIFICATION STRATEGY IN AGRICULTURE IN DEVELOPING COUNTRIES

It does make sense that the application of diversification strategy in the agriculture sector in developing countries is a mutually beneficial situation. On one hand, diversification strategy emphasizes the exploitation of multiple resources, rather than specialized in a single or few resources, leading to the long run coexistence among biological resources. On the other hand, proper diversification strategies in agriculture will support farmers' independence and increase their production. For example, mixed cropping, mixed crop and animal production, complex combinations of interactive farm enterprises, vertical integration through increased on-farm production of farm inputs and increased processing of farm outputs, are all considered to be methods of preserving biological diversity. It is reasoned that the diversification of practices in production and processing, tend to reduce farmers' needs for purchase inputs. Such practices, it is argued, reduce damage to the environment (Lin, 2011). One specific area where such an effect is said to be evident is in pest management. Plenty of the techniques named above, and others that are collectively included in "integrated pest management", have been shown to reduce pest incidence. Lin (2011) pointed out that crop diversification creates a more diverse agro ecosystem that will perform better in pest suppression and food production, when facing climate change. Another area of scientific effort is nutrient recycling, utilizing systems of composting, cover crops and other systems, which may substitute for chemical fertilizers. Other ongoing effort is to substitute water conservation methods such as no-till cultivation and soil mulches for irrigation. Yet another effort is directed to the use of animal powered equipment, to provide for beneficial interactions of livestock and cropping systems. These examples are cited because they substitute for the four key elements of modern farming technologies that have been most called into question by environmentalists: chemical pesticides, inorganic fertilizers, artificial irrigation and farm mechanization. A counter example is that landscape simplification, which transforms many landscapes into expansive monoculture with few natural habitats, results in greater pest pressure (Meehan et al., 2011).

Provided with a diversity of natural resources it was not easy for farmers to specialize according to the derived dictates of the law of comparative advantage. This is especially true for individual farmers in developing countries. Over even small spaces, natural processes had produced differences in soil fertility, water availability, vegetative cover that favored a variety of pests and diseases, and exposure to solar radiation and other factors that influence production. The superimposed variations of water, soils, pest and solar radiation regimes produced many small micro regimes, each of which, in its best use, required a different management technique - different crops or other technology. Of course, diversification strategy in agriculture requires higher management capacity and increases the cost of management per unit of output. One possibility, however, is that diversification for the purpose of biodiversity preservation might be combined with one or more of the strategies for diversification: self-sufficiency, off-setting gains and losses, or production complementarities. Combining the welfare of preserving biological diversity with the benefits obtained from complementarities in production may offer a way of offsetting the higher managerial costs of biologically diverse production systems. Complementarities among enterprises might even be sufficient to more than pay for higher management costs, and thus provide for some additional net increase in productivity. Ruttan (1989) suggested that the latter is necessary in order for sustainable agricultural systems to meet our needs as population and incomes increase. In addition, the accessibility of local processing facilities is a key factor that limits diversified agriculture.

DISCUSSION

The specialization strategy emphasizes welfare among nations, or spatial equity. Biodiversity preservation addresses the inter-temporal distribution of welfare. The locus of risk-taking is internal to the first strategy, and external to the latter. This review introduces a diversification strategy, which tries to combine the environmental objective with the internalized risk models. By internalizing risks, the diversification strategy is found to not only conserve natural resources, but also (a) self-sufficiency, (b) offsetting losses and gains, and (c) technical complementary. It is also argued that diversification strategies may play an active role in biodiversity preservation for developing countries, especially the fragile conflict-prone countries. A potential approach for developing countries to keep balance between biodiversity preservation and increasing productivity is to apply a risk-internalized diversification model that suits local circumstances.

A discussion of agricultural research is included as an extension of this article. The belief in specialization is especially strong among agricultural research organizations. Specialization induces a philosophy of management that suggests always investing most research resources where there has been greatest success in the past. The international research centers are a good example. Many have a commodity focus, and there is general belief that to stray from research on their mandated commodities will lead to unproductive research investment. The restriction goes even further to suggest that research on the commodity should focus on those aspects in which there has been greatest success, for example, lowland irrigated rice. The guiding principle is to continue to do what is believed to be done best. However, diversification strategy can be argued to deserve equal attention. Firstly, focusing on one commodity in research incurs the risk of total failure in case that the single commodity enterprise fails. Besides reducing the research organization's risk by spreading research across select commodities, diversification can increase benefits, utilizing and leveraging core facilities or administration across commodities enterprises. In fact, it is possible that the research organization's core commodity research would be benefited from studying other commodities, or cooperating with other institutions.

REFERENCES

Birgitta R, Lennartsson M, Davies G (2002). The use of mixed species cropping to manage pests and diseases – theory and practice. Presented at the UK Organic Research 2002 Conference in University of Wales Aberystwyth. pp. 207-210.

Bishop R (1993). Economic efficiency, sustainability, and biodiversity. A J. Hum. Environ. 22(2-3):69-73.

Champ P, Boyle K, Brown T (2003). A primer on nonmarket valuation. Kluwer Academic Publishers, Dordrecht, the Netherlands.

Chipman J (1965). A Survey of the theory of international trade: Part I, Classical Theory. Econometrica 33(3):477-519.

Chisholm M (1962). Tendencies in agricultural specialization and regional concentration of industry. Regional Sci. 10(1):157-162

Costanza R (1991). Ecological Economics: The Science and Management of Sustainability. New York, Columbia University Press.

Davis D, Weinstein D (1999). Economic geography and regional production structure: and empirical investigation. Eur. Econ. Rev. 43(2):379-407.

Day G (1981). The product life cycle: analysis and applications issues. J. Market. 45(4):60-67

Deming E (1994). The new economics for industry, government, education. The MIT Press, Cambridge, Massachusetts.

Fisher A, Krutilla J, Cicchetti C (1972). The economics of environmental preservation: a theoretical and empirical analysis. Am. Econ. Rev. 62(4):605-619.

Gowdy J, McDaniel C (1995). One world, one experiment: addressing the biodiversity- economics conflict. Ecol. Econ. 15:181-192

Gowdy J (1997). The Value of biodiversity: markets, society, and ecosystems. Land Economics, 73(1):25-41.

Gregson M (1996). Long-term trends in agricultural specialization in the United States: some preliminary results. Agric. History 70(1):90-101.

Hilborn R, Walters C, Ludwig D (1995). Sustainable exploitation of renewable resources. Ann. Rev. Ecol. Systemat. 26:45-67.

Jones R (1961). Comparative advantage and the theory tariffs: a multi-country, multi- commodity model. Rev. Econ. Stud. 77:161-175.

Lin B (2011). Resilience in agriculture through crop diversification: adaptive management for environmental change. BioScience 62(3):183-193

Lyons T (1988). Concentration and specialization in Chinese agriculture, 1979-1985. J. Dev. Areas 22(4):437-456.

McNeely J (2003). Conserving forest biodiversity in times of violent conflict. Int. J. Conserv. 37(2):142-152.

Meehan T, Werling B, Landis D, Gratton C (2011). Agricultural landscape simplification and insecticide use in the Midwestern United States. Proceedings of the national academy of sciences of the U.S.A. 108(28):11500-11505.

Norton B (1988). Commodity, amenity, and morality: the limits of quantification in valuing biodiversity. National Academy Press, Washington, DC.

Randall A (1991). The value of biodiversity. A.J. Hum.Environ.20(2):64-68.

Regev U, Gutierrez A, Schreiber S, Zilberman D (1998). Biological and economic foundations of renewable resource exploitation. Ecol. Econ. 26:227-242.

Ruttan V (1989). Sustainability is not enough. Am. J. Altern. Agric. 3(2-3):128-130.

Ricardo D (1817). On the principles of political economy and taxation. London.

Shiozawa Y (2007). A new construction of ricardian trade theory—A many-country, many- commodity case with intermediate goods and choice of production techniques. Evolution. Institut. Econ. Rev. 3(2):141-187.

Solow R (1993). Sustainability: an economist's perspective. Norton Press, New York.

Sull D, Escobari M (2004). Creating value in an unpredictable world. Bus. Strat. Rev. 15(3):14-20.

Thomas D, Zerbini E, Rao P, Vaidyanathan A (2002). Increasing animal productivity on small mixed farms in South Asia: a systems perspective. Agric. Sys. 71(1):41-57.

Tietenberg T (2009). Environmental Economics and Policy. Addison Wesley, Boston.

Torrens R (1815). An Essay on the External Corn Trade. London.

Unruh J, Gibson D, Woods K (2008). Biodiversity conservation and crisis. USAID.

7

Farm household level impacts of information communication technology (ICT)-based agricultural market information in Ghana

Ramatu M. Al-Hassan, Irene S. Egyir and James Abakah

Department of Agricultural Economics and Agribusiness, University of Ghana, Legon, Ghana.

This paper assesses the impact of the information communication technology (ICT)-based market information service (MIS) on farm households in the eastern corridor of Northern Ghana. Data was collected from 346 farm households in the eastern corridor of Northern Ghana where an ICT-based market information project was implemented between 2006 and 2009. The sample included 159 participants and 187 non-participants. The logit model was used to identify factors which influence the probability of participation in the ICT market information project, while propensity score matching (PSM) was used to assess the project impact. Results indicate that age and value of assets were negatively related to the likelihood of participation in ICT-based MIS. A unit increase in age reduced likelihood of participation by 0.5%, while a unit increase in value of assets reduces likelihood of participation by 7%. Previous participation in a development project increased likelihood of participation by more than 60%. Participation in the ICT-based project increased expenditure on pesticides, and food security by 11%. More participants (13%) used improved seed than non-participants. It is recommended that younger persons are made the target/focus group for ICT projects and that the private and public sectors collaborate to facilitate wide availability and expanded use of ICT MIS.

Key words: Information communication technology (ICT), agricultural market information, smallholders, Ghana.

INTRODUCTION

Development of the agricultural sector in many African countries hinges on the development of the smallholder systems that have sustained African agriculture to date but continue to face challenges of low productivity and limited access to remunerative markets. Non-competitive value chains and limited information about remunerative markets, and risk aversion of smallholders limit their integration into markets.

Ghana's agricultural growth has been erratic and at 5.2%, it has lagged behind Gross Domestic Product (GDP) growth of 5.9% [Institute of Statistical Social and Economic research (ISSER), 2005 - 2010]. Ghana's development strategies since 2001 have sought to turn the agricultural sector around through modernisation and commercialisation of smallholder agriculture (Republic of Ghana, 2003a, 2006). The Food and Agriculture policy has as one of its key objectives, integration of smallholders into domestic and international markets

Republic of Ghana, 2007) with improvement in access to market information as a key strategy. The information communication technology (ICT) for Accelerated Development (Republic of Ghana, 2003b) policy also aims among others, to facilitate the modernization of the agricultural sector through the deployment and exploitation of ICTs to improve the sector's efficiency and productivity. Initiatives taken to improve investments in ICT deployment include the privatisation of the telecommunications industry, establishment of a regulatory authority, and investments in infrastructure such as an internet backbone, and masts for mobile phones. Community Information Centres (CICs) have also been provided in several districts to facilitate training in ICT on a wide scale.

The focus on ICT-based methods of information provision is driven by the role they can play in communicating knowledge and information to rural farmers. Access to market information can enhance farmers' access to markets through better negotiation and meeting the demands of the market (Barrett, 2008; Moser et al., 2005), on condition that constraints to access to inputs are addressed. The ICT tools have included modern tools such as the internet, mobile telephony, and interactive video and CD-ROM programs, as well as traditional ICTs of the radio and television (Munyua, 2007).

There has been an increase in the use of ICT applications in rural Ghana in response to the enhanced policy environment, although issues of availability of the ICTs, electricity, literacy, telecommunications and content are still prevalent (Sampong et al., 2007; Alemna and Sam, 2006). An ICT-based agricultural market information service (MIS) was introduced by the private sector [Busy Lab through its TradeNet (now Esoko) platform]. In 2006, the Sustainable Enterprise Development Foundation (SEND Foundation), a non-governmental organisation operating in the north-east of Ghana, adopted the TradeNet market information platform to access market information for smallholder farmers. Specifically, farmers were trained to access and provide market information through text message alerts using cellular phones. Prior to this intervention, SEND Foundation had been promoting agricultural activities and cooperative credit unions with the goal of enhancing food security in the area. The objective of introducing the ICT-based MIS was to link the smallholder farmers to markets and promote their commercialisation process. The project was implemented for 3 years between 2006 and 2009. Initiatives such as this are widespread in several African countries (for example, Kenya Agricultural Commodity Exchange (KACE) and DRUMNET in Kenya; Malawi Agriculture Commodity Exchange (MACE) in Malawi, Market Information Systems and Traders' Organisation in West Africa (MISTOWA) (Tollens, 2006; Munyua, 2007; Okello and Ndiragu, 2009).

However, information on the impacts of the initiatives on smallholders and the markets they operate in is rather sparse. This paper assesses impact of the ICT-based MIS implemented in north-east of Ghana. Specifically, it identifies factors which influenced participation in the ICT-based market information project, and assesses the impact of the use of the ICT-based MIS on the level of commercialisation and food security, which are the SEND project goals.

METHODOLOGY

Methods of analysis

The decision to participate is modelled as a random utility function. The farmer decides to participate if the utility derived from participation exceeds that from not participating. That is, $U^a > U^b$; where *a* denotes participation and *b* denotes non-participation. U^a and U^b are modelled as:

$$U^a = x'\beta_a + \varepsilon_a \text{ and } U^b = x'\beta_b + \varepsilon_b$$

Since U^a and U^b are latent, it is the probability of the observed decision (participate or do not participate) that are modelled:
Let Y* be the binary response variable, $Y_i^* \in \{0, 1\}$, i = 1,...,n farm households and P is the probability that $Y_i^*=1$ given X;

$$\text{For } P(Y_i^*=1 \mid X) = \text{Prob}(U^a > U^b)$$
$$= \text{Prob}[x'\beta_a + \varepsilon_a - x'\beta_b - \varepsilon_b > 0 \mid x]$$
$$= \text{Prob}[x'(\beta_a - \beta_b) + \varepsilon_a - \varepsilon_b > 0]$$
$$= \text{Prob}[x'\beta + \varepsilon > 0]$$

The above binary choice model is estimated as a logistic model specified as follows:

$$\text{Log}\left(\frac{P}{1-P}\right) = \alpha + \beta_j X_j + \varepsilon \quad (1)$$

Where $\left(\frac{P}{1-P}\right)$ is the odds of Y* = 1 given that X_j and β_j is a vector of regression coefficients. β_j for variable, X_j is the log-odds and the exponential form of β_j is the multiplicative change in the odds for Y* = 1 when the jth variable increases by one unit, holding all other variables constant (Finger and El Benni, 2011). This model is used to explain farmers' decision to participation in the ICT-based market information system in the context of adoption.

Following the adoption literature, decisions to participate in an innovation programme depend on attributes of the innovation, channels of communicating the innovation and extent of promotion, as well as social systems (Rogers, 2003). In this paper, the explanatory factors are hypothesized to be observable personal characteristics of the farmer or their household which set the social context (for example, education, age and gender), awareness of ICT-based market information project (determined by promotion and communication), capital endowment, and level of output commercialization (indicator of economic empowerment as well as need for the innovation). Membership in a producer organization, and number of crop enterprises are also indicators of capital endowment. Being a member of producer organization is included because it creates opportunity for social interaction which the phone can facilitate. The radio and TV are substitutes for the mobile phone as a means of accessing information but they may also expose potential users, through marketing, to the capabilities of the mobile phone. Previous participation in an agriculture development project is both a measure of social capital and the farmer's appreciation for social interaction. Distance to main market is included to control for the trading environment, which influences need for information and therefore likelihood of participating in the

Table 1. Explanatory variables in Logit model participation in ICT-based MIS project.

Variable	Description	Expected sign of coefficient
Age	Age of farmer (number of years)	-
Gender	Gender dummy (1 if farmer is a male and 0 otherwise)	
Education	Number of years of schooling of farmer	+
Participation	Participation in ECAMIC project (1 if farmer is a participant and 0 otherwise)	
Awareness	Awareness of the ECAMIC project (1 if farmer is aware and 0 otherwise)	
Radio	Radio dummy (1 if farmer had a radio and 0 otherwise)	
TV	Television dummy (1 if farmer had a television and 0 otherwise)	
Radio/TV	Radio and TV dummy (1 if farmer had a radio and/or TV and 0 otherwise)	+/-
Land	Size of land cultivated before ECAMIC project inception (acres)	-
Member of FBO	Dummy for FBO membership (1 if farmer hold any membership a farmer-based organization and 0 otherwise)	
Past agricultural program	Ever participated in an agricultural program in the last 5 years (1 if farmer has ever participated and 0 otherwise)	+
Needed credit	Dummy for need for credit (1 if farmer needed credit for an agricultural activity and 0 otherwise)	+
Enterprises	Number of crops produced by the farmer (number)	+
Sold-out	Proportion of total output sold	
Distance to local market	Distance (in kilometers) to nearest local market (natural logged)	+
Distance to nearest phone service	Distance (in kilometers) to nearest mobile phone service (natural logged)	
Asset	Value of farmer's assets (natural logged)	+

project to access information. Although owning the cell phone is not a requirement for participation, those who own mobile phones are more likely to participate in the ICT-based market information project; the reverse is true because in both situations the mobile phone is the tool for accessing the information. The explanatory variables for participation in the ICT-based MIS project are summarized in Table 1.

The impact of participation in the ICT-based MIS project is assessed via the method of propensity score matching (PSM), which is used to correct selection bias in the assessment of impact based on a control group of non-participants (Becerril and Abdulai, 2009; Ali and Abdulai, 2010). The aim of matching is to line-up comparison individuals to sufficient observable factors to remove systematic differences in the outcome between treated and non-treated (Blundell and Costa Dias, 2008; Deheja and Wahba, 2002). PSM takes two individuals that are exactly similar in all characteristics except the treatment (in this case, participation in ECAMIC) and computes the difference in the outcome between them.

The outcome variables used in this study are, income from crop sales, the extent of commercialisation, and level of food security. Household commercialisation is measured by the share of crop output sold; this is a partial measure that captures commercialisation on the output side only. Food security is measured by the number of months the farmers' major staple last through the year and also a dummy variable with 1 if the household calorie intake per person meets the recommended daily allowance.

In this study, the observable characteristics by which participants and non-participants are paired are age, education, value of assets and distance to nearest local market and distance to centres with electricity. Kernel based matching (KBM) of participants and non-participants are used.

Data

The data is from a sample survey of 346 households, using a structured questionnaire. The study area was stratified into project and non-project communities. A total of 159 beneficiary and 112 non-beneficiary respondents were selected randomly from beneficiary communities. The remaining 75 households were randomly selected from non-project communities.

Household level data included household characteristics, awareness of existence of ICT-based MIS in the locality, number of ICT-based MIS used, other information sources used, participation in farmer organizations, access to formal and informal financial

Table 2. Estimates of logit model explaining participation.

Variable	Coefficient	S.E	z	p-value	Marginal effects (dy/dx)
Age	-0.04	0.02	-2.11	0.035**	-0.005
Youth	0.51	1.34	0.38	0.704	0.057
Education	0.03	0.05	0.52	0.601	0.002
Household size	0.13	0.11	1.18	0.239	0.014
Radio/TV	0.54	0.54	1.00	0.319	0.065
Land	0.10	0.35	0.31	0.755	0.012
Needed credit	0.70	0.63	1.11	0.265	0.093
Distance to local market	-0.64	0.31	-2.11	0.035**	-0.072
Past agricultural program	4.01	0.57	7.00	0.000***	0.617
Asset	-0.73	0.25	-2.94	0.003***	-0.081
Constant	3.63	1.67	2.16	0.031	

Dependent variable = Participation (dummy = 1 if a participant); Number of observation = 213; Pseudo R^2 = 0.43; Log likelihood = -68.43; LR $Chi^2(9)$ = 104.32; Prob > Chi^2 = 0.0000; S.E = Standard error; M.E = Marginal effects.

services, asset levels, crop and livestock production, participation in input and output markets, the types of markets used, and sources of income.

RESULTS AND DISCUSSION

Participation in ICT-based MIS project

The analysis of determinants of participation was conducted on the 187 households in beneficiary communities, who were aware of the project and therefore could choose to participate or not. Participants who were aware of the project were identified as ex-post (after the sampling) through their response to a question in the questionnaire, which asked if the respondent was aware of the project or not. Table 2 presents results of the logit model estimation for the identification of factors that influenced participation in the ICT-based market information intervention. The likelihood of participation decreased by 0.5 and 7%, respectively, for a unit increase in age and distance to a local market. A unit increase in the value of assets decreased the likelihood of participation by 8%. Those who had participated in an agricultural project were 61% more likely to participate in the MIS project. There is therefore a tendency for repeated participation in projects. This suggests a multiplicity of projects targeting same beneficiaries, and likely creation of dependency of some farmers on external interventions.

Contrary to expectations, farmers who accessed distant markets were less likely to participate in the project. Ordinarily one expects these farmers to have need for information; however the conduct of markets in the study area is such that those who sell or buy in distant markets usually have established trading partners (commonly called 'customers') and therefore will have less need for an MIS. Participation in such projects also demands time, and distant traders who are more likely to be frequently absent from the community are not likely to participate in the projects.

Impact of participation in ICT-market information project on smallholder farmers

The effect of participation in the ICT-MIS project on smallholder farmers was estimated with the KBM method and the results are presented in Table 3. The clustering of participants and non-participants in the propensity score range of 0.4 and 0.7 (Appendix 1) shows good matching based on the observed characteristics.

Thirteen percent more of participants than non-participants used improved seed and this was significant at 1%. More participants than non-participants also attained higher level of food security indicators of Recommended Daily Allowance (RDA) of caloric intake, for example, at the higher level of indicator based on IFPRI's RDA. Participants also spent more on pesticides but the level of significance was at 10% only. Since the use of pesticides is linked to growing of maize, it suggests that participants were growing more maize than non-participants.

Conclusions

This paper assessed the factors influencing the likelihood of participation in an ICT-based MIS, and the impact on farm households.

Participants in the ICT-based MIS project tended to be younger than non-participants and slightly more literate. The current study in Northern Ghana shows that participants of the ECAMIC project used more improved seed, spent more on pesticides and attained higher levels of food security. The enhanced access to market information has increased their orientation to produce for the market. Since the analysis explicitly considered the

Table 3. Average treatment effect on treated: effect of participation.

Outcome variable	ATT	Critical level of hidden bias (r)	Number treated	Number control
Commercialization index	0.03(1.02)	2.90 - 2.95	159	187
Food secure in maize	0.05(0.09)	3.30 - 3.35	159	187
Food secure (100% IFPRI RDA)	0.11(2.01)**	5.00 - 5.05	159	187
Food secure (80% IFPRI RDA)	0.09(1.54)	3.50 - 3.55	159	187
Food secure (100% WFP/UNHCR RDA)	0.13(2.12)**	2.80 - 2.85	159	187
Food secure (80% WFP/UNHCR RDA)	0.10(1.58)	2.20 - 2.25	159	187
Crop income	349.89(0.23)	3.15 - 3.20	159	187
Value of pesticides	1.38(1.71)*	3.90 - 3.95	159	187
Value of fertilizer	-1.85(-0.18)	2.25 - 2.30	159	187
Use of improved seed	0.13(2.80)***	5.65 - 5.70	159	187
Store for future sale	0.01(0.30)	1.00 - 1.05	159	187

Values in parentheses are t-values. ATT is the average treatment effect for the treated. *, ** and *** denote significance of t-statistics of mean difference at 10, 5 and 1%, respectively.

causal relationship between participation in ICT-MIS and household adoption of improved farm technology (seed, fertilizer and pesticides), it addressed the counterfactual questions that may be significant in predicting the impacts of policy change.

Policy implications

The policy implication of the findings is that promotion of high technology such as ICTs for communication with farmers should be targeted to youth in agriculture because they are already familiar with the communication tools and require little additional training for their use. Such programmes or interventions should be supported with complementary services such as financial services, because project participants tend to have multiple needs. In addition, access to production inputs should be improved because the use of such inputs is the route to commercialization and the ultimate realization of the benefits of enhanced market information.

There is a latent potential for ICTs, especially the mobile phone to facilitate transactions of rural farm households, which can be realized through positive partnership between the private and the public sectors; the latter has the responsibility to identify the needs of rural farm households that can be addressed most effectively by modern ICTs, while the former's role is to innovate to deliver the services cost effectively. Policies that improve farm level adoption of improved technology and facilities, and road surface conditions and networks will be needed to ensure that the farmers can produce more (through higher use of inputs and yields) and increase sales in more lucrative markets.

ACKNOWLEDGEMENT

Authors acknowledge the funding support provided by the International Development Research Centre (IDRC) to conduct the research project on the Role of ICTs in Linking Farmers to Markets in Africa. We also extend appreciation to all team members of the eARN project especially Dr Julius Okello and members of the advisory council for their counsel and support in various ways.

REFERENCES

Alemna AA, Sam J (2006). Critical issues in information and communication technologies for rural development in Ghana. Inf. Dev. 22:236-260.

Ali A, Abdulai A (2010). The adoption of genetically modified cotton and poverty reduction in Pakistan. J. Agric. Econ. 61:175-192.

Barrett C (2008). Smallholder market participation: Concepts and evidence from eastern and southern Africa. Food Policy 34:299-317

Becerril J, Abdulai A (2009). The Impact of Improved Maize varieties on poverty in Mexico: A propensity score matching approach. World Dev. 38(7):1024-1035.

Blundell R, Costa Dias M (2008). Alternative approaches to evaluation in empirical Microeconomics. J. Hum. Resour. 44(3):565-640.

Deheja R, Wahba S (2002). Propensity score matching methods for non-experimental causal studies. Rev. Econs. Stat. 84:151-161.

Finger R, El Benni N (2011). Farmers' Adoption of Extensive Wheat Production - Determinants and Implications. Paper prepared for the 122nd EAAE Seminar, "Evidence-Based Agricultural and Rural Policy Making: Methodological and Empirical Challenges of Policy Evaluation". Ancona, February 17-18, 2011.

Moser C, Barrett C, Minten B (2005). Missed opportunities and missing markets: Spatial-temporal arbitrage of rice in Madagascar. Selected paper presented at the American Agricultural Economists Association Annual meeting, Providence, Rhode Island, July 24-27, 2005.

Munyua H (2007). ICTs and small scale agriculture in Africa: a scoping study. Draft Report 1 Submitted to International Development Research Center.

Okello JJ, Ndirangu LK (2009). Background study of the status of ICT use in Kenyan agriculture. A paper presented at the proposal development workshop on "The role of ICT-based interventions in linking farmers to markets", Antananarivo, July 2008.

Republic of Ghana (2003a). The Ghana ICT for Accelerated Development (ICT4AD) Policy. Ministry of Communication, Accra.

Republic of Ghana (2003b). Ghana Poverty Reduction Strategy. National Development Planning Commission, Accra.

Republic of Ghana (2006). Growth Poverty Reduction Strategy. National Development Planning Commission, Accra.

Rogers EM (2003). Diffusion of Innovations, 5th edition. Free Press, New York, U.S.A.
Sampong DD, Irene Egyir S, Osei-Asare Y (2007). Assessing rural women food producers capacity to adopt modern ICTS- A case study of the Mfantsiman District of Ghana. 2nd Biennial Conference of African Association of Agricultural Economists (AAAE), La Palm Beach Hotel, Accra August 20-23, 2007.
Tollens EF (2006). Market information systems in sub-Sahara Africa challenges and opportunities. Poster paper prepared for presentation at the International Association of Agricultural Economists Conference, Gold Coast, Australia August12-18, 2006.

APPENDIX

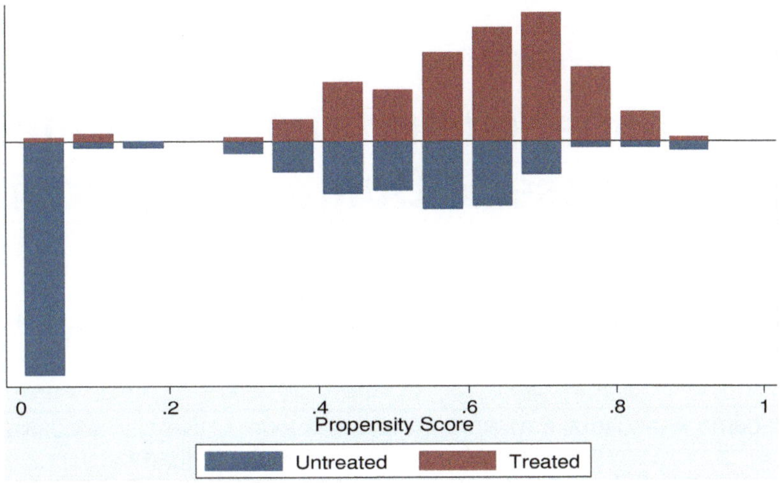

Appendix 1. Distribution of propensity scores.

Production efficiency and risks in limited resource farming: The case of Bulgarian peanut industry

Carel Ligeon[1], Curtis Jolly[2], Nelly Bencheva[3], Stanko Delikostadinov[4] and Naveen Puppala[5]

[1]Auburn University, Montgomery, Alabama, USA.
[2]Alabama Agricultural Experiment Station, Auburn University, Alabama, USA.
[3]Agricultural University in Plovdiv, Bulgaria.
[4]Institute of Plant Genetic Resources in Sadovo, Bulgaria.
[5]New Mexico State University, Las Cruces, New Mexico.

We used a stochastic production frontier and a Tobit model to evaluate technical efficiency and inefficiency of Bulgarian limited resource peanut farms for the period 2000/2002. We used a Just-Pope model to examine risks related to on-farm yield adjustments. Technical efficiency of farms ranged from 77 to 97%, with an average 92%, a median of 93%, mode of 90%, and skewness -1.49. The technical efficiency of these farms are largely influenced by on-farm decisions, the quantity of seeds, phosphate, fungicide, the amount investment capital, hours of manual labor used, and leased mechanized labor. The results indicate that efficiency is more crop specific than farm size dependent. Farm inefficiency was unrelated to size but rather to gender and age of operator. Farmers used less than the optimum levels of seeds and phosphate but there was risk associated with an increase in application rates of these inputs. These findings indicate that limited resource Bulgarian peanut farms are approaching efficiency, and adjustment to attain greater efficiency may be influenced by socio-demographic factors and limited government intervention.

Key words: Production, efficiency, peanuts, Bulgaria.

INTRODUCTION

Agriculture is the backbone of the Bulgarian economy. It employs about 26.2% of the labor force and contributes 17.3% to the gross domestic product (GDP) (Totev and Shahollari, 2001). However, most of the crops are produced on small farms that range in size from 0.9 to 11.3 ha, and a large proportion of the crops produced are consumed by the farm household. The small farm size has been suggested as one of the factors restricting agricultural development in post-adjustment periods since agriculture production fell since 1989. It has also been argued that subsistence agriculture is an impediment to agricultural growth in Central Eastern European (CEE) countries because of its lower technical and economic efficiencies (Hazel, 2004; Brent, 1999). Policy makers and researchers are worried that Bulgaria will not be competitive as a member of the World Trade Organization (WTO) because its small farm sizes limit the commercialization of agriculture. There are advocates who favor policies to expand farm size to enhance production efficiency (Lulcheva and Todolrova, 2005) without prior examination of the technical efficiency and resource substitution at the farm level. We investigate whether technical efficiency of Bulgarian peanut farms is farm-size specific and related to factor substitution or exogenous factors. We also examined the related risk associated with attainment of maximum yield efficiency. Peanut is the only crop in Bulgaria experiencing an increase in production over the transition period. Peanut production increased by 52% during the transition period, 1990 to 1999, and most of all the output is concentrated in the Plovdiv region where 73% of the production takes place. In 2000-2003, Bulgaria was responsible for 93% of the peanuts produced in Europe. Peanut seemed to be a logical crop choice for farm income enhancement during the post-transition period (period after 1989 there was a transformation in the society) because the soil and environmental conditions in southern Bulgaria and the Plovdiv region favor its production. The crop is produced by limited resource farmers in Bulgaria and has tremendous market potential. Given that peanut can be produced with limited external inputs and can be grown profitably under large or small scale conditions, it is important to examine the technical efficiency related to its production as new market opportunities open up for Bulgarian farm products, and the government is coerced to adopt new land tenure policies. Economic analyses conducted to measure the technical efficiency of limited resource peanut farms are non-existent. The factors influencing production efficiency and how the efficiency of small and large farms can be improved are yet to be analyzed separately.

Theoretical framework

A production function defines the technological relationship between inputs and output. An estimated production function from empirical data indicates average levels of outputs that can be produced from given levels of inputs. One of the basic assumptions in the estimation of a production function is that all firms are technically efficient and the representative firm defines the production frontier. Variations from the frontier are associated with mis-or un-measured production factors. Another assumption underlying the specification of a production frontier is that the firm engaged in production is applying "best management practices", and it receives the maximum potential output for a given set of inputs, employing a specific production process. Failure to attain this maximum is deemed inefficient; therefore, based on the observed output of a firm one is able to measure its level of efficiency (Greene, 1993). The measurement of firm specific technical efficiency is based upon deviations of observed output from the best production or efficient production frontier. If a firm's actual production point lies on the frontier, it is perfectly efficient. If it lies below the frontier then it is technically inefficient, with the ratio of the actual to potential production defining the level of efficiency (Herrero and Pascoe, 2002).

A production frontier is estimated through two distinct approaches, the data envelopment analysis (DEA) which is a non-parametric method whose main weakness is an inability to allow for stochastic shocks to the frontier. The stochastic frontier analysis (SFA), in contrast, is designed to incorporate stochastic disturbances, but requires strong parametric specifications in its implementation (Roy, 2002).

LITERATURE REVIEW

Studies which examine the technical efficiency of Bulgarian cooperative farms and that of smaller farm households indicate that there have not been clear indications that larger commercial or corporate farms are more efficient. Kopeva and Noev (2002) showed that cooperate farms in Bulgaria are more efficient in the production of vegetables and cereals, while family farms engaged in grape production perform better. They concluded that output was dependent on size of farm and use of inputs, no matter the type of production. Mathijs and Vranken (2001) measure and explain farm-specific technical efficiency of Bulgarian crop and dairy farms applying the Data Envelopment Analysis (DEA). They found that family farms were technically superior to corporate farms in crop farming, but rejected it for dairy farming. Totev and Shahollari (2001) indicated that even with small sized family holdings that Bulgaria had a revealed comparative advantage in most of its export crops when compared to its neighbors, Macedonia and Albania.

Empirical findings do not support the proposition that small-sized farms are generally inefficient. Mathijs and Swinnen (2001) argue that family size farms in Germany were technically more efficient than large scale successor organizations. They claimed that the difference in technical efficiency, however, declined during transition mainly due to structural changes in Agriculture. Mathijs et al. (1999) also used DEA to confirm that family farms in Czech and Slovak were technically efficient after transition.

In spite of the existing technical information, some researchers believe that the small-sized farms serve as an impediment to agricultural development in Bulgaria, and clearly emphasized that the major problem of agricultural commercialization in Bulgaria is embedded in fragmented land holdings (Lulcheva and Todorova, 2005). Todorova and Lulcheva (2005) are of the opinion that rural development and agricultural sustainability can be achieved through land consolidation and territorial planning. Some policy makers suggest that unless Bulgaria modernizes its agriculture there will be a continuous exodus of people from the rural sector. Policy discussions are to limit the growth of these small farms in order that Bulgaria becomes competitive on the global market (OECD, 1999).

Kostov and Lingard (2002) defended subsistence agriculture by stating that even with lower technical efficiency small-scale subsistence agriculture has positive economic effects both in terms of production and consumption.

They also indicated that in spite of low technical efficiency exhibited by small-scale farms, they display economic efficiency at the aggregate level as regards to their utility function. Limited resource farms that generate large marketable surpluses during the transition period of CEE countries are likely to engage in factor substitution in order to improve their technical efficiency. To evaluate the technical efficiency of limited resource farms in Bulgaria we employ a stochastic production frontier function. We evaluate whether farm characteristics, management decisions and resource use affect farm efficiency. We also investigate whether risks is a factor determining levels of input use.

Stochastic production frontier

The stochastic production frontier model was developed independently by Aigner et al. (1977), and Meeusen and van den Broeck (1977) to estimate the technical efficiency of the production process. Since then it has been employed to measure the production efficiency of various agricultural areas, such as factor productivity of industry, farms, crops and livestock. The technical efficiency of the UK potato industry was examined by Wilson et al. (1998). The technical production frontier was used by Coelli et al. (2003) to measure the total factor productivity for crop husbandry in Bangladesh for the period 1961 to 1992. Tzouvelekas et al. (2002) used the stochastic production frontier to examine the efficiency levels of organic and conventional farms in Greece. Karagiannis and Tzouvelekas (2001) also used the stochastic production function to look at efficiency of olive-growing farms in Greece. Seyoum et al. (1998) studied the technical efficiency and productivity of maize producers in Eastern Ethiopia. Yao and Liu (1998) used the same technique to examine the efficiency of grain production in China.

The stochastic production frontier model has also been widely used to study animal production efficiency. The technical efficiency of Dutch dairy farms was studied by Reinhard et al. (1999); the swine industry in Hawaii was evaluated by Sharma et al. (1997). Dey et al. (2000) used the production frontier to evaluate the technical efficiency of tilapia grow-out operations in the Philippines. Sharma and Leung (2000) used the production frontier to measure the technical efficiency of carp production in India.

Ali and Byerlee (1991) did an intensive literature review of studies that have focused on the economic efficiency of small farmers in developing countries. They summarized the studies by the method used, the countries where the studies were conducted, existing environmental factors, and the other factors that have resulted in farm inefficiencies. Bravo-Ureta and Pinheiro (1993) also did an intensive review of literature of the deterministic and stochastic production frontier approaches that have been employed in studies dealing with developing country agriculture. They organized the studies by the type of data used, the different countries, the different crops and the socio-economic factors that were applied in the different models.

METHODOLOGY

A stochastic production frontier model developed concurrently by Aigner et al. (1977) and Meeusen and van den Broeck (1977) and based on econometric specification of a production frontier is developed for peanut production in Bulgaria. The model can be written as:

$$Y = f(X_i;\beta)+\varepsilon \tag{1}$$

where Y is the output, X_i is ith input, and β is a vector of unknown parameters. The stochastic frontier model is based on the idea that the error term ε in Equation 1 is composed of two independent elements:

$$\varepsilon = v - u \tag{2}$$

The random factors outside the farmer's control (e.g. weather, disease, topography, luck etc.), measurement errors on the dependent variable, and other statistical noise are captured by v which is symmetric. This error is assumed to be independently, identically, and normally distributed as $N(0,\sigma_v^2)$. The term u, on the other hand, is a one-sided component ($u_i \geq 0$), and represents the technical efficiency relative to the stochastic frontier, and is independently and identically distributed, and it follows a half-normal distribution $|N(0,\sigma_u^2)|$. Thus u measures the deviation in the output Y from the maximum output given by the stochastic frontier $f(X_i;\beta)+v$. When a farm output lies on the frontier, u is equal to zero, and when a farm output is below the frontier, u is greater than zero.

The use of the maximum likelihood (ML) to estimate Equation (1) results in consistent estimators for β, λ and σ^2, where β is a vector of unknown parameters; $\lambda = \sigma_u / \sigma_v$ and $\sigma^2 = \sigma_u^2 + \sigma_v^2$ (Bravo-Ureta and Pinheiro, 1997). Inferences about the technical inefficiency for individual farmers can be made by using the conditional distribution of u, given the error term ε from Equation 1 (Jondrow et al., 1982). The formula for calculating the technical inefficiency for individual farmers is given in Equation (3).

$$E[u_i \mid \varepsilon_i] = \frac{\sigma_u \sigma_v}{\sigma}\left[\frac{f(\varepsilon_i \lambda/\sigma)}{1-F(\varepsilon_i\lambda/\sigma)} - \frac{\varepsilon_i \lambda}{\sigma}\right] \tag{3}$$

where $\sigma^2 = \sigma_v^2 + \sigma_u^2$ and $\lambda = \sigma_u / \sigma_v$. $F(.)$ and $f(.)$are the standard normal distribution and the standard normal density function, respectively, and are evaluated at $\varepsilon_i \lambda / \sigma$. Thus the technical efficiency TE_i for each individual farm is given by:

$$TE_i = \exp(-E[u_i \mid \varepsilon_i]) \tag{4}$$

The technical efficiency TE_i will be a number between zero and one (inclusive). A TE_i of one indicates that the farm is producing maximum output given the level of inputs. The coefficient $\lambda = \sigma_u / \sigma_v$ shows the relative variation of the standard error of u to the standard error of v. When λ is zero the symmetric error, v dominates the one-sided error u which is an indication that the difference between the observed output and the frontier output is due to factors outside the farmers control. When λ is large the one-sided error, u dominates the symmetric error v, and this implies that the difference between the observed and the frontier output is the consequence of technical inefficiency. Battese and Corra (1977) used γ, which is defined in Equation 5 as the ratio of variability of u to the total variability (u + v), to measure the difference in the observed output and the frontier output.

$$\gamma = \frac{\sigma_u^2}{\sigma^2} \tag{5}$$

where $\sigma^2 = \sigma_v^2 + \sigma_u^2$. The coefficient of γ gives the difference between the observed output and the frontier output due to the technical inefficiency. The mean technical efficiency for the population is given by Equation 6.

$$E[\exp(-u)] = 2\exp\left(\sigma_u^2 / 2\right)[1 - F(\sigma_u)] \tag{6}$$

Where $F(.)$ is the standard normal distribution function elevated at σ_u.

Production inefficiency

Production inefficiency is an unobserved variable. Hence to determine the level of inefficiency we employ a Tobit model. The Tobit model supposes that there is a latent (that is, unobservable) variable Y_i^*. This variable linearly depends on x_i. Under the Tobit model the relationship between the latent and observed variables for the $_i th$ farm is y^* if y is >0; otherwise 0. The inefficiency of peanut farms is described as:

$$Y = f(X_i;\beta) + \varepsilon \tag{7}$$

where ε_i ~N(0, d), x,

Production risk

While farmers may adjust farm inputs to increase farm efficiency there are risks associated with resource adjustments on a farm. A Just-Pope (J-P) production function is used to estimate the risk effects of a production function, since it relaxes the second moment of the production restrictions (Just and Pope, 1979; Traxler et al. 1995). The J-P function used in this study is given by:

$$Y_i = f(X_i,\beta) + g(X_i,\alpha) + \varepsilon_i \tag{8}$$

where Y_i is the yield or mean response output, $f(X_i,\beta)$ is the function of the explanatory variables, $g(X_i, \alpha)\varepsilon_i$ is the variance of output related to the explanatory variables, X_i is a vector of explanatory variables, β and α are parameter vectors, and ε_i is a random variable with zero mean. The J-P production function implies a multiplicative heteroskedastic model even if the production variances are the independent variables (Judge et al., 1985; Harvey, 1976). Therefore, the three-stage estimation method described by Judge et al. (1985) is used in this study. When the variance is an exponential function of K explanatory variables, the heteroskedastic error of the general model is expressed as:

$$Y_i = X_i'\beta + e_i, \text{ where } i = 1, 2, ... N \tag{9}$$

$$E(e_i^2) = \sigma_i^2 = \exp[Z_i'\alpha] \tag{10}$$

where $Z_i' = (z_{1i}, z_{2i}, ...z_{ki})$ is a vector of observations for K explanatory variables; $\alpha = (\alpha_1\ \alpha_1\ \alpha_1\ .....\alpha_K)$ is a (K x 1) vector of unknown coefficients, and $E(e_i) = 0$, $E(e_ie_s) = 0$ for $i \neq s$. Using the natural log transformation, Equation (10) can be rewritten as

$$\ln \sigma_i^2 = Z_i'\alpha .$$

Since σ_i^2 is unknown, the least square residuals from Equation (10) can be used to replace σ_i^2 in Equation (10) which then become

$$\ln e_i^{*2} = Z_i'\alpha^* + u_i \tag{11}$$

where $u_i = \ln(e_i^{*2} / \sigma_i^2)$

According to Harvey (1976) the u_i will be asymptotically independent with a mean of $E[u_i] = -1.2704$, and with an asymptotic covariance matrix $\Gamma = 4.9348\ (Z'Z)^{-1}$. This result is asymptotically valid in hypothesis tests for the risk effects. To obtain efficient coefficients the predicted values of Equation (11) are used as weights for Equation (8).

Empirical model specification

A semi log model is used to estimate the stochastic production function given in Equation 1:

$$\ln Y_i = \beta_0 + \beta_1 X_1 + \beta_2 X_1^2 + \beta_3 X_2 + \beta_4 X_2^2 + \beta_5 X_3 + \beta_6 X_3^2 + \beta_7 X_4 + \beta_8 X_4^2 + \beta_9 X_5 + \beta_{10} X_5^2 + \beta_{11} X_6 + \beta_{12} X_6^2 + \beta_{13} X_7 + \beta_{14} X_7^2 + \varepsilon_i \tag{12}$$

where $\ln Y_i$ is the log of yield in kg per ha for the ith farm. The explanatory variables for this model are, quantity of seed per hectare (X_1), quantity of seed per hectare squared $(X_1)^2$, quantity of phosphate per hectare (X_2), quantity of phosphate per hectare squared $(X_2)^2$, quantity of nitrogen per hectare (X_3), quantity of nitrogen per hectare squared $(X_3)^2$, quantity of fungicide per hectare (X_4), quantity of fungicide per hectare squared $(X_4)^2$, fixed investment costs per hectare (X_5), fixed investment costs per hectare squared $(X_5)^2$, amount of manual labor per hectare (X_6), amount of manual labor per hectare squared $(X_6)^2$, amount of mechanized labor per hectare (X_7) and amount of mechanized labor per hectare squared $(X_7)^2$. The semi-log model was considered superior after a likelihood test of the OLS, Cobb-Douglas, and the Quadratic functional forms, the semi-log was considered appropriate for model estimation. Maximum likelihood estimation is used to develop the model because of the low mean square error.
The inefficiencies in peanut production were calculated by subtracting the estimated technical efficiency from one. The inefficiencies are then regressed on the different explanatory variables. Since the inefficiencies are values that ranged from zero to one, a Tobit model is used. The advantages of a Tobit model are that it allows the dependent variable to be constrained between certain values; in this study between zero and one (Greene, 1993; Hossain, 1988; Bravo-Ureta and Pinheiro, 1997). The agriculture inefficiency model is given in Equation (13)

$$\delta_i = \alpha_0 + \sum_{j=1}^{6} \alpha_j Z_{ji} \tag{13}$$

Table 1. Number of farms and peanut area for the different villages in Bulgaria.

Village	Number of peanut farms	Peanut area (acres)	Peanut area (%)	Average size of peanut farm (acres)
Asenovgrad	21	49.50	8.70	2.4
P. Evtimovo	12	11.25	2.00	0.9
Kozanovo	11	41.25	7.30	3.8
Muldava	10	9.00	1.60	0.9
D.Voden	10	112.88	19.90	11.3
Zlatovrah	10	10.50	1.80	1.1
Konush	11	63.50	11.20	5.8
Izbegli	15	32.00	5.60	2.1
Karadzhovo	11	41.50	7.30	3.8
Hr. Milevo	10	8.75	1.50	0.9
Katunitsa	2	8.00	1.40	4.0
Kochevo	10	19.50	3.40	2.0
Popovitsa	10	19.50	3.40	2.0
Mominsko	14	32.25	5.70	2.3
Boljrtsi	10	36.75	6.50	3.7
D. Izvor	14	43.75	7.70	3.1
Debar	10	12.50	2.20	1.3
Gradina	14	15.50	2.70	1.1
Total	205	567.88	100.00	

where δ_i is the inefficiency in peanut production for the ith farm and Z_{ji} is the value of the jth explanatory variable associated with the technical inefficiency effect of farm i. The risk of related to peanut production is estimated using the yield variance of peanut production. After the specification of a parametric form for $f(X_i, \beta)$, the model can be consistently and efficiently estimated using a three step process. First the $Y_i = X_i' \beta + e_i$, expressed in Equation (8) is estimated , second the residuals $E(e_i^2) = \sigma_i^2 = \exp[Z_i' \alpha]$ as stated in Equation (9), and third the model is efficiently estimated taking into consideration the heteroscedasticity reflected by the estimated variance. The functional form for the mean yield is given as:

$$\ln Y_i = \beta_1 X_1 + \beta_2 X_1^2 + \beta_3 X_2 + \beta_4 X_2^2 + \beta_5 X_3 + \beta_6 X_3^2 + \beta_7 X_4 + \beta_8 X_4^2 + \beta_9 X_5 + \beta_{10} X_5^2 + \beta_{11} X_6 + \beta_{12} X_6^2 + \beta_{13} X_7 + \beta_{14} X_7^2 \quad (14)$$

While the functional form for the variance of the yield is given as:

$$\ln e_i^2 = \alpha_1 X_1 + \alpha_2 X_1^2 + \alpha_3 X_2 + \alpha_4 X_2^2 + \alpha_5 X_3 + \alpha_6 X_3^2 + \alpha_7 X_4 + \alpha_8 X_4^2 + \alpha_9 X_5 + \alpha_{10} X_5^2 + \alpha_{11} X_6 + \alpha_{12} X_6^2 + \alpha_{13} X_7 + \alpha_{14} X_7^2 \quad (15)$$

Data

The data for this study were gathered from 205 farmers from 18 villages during the period 2001 to 2002. Demographic, farm, crop production systems and marketing data were collected. Information on peanut production was also solicited from farmers. The data on the peanut farmers are summarized in Tables 1 and 2. The age distribution of farmers is skewed towards the age beyond retirement, with 24% belonging to this group. The education level is low with a majority of head of farm households only attaining a primary education level. A large portion of farmers depend on their experience in farming because most farmers have owned and managed their farms for over 20 years. The survey data show that about 95% of the household members participate in peanut growing activities. Of those who participate, about 50% work part-time of up to 4 h per day.

The average size of the peanut farms ranged from 0.9 to 11.3 ha. The peanut yield for the Bulgarian farmers ranged from 550 to 1690 kg/ha, with an average yield of 952 kg/ha. The average amount of seed used is 45 kg/ha, ranging from 32 to 76.8 kg/ha. The recommended quantity of seeds for Bulgarian farmers is 120 kg/ha, but given the factor price of $1.38 per kg, the optimal level is 130 kg/ha. This means that some farmers are using less than the recommended quantity of seeds. The average quantity of phosphate used by farmers is 35.71 kg/ha, with a minimum of 0 and a maximum of 200 kg/ha. The average quantity of fungicide used is 0.10 kg/ha, with a minimum of 0.00 and a maximum of 1.37 kg/ha. The statistical analytical software (SAS) and STATA were used for data analysis. The data on peanut production were used to examine the technical efficiency of peanut farmers by comparing the results of a stochastic production frontier using a maximum likelihood (ML) function and the average production function obtained through ordinary least squares function (OLS).

RESULTS

The estimated results for the mean response function for peanut production (average yield) in Bulgaria are given in Table 3. The factors affecting yield are quantity of seeds, quantity of seeds squared, quantity of phosphate squared, quantity of fungicide squared, investment capital, investment capital squared, mechanized labor, and mechanized labor squared. Farmers can increase yield by increasing the quantity of seeds used to a given

Table 2. Descriptive statistics for the factors use in the production functions.

Variable	Mean	Min	Max	CV (%)
Yield (kg/Acre)	952.14	550.00	1689.83	17.73
Seed (kg)	45.51	32.00	76.80	9.26
Qty. Phosphate (kg)	35.71	0.00	200.00	136.10
Qty. Nitrogen (kg)	169.76	48.00	400.00	25.07
Qty. Fungicide (kg)	0.10	0.00	1.37	171.85
Investment capital ($)	8.42	0.00	108.70	265.69
Manual labor (Hours)	37.45	4.00	126.00	31.77
Leased mechanized labor (Hours)	32.62	9.60	75.43	30.57
Peanut acreage	2.77	0.25	100.00	262.00

Table 3. The ordinarily least square (OLS) and maximum likelihood (ML) estimates for the semi-log model based on a sample of 205 farmers in Bulgaria.

Production factor		OLS	ML
Constant	β_0	5.414 (3.173) ***	5.793(19.156)***
Seed	β_1	0.0409(0.609)	0.034(2.938)**
Seed squared	β_2	-0.000382(-0.561)	-0.000316(-2.595)***
Qty. Phosphate	β_3	0.000555(0.228)	0.000375(0.953)
Qty. Phosphate squared	β_4	0.0000051(0.281)	0.00000651(2.283)***
Qty. Nitrogen	β_5	0.00092(0.260)	0.000262(0.218)
Qty. Nitrogen squared	β_6	-0.00000673(-0.665)	-0.0000039(-1.021)
Qty. fungicide	β_7	-0.283(-1.067)	0.299(2.108)**
Qty. Fungicide squared	β_8	0.429(3.800)***	-0.651(-4.302)***
Investment capital	β_9	-0.00263 (-0.378)	-0.0028(-2.182)*
Investment capital squared	β_{10}	0.0000355(0.394)	0.0000375(2.232)**
Manual labor	β_{11}	0.00956(1.053)	0.0104(3.112)***
Manual labor squared	β_{12}	-0.0000406(-0.484)	-0.0000564(-1.157)
Leased mechanized labor	β_{13}	1.479(0.638)	-0.0111(-2.950)***
Leased mechanized labor squared	β_{14}	0.000080(0.658)	0.000277(4.747)***
σ_u^2			0.0111
σ_v^2			0.01135
σ^2			0.02245(3.956)***
$\lambda = \sigma_u / \sigma_v$			0.99
$\gamma = \sigma_u^2 / \sigma^2$			0.496*** (2.14)
Log. Likelihood			-1280.3

*Significance level 10%; ** Significance level 5%; *** Significance level 1%; Values in brackets are t-statistics.

Table 4. Frequency distribution of farm's level technical efficiency for semi-log model.

Technical efficiency level	Number of farms	Percentage
0.00-0.10	0	0.00
0.10-0.20	0	0.00
0.20-0.30	0	0.00
0.30-0.40	0	0.00
0.40-0.50	0	0.00
0.50-0.60	0	0.00
0.60-0.70	0	0.00
0.70-0.80	1	0.49
0.80-0.90	32	15.61
0.90-1.00	172	83.90
Total number of farms and total percentage	205	100.00

Table 5. Descriptive statistics of the technical efficiency of the individual farms for semi-log model.

Statistical parameter	Value
Mean	92%
Median	93%
Mode	90%
Standard deviation	3%
Minimum	77%
Maximum	97%
Range	20%
First quartile	92%
Third quartile	94%
1st percentile	84%
5th percentile	87%
10th percentile	89%
90th percentile	95%
95th percentile	96%
99th percentile	97%
Skewness	-1.49

level. However, there is a point at which the optimum level is attained after which further increases in input may engender a decline in output. Farmers can also increase yield, manual labor, and reduce the amounts of fungicide, whereas increasing the investment level may have a decreasing effect on yield.

The quantity of seed used had a significant and positive effect on yield. An increase in the quantity of seed by 1% would increase yield by 1.6%. An increase in capital investment has a negative effect on yield suggesting that peanut producing farms in Bulgaria may be overcapitalized, or the available set of equipment may be incongruous to the farm sizes.

The total variability, (σ^2) for the stochastic model is 0.02245, while variability for *u*, (σ_u^2) is 0.0101 and the variability for *v*, (σ_v^2) is 0.01135. The coefficient λ is 0.99 indicating that the one sided error term dominates, and therefore, it is appropriate to use the stochastic frontier model to estimate the production function. The coefficient γ is 0.496, and is significantly different from zero. It shows that about 50% of the discrepancy between the production frontier and the observed output is due to the technical inefficiency at the farm level and this can be controlled by farmer decision. A frequency distribution of the individual farm's level of efficiency is given in Table 4, and shows that none of the 205 farms in this study are producing on the production frontier. Although none of the farms are producing the maximum output, 204 of the farms had technical efficiency that varied from 80 to 97%, and only one farm had technical efficiency that ranged from 70 to 80%. Table 5 gives the descriptive statistics of the technical efficiency of the farms in this study. The average technical efficiency of the farms is 92%. The mode of the technical efficiency for these farms is 90%. The minimum and maximum level of technical efficiency is 77 and 97%, respectively. The standard deviation is 3.0%. The skewness coefficient for the technical efficiency is -1.49.

The 4.0% inefficiency can be explained by gender and age of farmer. Increasing women managed farms increased the level of inefficiency. Increasing the ages of farm managers beyond 45 or below 30 years old increased inefficiency for farms employing more than 35 h of manual labor per ha. Physical farm size was not an issue in terms of farm inefficiency but the number of manual labor hours employed (Table 6).

Output variance response

The results of the variance response function estimation are given in Table 5. The R^2 for this model was 0.83. The joint F-test was used to test the hypothesis that each production factor did not affect the variance (Table 5). The F-test that the coefficients of quantity of seed and

Table 6. Tobit regression results for the technical inefficiency model for farms that use more than 35 h of manual labor per acre for the semi-log model.

Parameter		Coefficient	t-value	P>t
Constant	α_0	0.0679516***	11.92	0.000
Gender (0=male; 1=female)	α_1	0.013941***	2.51	0.014
Age (1= between 30 and 45 years; 0 = older than 45 years)	α_2	-0.0172938***	-2.95	0.004
Education (1= elementary School; 0 = secondary school)	α_3	-0.0039389	-0.83	0.409
Peanut area in acres	α_4	-0.0010011	-1.16	0.248
Number of observations		200[1]		
Log likelihood		242.72		

*Significance level 10%; **Significance level 5%; ***Significance level 1%; [1]The sample was divided into two; Farmers who use more than 35 h of manual labor=1 and those who use less=0. Two extreme observations were excluded.

quantity of seed squared are equal to zero shows that ($\beta_1 = \beta_2 = 0$) was rejected with an F-value of 4.43, indicating that quantity of seeds affected the variance of peanut yield. This means that increasing the quantity of seeds used increases the risk of the peanut farmers in Bulgaria. The F-test that the coefficients manual labor and phosphate squared were equal to zero ($\beta_3 = \beta_4 = 0$) was rejected with a F-value of 2.27; this implies increasing the quantity of manual labor employed increases the risk of peanut yield in Bulgaria. The F-tests for the other production factors are not rejected implying that they were not affecting the variance of the peanut yield and the risk of producing peanuts in Bulgaria (Table 7).

DISCUSSION

The results show that peanut production in Bulgaria is approaching full efficiency and that factors leading to efficiency may vary. Due to changes in farm structure during the transition period, there may be excess capacity at the farm level as farmers search for the optimum levels of input to produce a given quantity of peanuts. A study by Kopeva and Noev (2002) showed that cereal, vegetables and grape farms had average efficiency levels less than that of peanuts for both family and corporate farms. The mean efficiency levels obtained for cereal, vegetable and grape farms were 63.2, 42.9, and 43.1%, respectively, compared to the 92% for peanuts found in this study. Kopeva and Noev attributed the inefficiencies obtained for cereal, vegetables and grapes to the negative effects of the ongoing land, and structural reforms in Bulgarian agriculture. They stated that the loss of efficiency may be due to necessary changes resulting from shifts in the state agricultural policies. The technical efficiencies obtained from peanut farms in Bulgaria are also larger than that for small, medium and large rice farms in the Myanmar. The mean levels of efficiency for the small, medium and large rice farms were 81, 73 and 80%, respectively (Kyi and von Oppen, 1999). The results show, however, that the factors influencing efficiency may be crop specific, and about 50% of 8.0% (4.0%) of farmers considered technical inefficient can be reduced by government supportive programs and farmers adoption of best management practices.

The quantity of seeds used affected the levels of efficiency. Hence farmers can increase the quantity of seeds used to improve yields. The response of yield to quantity of seed used is elastic and a one percent increase in seed application results in a 1.17% increase in yield, *ceteris paribus.* Kyi and von Oppen (1999) found that seeds significantly influenced technical efficiency of small farms in the Myanmar. The average quantity of seeds used per hectare by Bulgarian farmers is 45.5 kg per ha whereas the recommended quantities in the US may vary from 65 to 180 kg/ha based on plant spacing, and that recommended for Bulgaria is 120 kg/ha. Therefore, there may be room for improving peanut yield by increasing the seeding rate, especially as farmers are using less than the optimal quantity of seeds. Feder (1980) indicated that farmers may be using less than the optimal quantity of input if they are risk averse. The results show that increasing seeding rate poses the greatest risk, and even if the seed cost shares are relatively small, the total seed cost and variation in quality are important budget items to limited resource farmers who are already financially constrained. In our case, the price of the seed to the farmer is the opportunity cost of consumption since farmers save their own planting materials.

Other factors that can be manipulated are pesticide treatment, investment and the use of manual labor. Additional amounts off manual labor may increase farm efficiency. However, increasing number of manual labor may increase yield risk.

Investment capital and mechanized labor are negatively related to yield, and there will be a reduction in yield if additional amount of inputs are employed. The negative relationship between investment capital and yield indicates that farms are overcapitalized and farmers may seek to make farm adjustments in the long run. An estimation of cross production elasticities would suggest that the present situation in peanut farming encourages

Table 7. Estimated coefficients for the mean and the variance of the peanut yield for the semi-log model based on a sample of 205 farmers in Bulgaria.

Production factor		Mean of yield	Variance of yield
Constant	β_0	5.414(3.173)***	-
Seed	β_1	0.0409(0.609)	-0.124(-2.25)**
Seed squared	β_2	-0.000382(-0.561)	0.00097(1.11)
Qty. Phosphate	β_3	0.000555(0.228)	-0.01011(-1.07)
Qty. Phosphate squared	β_4	0.0000051(0.281)	0.000125(1.79)*
Qty. Nitrogen	β_5	0.00092(0.260)	0.01026(0.657)
Qty. Nitrogen squared	β_6	-0.00000673(-0.665)	-0.000055(-0.76)
Qty. Fungicide	β_7	-0.283(-1.067)	0.1681(0.15)
Qty. Fungicide squared	β_8	0.429(3.800)***	-
Investment capital	β_9	-0.00263 (-0.378)	0.0259 (0.97)
Investment capital squared	β_{10}	0.0000355(0.394)	-0.00035(-1.01)
Manual labor	β_{11}	0.00956(1.053)	-0.0945(-1.64)*
Manual labor squared	β_{12}	-0.0000406(-0.484)	0.00093(1.52)
Leased mechanized labor	β_{13}	1.479(0.638)	-0.0256(-0.43)
Leased mechanized labor squared	β_{14}	0.000080(0.658)	0.00059(0.68)
R-squared		0.41	0.87
N		205	205

*Significance level 10%; ** Significance level 5%; *** Significance level 1%; values in brackets are t-statistics.

capital labor substitution in order to improve efficiency. The results show that manual labor can be increased while investment capital can be reduced to increase efficiency in yield, other factors remaining constant. Substitution of labor input for capital may encounter difficulty since Bulgaria and other CEE countries are facing a situation where the number of farm hands per hectare of cultivated land is decreasing as more employment opportunities open up in other sectors (Brent, 1999).

Peanut yield is also negatively influenced by fungicide use. Only a small number of farmers used fungicide in peanut production in Bulgaria. Fungicide use is not recommended for Bulgarian peanut farmers (Pan and AGROLINK, 2004). Hence the application of fungicide in peanut production may increase cost but not yield.

The need to improve farm efficiency through on-farm resource reallocation is imperative and can be accomplished at little or no cost to the public sector. The method of improvement of peanut production efficiency can be examined within a framework of modifying efficiency of other less efficient production enterprises at limited costs to the sector on a whole. In fact, there may be savings to be captured by the substitution of cheaper inputs for more costly factors of production while increasing economic efficiency of production in the agricultural sector.

Inefficiency was influenced by age. A large percentage of the farmers were above 45, and more farmers who are beyond this age, the greater the levels of inefficiency. As is observed in the survey results 69% of the farmers are older than 45 years old, and 1.9% were less than 30 years old. Draganova (2005) indicated there are large numbers of retired farmers beyond the ages of 70 who are involved in agricultural production, and it is difficult to recruit younger farmers. An increase in the number of farmers younger than 30 years old also increased the level of inefficiency while the opposite is true for farmers between 30 and 45 years. Hence, the government should implement policies that encourage farmers between the ages of 30 to 45 to remain in agriculture, while improving farm sizes as older farmers leave production agriculture.

Gender participation in farm operation and ownership influenced inefficiency. More women operators enhanced inefficiency. This seems contrary to the current literature on the efficiency of female farm operators. During the transition period women were less prepared educationally and had less experience in management of enterprises (Meurs, 1998, Mathijs and Vranken, 2001). Hence, this lack of experience is manifested in lower efficiency in farm management activities. Bulgarian women seem to have less access to farm investment capital and this may influence the level of farm efficiency. Farm operators in Bulgaria work on the average four hours per day out off farm to earn more money. However, women operators with their families may not have the opportunity to work outside and this may limit the non-agricultural resources they are able to transfer to on-farm activities. Also women farm operators have fewer avenues to obtain government assistance to improve their farm activities.

The results show that limited resource Bulgarian peanut farmers are approaching full efficiency. Efficiency is more crop-specific and is less dependent on farm size or levels of investment. Hence, the Bulgarian government, like all other CEE governments, should not rapidly merge farms to improve efficiency without technical information on the crop land requirements and the cropping system. To attain full efficiency farmers must be able to absorb risk and to make minor farm adjustments. The government can facilitate a drive towards full efficiency by putting into place policies that would target farmers between the ages of 30 to 45 to remain into peanut farming and provide farmers with a safety net. Assistance should be directed to women farm operators in terms of education and access towards government programs.

ACKNOWLEDGEMENTS

This study was funded by the Peanut Collaborative Research Support Program, USAID Grant No. LAG-G-00-96-90013-00, and supported by Auburn University and the University of Georgia.

REFERENCES

Aigner DJ, Lovell CAK, Schmidt P (1977). Formulation and estimation of stochastic frontier production function models. J. Economet. 6(1):21-37.

Ali M, Byerlee D (1991). "Economic efficiency of small farmers in a changing world: A survey of recent evidence. J. Int. Develop. 3(1):1-27.

Battese GE, Corra GS (1977). "Estimation of a production frontier model: with application to the pastoral zone of Eastern Australia. Australian J. Agric. Econ. 21:167-179.

Bravo-Ureta B, Pinheiro E (1997).Technical, economic and allocative efficiency in peasant farming: Evidence from the Dominican Republic". Dev. Econ. 35(1):48-67.

Bravo-Ureta B, Pinheiro E. (1993). Efficiency analysis of developing country agriculture: A review of the frontier function literature Agric. Resouc. Econ. Rev. 22(1):88–101.

Brent KJ (1999). Extension, research and farm competitiveness in Central and Eastern Europe, Part I, Sustainable Development (SD)." Food and Agriculture Organization of the United Nations (FAO). pp. 1-11.

Coelli T, Rahman S, Thirtle C (2003). A stochastic frontier approach to total factor productivity measurement in Bangladesh Crop agriculture, 1961-1992. J. Int. Develop.15:321-333.

Dey MM, Paraguas FJ, Bimbao GB, Regaspi PB (2000). "Technical efficiency of tilapia grows out pond operations in the Philippines. Aquac. Econ.Manag. 4(1-2):33-47.

Draganova M (2005). The restructuring of rural areas in Bulgaria: Needs must policy or policy based on the needs of rural people? Local development and innovation in rural areas and small towns in Europe; Chydenius Institute, Kokola, Findland, Nov. 26.

Feder G (1980) Farm size, risk aversion and the adoption of new technology under uncertainty, Oxford Economic Papers, New Ser. 32(2):263-283.

Greene WH (1993). Frontier Production Functions". EC-93-20. Stern School of Business, New York University.

Harvey AC (1976). Estimating Regression Models with Multiplicative Heteroscedasticity."Econometric 44:461-465.

Hazel P (2004). Last Chance for Small Farm?" IFPRI Forum pp.1-3 Considerations. Am. J. Agric. Econ. 61(2):276-284.

Herrero I, Pascoe S (2002). "Estimation of technical efficiency: a review of some of the stochastic frontier and DEA software". CHEER, Consumers in Higher Education Economics Review, Economics Network 15(1):1-19.

Hossain M (1988). Nature and Impact of the Green revolution in Bangladesh, Research Report 67. Washington, D.C.: International Food policy Research institute.

Jondrow J, Lovell CAK, Materov IS, Schmidt P (1982). On the estimation of technical inefficiency in the stochastic frontier production function model. J Econ. 19:233-238.

Judge GG, Griffiths WE, Hill RC, Lutkepohl H, Lee T (1985). The Theory and Practice of Econometrics," 2nd ed. New York: John Wiley and Sons.

Just R, Pope R (1979). Stochastic Representation of Production Functions and Econometric Implications. J. Econ. 7(1):111-118.

Karagiannis G, Tzouvelekas V (2001). Self-Dual Stochastic Production Frontiers and Decomposition of Output Growth: The Case of Olive-Growing Farms in Greece. Agric. Resour. Econ. Rev. 30(2):168-178.

Kopeva D, Noev N (2002). Technical efficiency and competitiveness of farms in transition: Lessons from Bulgaria." fourth coming IAMO series Agriculture and Food Industry in Central and Eastern Europe, IAMO, Halle/Saale, Germany.

Kostov P, Lingard L (2002). "Subsistence farming in transitional economies: Lessons from Bulgaria". J. Rural Stud. 18(1):83-94.

Kyi T, von Open M (1999). An Economic Analysis of Technical Efficiency of Rice Farmers at Delta Region in Myanmar", Institute of Agricultural Economics and Social Sciences in the Tropics and Sub-Tropics (Ed), Forshung zur EntwicklungsoKonomie und Politik-Research in Development Economics and Policy. June 1999. University of Hohenheim, Germany.

Lulcheva D, Todorova SA (2005). Actual Problems of the Structural Reform in Bulgarian Agriculture. J.Central Europ. Agric., 6:563-568.

Mathijs E, Swinnen JFM (2001). Production Organization and Efficiency during Transition: An Empirical Analysis of East German Agriculture. The Rev. Econom. Statist. 83(1):100-107.

Mathijs E, Dries L, Doucha T, Swinnen JFM (1999). "Production Efficiency and Organization of Chech Agriculture". Bulgarian J. Agric. Sci. 5(2):312-324.

Mathijs E, Vranken L (2001). Human Capital, Gender and Organization in Transition Agriculture: Measuring and Explaining the Technical Efficiency of Bulgarian and Hungarian Farms". Post-Communist Econ. 13(2):171-187.

Meeusen W, van den Broeck J (1977). Efficiency estimation from Cobb Douglas production functions with composed error. Int. Econ. Rev. 18:435-444.

Meurs M (1998). Imagined and imagining equality in East Central Europe: Gender differences in the economic transformation of Bulgaria. In Theorizing transition: the political economy of post-communist transformation; by Pickeles, J. A. Smith, Routledge. pp. 327- 329.

OECD (1999) (Organization for Economic Co-operation and Development). "Agricultural Policies in Bulgaria," http://www.oecd.org/document/25/0,2340,en_2649_2034649_1_1_1, 00.html

Pan G, AGROLINK (2004). "Pesticide Use in Bulgaria: Pesticide use, issues and how to promote sustainable agriculture in Bulgaria.Fact sheet, pp.1-4.
Reinhard S, Lovell CAK, Thijssen G (1999). "Econometric estimation of technical and environmental efficiency: An application to Dutch dairy farms. Am. J. Agric. Econ. 81(1):44-60.
Roy N (2002). A stochastic Production Frontier Model of the Newfoundland Snow Crab Fishery." Eleventh Biennial Conference of the International Institute for Fisheries Economics and Trade, Wellington, New Zealand, August. pp.19-22.
Seyoum ET, Battese GE, Fleming EM (1998). Technical efficiency and productivity of maize producers in eastern Ethiopia: A study of farmers within and outside the Sasakawa-Global 2000 project. Agric. Econ. 19(3):341-348.
Sharma KR, Leung PS (2000), Technical efficiency of carp production in India: a stochastic frontier production function analysis. Aquac. Res. 31:917-947.
Sharma KR, Leung P, Zaleski HM (1997). Productive efficiency of the swine industry in Hawaii: Stochastic Frontier vs. Data Envelopment Analysis. J. Prod. Anal. 8(4):447-459.
Todorova SA, Lulcheva D (2005). "Economic and Social Effects of Land Fragmentation on Bulgarian Agriculture. J. Central Eur. Agric. (4):555-562.
Totev S, Shahollari L (2001). Agriculture development and trade in Bulgaria, FYR of Macedonia and Albania in the context of Common Agriculture Policy." South-East Eur. Rev. 3(2001):S-51-70.
Traxler G, Falck-Zepeda J, Ortiz-Monasterio JL, Sayre K (1995). Production Risk and the Evolution of the Varietal Technol. Am. J. Agric. Econ. 77:1-7.
Tzouvelekas V, Pantzios CL, Fotopoulos C (2002). "Empirical evidence of technical efficiency levels in Greek Organic and conventional farms. Agric. Econ. Rev. 3(2):49-60.
Wilson P, Hadley D, Ramsden S, Kaltsas I (1998). Measuring and Explaining Technical Efficiency in UK Potato Production. J. Agric. Econ. 49(3):294-305.
Yao S, Liu Z (1998). Determinants of grain production and technical efficiency in China. J. Agric. Econ. 49(2):171-184.

9

Economic analysis of selected food crops in Ile-Ife of Osun State, Nigeria

Adisa O. D. and Sofoluwe N. A.

Department of Agricultural Economics, Obafemi Awolowo University, Ile-Ife, Osun State, Nigeria.

Studies showed that young farmers are not attracted to the production of food crops which is central to the reduction of poverty and food security among poverty ridden populace of Osun State. This study analyzes the economic factors responsible for productivity of selected food crops in Osun State. Factors influencing the level of production as well as problems militating against the production of the selected food crops in the study area were examined. Purposively, sampled data were collected from 100 farming households. Budgetary analysis and four functional forms of ordinary least square (OLS) were fitted to the data. Results showed that farm size allocated to crops imposes a positive and significant influence on return. The cost-benefit ratio to food crops are 0.65, 1.22 and 0.44 for cassava, maize and yam, respectively. Capital was the major constraints to food crop production in the study area for all the crops studied. It was concluded that yam production was more profitable but less cultivated compared to cassava. Farmers especially the young ones need to be financially encouraged to cultivate more of these crops and reduce the family size in order to sustain the productivity of these food crops.

Key words: Food crops, food security, productivity.

INTRODUCTION

Over 90% of Nigeria's agricultural output is produced by resource poor farmers who have, for centuries, sustained the national food supply by harnessing both natural and socio-economic factors of production (Adedipe et al., 2004). These small farmers produce about 80% of the total food in the country (IFAD, 2009). Despite the wide areas of arable land, productivity is still restricted (Mohammed, 2008). Decline in agricultural production in Nigeria began with the advent of the petroleum boom in the early 1970s which brought about a distortion of the labor market. The distortion in turn produced adverse effects on the production levels of the food crops. Idachaba (2004) argued that the dwindling agricultural production is a confirmation of the unattractiveness of agriculture as a result of low returns and compensation being paid to the farmers, which tend to discourage increased production. Over 90% of Nigeria's agricultural output is produced by resource poor farmers who have, for centuries, sustained the national food supply by harnessing both natural and socio-economic factors of production (Adedipe et al., 2004). These small farmers produce about 80% of the total food in the country (IFAD, 2009). Despite the wide areas of arable land, productivity is still restricted (Mohammed, 2008). Decline in agricultural production in Nigeria began with the advent of the petroleum boom in the early 1970s increased production. Government attitude to agriculture and adverse climatic change such as erratic rainfall, worsened the low productivity, both per unit of land and per farmer which in turn, made agricultural work unattractive. The problems also lead to loss of interest in farming and enhanced the lure of the cities for the young and able bodied. The interaction of these factors has heightened the problem of food insecurity in the country (Babatunde and Oyatoye, 2005; DFID, 2009).

As the food situation and the performance of the agricultural sector got worsened, a number of agricultural development institutions were set up to improve the performance of the sector to increase agricultural and food productivity. These institutions are National Accelerated Food Production (NAFPP), Agricultural Development Projects (ADPs), River Basin Development Authorities (RBDAs), National Seed Service (NSS), National Centre For Agricultural Mechanisation (NCAM), Agricultural And Rural Management Training Institute (ARMTI) and Agricultural Credit Guarantee Scheme Fund (ACGSF). Others were the Nigerian Agricultural Cooperative and Rural Development Bank (NACRDB)/agricultural bank, Operation Feed the Nation (OFN), Green Revolution Programme, Directorate of Foods, Roads and Rural Infrastructure (DFFRI), Nigerian agricultural insurance company (NAIC), National Agricultural Land Development Authority (NALDA), Specialized Universities for Agriculture, Root and Tuber Expansion Programme (RTEP) and rural banking scheme, etc. Despite all these, there is loss of interest in farming and consequently a reduction in food production (Babatunde and Oyatoye, 2005). This study therefore examines the economic factors responsible food production of selected food crops in Osun State. Specifically, the study seeks to:

i) Describe the socio-economic characteristics of selected food crop farmers in the study area;
ii) Examine the factors influencing the level of returns to selected food crops farmers;
iii) Identify the problems militating against the production of selected food crops in the study area.

MATERIALS AND METHODS

Study area

The study was carried out in Osun state, located in southwestern Nigeria, between latitudes 7.0° and 9.0°N, and longitudes 2.8° and 6.8°E. The topography is rolling hills and lies between 300 and 600 m above sea level. Average rainfall decreases from 1475 mm in the forest belt in the southern sections of the state to 1125 mm in the savannah section to the north. Mean annual temperature ranges from 27.2°C in June to 39.0°C in December. Soil types are varied but most contain a high proportion of clay and sand and are mainly dominated by the lateritic series.

The State is mainly agrarian. Food crops grown in the area include maize (*Zea mays*), yam (*Dioscorea* spp), cassava (*Manihot esculenta*), cocoyam (*colocasia* spp), rice (*Oryza sativa*) and vegetables (*Amaranthus* spp). The permanent crops cultivated include cocoa (*Theobroma cacao*), kolanut (*Cola nitida*) and oil palm (*Elaeis guinensis*). These crops are usually mixed or intercropped.

Data collection

The primary data used for this study were obtained from a cross sectional survey of farmers and was conducted in 2008/2009 cropping season. Data were collected using a well structured questionnaire to obtain data on the socio-economic characteristics of the respondents, scale of farming, inputs used, output and prices.

Sampling procedure and sample size

Multistage sampling technique was employed. Rural communities notable for agricultural activities in Ife central local government were purposively selected. The communities were stratified into major food and tree crop production zones. Out of the food crop producing communities, Kajola and Agbogbo villages were selected. 40 cassava farmers, 30 maize farmers and 30 yam farmers were randomly selected among the identified food crop farmers in the study area. Thus, a total of 100 farmers were sampled.

Methods of analysis

Data collected were analyzed using descriptive statistics (mean, frequency counts, and percentage), budgetary analysis and four functional forms of ordinary least square (OLS) regression analysis. The different forms of OLS regression analysis were used following the experimental approach to analysis to determine which model provides the best fit to the data (www.statpac.com/surveys/sampling.htm).

Descriptive statistics

Frequency counts, mean and percentage were used to analyze the socio-economic characteristics of the respondents such as age, educational level, farm size and problems militating against the food crop production.

Budgetary technique

An enterprise budget approach was undertaken to estimate costs and returns to each of the three enterprises. According to Alimi and Manyong (2000), a budget is a quantitative expression of total farm plan summarizing the income, costs and profit (a residue of total cost from total revenue). Revenue was computed as monetary value of the total farm output sold/consumed by the farmer's household, given out as gifts or used for other purposes. Costs and returns were computed on per hectare basis. Variable costs change with level of production and included fees, fertilizer, labor, interest charge and supervision charges. Fixed costs, which do not vary with level of production, included rent and cost of hand tools. The budgetary technique was used to evaluate levels of profitability of the enterprise by estimating the revenue, gross margin and net farm income at the end of the production process. The difference between the two parameters is a measure of net profit or return in food production. The following were computed for each category of food crop farmers:

1) Gross revenue (GR): $GR = P \times Y$
Where: P = output price and Y = yield,

2) Gross margin (GM): $GM = GR - TVC$
Where: GM = gross margin, TVC = total variable costs and GR = gross revenue,

3) Net farm income (NFI) = GM - Total Fixed Cost (TFC) or GR - TC,

4) Operating expense ratio = $\frac{TVC}{GR}$,

5) Net farm income ratio = $\frac{NFI}{GM}$,

6) Return/Naira outlay = $\frac{NFI}{TC}$

Where: TC = TVC + TFC and TC = Total cost.

Regression analysis

For this study, the OLS method of regression was employed to analyze each enterprise. Four functional forms (linear, semi-log, double log, and exponential forms) were used in order to determine which model provides the best fit to the data for each of the enterprises.

The implicit general from of the OLS is defined as:

Linear:

$$Q = a + b_1 X_1 + b_2 X_2 + b_3 X_3 + \mu \quad (1)$$

Exponential

$$\text{Log } Q = a + b_1 X_1 + b_2 X_2 + b_3 X_3 + \mu \quad (2)$$

Semi-log

$$Q = a + b_1 \log X_1 + b_2 \log X_2 + b_3 \log X_3 + \mu \quad (3)$$

Double-log

$$\text{Log } Q = Log a + b_1 \log X_1 + b_2 \log X_2 + b_3 \log X_3 + \mu \quad (4)$$

Where Y = output (₦); X_1= Age (years); X_2= Household size (number); X_3= Farm size (Ha); μ = error term.

The choice of a working functional equation was based on statistical measure of performance such as the adjusted R^2, the F-statistics, significance of the individual coefficients and the signs of the regression coefficients whether or not they conform to the a priori expectation. The higher the adjusted R^2, the better the equation fits the observed data.

The *a-priori* expectations of the signs of the coefficients are:

b_1: As a farmer grows older, it is expected that resource use and allocation would have been mastered for an optimum benefit. However, old age may also mean inability to manage resources beyond a little capacity that can be expected to cope with current market request. Hence, the variable can be positive or negative. Age is not necessarily synonymous with experience.

b_2: Larger family size is generally associated with a greater family labour being available to the household for the timely operation of farm activities. According to Akinbode (1973), as more labour is utilized, output is expected to increase and hence increase profit. However, as more family labour is used, increase in household expenditure is incurred. It is therefore difficult to predict this variable 'a priori' in this study.

b_3: The more the land area allocated to food crops production, the greater the expected output which translates to increased profit (Omotesho et al., 1993). Hence, the coefficient of farm size is expected to be positive.

RESULTS AND DISCUSSION

Demographic characteristics of the respondents indicated differences between respondents (Table 1). The result showed that most (77.5%) of the cassava farmers, 77.3% of the maize farmers and 83.3% of the yam farmers aged between of 30 and 50 years, respectively, showing that they are in active age brackets. The mean age was 42.4 years (cassava farmers), 40.9 years (maize farmers) and 41.3 years (yam farmers), respectively. Younger farmers may be more knowledgeable about better practices and may be more willing to bear risk and adopt improved technology because of their longer planning horizons. Older farmers may be less likely to understand inherent benefits in a given innovation.

The household size depicts that most of the farmers had a mean household size of 7.8 (cassava farmers), 7.3 (maize farmers) and 6 (yam farmers), respectively. The importance of this finding is that the large nature of household size could affect their food intake and food security of the household. This corroborates with the findings of Owu (1995) that the larger the sizes of the household, the more the food required within the household, this will in turn have a negative relationship with returns as a substantial portion of farm output is used to feed the family members.

Area of land cultivated to the food crops showed that 2.1, 0.9 and 0.4 acres were cultivated to cassava, maize and yam, respectively. The more the land area allocated to food crops production, the greater the expected yield which translates to increased profit (Omotesho et al., 1993). But as indicated, the land area allocated to the food crops is small which could translate into low aggregate output of food output hence, food insecurity for the people.

Literacy level analysis revealed that 70% of cassava farmers had no formal and incomplete basic primary education. Less than 30% of the maize farmers had basic education with the rest being illiterate. Fifty percent (50%) of the yam farmers had basic education.

Education plays important role in the optimal use of resources by farmers. As farmers acquire more education, their ability to obtain, analyze, interpret and use information improves. Education increases the ability of farmers to use their resources efficiently. Literacy level analysis revealed that 70% of cassava farmers had no formal and incomplete basic primary education. Less than 30% of the maize farmers had basic education with the rest being illiterate. 50% of the yam farmers had basic education.

The analysis of the problems (Table 4) militating against the food production showed that about (57.5%) of the cassava farmers had problems with capital. The main problems of the maize farmers were drought (43.3%) and labour cost (30%). For the yam farmers, theft and labour scarcity constituted the main problem. Access to capital enhances farmers' readiness to adopt technological innovations and increase production of food. It is hypothesized that the variable has a positive influence on the probability of adoption of land enhancing technologies (Bekele and Drake, 2003).

Table 1. Distribution of respondents by their socio-economic characteristics.

Variable	Cassava (n= 40)		Maize (n =30)		Yam (n = 30)	
	Frequency	%	Frequency	%	Frequency	%
Age (years)						
Below 30	3	7.50	3	10.00	4	13.33
31 - 40	12	30.00	12	40.00	11	36.67
41 - 50	16	40.00	7	23.34	7	33.33
Above 50	9	15.00	8	26.66	8	26.67
Mean	42.40		40.93		41.33	
Household size						
Below 6	14	35.00	12	40.00	15	50.00
7 - 9	14	35.00	12	40.00	11	36.67
Above 10	12	30.00	6	20.00	4	13.33
Mean	7.8		7.3		6	
Area of land cultivated (acre)						
Less than 2	28	70.00	28	93.34	15	50.00
2 - 3	11	27.00	2	6.66	8	26.67
Above 3	1	2.50	-	-	7	23.34
Mean	2.1		0.89		0.40	
Educational level (years)						
No schooling	17	42.50	12	40.00	11	36.67
Pry school incomplete	11	27.50	10	33.33	4	13.33
Secondary school incomplete	8	20.00	6	20.00	8	26.66
Post secondary	4	10.00	2	6.67	7	23.33

Source: Field survey (2011).

Table 2. Estimated costs and returns per hectare of selected food crops.

Source	Cassava (n = 40)	Maize (n = 30)	Yam (n = 30)
Total revenue	180, 655	81,467	236,400
Variable costs (₦)			
Labour	60,699.40	47,427	92,325
Fertilizer	10,714.28	14,158	39,583
Cuttings, seeds, setts	15,386.90	4,483	13,688
Supervision	8,746	6,639	20,531
Interest charge	13,125.60	12,208	-
Total variable cost	91,280	31,146	159,482
Fixed costs (₦)			
Tools	663.69	4,778	2,880
Rent on land	-	10,461	2,083
Total fixed cost	557.50	5486	4763
Total cost (TVC + TFC)	109,330.40	36,632	164,247
Net farm income (TR - TC)	71,324.60	44,835	72,153

Source: Computed with data obtained from field survey (2011).

Table 3. Profitability measures of the selected food crops.

Variable	Cassava	Maize	Yam
Profit (₦)[a]	71,324.60	44,835	72,153
Gross margin (₦)	89,325	50,321	76,918
Rate of return	0.65	1.22	0.44
Operating expenses ratio	0.51	0.38	0.67
Net income ratio	0.79	0.89	0.94

Source: Computed with data obtained from field survey (2011).

Table 4. Constraints to food production in the study area.

Problem	Cassava		Maize		Yam	
	Frequency	%	Frequency	%	Frequency	%
Capital	23	57.50	-	-	-	-
Low market value	7	17.50	-	-	-	-
Poor access road	2	2.50	-	-	-	-
High labour cost	8	20.00	9	30.00	10	33.33
Drought			13	43.33		
marketing			5	16.67		
Pests and diseases			3	10.00		
Theft	-	-	-	-	11	36.67
Yam sett	-	-	-	-	6	20.00
Harsh weather	-	-	-	-	3	10.00

Source: Computed with data obtained from field survey (2011).

Factors influencing level of return to food production

The regression results of the analysis (Table 5) showed that the double logarithm model gave the best fit for the analysis. For the cassava production, the result is explicitly stated as;

$$\text{Log}\, Q = Log a + b_1 \log X_1 + b_2 \log X_2 + b_3 \log X_3 + \mu$$

$$\text{Ln (TR)} = 4.87 - 0.00778\text{Ln age} - 0.02773\text{Ln household size} + 0.843\text{Ln farm size}$$

(8.225) (-0.251) (-0.103) (4.086)*

Adjusted $R^2 = 0.282$

The results showed that 28.2% of the variations in total revenue were explained by the explanatory variables included in the model. One of the three explanatory variables (farm size) was significant at 5% level. This means that a unit change in farm size has a strong influence on revenue. The positive sign of the coefficient of farm size conforms to *a priori* expectation that an increase in farm size leads to increased output and hence increased revenue. The negative sign of the coefficient of household size leads to a decrease in total revenue. This shows that a unit increase in family size decreases the revenue of the farmers.

The regression results of maize production showed that a large family size will cause a reduction to total revenue due to large consumption of what should have been sold.

The result is explicitly stated as:

$$\text{Ln (TR)} = 4.93 - 0.0533\text{Ln age} - 0.071\text{Lnhousehold size} + 0.882\text{Ln farm size}$$

(42.72) (0.702) (-1.062) (21.013)*

Adjusted $R^2 = 94.0$; F =151.96

The regression result of yam production is explicitly stated as:

$$\text{Ln (TR)} = 7.57 + 0.0256\text{Ln age} - 0.0120\text{Lnhousehold size} + 0.515\text{Ln farm size}$$

(16.799)* (3.811)* (-0.595) (14.416)*

Adjusted $R^2 = 94.0$; F =151.96

The positive sign of the coefficient of farm size conforms to the *a priori* expectation that an increase in farm size leads to increased output and hence increased revenue. The negative sign of the coefficient of household size leads to a decrease in total revenue. This shows that a unit increase in family size decreases the revenue of the yam farmers.

Table 5. Estimated OLS regression function of the selected food crops production.

Variable	Cassava (n = 40)				Maize (n = 30)				Yam (n = 30)			
(Code/Name)	Linear	Exponential	Semi-log	Double-log	Linear	Exponential	Semi-log	Double-log	Linear	Exponential	Semi-log	Double-log
Constant	8274.45 (0.223)	4.55 (18.672)*	96960.54 (1.054)	4.87 (8.225)*	9572.84 (1.187)	4.48 (68.016)	53809.57 (1.638)	4.93 (42.72)	101398.7 (1.324)	7.65 (16.658)*	-142009.3 (-2.710)*	7.57 (16.799)*
X_1- Age (years)	-204.03 (-0.397)	-3.15E-05 (-0.009)	-35531.91 (-0.737)	-7.78E-02 (-0.251)	115.06 (0.774)	-1.083E-03 (-0.891)	36707.66 (1.697)	5.33E-02 (0.702)	145.7 (0.983)	-3.15E-03 (-0.943)	1887.95 (0.522)	2.565E-02 (3.811)
X_2-Household size (No.)	1622.15 (0.668)	1.904E-03 (0.119)	10439.73 (0.252)	-2.73E-02 (-0.103)	-1110.45 (-1.377)	-3.55E-03 (0.538)	-21936.07 (-1.141)	-7.17E-02 (-1.062)	-1983.6 (-1.412)	2.04E-02 (0.919)	-3007.02 (-0.784)	-1.209E-02 (0.595)
X_3-Farm size (ha)	43575.23 (6.095)*	0.189 (4.015)*	188113.76 (5.863)*	0.843 (4.086)*	84396.18 (17.47)*	0.517 (13.093)	131597.87 (11.021)	0.882 (21.013)*	72577.3 (15.32)	0.514 (14.368)*	33761.04 (17.355)*	0.515 (14.416)*
Adj. R^2	47.9	27.4	40.0	28.2	92.1	86.1	43.60	94.0	84.9	88.4	91.8	84.8
F	12.95	5.89	12.09	6.12	114.15	61.12	81.5	151.96	37.8	56.24	81.96	75.01

Source: Computed with data obtained from field survey (2011). *Significant at 0.05%. Figures in parentheses are t-ratio.

Results of the budgetary analysis

Table 2 shows that net income was ₦59,912.5 for cassava farmers, ₦ 44,835 for maize farmers and ₦72,153 for yam farmers, respectively. The net income from the yam farmers was higher. There was a difference in cost of production for the different crop farmers. The cost-benefit ratio indicated that for every ₦1 spent by yam farmers, ₦0.55 profit was returned. For the maize farmers, for every ₦1 spent, ₦1.2 was returned. For the cassava farmers, for every ₦1 spent, ₦0.65 was returned (Table 3).

The operating expense ratio for cassava farmers indicated that 51% of gross revenue was used for operating expenses. The operating expenses ratio for maize farmers indicated that 38% of gross revenue was used for operating expenses, while 67% of gross revenue was used for operating expenses by the yam farmers. Net income ratio indicated that 94, 89 and 79% of gross revenue went to farmer equity for yam, maize and cassava farmers, respectively. Yam production was more profitable.

CONCLUSION AND RECOMMENDATION

Several factors are responsible for the decline in food crop production in Nigeria. Low farm size allocated to food production and high family size of the farming household negatively affects the return to food production in the study area. This has consequently led to loss of interest in farming activities, insufficient food production and food insecurity in the nation. Increased labour cost and scarcity, poor access to credit and theft are the other contributing factors.

It is therefore, recommended that farmers especially the young ones should be motivated financially, supported with farm machinery to reduce the labour problems. Institutions should also facilitate access to more cultivable land with adequate security to protect the effort of the farming household. The young farmers should be encouraged to add educational value to themselves and reduce their family size to benefit maximally and optimally from their farming activities. It is recommended that farmers should cultivate more of yam.

REFERENCES

Adedipe NO, Okunleye PA, and Ayinde IA (2004). The Relevance of Local and Indigenous Knowledge for Nigerian Agriculture. Presented at the International Conference on with Bridging Scales and Epistemologies: Linking

Local Knowledge Global Science in Multi Scale Assessments; March, 16-19. Alexandria, Egypt.
Akinbode A (1973). The Socio Economic Determinants of Farm Sizes. The Case of Ekiti Central Division in Western Nigeria. Niger. Agric. J. 10(2):192-207.
Alimi T, Manyong VM (2000). Partial budget analysis for on-farm research IITA research guide. P. 65.
Babatunde R, Oyatoye E (2005). Food security and Marketing Problems in Nigeria: the Case of Maize Marketing in Kwara State. Deutscher Tropentag, October 11-13, Hohenheim "The Global Food and Product Chain- Dynamics, innovations, Conflicts, Strategies".
Bekele W, Drake L (2003). Soil and water conservation decision behavior of subsistence farmers in the Eastern Highlands of Ethiopia a case study of the Hunde-Lafto area. Elsevier Science Ltd. Ecol. Econ. 46(3):437-451.
Department for International Development (DFID) (2009). One World A Million Stories Development. 45:23.
Idachaba F (2004). Food security in Nigeria: challenges under democratic dispensation'. Paper presented at ARMTI Lecture, Ilorin, March 24. pp. 1-23.
IFAD (2009). Who produces our Food? FAO, Rome, Italy. Assessed on 18 November 2009 on web page: http://www.ifad.org/ events/ op/ 2009/rt_4.htm.
Mohammed I (2008). Road Map to Attaining Food Security in Nigeria". Nigerian Tribune. Monday, 5th May.
Omotesho OA, Owe J, Oladeji SO (1993). Resource Productivity in Food Crop Production in Some Selected villages Oyi Local Government area, Kwara State, Nigeria. Ife J. Agric. P. 15.
Owu DO (1995). Farmer Adoption of Improved Soil Conservation Technologies Under International Agriculture in Imo State. Research Report to Fourth ARSSRN Programme.
Sampling Procedure on website www.statpac.com/surveys/sampling.htm. Assessed on 3rd February, 2010.

Effects of international fund for agricultural development (IFAD) credit supply on rural farmers in Rivers State, Nigeria

Tasie C. M.

Department of Agricultural Science, Ignatius Ajuru University of Education, Rumuolumeni, Port-Harcourt, Rivers state, Nigeria.

This study evaluated the effect of International Fund for Agricultural Development (IFAD) credit supply on rural farmers in Rivers state. Data for the study were collected using a structured questionnaire administered to 90 farmer's beneficiaries using a multi-stage sampling technique. The regression result shows that semi-log function gave the best fit with the highest value of coefficient of multiple determinations (R^2) of 0.8758 and seven explanatory variables were significant and there was a significant F-value. The significant variables are farm size, off-farm income, total household labour, educational level of farmer, gender, farm household size, and IFAD credit. The study also shows that IFAD credit impacted positively on the well-being of rural farmers. The IFAD programme has contributed to the increase in farm output and income. The study recommended that IFAD and their collaborating government agencies should expand their credit delivery in the study area to enable more farmers benefit from their services. Also, training programmes should be organized for all farmers in view of the fact that education produces significant influence on the income of the farmers.

Key words: Credit supply, rural farmers, international fund for agricultural development (IFAD), Rivers state.

INTRODUCTION

Investment expenditure provides the thrust for development. Agricultural development often involves expenditure on capital inputs. These expenditures require funding. Hence, the provision of funds is fundamental to agricultural development (Nwajiuba, 1989). Agriculture is a key sector in the Nigerian economy. Its importance is particularly glaring in a developing economy like Nigeria where land and labour resources are relatively abundant and the industrial sector is poorly developed. The contribution of agriculture to overall development especially in developing countries like Nigeria include provision of increased food supplies, provision of gainful employment, provision of capital and capital formation, increasing foreign exchange for development and increasing rural welfare. The agricultural sector's contribution to economic growth and sustained rural development remains to be fully exploited (FMARD, 2006). The contribution of agriculture to GDP was 64% in 1960, declined to 35% in 1988, and presently, the agricultural sector in Nigeria contributes less than 30% to GDP, with crop production accounting for an estimated 85% of this total, livestock 10%, with forestry and fisheries contributing the remaining 5% (Awotide and Akerele, 2010).

An examination of the Nigerian agricultural sector shows that it is not in a position to finance its own development. Nwagbo (1986) reports that emphasis on the financing problem is rightly founded on the belief that agriculture for various reasons is not in a strong competitive position in relation to other sectors to acquire or obtain investment and productive credit from the usual financial institutions. Farm credits are however important means for improving farm capital investment in Nigeria, with which there may be no progress in the agricultural sector to adequately fulfill its expected roles or millennium development goals (Musa et al., 2010).

These roles include achievement of self-sufficiency in the domestic production of food, revival of agricultural export crops production, generation of rural and agricultural employment and improvement of rural income and welfare (Nwaru, 2005). To attain agricultural policy objective, programmes such as the National Accelerated Food Production Programme, the Agricultural Development Programmes, River Basin and Rural Development Authorities, Operation Feed the Nation, the Green Revolution, the National Agricultural Land Development Authority (NALDA), Agricultural Credit Guarantee Scheme Fund (ACGSF), etc., were launched (ADP, 2005). Mention should also be made of NACB, now Nigerian Agricultural, Co-operative and Rural Development Bank (NACRDB). Yet, agricultural policy objectives have not been achieved, as evidenced by the general food scarcity in Rivers state and in the whole country.

Advancing reasons for this sordid situation, Balogun (1986) and Mejeha and Nnanna (2010) attributed this declaring trend of agricultural production to inefficient traditional practice which is being practiced by small-holder farmers. Supporting this view, Nwaru (2005) and Omeh (2006) stated that Nigerian small-scale farmers are known to be economically weak with little or no capital investment. Consequently, they use low technology tools and methods in their production activities, which in turn lead to reduced output and productivity. In its own contribution, IFAD (2002) opined that causes of food insecurity and famine were not so much failures in food production, but structural problems relating to poverty and to the fact that the majority of the developing world's poor population are concentrated in the rural areas.

Summing up all these views, Okerenta (2005) and Tasie (2008) identified insufficient extension or delivery of production credit to the poor farmers as the most critical factor responsible for the declining trend in agricultural production. It is therefore an irony of circumstance that the small-scale farmers who produce about 85% of food consumed in the country and the agricultural exports are perpetually handicapped by lack of production credit and bedeviled with poverty.

According to Nweze (1990), it was to obviate this sordid situation that successive Nigerian governments have attempted to ridge the credit gap in the agricultural sector through the establishment of various credit programmes. Those supply-led rural finance institutions include the Nigerian, Cooperative and Rural Development Bank (NACRDB), Community Banks (now Micro-Finance Banks), etc., which were established both to improve growth and equity and to neutralize or mitigate urban-biased macro-economic policies. Unfortunately, most of these financing programmes launched for over two decades have not had impressive impacts.

Among the factors responsible for this lack of significant effect of financing programmes, insufficient loan amount and poor loan repayment are considered the most critical (Okorie, 1986). Another is the inability and unsuitability of the formal credit institutions to adapt to the peculiar needs of the rural small-scale farmers in their socio-economic environment (Ijere, 1986). More frequently however, these problems simply expose and exacerbate more fundamental weaknesses within the credit system themselves.

To solve these problems, the Federal Government went into a funding agreement with International Fund for Agricultural Development for the funding of small-scale farmers in Nigeria. The International Fund for Agricultural Development (IFAD), a specialized agency of the United Nations, was established as an International Financial Institution in 1977 as one of the major outcomes of the 1974 World Food Conference. The conference was organized in response to the food crises of the early 1970s that primarily affected Sahelian countries of Africa. The conference resolved that an International fund for Agricultural Development should be established immediately to finance agricultural development projects primarily food production in developing countries.

In this context, IFAD was created to mobilize resources for programmes that alleviate rural poverty and improve nutrition. Unlike other International Financial Institutions which have a broad range of objectives, the Fund has a very specific mandate to combat hunger and rural poverty in developing countries. To achieve this objective, IFAD cooperates and collaborates with government agencies and parastatals such as Agricultural Development Programmes (ADP) and Ministries of Agriculture and Rural Development. IFAD-assisted programmes in Rivers state and Nigeria are generally deemed credible, highly relevant and effective, with positive impact. As a consequence, it is generally accepted that the Fund has a distinct and catalytic role in improving the livelihood of both subsistence and market-oriented small-holder farmers and producers. This belief is supported by ADP (2005), Mejeha and Nnanna (2010) and Tasie (2008).

Therefore, the broad objective of this study is to evaluate the effects of IFAD credit on the rural farmers of Rivers State. The specific objectives of this study are to examine factors that determine farm income amongst farmers in the IFAD credit programme and describe the effects of IFAD credit on these farmers. Based on these objectives, the hypothesis is that the credit supplies have not improved the farm income of the rural farmers in Rivers State.

METHODOLOGY

Rivers State is the study area. This is informed by the fact that agriculture is the major occupation of the people of Rivers Sate. This is induced by the rich soil, which stretches the length and breadth of the state. The climate is essentially tropical humid with an average annual rainfall of 220 to 250 mm, evenly distributed through its long wet season, which covers the period of eight months (March to November). The period is followed by the dry season spanning the months of November to March. The state is made of 23 Local Government Areas which are grouped into three agricultural zones.

A multistage sampling technique was used. Multi-stage sampling technique involves a procedure whereby the selection of units into the sample is organized into stages. It usually involves a combination of sampling methods. All the three agricultural zones were covered in this study. In stage one, all the IFAD credit beneficiaries were identified. The lists of these farmers form the sampling frame. For stage two, one Local Government area was randomly selected from each agricultural zone. In the third stage, 30 farmers each from the beneficiaries in the IFAD programme were randomly selected. This gave rise to 90 farmer-beneficiaries. Structured questionnaires were used to collect data from both farmers and ADP officials (IFAD credit facilitator). Other sources of data were publications in journals, textbooks, reports, and seminar materials. To achieve the objectives of the study, econometric techniques (regression analysis), and difference of means were applied on the data.

Factors that determine farm income were ascertained with the regression model. The four functional forms of the model, namely, linear, double-log, semi-log and exponential, was tried and the lead equation was selected based on the value of R^2, F-statistics and the conformity to a priori expectation. The implicit function was specified as:

GFI = f(LHA, OFI, HIR, HHL, PFI, EDU, GEN, HHS, ICS, U)

Where GFI = gross farm income. This was measured by the total amount of sales of farm produce in naira. LHA = farm size (in hectares). The total land the farmer has brought under cultivation. OFI = off-farm income. Total income to the farmer from sources other than his farm in naira. HIR = hired labour (man-days). HHL = total labour from household (man-days). PFI = purchase farm inputs like seeds, seedling, cuttings, agrochemicals including fertilizer in naira. EDU = educational level of the farmer. This was measured by the total number of years he spent in receiving formal education. GRN = gender of the farmer. A dummy variable (female = 1; male = 0). HHS = farm household size; total number of person that live and feed from the respondents. ICS = IFAD credit size. This is the total amount of credit extended to the farmers by IFAD and measured in naira farms. U = error term.

Test for difference between means

The test of the difference between two groups of data (gross farm income before IFAD credit supply (x_1) and gross farm income after IFAD credit supply (x_2)) was carried out. The range of years before and after IFAD financing was a maximum of two years before obtaining IFAD credit and maximum of two years after obtaining IFAD credit.

This analysis was focused on the difference D (when D = $x_2 - x_1$) between each matched pairs of observations and on the two groups of sample means, x_1 and x_2. The test was performed as a test of single mean D:

$$\bar{D} = \sum D/n$$

Where $\bar{D}$ = mean of the difference of the matched pairs; N = number of pairs of observation; D = $x_2 - x_1$ for each matched pair.

For large samples, D is approximately normally distributed, hence the test procedure is conducted the same way as for large independent samples using z – test. The test statistic is defined as:

$$Z = \frac{\bar{D}}{SD/n\sqrt{\ }}$$

D = sample mean difference between each pair of observation; SD = sample standard deviation of these differences and is defined as:

$$SD = \sqrt{\frac{\sum D^2 - n\bar{D}^2}{n-1}}$$

RESULTS AND DISCUSSION

In line with the objective of this study, the results are further discussed.

Factors that determine gross farm income

Analysis on the factors that determine gross farm income of farmers in IFAD credit programme in Rivers State as presented in Table 1 indicates that semi-log function gave the best fit with the highest value of co-efficient of multiple determination (R^2) of 0.8758 and significant variables and a significant F-value. The R^2 of 0.8758 implies that the significant explanatory variables explain or influence the criterion variables by 87.58. The significant F-statistics of 62.68 implies that the joint effect of all the included variables were significant.

From the result, the co-efficient of farm size, off-farm income, total household labour, educational level of farmer, gender, farm household size, and IFAD credit supply were all significant as shown in Table 1.

The co-efficient of farm size was significant at 1% and was positive, indicating that the gross farm name is directly related to farm size. This implies that the employment of more land resources would lead to higher income. The co-efficient of off-farm income was significant and negative, implying an inverse relationship with farm income.

This means that the farmers will most likely favour other employment that gives them higher financial returns. This seems to underpin the current high rate of rural-urban migration that has severely afflicted rural economics in Nigeria. Household labour was statistically significant and positive, indicating that the farm income is directly related to house hold labour. Farm operations in River State have remained labour intensive and farmers would rely more on their household labour than on hired labour. This is cost effective in relation to farm income.

Education is significant and positive, indicating a direct

Table 1. Regression result of factors that determine gross farm income.

Variable	Functional forms			
	Linear	Exponential	Semi log	Double log
Farm size	0.5317 2.4735 *	0.000024 2.2184 *	18052 3.1614	0.5700 1.9154 **
Off-farm income	-1.03562 -0.9525	- 0.000010 -0.4636	- 0.565 -9.3240 *	0.3171 0.7772
Hired labour	-0.4036 -0.9092	2.2736 1. 1152	-27.249 -0.38	4.6821 1.0253
Total household labour	1354.08 0.5145	0.2490 1.380	120.911 3.701 *	0.4161 2.4227 *
Input purchased	556.917 0.8672	1.17964 0. 3970	2227889 1.3412	1.486 2.2630 *
Educational level	1569.44 2.6013 *	4.5902 1.783 **	1582779 1.8405 **	9495.355 1.4825 **
Gender	1184.71 -1.675 **	-0.011545 -1.3643	4.9963 -1.9886 **	-11536.99 -2.4363 *
Household size	270.017 0.9645	0.07073 2.2952	4.0268 2.8676	2715326.3 1.495
IFAD credit supply	2540731 1.788 **	0.1898 1.687 **	5.654 3 .1420 *	3272425 1.6160 **
Constant term	2.2318	6.9234	4.6214	10.1135
R^2	0.5423	0 .5245	0.8758	0.8027
F-ration	10.5318	9.8048	62. 6803	36.1630
N	90	90	90	90

*Significant at 1%, **significant at 5% (Field survey, 2010).

relationship with farm income. Education and training produce a labour force that is more skilled and adaptable to the needs of a changing economy. It helps to unlock the natural talents and inherent enterprising qualities of the farmer. It enhances the farmer's ability to understand and evaluate new production techniques. This translates into higher farm income and productivity. The coefficient for gender is statistically significant and negative. Given that this is a dummy variable (female = 1 and male = O), the male gender generated farm income more than the females. Given the same conditions, male headed farm households will earn more than female headed ones. The coefficient of IFAD credit supply was highly significant and has positive relationship with farm income. This showed a direct relationship between farm income and IFAD credit supply. This implies that farm income increases with increase in credit supplied to farmers. Credit to rural farmers will generate in them the optimism and determination to venture into new fields, increase size of farm, increase productivity and farm income, facilitate adoption of improved farm practices, encourage capital formation, improve marketing efficiency, and improve the living conditions of the rural farmers.

The significant variables and their signs are in line or support the a-priori theoretical expectations. The implication of these findings for the rural area is that future policies on rural farmers should take adequate consideration of these variables which have significant affects on rural farmers.

Effects of IFAD credit on farmers beneficiaries

Table 2 shows that all the respondents (100%) farm income and output increased after IFAD credit supply. Also, 93.3% of the respondents had the hectarage of

Table 2. Distribution of farmers based on the effects of IFAD credit.

S/N	Effects	No of respondents	%	Ranking
1.	Increase in income/output	90	100	1
2.	Purchase of means of evacuation of farm produce	60	66.7	6
3.	Increase in farm holdings	84	93.3	2
4.	Increase in nutritional status	81	90	3
5.	Ability to meet short term expenditure	76	84.4	5
6.	Procurement of more working capital	80	88.9	4
7.	Building and repair of dilapidated buildings	53	58.9	7

Source: Field Survey (2010).

their farm holding (farm size) increased and 90% of the respondents had their nutritional status increased.

From this study, the important effect of IFAD credit on the farmers is increase in farm income and output which is ranked first, followed by increase in farm holdings which ranked second. Other lower ranked items such as increase in nutritional status, procurement of more working capital (fertilizers, farm implements and improved seeds), ability to meet short-term expenditure (payment of children's school fees and medical expenses), purchase of means of evacuation of farm produce, building and repair of dilapidated buildings, had also showed significant improvement after IFAD credit supply as shown in Table 2.

Testing of hypothesis

The credit supplies have not improved the farm income of the rural farmers in Rivers State, that is:

H_0: e = O
H_1: e > O

A t- test statistic cannot be used because the sampling distribution is no longer symmetrical but rather, skewed. The alternative hypothesis is stated on a greater than basis meaning a right tail test is required. This implies a one-tail test only. So, a Z-test statistic is used

Appendix 1

The result shows the zeal = 38.08. At 5% level of significance, since the zeal (38.08) is greater than the Z tab (1.658), we reject the null hypothesis and accept the alternative. Mean of deviation (D) is given thus:

$$\bar{D} = \frac{\sum = D}{n} \quad \frac{5165000}{90} = 57389$$

Sample standard deviation (SD):

$$= \sqrt{\frac{\sum D^2 - n\bar{D}^2}{n-1}}$$

$$SD = \sqrt{\frac{314615000000-90\ (57389)^2}{90-1}}$$

$$SD = \sqrt{\frac{314615000000 - 90\ (329349721)}{90-1}}$$

$$SD = \sqrt{\frac{314615000000 - 296414758900}{89}}$$

$$SD = \sqrt{\frac{182000241100}{89}}$$

$$SD = \sqrt{204497091}$$

$$SD = 14300$$

$$Zcal = \frac{\bar{D}}{SD/\sqrt{n}} = \frac{57389}{14300/\sqrt{90}} = \frac{57389}{14300/9.49}$$

$$Zcal = \frac{57389}{1507} = zcal = 38.08$$

CONCLUSION AND RECOMMENDATION

The findings of this study suggest that IFAD credit supply has produced positive impact on the income of farm in the study area. This was indicated from the regression analysis and test of difference of means. Some variables, namely, farm size, off-farm income, household labour, educational level of farmer, gender, farm household size, and IFAD credit supply were all significant in influencing the farm income of farmers. With increase in income and output, there will definitely be an increase in nutritional status and social development. In line with the results, it is recommended that the IFAD and their collaborating government agencies should expand their credit delivery in the study area to enable more farmers benefit from their services. Furthermore, training programmes should be organized for all farmers in view of the fact that education produces significant influence on the income of the farmers.

REFERENCES

Agricultural Development Programme (ADP) (2005). Annual Report of the Abia State Agricultural Development Programme, Umuahia, Abia State, Nigeria.

Awotide DO, Akerele EO (2010). Commercial Agriculture in Nigeria: Prospects, social impacts, Constraints and Policy Issues. In commercial Agriculture, Banking Reform and Economic Downturn: Setting a New Agenda for Agricultural Development in Nigeria. Proceedings of the 11th annual National conference of National Association of Agricultural Economics (NAAE). Nov. 30th-Dec.3rd 1-5.

Balogun ED (1986). Agricultural Development Strategies in Nigeria: past, present and future in Nigerian Agricultural Outlook (EDS) A. Okon and M. O. Ijere. NAAE, publication, Enugu Nigeria. 1:11-15.

Federal Ministry of Agriculture and Rural Development (FMARD) (2006). National Programme for Food Security(NPFS) Expansion Phase Project 2006 – 2010. (Main Report).

IFAD (2002). International Fund for Agricultural Development in www.ifad.com

Ijere MO (1987). Agricultural credit and Economic Development *in Reading in Agricultural Finance.* A. Okorie and M.O. Ijere (eds) CRDC, UNN.

Mejeha RO, Nnanna IN (2010). Effect of root and Tuber Expansion Programme (RTEP) on commercialization of staple food crops in Abia State Nigeria, *in commercial Agriculture, Banking reform and Economic downturn:* setting a New Agenda for Agricultural Development in Nigeria. Proceedings of the 11th annual National conference of National Association of Agricultural Economists (NAAE).

Musa SA, Hanisu AT, Yakubu SA (2010). Performance of the Agricultural credit guarantee scheme fund (ACGSF) in Kano State. In commercial Agriculture, Banking Reform and Economic Downturn: Setting a New Agenda for Agricultural Development in Nigeria. Proceedings of the 11th annual National conference of National Association of Agricultural Economics (NAAE).Nov.30th-Dec.3rd.pp. 66-71

Nwagbo EC (1986). The credit institution in managing Agricultural Development in Nigeria, ARMTI seminar series 1, Ijere, MO, Idachaba FS (eds).

Nwajiuba UC (1989). Factors in Commercial Banks lending to Agriculture in Nigeria: 1970-1988. An unpublished M.SC thesis. Department of Agric. Economics, UNN.

Nwaru JC (2005). Determinants of farm and off-farm incomes and savings of Food crop farmers in Imo State, Nigeria: Implication for poverty alleviation. Niger Agric. J. 36:26-42

Nweze NJ (1990). The structure, functioning and potentials of indigenous cooperative credit associations in financing agriculture. The case of Anambra and Benue States, Nigeria. An unpublished PhD dissertation. Dept of Agric. Economics, UNN.

Okerenta SI (2005). Evaluation of the Effects of Micro Finance Programmes on Rural Life of Farmers in the Niger Delta Region of Nigeria. Unpublished PhD dissertation submitted to the Department of Agricultural Economics, FUTO.

Okorie A (1986). Commercial Bank lending to agriculture. In the Performance of formal financial institutions in financing agriculture in Abia state. An unpublished PhD dissertation by Ezeh CI (1999). Agricultural Economics Dept, UNN.

Omeh NG (2006). Determinants of commercialization of cassava production in Abia State, Nigeria. Undergraduate project submitted to the department of Agricultural Economics, Michael Okpara University of agriculture, Umudike.

Tasie CM (2008). An Evaluation of the Effects of credit supply on Rural Farmers in River State; the case of International Fund for Agricultural Development (IFAD). Unpublished M.Sc Thesis submitted to the Department of Agricultural Economics, FUTO.

Determinants of income diversification among rural households: The case of smallholder farmers in Fedis district, Eastern hararghe zone, Ethiopia

Amare Demissie[1] and Belaineh Legesse[2]*

[1]International Committee for the Development of Peoples (CISP), P. O. Box 888, Harar, Ethiopia.
[2]Agricultural Economics, Haramaya University, P. O. Box 138, Haramaya, Ethiopia.

This paper examines the determinants of income diversification among rural households using cross sectional data collected from Fedis District of Eastern Hararghe Zone, Ethiopia. Both descriptive statistics and rigorous econometric models are used to analyze the data. Multinomial logit model was used to pinpoint factors influencing households' participation in non/off-farm activities while the determinants of non/off-farm income were analyzed by Tobit model. The descriptive statistics result shows that agricultural activities are the most important source of income for rural households in Fedis District contributing 77% of total household income with the remaining 23% originating from non-agricultural activities. About 84% of the sample households involved in non/off-farm activities and only 16% did not participate in any non/off-farm activity. Participation in non/off-farm employment activities and the level of income derived are found to be influenced by human capital related variables (gender and age of household head, number of economically active family members, education level of household head and presence of children attending school), livelihood assets (livestock holding, size of cultivated land), livelihood diversifying strategy (crop based diversification through number of crops grown and harvested) and infrastructure related variable (proximity to market). The results imply that these factors need to be considered by policy makers in the planning of agricultural and non agricultural initiatives in this study area.

Key words: Smallholder, livelihood diversification, participation decision, household income, determinants of economic choices, Ethiopia.

INTRODUCTION

In Ethiopia, the policy focus is to increase agricultural productivity and farm income so as to attain food self sufficiency at a national, regional and household levels. While substantial resources have been spent on agricultural research and extension to alleviate food shortage in the nation, research and extension activities have not been done adequately on the issues related to off/non-farm employment.

Despite this fact, farmers are engaged in a variety of off/non-farm activities to diversify their income with a view to feed and sustain themselves during crop failures. The main question and concerns of policy makers is to make sure whether or not it is possible to support farmers to engage in off/non-farm activities without sacrificing the farm productivity and food self sufficiency objectives. Hence, looking into the link between farm and non/off-farm activities and their determinants is necessary before policy measures are taken to promote non/off-farm activities (Tassew, 2000). In view of these outstanding issues, various empirical studies have pinpointed the socio-economic rationale of rural livelihoods for pursuing differentiated and contextual livelihood strategies.

According to Ellis (2000), participation in multiple activities by farm families is not new or only confined to

*Corresponding author. E-mail: belaineh.legesse@yahoo.com

the rural sectors of developing countries. Most rural families have truly multiple income sources which may indeed include off-farm wage work in agriculture and wage from non-farm activities, rural non-farm self employment (e.g. trading), and remittances from urban areas and from abroad.

In assessing diversification strategies of households, Barrett et al. (2005) and Reardon et al. (2001) indicate that because of differences in initial asset endowments, rich and poor households diversify differently. The rich typically engage in more capital intensive and more remunerative activities, while leaving the poor confined to labour intensive, highly contested niches with low barrier to entry and low returns. Wealthier households often mention "profit maximization" as their motive for entering in to rural non-farm activities, whereas lower income households emphasize "risk minimization" and "income stabilization". Poorer households have a high incentive but a low capacity to diversify successfully, even if in some cases they rely more on non-farm activity in percentage terms.

Hoogeveen (2001) and Sumberg et al. (2004) also argue that most attractive non/off-farm employment opportunities have the highest entry barriers and that the poor, who have fewer buffer stocks, less access to credit and greater interest in risk management strategies, are often not able to access the safest and most rewarding income opportunities because of entry barriers. Richer people thus have greater freedom to choose among a wider range of non-farm options than do the poor. On the other hand, the poor have little choice when diversifying out of farming: they go in to unskilled off-farm labour and other activities with low barriers and therefore generally poor returns.

Barrett and Reardon (2000), based on review of 27 case studies in Africa, showed that the average share of non-farm income in total rural household income is around 45% with non-farm wage labor income exceeding self-employment income; and non-farm earnings are substantially greater than either agricultural wage employment earnings or migration earnings.

In Latin America, Deininger and Olinto (2001) showed that in the Colombian case, non-farm income contributes between 30 to 40% of total household income. Escobal (2001) shows that about 51% of the net income of Peruvian rural households comes from off-farm activities while in Mexico, off-farm activities generate more than half of farm households' incomes.

Despite the fact that agriculture is the main source of livelihood in rural Ethiopia, farmers are engaged in a variety of off/non-farm activities to diversify their income and enable them cope with the risk of crop failures. However, there is a wide difference between literatures regarding the share of non/off-farm income in total household income in Ethiopia. Barrett and Reardon (2000) reported that the share of non/off-farm income in rural Ethiopia averaged about 36% in 1989/90. On the other hand, Reardon et al. (2006) found that non/off-farm share of total income in rural Ethiopia was about 20% in 1999 fiscal year. Despite the differences in the percentage of income share derived from non-farm employment, the role of non-farm income in total household income is significant. Sara (2007) found that about 75% of rural households engage in off-farm activities and approximately 31% of their livelihood income is generated from off-farm employment indicating that income from farming is not sufficient to support the household economy. The working hypothesis of this study is that motivation for and participation in livelihood diversification strategies is contextually varied.

Thus, this present study tries to explore the determinants of income diversification among rural households in one of the most drought risk prone districts in eastern Ethiopia. This study attempts to address the following questions. What are the types and characteristics of non/off-farm activities existing in this study area? What inspires participation in rural non/off-farm activities? What contributes to the variation in non/off-farm income level of households?

DATA AND SAMPLING DESIGN

This study is conducted in Fedis district of Eastern Hararghe Zone of Oromia National Regional State in Ethiopia. Fedis, one of the 18 districts of East Hararghe Zone, is among the food insecure districts of the Zone where rain-fed agriculture is the mainstay of the local economy. A simplified formula provided by Yamane (1967) is employed to determine the required sample size. Data for the study was collected from 120 sample households residing in four farmers' associations (*'kebeles'*) using structured questionnaire. Sample households were selected using a two-stage random sampling technique. In the first stage, four farmers' associations were randomly selected out of the 19 farmers' associations of Fedis district. In the second stage, using the fresh household list obtained from each sampled farmers' association, households were classified into participant and non-participant categories. Data collected by field personnel of International Committee for the Development of People (CISP) during the 2009/2010 production year had been utilized as an input in the stratification process. Then sample households were randomly selected from each category with probability proportional to size (PPS).

Method of analysis

This study intended to identify the determinants of households' participation into non/off-farm activities using multinomial logistic regression model with a particular interest to explore factors influencing households' choices between varying income diversifying activities. This study also analyzed factors contributing to variations in participating households' income. To this effect, Tobit model is used to understand why some households were able to derive better/lower income from specific non/off-farm activities than others. The characteristics of multiple non/off-farm activities in this study area were explored using the different techniques of descriptive statistics. Following the work of Tassew and Oskam (2001), the multinomial logit model can be specified as follows:

Let U_{ij} denote the utility that the household i gets from choosing alternative activity j and

$$U_{ij} = u_{ij} + e_{ij} = X_{ij}\gamma_j + e_{ij} \quad (1)$$

Where γ_j varies and X_i remains constant across alternatives; and e_{ij} is a random disturbance term reflecting intrinsically random choice behavior, measurement or specification error and unobserved attributes of alternatives.

Let also $P_{ij\,(j=0,1,2,3)}$ denote the probability associated with the off/non-farm activities choices of household i with; j = 0 if the household does not participate in non/off-farm work, j= 1 if the household participates in non/off-farm wage employment, j= 2 if the household participates in non/off-farm self employment, andj= 3 if the household participates in both non/off-farm wage and self employment.

Then the multinomial logit model can be given by:

$$P_{ij} = \frac{\exp(X_i\gamma_j)}{\sum_{j=0}^{3}\exp(X_i\gamma_j)} \quad (2)$$

Setting $\gamma_0 = 0$, the multinomial logit model can be written as:

$$P_{ij} = \frac{\exp(X_i'\gamma_j)}{1+\sum_{j=1}^{3}\exp(X_i'\gamma_j)}(j = 1,2,3)\,and\ P_{i0} = \frac{1}{1+\sum_{j=1}^{3}\exp(X_i'\gamma_j)} \quad (3)$$

This can be estimated using the maximum likelihood method;

Similarly, Based on Sesabo and Tol (2005), the Tobit model can be specified as follows:

$$y_i^* = X_i'\beta' + \varepsilon_i'$$

For a Tobit model, the dependent variable (activity income) can take the value of zero or positive values as follows:

$$y_i = y_i^* \text{ if } X_i'\beta' + \varepsilon_i' > 0 \text{ and } y_i = 0 \text{ if } X_i'\beta' + \varepsilon_i' \leq 0 \quad (4)$$

Where y_i^* is a latent dependent variable that captures the ith household propensity to earn income from a certain source, X_i' is a matrix of variables such as household asset endowments, household characteristics, institutions and location characteristics, which describe the potential benefits of participating in various activities, β' is a parameter vector to be estimated, ε_Ω is a random disturbance term. The model assumes that $\varepsilon_\Omega \sim N(0,\sigma)$.

The coefficients of activity income are estimated by the maximum likelihood estimation and the log-likelihood function for the Tobit model is expressed as follows:

$$\ln L = \sum_{yi>0} -1/2\left[\ln(2\pi) + ln\sigma^2 + \frac{(y_i - \beta' X)^2}{\sigma^2}\right] + \sum_{yi>0} ln\left[1 - \Phi(\frac{\beta' X}{\sigma})\right] \quad (5)$$

Where, Φ is the Cumulative Density Function (CDF) of the standard normal distribution function; Here the first part of the likelihood function is essentially the classical regression model for the non-zero observations, while the second half represents the probabilities for the censored observations. The maximum likelihood estimator has the desirable properties for being both consistent and asymptotically efficient (Greene, 2003; Sesabo and Tol, 2005).

The explanatory variables used for the analyses are grouped into household demographic characteristics, assets and mediating factors. The demographic characteristics include age, sex, and family size and number of economically active family members. The asset variables include size of cultivated land, livestock wealth, education of household head and number of children attending school. Mediating factors include variables such as perceived quality of land, number of crops harvested per year, access to credit, proximity to market and agro-ecology in which rural households make and earn their livelihoods.

RESULTS AND DISCUSSION

Descriptive statistics

The descriptive statistics results show that 84% of sample households have participated in non/off-farm activities while 16% did not engage in any form of non/off-farm employment and their sole employment was only farming. Of the total sample households, about 86% were male and 14% were female headed households. From the group of participants, 58% of sample households involved in sole wage employment, 12% in self employment and 30% in both self and wage employment activities. Non/off-farm employment income contributed to about 23% of total household income of participants. Of those participating in non/off-farm wage employment, 70, 16 and 14% were involving in the government productive safety net program (PSNP), agricultural daily wage and other casual non/off-farm wage activities, respectively.

Determinants of participation in non/off-farm activities

The result of multinomial logit model showed that there is disparity in the effect of explanatory variables on participation decision of households in non/off-farm wage and self employment activities. Sex, age and education of household head, number of economically active family members, and number of children attending school significantly affected participation into wage employment activities. Whereas participation in non/off-farm self employment was significantly influenced by sex, age, education of household head, number of economically active family members, distance to market, livestock holding and crop based diversification strategy, that is, number of crops grown and harvested in a season. Similarly, the variables: sex, age, education, number of economically active family members, distance to market and livestock holding have significantly influenced participation in multiple self-wage employment activities.

The age of household head has significantly and

Table 1. Multinomial Logit estimation result of determinants of participation.

Variable	Non/off-farm wage employment		Non/off-farm self employment		Non/off-farm wage + self employment	
	Coefficient	t-ratio	Coefficient	t-ratio	Coefficient	t-ratio
Intercept	0.6279	0.5430	0.0703	0.9189	0.7621	0.8703
Age	-0.2486	-3.0271***	-0.2377	-2.3485**	-0.2753	-3.0725***
Sex	4.5047	3.3780***	3.1444	2.2431**	3.6672	2.6894***
Family size	0.4059	0.7649	0.0049	0.1886	0.7164	1.2490
Education	-2.4452	-2.7692***	0.0091	2.1802**	2.5173	2.4069**
Ec. active members	3.3062	2.3551**	3.0583	2.0593**	2.0558	2.8960**
Children in school	-2.2489	-2.2800**	1.7879	2.6120***	2.6204	1.2584
Credit amount	0.1305	0.9807	0.0733	0.9981	0.1326	0.8961
Cultivated land	3.5584	0.7492	2.7342	0.6174	0.5081	0.9946
Livestock holding	1.6658	1.6315	0.5402	2.2447**	1.5748	2.4512**
Number of crops	0.2329	0.0647	-0.9521	-2.1078**	0.5622	1.2221
Distance to market	-0.8459	-1.7544	-1.2668	-2.0476**	-1.0654	-2.7081***
Agro-ecology	-3.1444	-1.4928	0.8914	0.4633	1.0970	0.0753
Soil fertility	-0.6098	-0.5264	1.0069	0.6334	0.6706	0.6372

***, **and * indicate statistical significance at 1, 5 and 10% probability levels, respectively. Dependent variable: Participation. Weighting variable: one; Number of observations: 120; Log likelihood = - 79.66; Restricted log likelihood = -117.67; Chi-square = 105.62.

negatively affected participation in sole wage, self employment and multiple self-wage employment activities at 1, 5 and 1% levels of significance, respectively (Table 1). This implies that supply of labor to sole and combined non/off-farm employment activities was higher for younger households than older households. Hence, younger households rely on non/off-farm employment to support their livelihoods while the older ones concentrate on farming instead of opting for engagement in non/off-farm work. Existence of entry barriers and lack of *a priori* exposure might be the push factors for the elderly, while shortage of arable land and ability to meet graduation requirements are the pull factors, for the rural youth.

Sex of household head became a significant and positive determinant of participation in non/off-farm wage, self and mixed self-wage employment activities at 1, 5 and 1% levels of significance, respectively revealing that the male headed households were able to participate in all non/off-farm employment activities compared to female headed households.

The presence of large number of economically active members in the household has a significant and positive influence on participation in all the three forms of non/off-farm activities (wage, self and mixed employment) at 5% significance level in agreement with *a priori* expectations. A possible explanation is that households with abundant economically active and working age members could participate in non/off-farm employment activities with a view to generate more income by absorbing the available extra labor force from the farm work. Educational level of household head is found to significantly and negatively affect participation in non/off-farm wage employment at 1% significance level while this variable influences participation in sole self and multiple self -wage employment opportunities significantly and positively at 5% significance level.

It is understood through this research that number of children attending school do significantly and negatively affect the households' participation in wage employment at 5% significance level but its effect on non/off-farm self employment participation is significant and positive at 1% level. The interpretation could be that households with more number of children going to school have high probability and preference to participate into self employment activities than the wage employment activities weighed against households with no or small number of children in school.

Distance to the main market center appears to determine participation into non/off-farm self employment and mixed self-wage activities significantly and negatively at 5 and 1% levels of significance, respectively while its influence on participation in wage employment activities is insignificant. The possible reasons for obtained results might be the following. Since most of the wage employment activities are dominated by the Government rationing productive safety net program (PSNP) public works, which are carried out within the vicinity of the villages, participants may not be obliged to travel to far distant areas to access and undertake safety net public works. Conversely, farmers residing at far distant locations from market centers are less likely to participate in non/off-farm self employment and the mixed self-wage activities.

The multinomial logistic regression analysis result also

showed that the livestock holding significantly and positively influence participation in sole self and mixed self-wage activities at 5% level of significance. Households with more livestock holding do have the capacity to participate in lucrative non/off-farm employment activities, putting them in a better position than those households with no or small size livestock holding.

The insignificant effect of livestock ownership on participation into wage employment could be attributed to the fact that households with no or small livestock holding are pushed towards the less rewarding wage employment activities like the Government rationing programs (PSNP) in which the resource poor households are targeted.

The number of crops grown and harvested per year which can be taken as a proxy for temporal diversification has a statistically significant and negative effect on participation in non/off-farm self employment activities at 5% significance level. But this variable appears to be insignificant with regard to sole wage and mixed self-wage employment activities participation. The result confirms the argument that farmers who cultivate diverse crops are less likely to participate in non/off-farm activities. A possible explanation is that such farmers devote more of their time to farm operation to sustain their livelihoods.

Censored regression results of determinants of non/off-farm income

Non/off-farm income is calculated as income derived from a certain non/off-farm activity as the result of participation of the household into the specific income diversifying activity. Tobit model was applied for the analysis of determinants of non/off-farm activity income using

LIMDEP 7.0 software package and the parameter estimation were done through MLE procedure. Tobit model was applied because of its superiority in identifying the intensity of explanatory variables on non/off-farm self and wage employment incomes of participant households.

Non/off-farm self employment income

The Tobit model result showed that the variables age, sex and education of household head, number of economically active family members, size of cultivated land, number of crops grown and harvested, livestock holding and distance to market have statistically significant effects on off-farm self employment income.

Sex of the household head has significantly and positively affected off-farm self employment income at 5% level of significance revealing that male household heads earned more income than the female headed ones. Thus the male headed households have obtained better self employment income because of their ability to access more lucrative non/off-farm self employment activities than the female headed households.

The number of economically active family members in the household had also significantly and positively influenced non/off-farm self employment income at 1% significance level. The positive relationship indicates that as the number of working age family members increases, the probability of the household to earn non/off-farm self employment income also increases. This could mean that households with large economically active labour force were able to participate in different self employment activities and earn more income compared to households with small number of working labour force.

Size of cultivated land operated by the household had significant and positive effect on non/off-farm self employment income at 1% probability level. Though result obtained from multinomial logit analysis hinted that size of cultivated land does not significantly affect participation, the Tobit analysis revealed that, for participant households, non/off-farm self employment income is significantly and positively influenced by the size of cultivated land. The justification for this finding could be that farmers who cultivate large area of plots have the capacity to produce more and that would enable them to accumulate startup capital for participation in non/off-farm self employment.

Education level of household head was a significant and positive determinant of non/off-farm self employment income at 5% significance level (Table 2). It is presumed that household heads with formal education are better in their perceptive values, knowledge and decision making ability to participate into rewarding self employment activities and earn better income than the illiterate households. The size of livestock holding measured in TLU per sample household is found to be the significant and positive determinant of off-farm self employment income at 5% level of significance. Households owning large livestock holding have the capacity to accumulate capital and overcome cash constraints for participation and obtain non/off-farm self employment income as compared to those households with small or no livestock holding. Age of the household head has a significant and negative relation with non/off-farm self employment income at 1% level of significance, indicating that as age increases the likelihood of obtaining non/off-farm self employment income declines. Results show that younger heads of households could obtain more income from non/off-farm self employment compared to their older fellow farmers. Ageing heads of households concentrate and devote more of their time on farm operation instead of searching for income diversifying non/off-farm activities.

The effect of number of crops grown and harvested (temporal crop based diversification strategy pursued) is significantly and negatively related to non/off-farm self employment income at 1% significance level implying that

Table 2. Tobit Maximum Likelihood estimates of determinants of off-farm employment incomes.

Variable	Non/off-farm wage employment income		Non/off-farm self employment income		Non/ off-farm wage + self employment income	
	Coefficient	t-ratio	Coefficient	t-ratio	Coefficient	t-ratio
Intercept	0.6279	1.5430	0.0703	1.9189*	0.4700	1.2365
Age	-0.0034	-2.4720**	0.0051	-7.4721***	-0.0232	-3.2457***
Sex	-0.0922	-2.7512***	0.0204	2.5014**	0.0085	2.3021**
Family size	0.0022	0.8111	0.0203	1.0112	0.0742	1.2540
Education	-0.0489	-2.6325***	0.0339	2.2783**	0.0066	2.4953**
Ec. active members	0.0642	2.4305**	0.0227	2.8112***	0.0455	3.2052***
Children in school	-0.0112	-2.2423**	0.0355	1.1163	0.0321	1.2541
Credit amount	0.0056	0.1837	0.0005	0.0811	0.0029	0.9624
Cultivated land	-0.0357	-1.9814*	0.0697	4.4551***	0.0015	2.3215**
Livestock holding	-0.0074	-1.3822	0.0049	2.5009**	0.0094	1.6355
Number of crops	0.0108	-1.0782	0.0477	-5.7195***	-0.0028	-1.4126
Distance to market	-0.0086	-1.0155	0.0210	-3.5742***	-0.0245	-2.8743***
Agro-ecology	-0.0004	-1.0046	0.0014	0.6633	0.0038	1.0214
Soil fertility	-0.0098	-1.0991	0.0069	0.5334	0.0071	0.6427

*, **and *** indicate statistical significance at 10, 5 and 1% probability levels, respectively. Dependent variable: Activity income; Threshold values for the model: lower = 0 and upper = +infinity; Weighting variable: one; Number of observations: 120; Log likelihood for self employment income: -73.15; Log likelihood for wage employment income: - 49.78; Log likelihood for wage + self employment income: - 88.51.

farmers who decide to diversify the cropping system have less probability of obtaining self employment income as they invest more time on their farm work in order to ensure food availability for their household.

Distance to market influences non/off-farm self employment income significantly and negatively. The negative sign, at conventional level of significance, shows that as the distance from homestead to market center increases, the likelihood of the household to earn non/off-farm self employment income declines. The implication is that households residing far from market centers have less probability to access and participate into opportunistic non/off-farm self employment activities.

Non/off-farm wage employment income

Non/off-farm self employment income is significantly influenced by age and sex of household heads, number of economically active family members, education, number of children in school and size of cultivated land. In agreement to the result obtained in analyzing self employment income, age of the household head is found to be significantly and negatively affecting non/off-farm wage employment income at 5% level of significance which showed that as age increases, the intensity of non/off-farm wage employment income decreases (Table 2). Thus, the result reveals that younger household heads derived more wage employment income compared to their older fellow farmers. The findings of Mohammed (2008) and Sosina et al. (2009) is similar to our result, while Escobal (2001) reported that age is not significant determining factor of off-farm wage employment income.

The Tobit analysis also yielded that sex of household head has a significant and negative relationship with non/off-farm wage employment income at 1% level of significance indicating that male headed households were able to generate better income from non/off-farm wage employment compared to female headed households. As the wage employment activities are physically demanding and less preferable to women, the likelihood of their participation in income diversifying livelihood strategies is minimal.

The number of economically active members has significant and positive relationship with non/off-farm wage employment income, as expected. The presence of large number of economically active members in the household improves the capacity and ability of household to participate in different income earning activities such as non/off-farm wage employment.

The effect of education of household head on non/off-farm wage employment income is found to be significant and negative at 1% level of significance (Table 2) showing disparity on the effect of education between self employment (positively related) and wage employment incomes. The result shows that heads of households with little formal education, the probability and willingness to search for wage labour may be low as they are pulled towards the more profitable non/off-farm self employment activities. Hence, the illiterate heads of households are mostly pushed to the less attractive wage employment activities. De Janvry and Sadoulet (2001), in their study conducted in Latin America, reported that education has no role to play on agricultural wage employment income but it was a key factor in determining the income level derived from the more remunerative off-farm self employment activities.

Number of children attending formal school is also a significant and negative determining factor for non/off-farm wage employment income at 5% level of significance (Table 2). Households who have more children in school have less preference to participate and obtain income from non/off-farm wage employment activities. A possible explanation is that such households might opt to involve in self employment activities than wage employment activities to supplement their on-farm income.

The effect of size of cultivated land on non/off-farm wage employment income is also significant and negative at 10% significance level. The result shows that cultivation of more land increases households' crop production level and hence better cash for investing in self employment activities which tends to reduce the willingness to search for and participate in less remunerating daily wage activities.

Non/off-farm wage + self employment income

Determinants of multiple non/off-farm self-wage employment income were analyzed by employing Tobit model. The income from the combined non/off-farm self + wage employment activity was significantly influenced by age and sex of household heads, number of economically active family members, education, size of cultivated land and distance to the nearest market center.

The age of the household head has significantly and negatively affected non/off-farm self + wage employment income at 1% level of significance showing that as age increases, the level of non/off-farm self-wage employment income decreases (Table 2). The result reveals that younger household heads derived more non/off-farm wage with self employment income as compared to the older farmers.

The sex of household head has significantly and positively influenced non/off-farm (wage + self) employment income at 5% level of significance indicating that it was the male headed households who were able to generate better income from the multiple non/off-farm self and wage employment activities than the female headed households. As women have less access to different non/off-farm activities due to different factors such as cultural, religious and financial constraints, the likelihood of their participation and/or income level derived from such mixed income diversifying livelihood strategies is less.

The effect of educational attainment of household heads on the multiple non/off-farm self and wage employment income is found to be significant and positive at 5% level of significance (Table 2). The result confirms that households with formal education had the possibility and capacity to participate into both wage and self employment activities than the households with no formal education. Based on this result, one may argue that literate households have willingness and knowledge to participate in multiple non/off-farm self and wage employment activities as compared to those households with no formal education.

The number of economically active members has significantly and positively influenced non/off-farm self + wage employment income. The presence of large number of economically active members in the household improves the capacity and ability of household to participate in multiple non/off-farm self and wage activities. This enables to generate better income than those households constrained by availability of working age family members. The result is similar to the analysis found with respect to the sole wage and self employment activities.

Distance to market has significantly and negatively affected non/off-farm self + wage employment income. The interpretation is that as the distance to reach the nearest market center increases, the probability of the household to participate in and generate income from multiple non/off-farm self and wage employment activities decreases. This could be attributed to the fact that households residing in the villages distant from market centers have less access and opportunity to engage in multiple self and wage employment activities.

The size of cultivated land was a significant and positive determinant of non/off-farm self with wage employment income at 5% significance level (Table 2). This means that farmers who cultivate more land have the capacity to obtain high level of crop production which might enable them generate capital for investment in multiple non/off-farm activities.

SUMMARY AND CONCLUSION

Study concluded in 2009 depicts that activities outside of the agricultural sector play an important role in this study area contributing about 23% of the total household income. In terms of participation, about 84% of sample households are participants of non/off-farm employment and of these households, 58% involved in sole wage employment, 12% in self employment and 30% were involved in both self and wage employment activities. The descriptive statistics result shows that from the wage employment participants, 70, 16 and 14% were participants of productive safety net program (PSNP), agricultural daily wage and other casual non/off-farm wage activities, respectively.

Econometric results obtained go somewhat towards establishing and clarifying relationships between various explaining variables and livelihood diversifying options.

Based on these results the following concluding remarks are made. Various human capital related variables (gender, education, number of children, number of economically active family members and age of household head) are found to be strongly associated with non/off farm income employment decisions. Age of household head has significantly and negatively affected participation in sole wage and self employment as well as the mixed self-wage employment activities implying that younger households rely on non/off-farm employment to support their livelihood. In contrary to this, ageing heads of households concentrate and devote more of their time on farm operation instead of searching for income diversifying off-farm activities.

Gender of household head and the number of economically active family members significantly and positively influenced households' participation in non/off-farm employment activities. This indicated that the male household heads and households with abundant working age members were more able to participate and derive income from sole wage, self employment and multiple self and wage employment activities. One policy implication is that entry barriers for female household heads and disadvantaged groups who are lacking working age family members to participate in non/off-farm activities need to be overcome.

Educational status of household head significantly and positively affected participation and income from sole non/off-farm self employment and mixed self and wage activities while its effect was significant and negative on participation and income that households derived from wage employment activities. This implies that household heads with formal education were found to have better information and knowledge to participate and earn better income from non/off-farm self and self with wage employment activities than the illiterate households. The negative effect of education on wage employment could be justified that due to lack of skill and knowhow, the illiterate households are mostly pushed to the less attractive wage employment activities. Therefore, efforts should be made in improving skill and knowledge of farmers through provision of training.

Number of children in school has significantly and negatively affected participation in off-farm wage employment activities but significantly and positively affected participation in sole self and self with wage employment activities. This implies that households with more number of children in school have low probability and preference to participate into wage employment activities compared to households with no or small number of children in school. Hence development efforts shall be made to the education sector in order to pave the ground for the rural people so that they access the opportunity to participate in the best paying self employment activities.

Infrastructure related variable is found also to determine participation in livelihood diversifying strategy. Accordingly, travel distance to reach the main market center significantly and negatively affected participation and income level derived from non/off-farm employment activities. Here, it is mirrored that as the distance from homestead to market center increases, the likelihood of the household for participating and earning income from income diversifying activity declines. This is possibly because households residing far from market centers have less probability to access and participate into non/off-farm self employment activities. Therefore, the policy should give due emphasis for the development of rural infrastructure and improve transport services in the area.

Akin, crop based diversification strategy, that is, number of crops grown and harvested (temporally and spatially) is found to significantly and negatively influence income derived from non/off-farm employment activities. Farmers living in drought prone and risky areas like Fedis tend to cultivate diverse number of crops as a strategy to minimize the risk of staple crops failure and to maintain food availability. Hence supports to improve the farm production and income of farmers through provision of extension services, yield enhancing inputs should be a possible area of intervention to support rural households produce beyond the subsistence level.

Access to livelihood asset endowment such as livestock and arable land are found to affect involvement in rural non/off-farm sector. In line with this, number of livestock owned has a significant and positive influence on participation and income derived from sole self employment and multiple self and wage employment activities implying that households with more livestock holding had the capacity to participate and generate income from non/off-farm self and self-wage employment activities than those households with no or small livestock holding. More number of livestock ownership, particularly oxen, provides an opportunity for increased crop production and better capacity to generate capital needed for non/off-farm activity participation and thereby increased total household income. Livestock wealth is a key asset of rural livelihood in the districts like Fedis because livestock, particularly oxen, serve as means of draught power, source of capital and serve as prestige. Apart from these, livestock are source of food (animal products and by-products), enable farmers to absorb risks through disposal (destocking) and play a vital role in livelihood diversification (option). Hence the policy environment shall aim at supporting the livestock sub-sector development, in this study area.

The size of cultivated land significantly and positively influenced income obtained from self employment and self-wage employment activities while its effect was significant and negative with respect to off-farm wage employment income. The justification could be that farmers who cultivated large area of plots had the capacity to increase farm production and income which enabled them to accumulate startup capital for off-farm self and self-wage employment participation which contributed to increased total household income than those households constrained by land shortage. The negative effect of size of cultivated land on wage employment income could be justified that households deriving better income from their farm operations are pulled towards self employment than the wage employment activities. The result shows that there are important complementarities between farm and off-farm activities suggesting for appropriate policy instruments that can actually serve both purposes.

This study result shows that despite the high level of participation in non/off-farm activities, the contribution of non/off-farm income to total household income is small compared to farm income. If success is to be achieved in terms of poverty reduction and food security, it is important to support both agricultural and non agricultural sectors. Factors deterring farmers' participation in lucrative non-farm activities need to be addressed through provision of training, credit, improving infrastructure and promotion of employment opportunities with emphasis of targeting the poor and women household heads.

REFERENCES

Barrett CB, Reardon T (2000). Asset, Activity and Income Diversification among African Agriculturalists, Some Practical Issues. Project report to the USAID BASIS CRSP, March 2000.

Barrett C, Mesfin-Bezuneh B, Clay D, Reardon T (2005). Heterogeneous constraints, incentives and income diversification strategies in rural Africa. Q. J. Int. Agric. 44(1):37-60.

De Janvry A, Sadoulet E (2001). Income Strategies among Rural Households in Mexico. The Role of Off-farm Activities. World Develop. 29(3):467-480.

Deininger K, Olinto P (2001). Rural Nonfarm Employment and Income Diversification in Colombia. World Development The World Bank, Washington DC, USA 29(3):455-465.

Ellis F (2000). The Determinants of Rural Livelihood Diversification in Developing Countries. J. Agric. Econ. 51(2):289-302

Escobal J (2001). The Determinants of Non-farm Income Diversification in Rural Peru. World Develop. 29(3):497-508.

Greene WH (2003). Econometric Analysis. 5th edition, Prentice Hall, Upper Saddle River, New Jersey.

Hoogeveen JGM (2001). Income Risk, Consumption Security and the Poor. Oxford Develop. Stud. 30(1):105-121

Mohammed AK (2008). Livelihood strategies and employment structure in Northwest Pakistan. University of Kassel-Witzenhausen and University of Gottingen. PhD, thesis.

Reardon T, Julio A, Berdegue A,. Escobar G, (2001). Rural Nonfarm Employment and Incomes in Latin America. Overview and Policy Implications. World Develop. 29(3):393-409.

Reardon T, Bedegue. J, Barrett CB, Stamoulis K (2006). Household Income Diversification into Rural Non-Farm Activities. (Forthcoming in Haggblade, S., P. Hazell, and T. Reardon, editors). Transforming the Rural Non–Farm Economy. Johns Hopkins University Press, Baltimore, 2006.

Sara W (2007). Livelihood strategies of rural women with emphasis on income diversification and demographic adjustment. The case of Wolonkomi, Oromia Region. Working papers on population and land use change in central Ethiopia, p. 9, Addis Ababa University

Sesabo JK, Tol RS (2005). Factors Affecting Income Strategies among Households in Tanzanian Coastal Villages. Implications for Development Conservation Initiatives. Working paper FNU-70.

Sosina Bezu S, Holden B, Barrett C (2009). Activity Choice in Rural Non-farm Employment (RNFE): Survival versus accumulative strategy.

Sumberg J, Gilbert E, Blackie M (2004). Income Diversity, Technology Choice and Agricultural Research Policy in Sub-Saharan African. Develop. Policy Rev. 22(2):131-146.

Tassew W (2000). Economic analysis and policy implications of farm and off-farm employment. A case study in the Tigray Region of Northern Ethiopia. Mansholt Studies 18 and Ph.D. thesis, Agricultural Economics and Rural Policy Group, Wageningen University.

Tassew W, Oskam A (2001). Income diversification and entry barriers: evidence from the Tigray Region of Northern Ethiopia. Food Policy 26(4):351–365.

On-farm evaluation of maize varieties in the transitional and savannah zones of Ghana: Determinants of farmer preferences

Prince Maxwell Etwire[1], Tahirou Abdoulaye[2], Kwadwo Obeng-Antwi[3], Samuel S. J. Buah[1], Roger A. L. Kanton[1], Henry Asumadu[3], Mashark S. Abdulai[1], Alidu Haruna[1] and John C. Etwire[4]

[1]CSIR-Savanna Agricultural Research Institute (CSIR-SARI), P. O. Box TL 52, Tamale, Ghana.
[2]International Institute of Tropical Agriculture (IITA), Ibadan, Nigeria.
[3]CSIR-Crops Research Institute (CSIR-CRI), Kumasi, Ghana.
[4]Faculty of Mathematical Sciences, University for Development Studies, Navrongo, Ghana.

Maize is one of the most important food crops in Ghana even though its production has not reached self sufficiency levels. Drought and striga infestation are among the most important production constraints of maize in Ghana. Promising high yielding, drought and striga tolerant maize varieties are being evaluated by CSIR and IITA in participatory on-farm trials and demonstrations. These varieties however need to meet farmers' varietal preferences in order for them to adopt. This study therefore sought to assess farmers' preference for the different drought tolerant maize varieties, and determine factors that influence their choices. Kendall's coefficient of concordance was used to test the level of agreement between farmers on their preferences. The ordered logistic regression was used to estimate the determinants of farmer's preference using cross-sectional observations from 120 maize farmers in the Transitional and Savannah zones of Ghana. Results indicate that maize varieties that are early maturing (2.38) and drought tolerant (2.45) were most preferred by farmers. Area under maize cultivation, fertilizer usage and family size are the factors that were found to influence farmers' preference for improved maize varieties. These factors should therefore be considered in varietal promotion.

Key words: Drought tolerant, farmers' preferences, determinant, savannah and transition zones of Ghana, ordered logit.

INTRODUCTION

Maize (*Zea mays* L.) is one of the most important food cereals in the developing world. It is not only the largest staple crop in Ghana, but also the most widely cultivated crop accounting for 50 to 60% of total cereal production. Additionally, maize is the largest commodity crop in the country second only to cocoa (Millennium Development Authority, 2010). Therefore, the importance of maize to the economy of Ghana cannot be overemphasized. The crop is a major source of food, feed and cash for many households in Ghana. Ghana is, however, not self sufficient in the production of maize because its production is faced with a number of constraints including low soil fertility, erratic rainfall pattern, drought, inadequate access to certified seed, poor agronomic practices, among others. It has therefore become important to develop and disseminate maize varieties that are high yielding and tolerant to the major stresses in the country in order to boost its production. The Council for Scientific and Industrial Research of Ghana (CSIR) has been collaborating with the International Institute of Tropical Agriculture (IITA) in developing and evaluating improved maize varieties suitable for various agro-ecological systems in Ghana. Two CSIR institutes, namely Savanna Agricultural Research Institute (CSIR-SARI) and Crops Research Institute (CSIR-CRI) are responsible for developing improved maize varieties for the northern Guinea savannah, and forest and transitional zones, respectively.

Since 2008, promising high yielding, drought and striga tolerant maize varieties are being evaluated in participatory on-farm trials and demonstrations. These trials serve as important tool to showcasing the potential of new maize varieties to farmers. Additionally, the participatory on-farm testing of the varieties is also a means of facilitating rapid dissemination and adoption of these improved maize varieties by farmers. This study therefore sought to assess farmer preference for the different maize varieties that have been developed and are being evaluated. The findings from the study have important policy implications in terms of guiding breeders and agronomists in prioritizing potential varieties to concentrate on, as means of maximising scarce resources. Most studies have either used contingent valuation to identify factors influencing preference (Chern et al., 2002; Nouhoheflin et al., 2004; Parcell and Gedikoglu, 2012; Chen and Gao, 2013) or product characteristics to explain preference (Tetteh et al., 2011). This study attempts to explain preference using farmer characteristics and the determinants of preference were estimated using ordered logistic regression based on random utility theory.

Study areas

According to the Ministry of Food and Agriculture (2011), rain forest, deciduous forest, transitional zone, coastal savanna and Northern Savannah are the 5 main agro-ecological zones in Ghana. This classification takes into consideration climate, natural vegetation and soil types. This study was undertaken in the transitional and northern savannah zones with mean annual rainfall of 1300 and 1100 mm, respectively. The northern savannah with a uni-modal rainfall pattern is located in northern Ghana and the transitional zone with a bi-modal rainfall pattern is located in the middle belt of the country. Differences in mean annual rainfall and distribution as well as soil types have implications for agricultural production in the two zones (Codjoe and Bilsborrow, 2011). The study district and community in the transitional zone was Techiman and Pramposo, respectively. The study districts in the savannah zone were Sissala West (Silbelle community), Wa Municipality (Kpongu community) and Garu-Tempani (Garu community).

METHODOLOGY

Data

The study used primary data collected through farmer interviews. Questionnaires were developed, tested and administered to maize farmers. Information captured by the questionnaire was mainly on socio-demographic characteristics, preferred maize varieties and reasons for preference. A multi-stage sampling technique was employed by the study. All the districts and communities were purposively sampled because they hosted an on-farm trial of the Drought Tolerant Maize for Africa (DTMA) Project Phase III. At the community level, farmers were grouped by sex and 30 farmers were sampled through simple random technique. A total of 120 farmers were interviewed in 4 communities in 4 districts. In terms of agro-ecology, 30 farmers were sampled in the transitional zone.

Models

Kendell's coefficient of concordance

Evaluation of the maize varieties was done when the crop was still on the field and almost ready for harvesting. The objective of the on-farm trial and the purpose of the variety preference ranking were explained to the sampled farmers. Farmers were asked to view all the plots before selecting the variety they liked most after which a questionnaire was administered. Farmers showed preferences for 17 varieties. These varieties were then categorized into 5 for the purpose of ranking. The highest ranked category was assigned a value of one. Categorisation of the varieties and the levels of preference are presented in Table 1. The rankings were then subjected to the Kendall's coefficient of concordance in order to measure the degree of agreement of rankings by different farmers. Following from Tetteh et al. (2011), the total rank score for each factor was computed and the factor with the least score was interpreted as the most preferred. The coefficient of concordance (equation 1) was computed using a statistical software, SPSS:

$$W = \frac{12\left[\sum T^2 - \frac{(\sum T)^2}{n}\right]}{nm^2(n^2-1)} \tag{1}$$

Where T = sum of ranks for factors being ranked; m = number of farmers; and n = number of factors being ranked.

The Coefficient of concordance *W* was tested for significance using the F distribution (Tetteh et al., 2011). The hypotheses tested were:

H_0: There is no agreement among farmers in ranking of maize varieties.

H_1: There is agreement among farmers in ranking of maize varieties.

Ordered logit

In modeling, the factors that influence the choice of a preferred maize variety, the study employs qualitative response regression models because the dependent variable, which is preference for a set of maize varieties, is not measured quantitatively. When the dependent variable is polytomous and is unordered, then the multinomial logit is appropriate and mostly used (Hassan and Nhemachena, 2008; Gbetibouo, 2009; Deressa et al., 2010). The multinomial logit is however inappropriate in instances when the dependent variable is ordered. It fails to account for the ordinal nature of the dependent variable (Greene, 2003). The ordered logit is appropriate in this situation. Empirical works that employ the ordered logit include Curtis et al. (2003) and Geda et al. (2005). The ordered logit is considered for this study because farmers were asked to rank their most preferred maize variety which enters the model as the dependent variables. The theoretical framework adopted for the qualitative response model for this study is based on the random utility model as espoused by Greene (2003).

Following from Greene (2003), the choice of a preferred maize variety may be modelled as a linear function of a vector of observable exogenous variables, X and an unobservable stochastic term, e:

$$Y = X\beta + e \tag{2}$$

The mean and variance of the random term is normalized to zero and one with an independent and identical logistic distribution. The model therefore has only one unknown threshold parameter and the probabilities are such that:

$$\begin{aligned} \Pr ob(Y=0|X) &= 1-\phi(X'\beta) \\ \Pr ob(Y=1|X) &= \phi(\mu - X'\beta) - \phi(-X'\beta) \\ \Pr ob(Y=2|X) &= 1-\phi(\mu - X'\beta) \end{aligned} \tag{3}$$

The coefficients of the exogenous variables are not the marginal effects; the marginal effects of X on the probabilities are estimated as:

$$\begin{aligned} \frac{\partial \Pr ob(Y=0|X)}{\partial X} &= -\phi(X'\beta)\beta \\ \frac{\partial \Pr ob(Y=1|X)}{\partial X} &= \left[\phi(-X'\beta) - \phi(\mu - X'\beta)\right]\beta \\ \frac{\partial \Pr ob(Y=2|X)}{\partial X} &= \phi(\mu - X'\beta)\beta \end{aligned} \tag{4}$$

For this study, ϕ is defined as the standard logistic distribution with mean zero and standard deviation $\delta = \pi/\sqrt{3}$. From the marginal effects equations aforementioned, it can be seen that a change in the prob $(Y=0|X)$ with respect to a change in the exogenous variable X, has the opposite sign of beta. A change in the prob $(Y=2|X)$ by a similar analogy has the same sign as beta. Hence, the marginal effects of these two probabilities are not ambiguous. A change in prob $(Y=1|X)$ is however dependent on the two densities, the sign of the beta is therefore ambiguous, thus, the sign of the marginal effects is also not interpretable (Greene, 2003). This study therefore considers the effects of the exogenous variables on the endogenous variables in terms of log odds.

The parameters for the model were estimated using the maximum likelihood method. The likelihood ratio (LR) statistic was used to test the overall significance of the model (Gujarati, 2004). The model was tested for the major econometric problem associated with ordered logistic regressions. According to Green and Hensher (2009), the parallel regression assumption also known as the proportional odds assumption is a very important assumption underlying the use of the ordered logistic regression. It states that the 'independent variables' effect on the cumulative odds does not change from one cumulative odds to the next. The only thing that changes is the constant term". This implies that, the ordered logit unlike the multinomial logit model does not provide different parameter estimates for each dependent variable. Only one set of parameter estimates are provided for all the dependent variables because the cumulative odds do not change, only the constant term changes. According to Greene and Hensher (2009), Brant test can be used to determine if the proportional odds assumption has been violated or not.

STATA software was used for the estimation. A significant Brant test statistic indicates that the parallel regression assumption has been violated.

Description of variables

The dependent variables for this study are maize varieties that have been broadly categorized into 5 as presented in Table 1. Independent variables that were considered to have an effect on farmer's choice of a preferred maize variety are sex, household headship, family size, area under maize cultivation and quantity of fertilizer used. Preference for maize varieties by men may differ from that of women. Sex is included in the model as an independent variable to capture such differences. Sex may have a negative or positive effect on preference. Several authors have posited that males have more access to, and control of resources (Hassan and Nhemachena, 2008; Deressa et al., 2010; Mandleni and Anim, 2011). Male farmers are therefore more likely to have access to different maize varieties and as a result make choices between these different maize varieties. Nhemachena and Hassan (2007) however disagree opining that, female farmers undertake most of the agricultural activities. Female farmers may therefore have an influence on the choice of maize variety that is cultivated. A household head is usually the key decision maker of the household. Household headship is therefore expected to have a positive influence on the choice of a maize variety that is cultivated by the household. As a result of cash constraints, it is not uncommon for farmers to apply less than the recommended fertilizer rates. The average quantity of fertilizer applied, which can be used as a proxy for wealth, is expected to have a positive effect on the choice of a maize variety by a farmer. Increase in family size means increased consumption which consequently has a negative effect on household income (Shehu and Mshelia, 2007; Maganga, 2012). Family size may therefore have a limiting effect on the number of options of maize varieties available; hence, family size may have a negative influence on the decision to choose between different maize varieties.

According to Simonyan et al. (2011) and Ezeh et al. (2012), an increase in family size implies an increase in family labour. The additional labour can be channelled into production of different maize varieties; hence, family size could also have a positive influence on the decision to choose between different maize varieties. Farm size is expected to have a positive effect on the decision to choose between different maize varieties.

Table 1. Categorisation and levels of preference for maize varieties.

Variety	Most preferred variety selected by sample (%)	Category
2004 TZEE-W Pop STR C4	9.2	TZE (34.2%)
TZEEI 29 × TZEEI 21	7.5	
TZEE-W Pop STR QPM C0	5.0	
TZE Comp 3 DT C2F2	5.0	
TZEE-Y Pop STR C5 × TZEEI 82	3.3	
99 TZEE Y STR C1	1.7	
99 TZ WW Y STR	1.7	
TZEEI 49 x TZEEI 29	0.8	
DT SR W CO F2	11.4	DT (17.4%)
DT STR SYN	4.7	
DT SR W C2	1.3	
2000 Syn EE-W STR	7.5	SYN (11.7%)
2008 Syn EE-W DT STR	4.2	
PAN 53	10.0	Others (26.7%)
L0904 27	8.3	
IWD C3 SYN F2	4.2	
M0926 8	4.2	
Local check (various)	10.0	Local (10.0%)

Deressa et al. (2009) opines that farmers with large farm sizes are also wealthier farmers and therefore have the resources to invest into different farming options.

RESULTS AND DISCUSSION

Socio-demographic characteristics of respondents

The use of improved maize varieties by farmers in northern Ghana and to a large extent Ghana as a whole is pervasive (Wiredu et al., 2010). All the farmers interviewed were found to have been using improved maize varieties for at least 10 years. It was observed that most varieties perceived by farmers to be traditional were in fact developed or introduced by research even though in some cases they were developed a long time ago and have been recycled by farmers over the years. Even though majority (80%) of the sample were still cultivating an improved maize variety, only 25% reported knowing the difference between hybrids and openly pollinated varieties (OPVs). Farmers therefore tend to give all maize varieties the same treatment, selecting seeds from previous harvest of both OPVs and hybrids thereby culminating in yield loss as a result of loss in hybrid vigour. The average age of the sampled farmers was 41 years. The experience of the farmers in either cultivate improved maize varieties in particular or engaging in maize production in general, could have had an influencing factor on farmer's selection of their most preferred maize variety. The summary statistics of the independent variables that were considered to have an effect on farmers' preferences for maize varieties are presented in Table 2. On the average, each farmer interviewed had a family size of 10 people.

According to the Ghana Statistical Service (2008), the mean household size in rural savannah of Ghana is 5 people. Family labour is a significant contributor to the labour requirements of most cropping systems in Ghana; hence, farmers with larger family sizes are less likely to be labour constrained during the peak cropping season. About 58% of the farmers interviewed were heads of their household. The average maize area cultivated was estimated to be 1 ha. Agriculture in Ghana is on a smallholder basis with majority (90%) of farmers cultivating less than 2 ha (MoFA, 2011). A large family and farm size is normally considered as an indication of prosperity and may affect one's social status in most farming communities in Ghana. The average inorganic fertilizer usage was estimated to be 283 kg/ha. With declining soil fertility, most farmers will record low or no yield without fertilizer application. About 48% of the respondents were females.

Farmer's ranking of maize varieties

The results of farmers' rankings of their preferred maize

Table 2. Summary statistics.

Variable	Minimum	Maximum	Mean	Standard deviation
Family size	1	28	9.975	5.173
Maize area (Ha)	0.10	6.00	1.054	0.992
Fertilizer quantity (kg)	0	800	283	200.2
Relative frequency				
Sex	Female (47.5%) Male (52.5%)			
Household head	Yes (58.3%) No (41.7%)			

Table 3. Farmers ranking of preferred maize varieties.

Category	Mean rank		
	Transition	Savannah	Overall
TZE	2.65	2.29	2.38
DT	1.40	2.80	2.45
SYN	3.37	2.96	3.06
Other	2.92	3.42	3.30
Local	4.67	3.53	3.81
Test statistics			
N	30	90	120
Kendall's W[a]	0.56	0.10	0.14
Chi-Square	67.33	36.17	69.15
df	4	4	4
Asymp. Sig.	0.00***	0.00***	0.00***

a. Kendall's coefficient of concordance. *** Significant at 1% level.

varieties are presented in Table 3. Analysis of the results indicated that earliness is the most preferred trait required by farmers for their maize varieties. Those varieties with TZE (meaning early maize) in their pedigree received the highest ranking (2.38) in terms of farmer preference. Early and extra early maize varieties enable farmers to have an early harvest which is critical considering the fact that the harvesting period is usually the lean season. These varieties are also usually harvested before the end of the regular season thereby enabling the crop to escape some biotic and abiotic stresses that may affect regular maize production. The second most preferred varieties were the DT category (that is, drought tolerant varieties) with the various local checks been ranked as the least preferred. Drought tolerant varieties are increasingly becoming important with increasing unreliability and unpredictability of the weather pattern. The Kendell's coefficient of concordance indicates that about 14% of the sample agrees with each other on the rankings. Analysis of the rankings by ecology reveals that the various local checks are the least preferred varieties. The local checks were perceived to be drought and striga susceptible and therefore low yielding.

Farmers however revealed different preferences for their most preferred variety. Whereas, farmers in the savannah zone preferred early and extra early varieties, farmers in the transitional zone preferred drought tolerant varieties. The level of agreement among farmers in terms of the rankings, were 56 and 10% for the transitional and savannah zones respectively.

Determinants of preference

The overall model is statistically significant suggesting that the model with predictors is to be chosen over a model without repressors in explaining determinants of

Table 4. Determinants of preference.

Variable	Odds ratio	Robust standard error	P > |Z|
Sex	0.839	0.434	0.734
Household head	1.054	0.514	0.914
Family size	1.057*	0.349	0.091
Maize area	0.638**	0.114	0.012
Fertilizer quantity	1.002**	0.001	0.025
Cut 1	-0.218	0.406	
Cut 2	0.552	0.409	
Cut 3	0.991	0.412	
Cut 4	1.596	0.422	
Number of obs	120		
Wald Chi^2 (5)	11.94		
Prob > Chi^2	0.036		
Pseudo R^2	0.024		
Log pseudo likelihood	-177.098		

*, ** represents statistical significance at 10 and 5%, respectively.

Table 5. Brant test of parallel regression assumption.

Variable	Chi^2	Prob > Chi^2	df
All	14.34	0.500	15
Sex	2.59	0.460	3
Household head	5.09	0.165	3
Family size	3.80	0.284	3
Maize area	5.15	0.161	3
Fertilizer quantity	3.53	0.316	3

A significant test statistic provides evidence that the parallel regression assumption has been violated.

farmer preference. Parameter estimates of the model showed that area under maize cultivation, fertilizer usage and family size are the factors that influence farmer's preference for improved maize varieties as shown in Table 4. Results of the Brant test presented in Table 5 also indicated that the parallel regression assumption has not been violated justifying the use of the ordered logit regression technique. Area under maize was found to have a positive and significant influence on farmer's choice of a maize variety. The log odds in favour of selecting a preferred maize variety increases by 0.638 units if a farmer puts an additional hectare of land under maize cultivation. Farmers usually try new varieties on a small piece of land before wide scale adoption takes place after they have been found to be preferable. It is not uncommon for farmers to cultivate more than one variety on a parcel of land if it is large. Farmers with larger farm sizes are therefore more likely to show preference for different maize varieties unlike those with only a small piece of land that may not be able to effectively support several varieties of maize. Farm size is mostly considered as an indication of manliness or wealth in most farming communities in Ghana.

Farmers who are therefore able to manage an additional hectare of maize are more likely to be able to buy newly improved varieties as compared to resource poor farmers. The choice of an improved maize variety could therefore be explained by the amount of land kept under maize cultivation. It is not unusual for farmers with large maize farms to have several plots at different locations. Different plots may be suitable for different varieties thus farmers with large farm sizes are more likely to show preference for different maize varieties as compared to farmers with small farm size who may be constrained to utilize only one maize variety. The findings of this study is consistent with Kankwamba et al. (2012) who also reported a positive and significant relationship between farm size and farmers choice to participate in improved seed technology transfer. Application of inorganic fertilizer was found to have a positive and significant relationship on farmer's preference for a maize variety. Application of an additional unit of inorganic fertilizer increased the odds ratio in favour of choosing any preferred maize variety by 1.00 units. In Ghana, inorganic fertilizers and certified seeds are sold by agro-input dealers; hence, farmers are likely to purchase certified seeds and inorganic fertilizers from the same agro-input shop.

Farmers who are able to afford additional inorganic fertilizers are more likely to be able to purchase certified seeds. There is therefore a high likelihood of such farmers showing preference for different maize varieties. Inorganic fertilizers have shown to be more effective on improved seed as compared to traditional varieties. Farmers who apply additional inorganic fertilizers are more likely to be cultivating improved varieties and are therefore more likely to show preference for different maize varieties. Most agricultural related interventions in Ghana with input support usually provide farmers with both inorganic fertilizers and certified seed. The farmers upon imbibing the importance of using certified seeds and inorganic fertilizers are more likely to use both inputs together; hence, farmers with access to inorganic fertilizers are more likely to be exposed to several maize varieties and therefore show preference for different maize varieties. Parameter estimates of the model shows a positive relationship between preference for maize varieties and family size. Increasing the family size by an additional unit increases the odds ratio in favour of choosing a preferred maize variety by 1.06 units. Farmers with large families usually have members with diverse taste. Whereas, some members may prefer maize varieties that have good roasting properties when green, others may prefer maize varieties that have good cooking or milling properties. A farmer with a large family size is therefore more likely to show preference for different maize varieties. The extended family system is still pervasive in most farming communities in Ghana. It is common for grown up or married children to live and work with their parents in the same household. Such large families may cultivate different varieties of maize; hence, the likelihood of a farmer experiencing and showing preference for different maize varieties increases with family size.

Access to labour is not usually a problem for farmers with large families. Such farmers are therefore more likely to either try new varieties or cultivate multiple varieties since labour may not be a limiting factor. Family size therefore seems to be important in determining the choice of any preferred maize variety. Kankwamba et al. (2012) also reported a positive and significant relationship between household size and farmer' decision to use improved seed of groundnut and pigeon pea.

Conclusion

Factors that were found to influence farmer preference for improved maize varieties are area under maize cultivation, fertilizer usage and family size. Maize varieties that are early maturing and drought tolerant are most preferred by farmers in the savannah and transition zones of Ghana, respectively. Breeding for other important maize traits such as striga tolerance, high yielding, resistance to lodging among others, should also take into consideration earliness and drought tolerance. For maximum adoption and impact, access to land, fertilizer and labour are important factors that need to be taken into consideration for effective promotion and dissemination of improved maize varieties.

ACKNOWLEDGEMENTS

The authors are grateful to the Drought Tolerant Maize for Africa Project, Phase III, for funding this study. The authors also acknowledge Zakaria Achulo, Haruna Ali, Solomon Antuona and Emmanuel Ansoba for administering the questionnaire.

ABBREVIATIONS

CRI, Crops Research Institute; **CSIR,** Council for Scientific and Industrial Research; **DTMA,** Drought Tolerant Maize for Africa; **GSS,** Ghana Statistical Service; **IITA,** International Institute of Tropical Agriculture; **MiDA,** Millennium Development Authority; **MoFA,** Ministry of Food and Agriculture; **SARI,** Savanna Agricultural Research Institute.

REFERENCES

Chen X, Gao Z (2013). Chinese Consumer Knowledge, Perception and Willingness to Pay (WTP) for Orange Juice Products: Any Opportunities for the U.S. Juice Producers? Selected Paper prepared for presentation at the Southern Agricultural Economics Association (SAEA) Annual Meeting, Orlando, Florida, 3-5 February. Available at http://ageconsearch.umn.edu/bitstream/143104/2/SAEA-Xuqi_Gao.pdf.

Chern WS, Rickertsen K, Tsuboi N, Fu T (2002). Consumer Acceptance and Willingness to Pay for Genetically Modified Vegetable Oil and Salmon: A Multiple-Country Assessment. Ag Bio Forum. 5(3):05-112. Available at http://agbioforum.org/v5n3/v5n3a05-chern.htm.

Codjoe NAS, Bilsborrow RE (2011). Population and Agriculture in the Dry and Derived Savannah zones of Ghana. Popul Environ. Available at http://www.ug.edu.gh/rips/pub/Codjoe_SN_April_2011PopEnvt.pdf 23-1-13.

Curtis RK, McCluskey JJ, Wahl TI (2003). Westernization in China: A Case Study in Processed Potatoes. American Agricultural Economics Association Annual Meeting, Montreal, Canada, July 27-30.

Deressa TT, Hassan RM, Ringler C (2009). Assessing Household Vulnerability to Climate Change: The Case of Farmers in the Nile Basin of Ethiopia. IFPRI Discussion Paper No. 00935. International Food Policy Research Institute, Washington, D.C.

Deressa TT, Ringler C, Hassan RM (2010). Factors Affecting the Choices of Coping Strategies for Climate Extremes: The Case of Farmers in the Nile Basin of Ethiopia. IFPRI Discussion Paper No. 01032. International Food Policy Research Institute, Washington, D.C.

Ezeh CI, Anyiro CO, Chukwu JA (2012). Technical Efficiency in Poultry Broiler Production in Umuahia Capital Territory of Abia State, Nigeria. Greener J. Agric. Sci. 2(1):001-007.

Gbetibouo AG (2009). Understanding Farmers' Perceptions and Adaptations to Climate Change and Variability: The Case of the Limpopo Basin, South Africa. IFPRI Discussion Paper No. 00849. International Food Policy Research Institute, Washington, D.C.
Geda A, Niek de J, Kimenyi MS, Germano M (2005). Determinants ofPoverty in Kenya: A Household Level Analysis. Economics Working Papers. Paper 200544. Available at http://digitalcommons.uconn.edu/econ_wpapers/200544.
Ghana Statistical Service, GSS (2008). Ghana Living Standards Survey Report of the Fifth Round (GLSS 5). Accra, Ghana. P. 131.
Greene HW (2003). Econometric Analysis. 5th Edition, Pearson Education, Inc., Upper Saddle River, New Jersey, USA. P. 1026.
Greene HW, Hensher DA (2009). Modeling Ordered Choices. Available at http://pages.stern.nyu.edu/~wgreene/DiscreteChoice/Readings/OrderedChoiceSurvey.pdf.
Gujarati ND (2004). Basic Econometric. 4th Edition, The McGraw–Hill Companies. P. 1002.
Hassan R, Nhemachena C (2008). Determinants of African farmers' Strategies for Adapting to Climate Change: Multinomial Choice Analysis. Afr. J. Agric. Resour. Econ. 2(1):83-104.
Kankwamba H, Mangisoni J, Simtowe F, Mausch K, Siambi M (2012). Improved Legume Seed Demand Systems in Central Malawi: What do Farmers' Seed Expenditures say about their Preferences? International Association of Agricultural Economists Triennial Conference. Foz do Iguaçu, Brazil, 18-24 August. Available at http://ageconsearch.umn.edu/bitstream/131684/2/H_kankwamba_IAAEx.pdf.
Maganga AM (2012). Technical Efficiency and its Determinants in Irish Potato Production: Evidence from Dedza District, Central Malawi. J. Agric. Envion. Sci. 12(2):192-197.
Mandleni B, Anim FDK (2011). Climate Change Awareness and Decision on Adaptation Measures by Livestock Farmers. 85rd Annual Conference of the Agricultural Economics Society. Warwick University, 18 - 20 April.
Millennium Development Authority, MiDA (2012). Available at http://www.mcc.gov/documents/investmentopps/bom-ghana-english-grain.pdf.
Ministry of Food and Agriculture, MoFA (2011). Agriculture in Ghana: Facts and Figures (2010). Accra, Ghana. P. 53.
Nhemachena C, Hassan R (2007). Micro-Level Analysis of Farmers' Adaptation to Climate Change in Southern Africa. IFPRI Discussion Paper No. 00714. International Food Policy Research Institute, Washington, D.C.
Nouhoheflin T, Coulibaly O, Cherry AJ, Al-Hassan R, Adegbola PY (2004). Consumers' Perceptions and Willingness to Pay for organic vegetable in Benin and Ghana, African Association of Agricultural Economists, Shaping the Future of African Agriculture for Development: The Role of Social Scientists. Proceedings of the Inaugural Symposium. 6 to 8 December, Grand Regency Hotel, Nairobi, Kenya. Available at http://ageconsearch.umn.edu/bitstream/9525/1/cp04no01.pdf.
Parcell JL, Gedikoglu H (2012). Consumer Preference Variation between Domestic and Imported Food. Selected paper prepared for presentation at the Agricultural and Applied Economics Association's 2012 AAEA Annual Meeting, Seattle, Washington. August 12-14. Available at http://ageconsearch.umn.edu/bitstream/124183/2/AAEA%202012%20Artisan%20Cheese%20Final.pdf.
Shehu JF, Mshelia SI (2007). Productivity and Technical Efficiency of Small-scale Rice Farmers in Adamawa State, Nigeria. J. Agric. Soc. Sci. 3(4):117-120.
Simonyan JB, Umoren BD, Okoye BC (2011). Gender Differentials in Technical Efficiency among Maize Farmers in Essien Udim Local Government Area, Nigeria. Int. J. Econs. Manage. Sci. 1(2):17-23.
Tetteh AB, Adjetey NAS, Abiriwe SA (2011). Consumer Preferences for Rice Quality Characteristics and the Effects on Price in the Tamale Metropolis, Northern Region, Ghana. Int. J. Agric. Sci. 1(2):67-74.
Wiredu NA, Gyasi KO, Abdoulaye T, Sanogo D, Langyintuo A (2010). Characterization of Maize Producing Households in the Northern Region of Ghana. Country Report – Ghana: CSRI/SARI – IITA, Ibadan, Nigeria. P. 24.

13

The performance of potato markets in South Western Uganda

J. Bonabana-Wabbi[1], S. Ayo[1], B. Mugonola[2], D. B. Taylor[3], J. Kirinya[1] and M. Tenywa[4]

[1]Department of Agribusiness and Natural Resource Economics, Makerere University, P. O. Box 7062 Kampala, Uganda.
[2]Department of Agribusiness and Rural Development, Gulu University. P. O. Box 166, Gulu, Uganda.
[3]Department of Agricultural and Applied Economics, VirginiaTech, Blacksburg, VA, 24060, USA.
[4]Department of Agricultural Production, Makerere University, P.O. Box 7062 Kampala, Uganda.

Market-oriented farmers play a significant role in the rural agricultural sector in Uganda. However, these trader-farmers are often disadvantaged by limited access to information, services, appropriate technology and capital. These factors restrict their capacity to effectively participate in the marketing of their produce. In many instances farmers, including those in the potato innovation platforms (IPs) of Southwestern (SW) Uganda are relegated to the lower end of value chains where they are price takers with little bargaining power. Therefore they end up earning little margins while giant chain actors along the chain like middlemen have the power to determine prices paid by the final consumer and thus extract huge marketing margins. This study aimed at identifying marketing channels of potatoes from the farm to consumption; identifying marketing constraints faced by farmers and traders in the potato marketing chain in SW Uganda; and assessing the marketing performance of potato markets. Data were collected in July 2010 from 291 respondents in 2 IPs in SW Uganda. Descriptive statistics were used to characterise potato traders and farmers while Gross Margin Analysis was used to determine the market performance for different potato varieties. Purchase prices varied by district between UgSh350/kg for Rwashaki variety and UgSh531/kg for Kinigi variety. Sutama and Victoria varieties had the highest marketing margins in Kabale and Kisoro districts respectively.

Key words: Marketing channels, marketing chains, marketing margin analysis, innovation platform, ware potato.

INTRODUCTION

In Uganda, potatoes[1] are grown in the highland areas (1,500 - 3,000 masl). The major production areas are Kabale and Kisoro districts in SW Uganda which produce the bulk of the crop. Kabale alone produces between 50 and 60% of the total annual ware potatoes[2] consumed in Uganda. The 2 districts are followed in production by Mbale and Kapchorwa on the slopes of Mount Elgon in Eastern Uganda. However, production of potatoes has

[1] The term potato is widely used to refer to the solanum potato, differentiating it from sweet potato the tuber

[2] Ware potatoes are those potatoes that are grown for human consumption as fresh or processed products.

now spread to other districts of Uganda (Ferris et al., 2002).

Potatoes are considered a source of both food and income, thus development of the potato sector can improve livelihoods of rural dwellers in Uganda in the context of urbanization and market integration (Horton, 2008). Urbanization, increasing incomes, market liberalization and direct foreign investment are causing changes in the food marketing systems (Kennedy et al., 2004). In addition, increased participation of women in the labour force has led to transition from traditional staple foods to convenience foods. The changes in the marketing of food products have thus led to a shift from informal to formal market channels.

Formal marketing channels are characterized by standardized branded products, use of efficient, integrated marketing, logistical, and financing processes. In addition, the terms of production, processing, procurement, payment and product type are set by buyers and not producers. This is due to the demand by urban consumers who require high quality products at consistent prices throughout the year. Supermarkets are becoming significant players in vertically integrated food marketing systems. Other market players include hotels and fast-food outlets. These trends confront smallholder farmers with market challenges and opportunities. As a result of such trends, the livelihoods of smallholder farmers are being influenced by the demands of urban consumers, market intermediaries, and agricultural (food) industries (Horton, 2008).

Often smallholder farmers have limited access to marketing information, services, technology and capital. This lack of access restricts the capacity of the smallholder farmers to participate efficiently in the marketing of their produce (Horton, 2008). Consumers need standardized products, yet these farmers have little knowledge of consumers demand and hence cannot produce what the market needs. Even if they produce what the market needs, they may have little information of reliable and profitable markets. In such circumstances, there is potential exploitation of farmers by the middlemen and wholesalers in the chain because the market value of the potatoes is subject to very limited negotiation, given that almost all farmers sell to middlemen at the farm gate. The exploitation is further exacerbated by absence of standardized packing and weighing scales (Hoffler and Maingi, 2005). The growing demand for potatoes in urban areas could therefore contribute positively to the development of the rural areas and the overall economy of Uganda if there is 2 way efficient flow of market information.

High marketing margins exist either because of monopolistic elements in the marketing chain or because the real costs of marketing are high. High marketing costs may be due to poor marketing services and infrastructure. Thus, improving the marketing services such as storage, transportation, and processing can lead to improvement of rural income by reducing marketing costs (Fuglie, 1993). Farmer collective action has also been proposed as a way of improving the welfare of smallholder farmers in the emerging high-value agricultural markets (Horton, 2008) as it can improve the bargaining power.

Production of potatoes in Uganda is basically for subsistence use (mainly household consumption) with limited surplus for sale in order to earn income despite enormous opportunities for national, regional and global trade. There is limited cross-border trade with Rwanda, but this only occurs to meet the very short term potato supply shortages (Okoboi, 2001; Ferris et al., 2002). The potato value chain is not well organized or integrated because producers, transporters, marketers, wholesalers and retailers are fragmented. This lack of organization is one factor that isolates the potato sub-sector from regional and global markets. There are few initiatives for collective action in potato production and marketing and those existing are in their infancy and widely scattered (Ferris et al., 2002) leading to limited or no integration of stakeholders along the potato value chain.

With increasing population and urbanization and thus growing demand coupled with the increase in fast food restaurants and supermarkets, the potato sub-sector in Uganda is bound to expand. This was noted by Ferris et al. (2002) who estimated the demand for potatoes to be approximately 850,000 to 1,000,000MT per year by 2015. Production volumes increased from 478,000MT in 2000 to 695,000MT in 2010 (FAOSTAT, 2012). This is an opportunity for potato farmers to increase production and productivity of improved and suitable potato varieties, which will in turn increase their income and improve food security and livelihood.

Given the challenges and opportunities that smallholder farmers face in Uganda, it is important to identify and compare marketing challenges as faced by potato farmers and traders in selected Innovation Platforms[3] in addition to estimating the marketing margins in the potato trade. An important contribution of this paper is the comparative assessment of constraints at different parts of the potato market chain hoping to identify common constraints and possible solutions and concerns along the length of the potato marketing chain.

The objectives of this study were to: (a) Characterize potato farmers and traders in selected Innovation Platforms (IPs) on the basis of potato volumes and varieties traded, type of trading business, sources of financing, customer base and pricing strategies;

[3]An Innovation Platform is a tool for bringing together multiple stakeholders for visioning, planning and implementing or application of new ideas, practices, services which arise through interaction, creativity, insight, and empowerment. The aim of the Innovation Platform is to improve the existing situation /conditions around a common interest/challenge and thereby bring about a desired change. It is a forum for sharing and creation of new knowledge gaps relevant for planning explicit systemic innovation agricultural development strategies (Tenywa et al., 2011).

Table 1. Number of Respondents.

District	Traders	Farmers
Kisoro	37	51
Kabale	103	100
Total	140	151

(b) Identify and compare marketing constraints across different actors in the potato trade; and (c) Assess the marketing margins and performance of potato varieties traded in the area (d) Identify and propose solutions for improving performance in potato marketing.

Conceptual framework

Farmers seek to maximise profits from the sale of their agricultural produce by exploiting the margin or the difference between what the consumer pays and the amount producers receive. Margins represent the price charged for one or a collection of marketing services. Along the market chain, margins reflect the value added or the price of all utility adding activities at each level of the market chain. The concept of marketing margins has been defined variously by a number of researchers (FEWS NET 2009, Arene, 2003; Minot and Goletti, 2001; Abbott and Makeham, 1986) all of whom contend that marketing margins are an important tool in analysing the performance of marketing systems. They measure the portion of the consumer's food expenditure that goes to food marketing. Marketing costs and profit margins which make up marketing margins can be both indicators of efficiency or inefficiency of marketing systems and act as incentives or disincentives in business sustainability (Achike and Anzaku, 2010). The concept of margin analysis offers a chance to understand market conditions made by different decision makers in the market.

When looking at the concept of margins, it is imperative to consider the flow of agricultural commodities along the marketing chain. This angle enables the consideration of the relationship between and amongst the different players and the share of the total margin claimed at each stage. This enables the determination of margins at various stages given that margins are likely to differ for different stages. Therefore in order to make a complete analysis of marketing margins, one needs to analyse the complete flow of goods. This can be achieved through use of market-behavior equations (Carambas, 2005). For example Smith (1992) elaborates the concept of costs and margins using the farmers' share of retail price thus:

$$\text{Farmers' share (\%)} = \frac{\text{Farmgate price}}{\text{Retail price}} \times 100$$

And further uses the above to obtain the Total Gross Margin as:

$$\text{Total gross margin (\%)} = \frac{\text{Retail price} - \text{Farmgate price}}{\text{Retail price}} \text{ X } 100$$

The Ugandan potato marketing chain has various actors including collectors, retailers, transporters, wholesalers and farmers. In this study we adopt the Gross Marketing Margin model as suggested by Mendoza and Rosegrant (1995) to analyse the margins of traders because of their central position in the potato marketing chain.

$$\text{Marketing margin (\%)} = \frac{\text{Selling price} - \text{Supply price}}{\text{Selling price}} \text{ X } 100$$

Data and methods

Study area

The study was conducted in Kabale and Kisoro districts which are situated in South-Western Uganda. The districts are currently the leading potato producers in Uganda. The 2 districts have various types of traders in the potato market chain and thus were purposively sampled for the study. Both districts were part of the Sub-Saharan Africa Challenge Program (SSA-CP) as pilot sites where an Integrated Agricultural Research for Development (IAR4D) initiative was conducted. The road ne2rk in these districts is all weather and thus transporting potatoes to urban areas for marketing is feasible all year round.

Data collection

Data were collected in July 2010 using pre-tested structured questionnaires. Information collected included household characteristics, communication equipment, sources of business finance, type of traders and their customers, types of potato varieties traded, quality characteristics considered, pricing strategy, produce grading and mode of transport. Direct one-on-one interview sessions were held with 291 randomly selected respondents, including 140 potato traders and 151 potato farmers in 48 villages in 3 sub counties (Table 1).

Data analysis

Descriptive statistics (frequencies and percentages) were generated from the data and used to characterize potato farmers and traders in the IPs and to identify and compare marketing constraints across different actors in the potato trade. The analyses were conducted using a standard Statistical Package for Social Scientists (SPSS) version 17.0.

Empirical model

Gross marketing margins made by the traders were calculated for each potato variety as follows:

$$V_1 = \sum_{i-1}^{s} (Q_{1i} * P_{1i}) \quad (1)$$

Table 2. Demographic characteristics of traders and farmers in Kabale and Kisoro.

Variable	Response	Overall sample		Kabale		Kisoro	
		Farmers (%)	Traders (%)	Farmers (%)	Traders (%)	Farmers (%)	Traders (%)
Gender of Respondent	Male	64.9	60.7	63	50.5	68.6	89.2
	Female	35.1	39.3	37	49.5	31.4	10.8
Marital status	Married	78.1	85.0	78	82.5	78.4	91.9
	Widowed	13.3	3.6	11	4.9	17.7	0
	Single/divorced/separated	8.6	11.4	11	12.6	3.9	8.1
Age group	18-30	17.9	23.3	22	23.0	10	24.0
	31-50	48.3	62.2	49	63.0	47	60.0
	51+	33.7	14.5	29	14.0	43	16.0
Years of schooling	0-7	73.2	76.4	73.3	78.6	73	70.3
	8-14	18.2	21.4	19.8	20.4	15	24.3
	> 14	8.6	2.2	6.9	1.0	12	5.4
Experience in trading (Years)	0-10		69.2		65		81
	11-20		22.2		26.3		10.8
	21-30		8.6		8.7		8.2

Source: Survey data, 2010.

Where; Q_{1i} is the average quantity (kgs) of potatoes supplied to the ith trader, P_{1i} is average purchase price (UShs) paid by the ith trader to the farmers and V_1 is the average value of potatoes at the farm gate. To determine the average quantity of potatoes reaching market, Q_{2i}:

$$Q_{2i} = Q_{1i} - Q_{Li} \text{, and } Q_{Li} = Q_{1i} * P_L . \quad (2)$$

Where P_L is the percentage loss, Q_{Li} is the quantity lost by the ith trader before selling potatoes (obtained by multiplying quantity supplied by percentage loss).

In order to determine the average value of potatoes after adjusting for the losses, V_2 :

$$V_2 = \sum_{i-1}^{s}(Q_{2i} * P_{2i}) \quad (3)$$

Where P_{2i} is the average selling price by each trader in the market. Thus:

$$GMM = (\frac{V_2 - V_1}{V_2}) * 100 \quad (4)$$

Where GMM is the gross marketing margin. It is the proportion of revenue that traders retain as gross profit, expressed as a percentage.

RESULTS AND DISCUSSION

Respondent characteristics

The majority of the respondents in both districts were males (for traders, 50.5% in Kabale and 89.2% in Kisoro, and for farmers, 63% in Kabale and 68.6% in Kisoro (Table 2). In Kabale about equal numbers of males and females were involved in potato trading reflecting equal proportion of gender types in potato production and trade further reflecting the crop's importance as a source of livelihood for people in this area. In both districts, the majority of respondents in the selected sample were in the 31 to 50 year age-group. In Kabale, 63% of the traders and 49% of the farmers were in this age-group while for Kisoro, this age group formed 47% of the trader sub-sample and 60% of the farmer sub-sample (Table 2). The majority of respondents in both districts had only attained primary level of formal education or no education at all, that is from 0 to 7 years of schooling. Few traders and farmers had attained higher than secondary level education. Most traders in both districts had up to 10 years of trading experience. Overall retailers formed over half the trader sample with proportions as shown in Table 3.

The second most common type of trade was the wholesale business as indicated by the percentages of 32.9% for the overall sample (and 31 and 38% in Kabale and Kisoro, respectively). Traders did not significantly

Table 3. Types of potato traders and sources of income.

Variable	Types of traders	Overall sample (%) N = 140	Kabale (%) N = 103	Kisoro (%) N = 37
Types of traders	Wholesaler	32.9	31	38
	Retailer	56.2	61	43
	Transporters	2.2	3	0
	Collectors	6.4	3	16
	Agents/Brokers	2.3	2	3
Sources of income	Sale of crops	36.4	36.9	35.1
	Sale of livestock	1.4	1.9	0.0
	Salary	0.7	1.0	0.0
	Wages	2.1	1.9	2.7
	Trading business	57.9	56.3	62.2
	Farmers groups (proceeds from group activities)	0.7	1.0	0.0
	Loans	0.7	1.0	0.0

Source: Survey data, 2010. [1] According to Ferris et al. (2002), the potato value chain contains a number of participants/traders. They are Wholesalers-these are traders who get their supplies from travelling traders. They rarely buy from farmers; Retailers-these range from supermarkets to roadside sellers and buy about 1-5 bags from the wholesalers and sell them in heaps of various sizes; Transporters (at times called travelling traders)- are traders who either own trucks or hire them for buying potatoes from farmers or village traders and then transport and sell to wholesalers and urban traders in other district markets; Collectors- these are village traders who know farmers in their village and surrounding areas and make it their business to know which farmers are ready to harvest. They are in contact with transporters, wholesale buyers and financial service providers; Agents/Brokers- are one of the prominent participants in the potato marketing chain and are a contact for travelling traders and wholesale buyers to farmers, as well as the key link between farmers and traders.

process the product either in order to lengthen the shelf life or add value to the final product. As such few other trader types are observed in both districts. These findings suggest that the potato marketing system is still in its infancy and a lot needs to be done in order to add value in the market chain. Respondents reported several sources of income, the commonest being income from trading businesses.

In both districts, traders also obtained income from growing and selling their own crops, hence the potential for expanding potato production. Other sources of income for traders were from sale of livestock, and from loans, wages or salaries. This broad range of activities reflects the desires of the traders to diversify their income sources in order to manage risks. However, this also may imply that these traders and farmers may not be able to commit all their time to one activity which may hinder intensification of potato production and marketing.

Quality characteristics in potato trade

Traders look for a number of quality characteristics and attributes when buying potatoes. The study revealed that traders in both Kisoro and Kabale preferred large-sized potatoes because of the suitability for making chips (French fries) (Table 4) and hence can penetrate the market easily. Traders had little preference for other potato characteristics. This implies those characteristics are of less importance to determine marketing of potatoes. Therefore any improvement in the potato production for attracting consumers/market should first consider size. When the marketing system develops further and becomes more integrated, information flow should be greatly enhanced such that the producers get feedback about the consumer desires and hence production becomes demand driven. This demand responsive production would be a great stride towards more effective resource use in the marketing system. The most important criteria for grading potato were variety, size and colour as determined by consumer preferences in the market. Overall 73.5% of the traders considered variety as the most important grading criterion followed by size (66.4%) and color (48.6%). In both districts this order of grading preference was maintained. However, there was an additional important criterion - cleanliness, for Kisoro district where it scored 32.4% as opposed to Kabale district where it scored 1.9%. These grading standards are indications of moment towards a well functioning marketing system. However, it was not clear whether consumers were willing to pay premium prices for the best quality and grades.

Types of potato markets, pricing strategies and product transportation

Traders operate in a number of markets (Table 5). Shops constructed in the trading centers, stalls constructed in markets and trading spaces in open markets formed the

Table 4. Characteristics of potatoes used for purchase decisions and grading criteria.

Variable		Overall sample (%)	Kabale (%)	Kisoro (%)
Characteristics	Large size	34.3	35	32.4
	Small size	1.4	1	2.7
	Round shape	2.8	2.9	2.7
	Good for chips	8.6	7.8	10.8
	Good for crisps	0.7	1	0
	Good for mashing	1.4	1.9	0
	Kind of customer	2.8	2.9	2.7
	Maturity	3.6	4.9	0
Grading criteria	Variety	73.5	71.8	78.4
	Color	48.6	45.6	56.8
	Size	66.4	62.1	78.4
	Taste	7.9	10.7	0
	Origin	5.0	2.9	10.8
	Cleanliness	10.0	1.9	32.4

Source: Survey data, 2010.

Table 5. Traders by type of potato market by District*.

Variable	Stall constructed in the market	Overall sample (%)	Kabale (%)	Kisoro (%)
		20.7	24.3	10.8
Type of market	Shop constructed in the trading center	50.0	52.4	43.2
	Space in open market	28.5	25.2	37.8
	Roadside market	17.9	20.4	10.8
	Selling to stores	4.3	0.0	16.2
	Others (Mobile (wheelbarrow/bike) market, home market)	10.1	12.7	2.7
Pricing strategy	Cost – plus	43.5	41.7	48.6
	Target profit	22.1	22.3	21.6
	Going rate	25.0	24	27
	Negotiation	7.9	6.8	10.8

* Multiple responses. Source: Survey data (2010).

commonest points for marketing produce. In Kabale, the majority of traders (52.4%) operated from shops constructed in the trading centers, followed by open spaces in markets (25.2%), stalls in markets (24.3%) and road side markets (20.4%). In Kisoro the majority of traders interviewed (37.8%) operated from shops constructed in the trading centers (40.5%) and in open spaces within the market.

Traders had various ways of determining prices of their produce (Table 5). Overall, the major pricing strategy in both districts was the cost plus strategy (43.5%) followed by going rate at 25% and target profit (22.1%). Negotiation was not a major pricing strategy. The ordering of importance of these pricing strategies is not different across the 2 districts. The cost-plus strategy is important given that traders incur costs during their trading activities and simply look to meeting the costs of trading without necessarily making profit. In addition, the fact that about a quarter of traders priced at the going rate confirms why negotiation was not a major pricing strategy.

Potato varieties traded in Kisoro and Kabale Districts

Rwangume was the most traded (47.6%) potato variety in Kabale followed by Kinigi (40.8%), and Rwashaki (18.4%) (Figure 1). In Kisoro however, the most traded variety was Kinigi (51.4%). In general potato traders did not trade the other varieties including Rwashaki, Victoria and

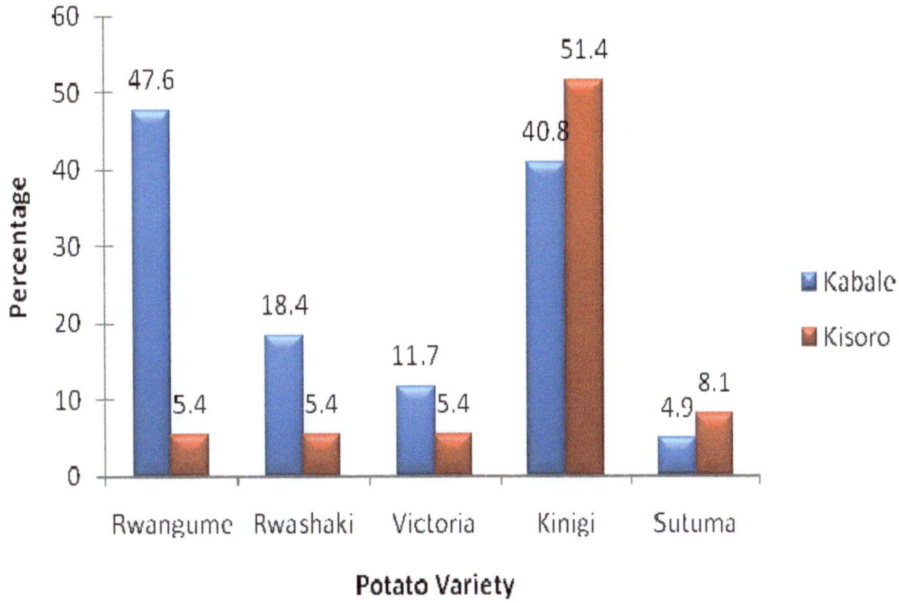

Figure 1. Varieties traded by District.

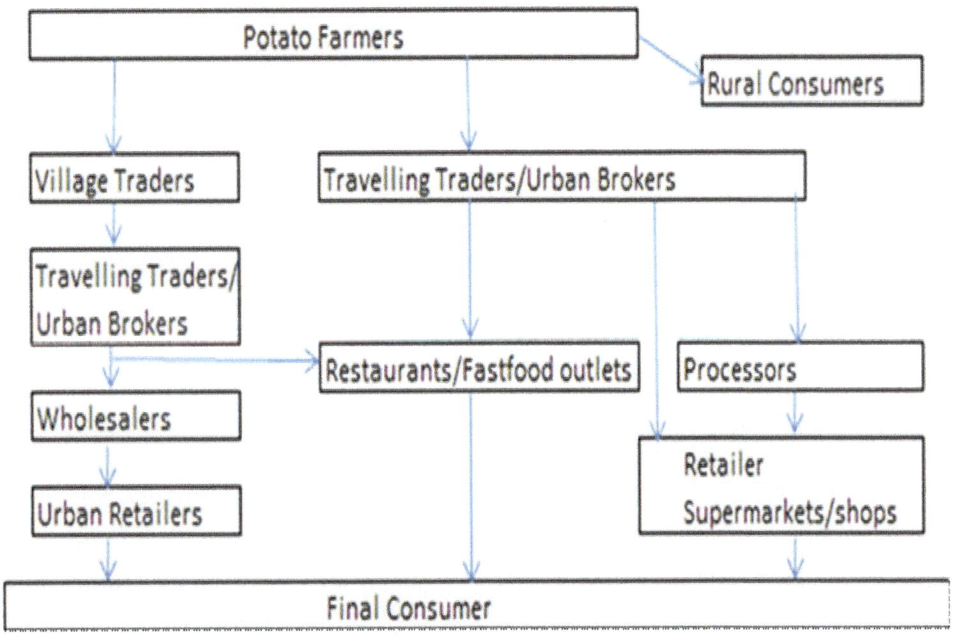

Figure 2. Ware potato marketing chain in Kabale and Kisoro. Source: Survey Data (2010).

Sutama. The observed differences are due to differences in tastes and preferences for the varieties.

Potato marketing channels and transportation modes

Ware potatoes are relatively highly perishable and due to limited on-farm storage facilities thus they are normally moved quickly to buyers/consumers. There are various participants in the ware potato market chain namely: farmers, village traders, brokers, transporters, mobile traders, wholesalers, retailers, processors, restaurants/ fast food outlets and consumers as displayed in Figure 2.

In ware potato marketing, farmers form the first link in the supply chain. They are also consumers of their own produce as well as other farmers' produce. Some farmers

Table 6. Transportation modes.

Variable		Overall sample (%)	Kabale (%)	Kisoro (%)
Transportation mode	Hired vehicle	19.2	18.4	21.6
	Motorbike	1.4	1	2.7
	Non-motorized bike, cart etc	5.0	3.9	8.1
	On head/shoulder/back	10.7	12.6	5.4
	Brought in by suppliers	1.4	1.9	0
	Own transport	1.4	1	2.7

Source: Survey Data (2010).

Table 7. Constraints to potato marketing.

Variable	Institutional factors		Product-related factors	
	Kabale (%)	Kisoro (%)	Kabale (%)	Kisoro (%)
Traders	42.7	43.2	12.7	16.2
Farmers	31	33.3	-	-

sell their potatoes to the rural consumers on-farm or by the road side or in the weekly village markets. However, the main channel through which farmers sell their produce is through the village traders. Farmers also sell directly to travelling traders or urban brokers. Village traders sell potatoes directly to the travelling traders or urban brokers mainly from Kampala and also from other towns. Some village traders also double as brokers. Brokers provide the link for other market participants to each other especially to farmers.

Travelling traders may own trucks or hire them for buying potatoes from farmers or village traders. Travelling traders then transport and sell the potatoes to wholesalers, processors and retailers like supermarkets, shops and restaurants, fast food outlets in other towns mainly Kampala. Travelling traders know the best varieties required especially for chips and crisps (potato chips) and so tend to dominate the chain. Wholesalers sell their potatoes mainly to the urban retailers. Urban retailers (supermarkets, shops, local markets and restaurants or fast food outlets) then sell them in the various grades (heaps, tins, kilograms or as prepared food).

Potato processors obtain their supply directly from travelling traders or urban brokers. They mainly sell their products to supermarkets or shops. Consumers are the final link in the potato market chain. They obtain their raw produce and processed products from urban retailers, restaurants, fast food outlets, supermarkets and shops.

Table 6 shows the percentages of farmers in the district that had access to the different transportation/ delivery modes. The most common means of transport for potatoes in both districts was hired vehicles (19.2%) with Kisoro having a higher percentage (21.6%) than Kabale (18.4%). The second most common means of delivering produce to markets in Kabale was carrying it on heads/shoulders/backs (12.6%) followed by non-motorized bikes (3.9%). In Kisoro the second major mode of transport used was the non-motorized bikes (8.1%) followed by people carrying produce on their heads/shoulders/backs (5.4%). The other means of transport were not used much in both districts (Table 6). A good number of traders in both districts (on average 39.2%) sold their produce either by the roadside or at home markets and therefore did not incur transportation costs. It is worth noting that the steep terrain in these districts largely prohibits motorized transport especially in the rural areas making collection of produce problematic.

Constraints in potato marketing

Traders in both districts encountered various constraints in the marketing of their produce (Table 7). The most serious constraint faced by Kabale traders was that of low prices at the time of sale. This could be due to the nature of the supply chain where middlemen determine the price the consumer pays and other chain actors become price takers. It could also be due to the perishable nature of potatoes and lack of proper storage facilities, so potatoes cannot be stored long after harvest until prices go up. In Kisoro traders reported high cost of transport as the main marketing constraint. Farmers and traders linked marketing constraints to institutional inadequacies (Table 7). For example for traders, institutional constraints included low prices at the time of sale, high prices at the time of purchase, delayed delivery of produce, and long distances to markets.

Table 8. Potato marketing margins for traders in Kabale District.

Variety	Q_1	P_1	P_2	V_1	Q_2	V_2	V_2-V_1	GMM (%)
Rwangume	1,290	411	513	530,190	1148	588,975	58,785	11.1
Rwashaki	431	457	556	196,967	384	213,276	16,309	8.3
Victoria	591	413	514	244,083	526	270,359	26,276	10.8
Kinigi	663	460	563	304,980	590	332,209	27,229	8.9
Sutama	1,105	400	560	442,000	983	550,732	108,732	24.6
Total	4,080	2,141	2,706					
Average prices		428	541					
Percentage loss = 0.11								
Average price paid by final consumer for all the varieties = UgShs1,200/kg								

Table 9. Potato marketing margins for traders in Kisoro District.

Variety	Q_1	P_1	P_2	V_1	Q_2	V_2	V_2-V_1	GMM (%)
Rwangume	650	375	550	243,750	572	314,600	70,850	29.1
Rwashaki	150	350	450	52,500	132	59,400	6,900	13.1
Victoria	465	500	850	232,500	409.2	347,820	115,320	49.6
Kinigi	829	531	723	440,199	729.52	527,443	87,244	19.8
Sutama	400	483	567	193,200	352	199,584	6,384	3.3
Total	2,494	2,239	3,140					
Average prices		447	628					
Percentage loss = 0.12								
Average price paid by final consumer for all the varieties = UgShs1,200 per kg								

The constraints are heightened by high transportation charges, lack of credit arrangements, lack of ready markets, and failure of consumers to pay in time thereby impairing the traders' wellbeing due to delayed incomes which are also low. For farmers, institutional constraints revolved around low prices at the time of sale, inaccessible markets and lack of packaging materials. Traders also mentioned product-related constraints including the bulky nature of potatoes, low supply volumes, high perishability and high incidence of pests/diseases that lead to easy spoilage. About the same proportion of traders in Kabale (42.7%) and Kisoro (43.2%) mentioned institutional factors compared with Kabale (31%) while in Kisoro (33%) of farmers mention the same factors. Similarly, product-related constraints were mentioned by about the same percentages of traders across districts. Product-related factors were not mentioned by farmers. Institutional and product-related interventions would have similar effects in both districts.

Marketing margins in potato trade

In Tables 8 and 9, the buying and selling prices and margins of different varieties of potato are presented. In Kabale, purchase prices (P_1) varied between UgShs400/kg for the Sutama potato variety and UgShs460/kg for the Kinigi variety, while the average selling price (P_2) was UgShs541/kg.

In both Kabale and Kisoro the lowest buying price is for Rwangume which traders buy at an average of UgShs411 per kg in Kabale and UgShs350 per kg in Kisoro. In Kisoro, the highest margins are obtained from selling Victoria while the lowest are obtained from Sutama. The opposite is true however for Kabale where Sutama has the highest margins (Figure 3). These differences in margins point to variations in market suitability for the varieties in both districts. The variations are mainly brought about by out of district sales and the marketing channels through which potato passes. Sutama variety had the highest margins in Kabale while in Kisoro, given the prevailing market prices and supply volumes marketing margins were highest with Victoria variety. Marketing margins span from a low of 18% for Sutama to a high of 39% for Victoria in Kisoro and from a high of 72% (for Sutama) to a low of 37% (for Rwashaki) in Kabale. In both districts, the highly traded Rwangume and Kinigi varieties had medium margins while the least traded varieties lay on the opposite end of the spectrum. Thus, even though Sutama and Victoria were among the least traded varieties, if traded they would result in higher margins for Kabale and Kisoro traders respectively. These results may point to the need for trade diversification within the different potato varieties.

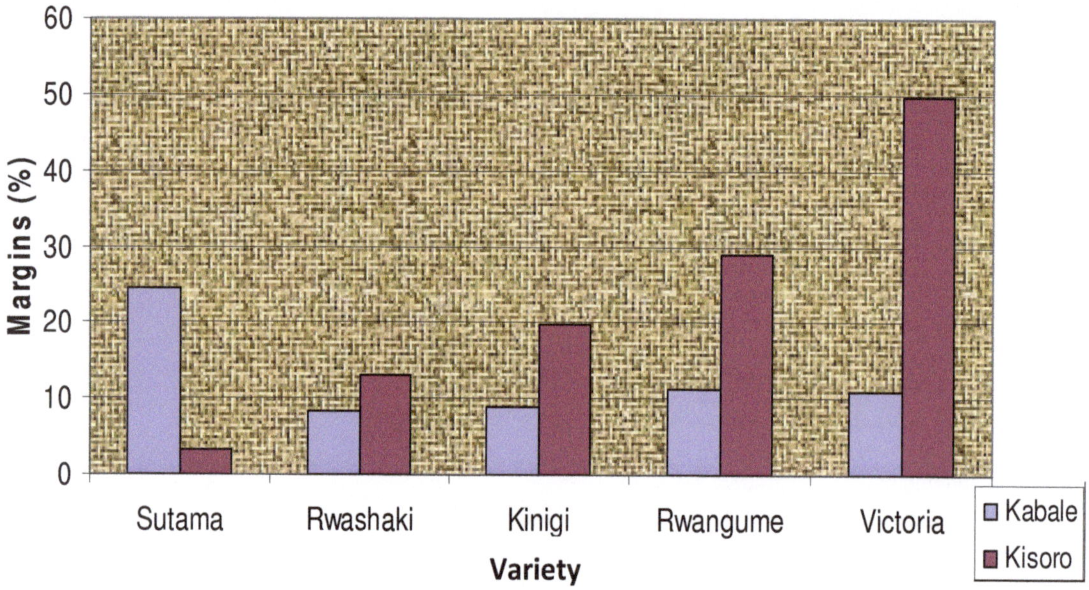

Figure 3. Potato marketing margins (%) in Kabale and Kisoro districts.

SUMMARY AND CONCLUSIONS

In the 2 major potato producing districts of Uganda, a number of potato varieties were traded. A typical trader was a male in the 31 to 50 year age bracket with less than secondary education, whose major source of income was the trading business and occasional sale of crops but with little or no other source of income. Rwangume was the most traded potato variety in Kabale while Kinigi was most traded in Kisoro. The major pricing strategy employed by traders was the cost-plus strategy. Low prices at the time of sale were the most significant marketing constraint in Kabale while in Kisoro the high cost of transport limited trade. Purchase prices varied widely across varieties with implications on the overall marketing margins for traders of the different varieties. Higher prices in Kisoro could be due to high transport costs there that translated into high selling prices.

From the results, institutional and product-related constraints were the major inhibitors of the development of the potato value chains. These included low prices at the time of sell, high perishability of the product and poor market access due to bad road conditions leading to high transport costs. Therefore interventions towards addressing such challenges would help improve /enhance the performance of potato markets most especially in terms of increasing the marketing margins. More importantly, improvements in potato post-harvest handling operations specifically in processing and storage are urgently needed to curb on the wide price swings that negatively impact the stability of incomes for both traders and farmers. This calls for designing and development of cost-effective farmer/trader user-friendly technologies.

ACKNOWLEDGEMENTS

This research was conducted with funding from the Forum for Agricultural Research in Africa (FARA) through the Sub-Saharan Africa Challenge Program (SSA-CP). The authors appreciate the helpful comments of 2 anonymous reviewers.

REFERENCES

Abbott JC, Makeham JP (1986). Agricultural Economics and Marketing in the Tropics, 5th Impression. Hong Kong: Longman (FE) Ltd.

Achike AI, Anzaku TAK (2010). Economic Analysis of the Marketing Margin of Benniseed in Nasarawa State, Nigeria. ISSN 1119-7455. Agro-Sci. J. Trop. Agric. Food Environ. Ext. 9(1):47-55.

Arene CJ (2003). An Introduction to Agricultural Marketing Analysis and Policy. Enugu: Fulladu Publishing Company.

Carambas MC (2005). Analysis of Marketing Margins in ECO-Labelled Products. Paper prepared for presentation at the XIth Congress of the EAAE (European Association of Agricultural Economists), ‗The Future of Rural Europe in the Global Agri-Food System', Copenhagen, Denmark, August 24-27, 2005. Center for Development Research (ZEF), University of Bonn, Germany email: mcaramba@uni-bonn.de.

Ferris RSB, Okoboi G, Crissman C, Ewell P, Lemaga G (2002). Uganda's Irish Potato Sector. Report Prepared for the Government of Uganda's Conference on Competitiveness of Selected Strategic Exports.

FEWS NET (2009). Commodity Market Maps and Price Bulletins: Tools for Food Security Analysis and Reporting FEWS NET Markets Guidance, No 4. July 2009. FEWS NET Washington 1717 H St NW Washington DC 20006 info@fews.net.

FAOSTAT (2012). Statistics division. www.faostat3.fao.org/home/index.html#DOWNLOAD_STANDARD.

Fuglie KO (1993). Measuring Welfare Benefits from Marketing Improvements: Potato Storage in Tunisia. Soc. Sci. Dept. Working Paper Series. P. 2.

Hoffler H, Maingi G (2005). Rural-urban Linkages in practice. Promoting agricultural value chains.

Horton H (2008). Facilitating Pro-poor Market Chain Innovation: An Assessment of the Participatory Market Chain Approach in Uganda. Working Paper. P. 1.

Kennedy G, Nantel G, Shetty P (2004). Globalization of Food Systems in Developing Countries: A Synthesis of Country Case Studies. FAO Food Nutr. Paper. P. 83.

Mendoza MS, Rosegrant MW (1995). Pricing Behavior in Philippine Corn Markets: Implications for Market Efficiency. Int. Food Policy Res. Inst. P. 101.

Minot N, Goletti F (2001). Rice Market Liberalization and Poverty in VietNam. International Food Policy Research Institute, Washington. D. C. Res. Report. P. 114.

Okoboi G (2001). The Marketing Potential of Potatoes in Uganda and Market Opportunities for Rwanda. Draft Report. Int. Inst. Trop. Agric. pp. 4-25

Smith LD (1992). Costs, Margins and Returns in Agricultural Marketing. Marketing and Agricultural Development Paper No. 1. Department of Political Economy University of Glasgow. A paper prepared for marketing and rural finance service. Food Agric. Organ. pp. 3-9

Tenywa MM, Rao KPC, Tukahirwa JB, Buruchara R, Adekunle AA, Mugabe J, Wanjiku C, Mutabazi S, Fungo B, Kashaija NI, Pali P, Mapatano S, Ngaboyisonga C, Farrow A, Njuki J, Abenakyo A (2011). Agricultural Innovation Platform As a Tool for Development Oriented Research: Lessons and Challenges in the Formation and Operationalization. Sub-Saharan Africa Challenge Program, Forum for Agricultural Research in Africa, P.M.B Ct 173 Cantonments Accra Ghana. ISSN-L: 2026-5654. Learn. Publics J. Agric. Environ. Stud. 2(1):117-146.

14

Regional differences of technology modernization in agriculture of Pará State

Mario Miguel Amin Garcia Herreros[1], Heriberto Wagner Amanajás Pena[1], Allan Klynger da Silva Lobato[2] and Raimundo Thiago Lima da Silva[3]

[1]Centro de Ciências Naturais e Tecnologia, Universidade do Estado do Pará, Belém, Brazil.
[2]Núcleo de Pesquisa Vegetal Básica e Aplicada, Universidade Federal Rural da Amazônia, Paragominas, Brazil.
3Centro de Ciências Rurais, Universidade Federal de Santa Maria, Santa Maria, Brazil.

The objective of this study is to identify, through the application of factor analysis and cluster analysis, the spatial organization of the micro hydrographic regions (MRHs) of the State of Pará, as a result of the technology level used in agriculture, the volume of investments, financed credit and by the incorporation of modern means of production. To meet the objectives of the study, information from the Agricultural Census of 2006, published in 2007 by the Brazilian Institute of Geography and Statistics (IBGE) was used. The results indicate that the structural changes occurring within the State's agriculture, during the last two decades have increased considerably in certain MRHs, the regional income by directing most of the production for the export market. However, this led at the same time to a significant reorganization of agrarian space, in other MRHs, where high rates of unemployment and poverty increasing the regional inequalities prevail.

Key words: Technological modernization, factor analysis, clusters analysis.

INTRODUCTION

The investment needed for economic expansion and social development of a region can only be achieved through the accumulation of capital. Their availability makes it possible to move from low to high levels of productivity. Capital formation is therefore of vital importance to ensure the ability to break the "vicious circle of poverty" that, according Nurske (1957), becomes the main obstacle to economic development.

After seven decades, problems linked to insufficient investments still appear in several publications, which are explained in works of Diederen (2003) and Bahiigwa et al. (2005), when explaining why the producer must use technology. Therefore, the modernization of agriculture can reduce poverty. In addition, different methodologies suggests that the innovation is positively related to work level, increase in productivity, market position, access to information, and accumulation of capital.

The new productivity gains for agricultural products supplied by technological advancement were considered as the main driving force for the development of this sector, and public agents have promoted technological change through investments in research, infrastructure and institutional support, capable of regulating the market failures. However, the economic power in the study of the interactions between technology, markets, agricultural incomes and public policies involving farmers, reveal that the dimensions of those trends remain determinants, and in essence imperfectly understood (Gardner, 2001).

In this sense, Pastore (1973) also indicates that during stages of economic development, agriculture is a dominant sector in the use of resources and generation of income. Therefore, the development of society is considered to depend fundamentally on the process of modernization of agriculture.

In comparison, the three economies of Latin America (Brazil, Chile and Mexico), present similar problems, because production and productivity increased, however employment and cultivated area reduced within situation of change from traditional to non-traditional products, the amount of family farmers, accompanied by significantly reduced greater concentration of land and social inequality (David et al., 2000).

A large agricultural sector growth does not define its leadership for economic growth, because in poor countries and in development, the proportion linked to agricultural products is very low, compared to the rest of the economy, which already admits that its expansion may not be as good for growth, because there are problems of diverse causes. After fifty years of scientific research on development of agriculture, there are clear positive associations between substantial increase in agricultural productivity and economic growth (Gollin, 2010).

In the perspective region, the dynamics of modernization in agriculture of Pará is in sharp contrast with the concepts and approaches that underscore the importance of technological innovation for economic development that represents a region. In a study about the new standard of Brazilian agriculture, Kageyama et al. (1990) described that the modernization of agriculture means primarily the change in technical basis of agricultural production. It is a change in the technical basis of production that turns the craft production of the peasant-based hoe, in a modern agriculture, intensive, mechanized, and finally, a new way of producing agricultural products.

The incentive for centuries has been given to the extractive economy over the market economy; it was a factor in the process of occupation and use of the geographical area of the State of Pará. The extraction conditioned the region to a position of a typical manufacturer-exporter of raw minerals, animal and plant, reducing its ability to generate the capital needed for the structural transformation of the regional economy.

Other factors were appointed by the analysis of export structure of the State of Pará, in which the image visualized is extractives, reflexing it due to 25 products studied respond for 97.2% of the total value, composed basically by commodities, if compared with international scenario, and these facts represent loss of business opportunities or market exceeding 80%, being large portion of this amount transferred to other countries that maintain business relationship with Pará State (Pena et al., 2008).

Technological advances have accelerated the food production and the consolidation of agribusiness as the main income generating sector in the national economy. Similarly, the process of modernization and expansion of the horizontal and vertical growth of agriculture in other states of the federation has been a key factor for maximizing the profitability of the agricultural sector. This process seems to have been very well incorporated through the economy of the State of Pará. While maintaining a close relationship between public policies with the old economic model of exploitation, these regional inequalities in social and economic terms have become substantial. The current growth of the agriculture of Pará presents a spatial differentiation that comes to reinforce a trend that began in earlier decades, when the projects come to occupy well-defined regional areas making it clear not only in the form of occupation, but as well as the influence of the implementation of large projects responsible for micro-regional agricultural focus of economic development of the state.

The importance and contribution to the modernization of agriculture for capital formation is expressed by Johnston and Mellor (1972) considering five principles: (i) The increase in agricultural production and agricultural productivity contribute substantially to meet the demand for food; the lack of an expansion in the food supply can seriously damage the region's economic growth; (ii) the exportation of agricultural surpluses is also one of the fastest ways to raise incomes and accelerate regional economic development; (iii) once you overcome the first stages of development, the agricultural sector passes to supply the processing industry and other sectors with a large contingent of skilled labor; (iv) agriculture has an important role in contributing to the formation of capital (savings), and capital accumulation is considered as a factor of vital importance to the expansion of other sectors of the economy; (v) it has been established as a source of food production and capital, that the agricultural sector becomes a major center of encouraging the expansion of the industrial sector.

By have remained in merely extractive activities: plants, minerals and animals, it was not possible to develop the purchasing power of the state population. It avoided, of course, the generation and consolidation of sufficiently large market to stimulate investment. Part of the stagnation of the economy of the State of Para may be explained also by the size of its market.

Regarding the importance of developing a market economy in which the interaction of the forces of supply and demand are factors that help in determining the size of the market and at the same time streamline their behavior, Nurske (1957) is very clear to state that:

... the common observation shows that, in the poorest countries, the use of equipment capital in the production of goods and services for production is constrained by market size, lack of purchasing power... In the economy of exchanges in the real world, is not difficult to illustrate the way in which the small market size of a country can discourage or even prevent the profitable application of modern capital equipment for any particular individual entrepreneur or industry ... The limited domestic market size in a low-income country may thus constitute an obstacle to application of capital by any private company. In this sense, the smallness of the internal market is usually an obstacle to its development".

The regional characteristics have shaped the spatial concentration of investment and capital formation that contributes most to increasing regional imbalances than to improve the economic dimension of most agricultural establishments. This study aims to identify, through the application of factorial analysis and cluster analysis, the spatial organization of micro regions of the State of Pará, the resulting level of technology used in agriculture, the volume of investment, credit financed and the incorporation of modern means of production.

MATERIALS AND METHODS

Source of data

To answer the objectives of the study, it was necessary to choose, first, the level of aggregation of information. This phase is very important in studies of regional character in which one seeks to identify the dynamics and the relationship between a certain group of variables and a selected group of locations. For this work, we used the information presented in the 2006 Agricultural Census of the Brazilian Institute of Geography and Statistics - IBGE. Three options were available: the level of municipalities, micro and meso regions. Given the peculiar characteristics of the agricultural economy of the State of Pará, it was deemed appropriate to choose the statistics presented in micro hydrographic regions (MRHs), since the level of aggregation used by the IBGE allows identifying regional inequality without compromising local specificities. The Agricultural Census of IBGE of 2006 identifies 22 MRHs Homogeneous in Para State, including 138 municipalities out of 143. The difference is explained by the lack of regional information on some of the municipalities visited.

Selection of variables

Given the marked regional differences in Brazilian agriculture regarding the use of productive factors and the incorporation of technological innovations, it was necessary to select variables which, although different from those incorporated in the studies of Hoffmann (1992), Figueiredo and Hoffmann (1998), as well as Souza and Khan (2001), aiming to characterize the level of technological modernization between different regions, allowing to identify, given the spatial location of administrative and economic activities of farmers, the specific process of technological specialization of each micro-region.

The characterization of the regional process of technological incorporation of farmers of Para was obtained by selecting four groups of indicative variables of the intensity of capital used in the modernization of regional agriculture. The variables selected are related to expenses in productive activities, with the value of production of permanent and temporary crops, and quarrying, with funding and, finally, with investments. Given the importance that livestock has reached in recent years in the State of Pará, it was included as a variable of analysis, the pasture of planted area. The variables used as indicators of the degree of technological incorporation in MRHs selected are listed as follows:

X1 = Expenses with payroll,
X2 = Expenses with fertilizers,
X3 = Expenses with seeds,
X4 = Expenses with pesticides,
X5 = Expenses with production transportation,
X6 = Expense with interest / bank,
X7 = Expenses with taxes,
X8 = Expenses with fuel / lubricants,
X9 = Expenses with energy,
X10 = Value of large livestock,
X11 = Value of production of permanent culture,
X12 = Value of production of temporary culture,
X13 = Value of extractive production,
X14 = Financing for cost,
X15 = Financing for investment,
X16 = Financing for commercialization,
X17 = Investments in land,
X18 = Investment in improvements,
X19 = Permanent investment in culture,
X20 = Investment in machinery,
X21 = Investment in vehicles,
X22 = Planted area pasture.

It can be inferred that, given the rational behavior of producers, each of the 22 selected variables is sufficiently a representative of the technical relations that determine the application of capital and the strategies adopted in the selection of agricultural practices that, within the limitations and specificities, are considered the most profitable.

Method of analysis

The availability of a large number of statistical methods, as well as several software programs that enable the rapid implementation, has helped to gain a high degree of speed in identifying georeferenced regional of socioeconomic differences. In this study, as it will be used, initially, the multivariate analysis is known as factor analysis and cluster analysis. Subsequently, the results will be georeferenced using the technique of the Geographic Information System (GIS), thus allowing better visualization of regional differences in the use of technological factors.

Factorial analysis

Factor analysis can identify the pattern of dependence or interdependence between a certain groups of variables. This relation is given initially by the degree of correlation between variables, passing later to be grouped on factors that highlight the "real relationship" between the structural variables (Hartman, 1976; Boyd et al., 1981; Mardia et al., 1980; Hair et al., 1998).

Otherwise, the correlation structure used in statistical technique involves how the database search will respond to the research problem in general, provoking reduction in number of variables, without causing loss of significance (Rodrigues and Paulo, 2009). The application of factorial analysis requires several steps. The first is related to determining the level of association between variables by calculating the correlation matrix. The degree of association between the variables determines whether it is feasible to use factor analysis. In the second step, it is necessary to determine the number of factors that must be estimated.

There are several alternatives depending on the software used. In this paper, we used the method described by Boyd et al. (1981) and Aaker and Day (1983), which is based on the criterion of Kaiser. The criterion states that those factors that have eigenvalues with a value greater than 1, guide the decision to advise or advice regarding the replacement of the variables set by the factors observed (Pison et al., 2003; Jin, 2011). This process can be validated through the generation of the graph called the Scree Plot, in which all eigenvalues are presented.

Once selected, the number of factors, it passes to extract the most significant factors that represent the data. When performed this step, it is necessary to apply the technique of rotation of factors to make them statistically more usable. The option used in this study was to Varimax normalize. Normalizing the data is necessary, since some of the variables used in the research are expressed in different units. Estimated factors and factor loadings (factor loadings) indicate the correlation of variables within each factor, thus it becomes necessary to estimate the factor scores (factor scores) that will be used later in the cluster analysis.

Cluster analysis

Cluster analysis is widely used to identify the degree of interdependence between numbers of variables. The technique allows identifying and grouping those variables that have a strong similarity to its basic structure. Significant differences in the structural composition of the cluster to cluster also can be identified (Boyd et al., 1981; Mardia et al., 1980; Hair et al., 1998).

The identification of clusters can be accomplished through various algorithms. The most commonly used in the literature of multivariate analysis are the nodal linkage procedures and nodal procedures. The first alternative has the simple, complete and average linkage. The second algorithm, hardly used because of its complexity analysis, refers to the choice of a cluster object that will serve as the basis for the identification of other clusters (Churchill Jr., 1982). In this study, we used the complete linkage method, viewed as a hierarchical clustering technique, which identifies, through the dendogram, regional differences in the use of technological factors. A more detailed mathematical description of the application of methods of factorial analysis and cluster can be found in the writings of Hartman (1976), Mardia et al. (1980) and Hair et al. (1998).

RESULTS AND DISCUSSION

Identification of technological modernization

As a first step in the application of factor analysis, it was estimated in the correlation matrix. It can be seen in Table 1, a strong correlation between most variables. Only four variables, X4 (cost of pesticides), X9 (energy costs) and X13 (value of extraction production) had very low correlations with other variables.

The result is quite surprising. The low correlation shown between expenditure on pesticides and other variables, especially with temporary and permanent crops, crops, occupy a large part of the economy in Pará in the area of mineral extraction, animal or plant character. The low use of pesticides is related also to the kind of climate that exists in the Amazon region. Likewise, extraction and production are basically for family meals and with a much localized trade, the value of extractive production does not present a very significant contribution to overall economic growth of the state.

The lack of energy in cities farther away from the capital, Belém, is a serious problem, passing it to influence negatively the productive capacity and slow the process of incorporating the most modern techniques in agriculture in the state. The funding generally appears as the main inducer of the modernization process of agriculture in Para. Permanent crops depend significantly on the cost of financing, investment and trade. The agricultural credit, especially the Institutional Fund for Financing the North (FNO), plays a very important social and economic development by financing large part of agricultural projects implemented in the Amazon region. Amazon shows how the allocation of resources to productive sectors was essential for the generation of regional income and employment (Amin, 2002; Santana, 2002).

The costs are strongly correlated with most agricultural activities. It indicates that producers are reinvesting part of the capital raised in machinery, improvements, livestock, seed, pasture area, fertilizers and, especially, for payment of the financing arranged.

Determination of groups of factors

After the implementation of statistical tests and verification that the correlation matrix was not an identity, eigenvalues were estimated to help in identifying a number of factors that can be extracted from the variables used. The criterion of extraction is somewhat arbitrary, but modern statistical algorithms already allow obtaining preliminary information that facilitates the selection of the final number of factors. To answer the objectives of the study, three factors were selected showing eigenvalues of 7.73 for F1, 5.36 for F2 and 1.84 for F3. The first factor explains 35.16% of the total variance of 22 variables, the second factor explains 24.38% and the third factor explains 8.38% of total variance. It means that three factors explain 68% of the total variance.

From this information, the rotation was carried out by the VARIMAX method to obtain the normalized factorial loadings (factors loadings) that indicates a correlation between the respective variables and factors, representing the most important information of the factor analysis. The normalization or standardization of variables is of great importance, given that the main objective of the study is to identify, through the application of cluster analysis, the spatial distribution of

Table 1. Correlation matrix of selected variables.

	X1	X2	X3	X4	X5	X6	X7	X8	X9	X10	X11	X12	X13	X14	X15	X16	X17	X18	X19	X20	X21	X22
X1	1.00																					
X2	0.59	1.00																				
X3	0.52	-0.06	1.00																			
X4	0.12	0.19	0.41	1.00																		
X5	0.15	0.50	-0.45	-0.15	1.00																	
X6	0.78	0.50	0.31	0.08	0.15	1.00																
X7	0.75	0.44	0.37	0.04	-0.18	0.56	1.00															
X8	0.63	0.13	0.44	0.08	0.04	0.55	0.38	1.00														
X9	0.23	0.37	-0.18	-0.10	0.08	0.00	0.17	-0.09	1.00													
X10	0.53	-0.29	0.79	0.07	-0.10	0.37	0.35	0.58	0.00	1.00												
X11	0.52	0.67	-0.12	-0.09	0.07	0.40	0.27	0.39	0.15	-0.12	1.00											
X12	0.39	0.49	-0.12	-0.03	-0.09	0.23	0.13	0.44	0.15	-0.04	0.75	1.00										
X13	-0.19	0.00	-0.28	0.013	-0.13	-0.06	-0.15	0.24	-0.20	-0.37	0.05	0.00	1.00									
X14	0.76	0.69	0.10	0.01	0.01	0.82	0.49	0.45	0.05	0.09	0.56	0.38	0.04	1.00								
X15	0.50	0.31	0.18	0.01	0.01	0.27	0.60	0.38	0.10	0.21	0.52	0.36	-0.21	0.27	1.00							
X16	0.65	0.73	-0.01	0.02	0.02	0.77	0.47	0.29	0.00	-0.11	0.56	0.28	0.15	0.94	0.26	1.00						
X17	0.26	-0.27	0.45	-0.07	-0.07	0.39	0.24	0.29	-0.17	0.57	-0.24	-0.15	-0.30	-0.02	0.18	-0.09	1.00					
X18	0.57	0.16	0.42	0.07	0.07	0.35	0.77	0.40	0.09	0.49	0.14	0.20	-0.27	0.23	0.79	0.14	0.42	1.00				
X19	0.52	0.65	-0.07	-0.07	-0.07	0.45	0.08	0.33	0.14	-0.05	0.86	0.68	-0.08	0.64	0.34	0.57	-0.14	0.01	1.00			
X20	0.81	0.65	0.24	0.12	0.12	0.91	0.60	0.48	-0.05	0.18	0.52	0.30	0.02	0.93	0.36	0.93	0.17	0.34	0.54	1.00		
X21	0.55	-0.12	0.63	0.00	0.00	0.39	0.52	0.57	-0.12	0.75	-0.08	0.01	-0.24	0.17	0.24	0.01	0.69	0.57	-0.01	0.29	1.00	
X22	0.51	-0.29	0.83	0.01	0.01	0.37	0.39	0.53	-0.22	0.91	-0.16	0.09	-0.31	0.06	0.21	-0.09	0.75	0.51	-0.08	0.22	0.85	1.00

the technological modernization of agriculture in the State of Pará, in 22 selected MRHs.

Table 2 presents the estimated factorial loadings after orthogonal rotation. The analysis is based on values greater than 0.50, in order to identify the extent to which technological changes in agriculture have been incorporated into national regional level. The estimated communalities indicate that the variables selected have their variability very well captured by the three selected factors.

Factor 1 is positively related to the indicators X1, X2, X6, X11, X14, X16, X19 and X20 which represent, in its entirety, the decisions taken by farmers in relation to expenditure, investments and financing in processes. This factor therefore represents all the "financial activities".

The high values of X14 (0.94) and X16 (0.96) indicate that the state's agricultural production depends significantly to the cost of financing and marketing of production. Given the poor conditions of road infrastructure in the state, farmers need to make heavy investments in the purchase or lease of means of transport to the flow of production. This is especially evident in those municipalities located near the Transamazônica highway. Similarly, credit is a key role in encouraging and strengthening the agricultural economy of the state.

In Factor 2, on the other hand, all those activities related to the livestock industry are identified. The factor is positively related to the indicators X1, X3, X7, X8, X10, X17, X18, X21 and X22. The high values of X10 (0.88) and X22 (0.91) reflect those in which MRHs predominate the beef cattle industry and milk as the main

Table 2. Factor loadings and commonalities.

Indicators	Factor			Communal
	1	2	3	
X1 - Wages exp.	0.73	0.57	0.25	0.992
X2 – Fertilizers exp.	0.72	-0.17	0.45	0.990
X3 - Seeds exp.	0.15	0.76	-0.31	0.971
X4 - Pesticides exp.	0.15	0.07	-0.21	0.638
X5 - Exp. trans. prod.	0.30	-0.39	0.68	0.997
X6 - Interest / bank exp.	0.86	0.34	-0.05	0.983
X7 – Taxes exp.	0.49	0.57	0.15	0.886
X8 - Fuel / lubric. exp.	0.44	0.53	0.22	0.924
X9 – Energy exp.	-0.05	-0.01	0.49	0.851
X10 - Value of large livestock	0.03	0.89	-0.14	0.996
X11 - Perm. value	0.55	-0.05	0.72	0,866
X12 - Temp. value	0.29	0.03	0.81	0.804
X13 -Extra. value	0.16	-0.44	-0.09	0.891
X14 - Fin. cost	0.95	0.05	0.13	0.992
X15 - Fin. invest.	0.21	0.49	0.54	0.998
X16 - Fin. comer.	0.96	-0.11	0.06	0.997
X17 - Inv. in land	0.00	0.71	-0.28	0.876
X18 - Inv. in improvements	0.16	0.76	0.28	0.890
X19 - Inv. in perm. culture	0.58	-0.06	0.59	0.995
X20 - Inv. in machines	0.96	0.20	0.02	0.997
X21 - Inv. in vehicles	0.14	0.85	-0.13	0.990
X22 - Pasture planted area	0.06	0.91	-0.25	0.982

economic activity. This factor is basically the "farming activity." Factor 3 contrasts with the other two factors. Here, the focus indicators related to the production of permanent crops (X11) and temporary crops (X12), financing for investment (X15), investment in permanent crops (X19) and transportation costs of production (X5). Again, it is evident that the importance of the cost of shipping into the productive matrix represents the State of Pará. This factor reflects the MRHs, where agriculture is the main economic activity. This factor therefore represents the "agricultural activity".

Determination of technological dynamics of micro regions

As a complementary result to the identification of factors and factorial loadings, it is possible to estimate the factor scores. The scores can be used to identify the dynamics of technological incorporation in each of the selected MRHs, as well as to verify, through cluster analysis, the spatial grouping MRHs according to their productive and technological characteristics. Table 3 presents the results identifying the level of technological incorporation in 22 MRHs analyzed. Building on the analysis of the 22 indicators shown earlier, are highlighted in bold, those MRHs that have a more advanced technological standard. The results show strong regional disparities in the level of technological modernization of agriculture in the State of Pará. It is observed not only in a well-defined stratification of productive processes, but the low technological level of agriculture that still exists in most MRHs state.

The micro region of Tomé-Açu presents the highest level of technology due to the significant use of intensive, the capital / labor ratio. The micro region incorporated, from the arrival of Japanese farmers in the 50's, modernizing agricultural practices which have contributed to this region, at present is responsible for a significant portion of the state production of cocoa, black pepper and kingdom tropical fruits. The production orientation of regional products, mostly targeted for the export market, served to ensure a structural change in the growth of agriculture induced by the implementation of modern technologies.

It is worth mentioning also that the process of technological modernization of the micro region of Tome-Açu was highly benefited by the location in relation to its proximity to the main ports of export, as well as the available infrastructure. The MRHs of Redenção, Paragominas, São Felix do Xingu, Parauapebas, Conceição do Araguaia and Maraba have developed an economy oriented toward specialization in livestock. These MRHs respond for about 65% of the effective cattle in the state of Pará estimated in the year of 2000 in 10.2 million heads (IBGE, 2008). In another group, we

Table 3. Composition of factor scores.

Micro regions	Factor		
	1	2	3
Almeirim	0.058071	-0.77443	-1.40348
Altamira	0.092584	0.97213	0.96400
Arari	-0.798371	-0.56678	0.08455
Belém	-0.265567	-0.76541	-0.27299
Bragantina	-0.128431	0.93009	2.69777
Conceição do Araguaia	-0.058509	1.80241	-0.62966
Cametá	-0.312431	-1.00145	0.33003
Castanhal	-0.012178	-0.66645	0.87475
Furo de Breves	-0.159793	-1.32652	-0.50526
Guamá	0.058034	-0.12642	2.29103
Itaituba	-0.512613	-0.45834	-0.29093
Marabá	-0.346820	0.37682	-0.69764
Óbidos	-0.537355	-0.70472	-0.27265
Paragominas	-0.055366	1.23860	-0.37433
Parauapebas	-0.187892	0.72893	-0.50797
Portel	-0.521210	-1.05315	-0.31951
Redenção	0.454242	2.19872	-0.87229
Salgado	-0.365600	-0.98510	-0.22399
Santarém	-0.104096	-0.12307	0.85066
São Felix do Xingu	-0.194987	0.93994	-0.91422
Tomé-Açu	4.311826	-0.69396	-0.15508
Tucuruí	-0.413540	0.05815	-0.65281

highlighted the MRHs of Bragantina and Guamá with an oriented agriculture to supply the regional market. The return of capital from agriculture contributes significantly to these MRHs, the most important centers of employment and income generation in the State. As in the case of micro region of Tome-Açu, these MRHs have also been privileged with a great infrastructure. Micro regions as Altamira, Cametá, Santarém and Castanhal occupy an intermediate position regarding the level of technology embedded in their productive activities. The micro region of Tucuruí has taken, in recent years, an important position in the agricultural state, accounting today for 3.5% and the actual state of the beef cattle (IBGE, 2008). The micro regions Almeirim, Arari, Belém, Furo de Breves, Itaituba, Óbidos, Portel and Salgado present no score for identifying the development of farming practices that are being incorporated into technological modernization. Low scores represent the existence of a low technological level due to the strong presence of agriculture and quarrying.

Cluster analysis

In possession of scores of factors, it is possible to perform additional statistical analysis. In particular, the resulting information can be exploited to define groups that characterize cases (MRHs) with a certain degree of homogeneity. Using the three estimated factors scores, the 22 MRHs were hierarchically grouped into six clusters.

According to the results of dendrogram (Figure 1), there are well-defined groups according to the level and technological use in agriculture. The first cluster is formed by micro region of Tome-Açu. The importance of this micro region in the state economy is evident from the rank observed in the dendogram estimated. The second cluster, where commercial agriculture is the basis of the regional economy, is composed of five MRHs: Guama, Bragantina, Santarem, Conceicao de Araguaia and Altamira. These are cluster grouped areas where capital flows and the incorporation of technology have been instrumental in the development and structural transformation of agriculture in the state.

The third cluster is formed by micro region of Redenção participating at the time, with 1.9 million head of sheep and 18.63% effective in the State of Pará (2008). The pattern of technological factors adopted from 1995, has converted this area into a major beef-producing center of the North. Completing this cluster were the MRHs Paragominas and Cameta. The fourth cluster is formed by MRHs of Sao Felix do Xingu, Paraupebas Tucuruí and Maraba. The MRHs of this group and the previous one forms the largest producer and beef cattle in the state. The significant increase in exports of beef from Brazil has been a conditioning factor for the adoption of advanced technology initiatives that are transforming the pattern of development in this region.

The MRHs of Salgado, Portel, Belém, Castanhal, Óbidos, Itaituba and Arari form the fifth cluster. This group of MRHs is characterized by an agriculture that coexist a low technological level of production and a regional economy geared basically for mining activities. The sixth cluster is formed by micro-Almeirim. The segmentation of this region in a well isolated cluster reflects very clearly the inequality in the use of regional factors and technological modernization. Characterized by a regional economy and quarrying, their contribution to the state's economy is almost marginal. The six clusters identified by the 22 indicators used and agricultural modernization, allow three basic groupings of economic activity, showing the regional potential, to be formed: agriculture (Clusters 1 and 2), livestock (Clusters 3 and 4) and extraction (Clusters 5 and 6).

The significant heterogeneity in the process of modernization of agriculture can be explained also by how the integration of micro-regions, given by the Euclidean distances (similarity coefficient), is performed (Figure 1). Regional differences can be characterized in terms of the information used and the explanatory power of each factor. The micro-regions, for example, where agriculture is predominant extraction, have been grouped close to the Euclidean distances of 1, showing therefore a relatively lower agriculture in the level of modernization and

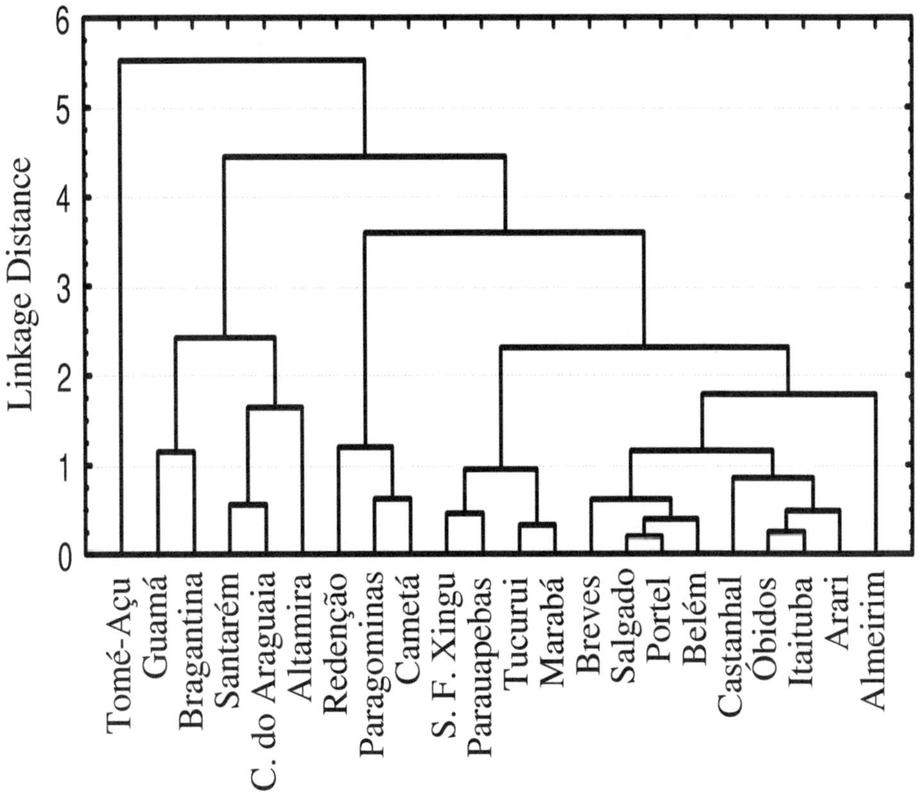

Figure 1. Dendrogram for the clustering of the State of Pará MRHs.

technological expertise. This cluster is integrated into the group formed by the micro-regions of Sao Felix do Xingu, Paraupebas Tucuruí and Maraba, where ranching is the main activity, at a distance of 2.25.

The micro-regions of Redenção, Paragominas and Cameta, on the other hand, where ranching is the predominant basis of the regional economy, integrate the general system, at a distance of 3.75. Following by importance are the micro regions of Guama, Bragantina, Santarem, Conceição de Araguaia and Altamira where commercial agriculture is dominant and most representative industrial. The integration of these micro regions is made at a distance of 4.25.

The micro region of Tome-Açu is the last to integrate the system due to its large Euclidean distance of 5.5, emphasizing thus the importance that the implementation of advanced standards of technology and modernization of the capital / labor ratio in agriculture has regional level. By georeferencing the results presented earlier, it is possible to identify spatially regional inequalities, and the use of modern technologies in agriculture and regional development. Figure 2 shows the regional differences in the level of use of technological factors and the modernization of agriculture in the State of Pará.

At first glance, it emphasizes the heterogeneity of an agriculture state. This is reflected as an indication of the pattern of technological and productive economic activities in MRHs.

The observed clusters follow a geographical contiguity, indicating a regionally well defined, in relation to the technological modernization of agriculture in the state. In the northern region of the state (Clusters 5 and 6), the extractive agriculture, whether by animal or plant, is the main source of employment and income. The remarkable lack of energy, transport infrastructure and services in general, has served to that in terms of technology investments, this part of the state remains on the fringes of economic development.

Those MRHs in which the predominate agriculture has a more advanced technological level are located in the center and northeastern parts of the state (Clusters 1 and 2). This region is characterized by a level of specialization and diversification of productive processes reasonably modern making it the leading supplier of food both for regional consumption and for export.

In the southeast of the state (Clusters 3 and 4), it is a distinguished set of MRHs characterized by a high technological standard and by a significant increase in agriculture, which became the main activity of regional development. The proximity of the States of Mato Grosso, Goias and Tocantins, major producers of beef cattle have served to create a favorable environment for

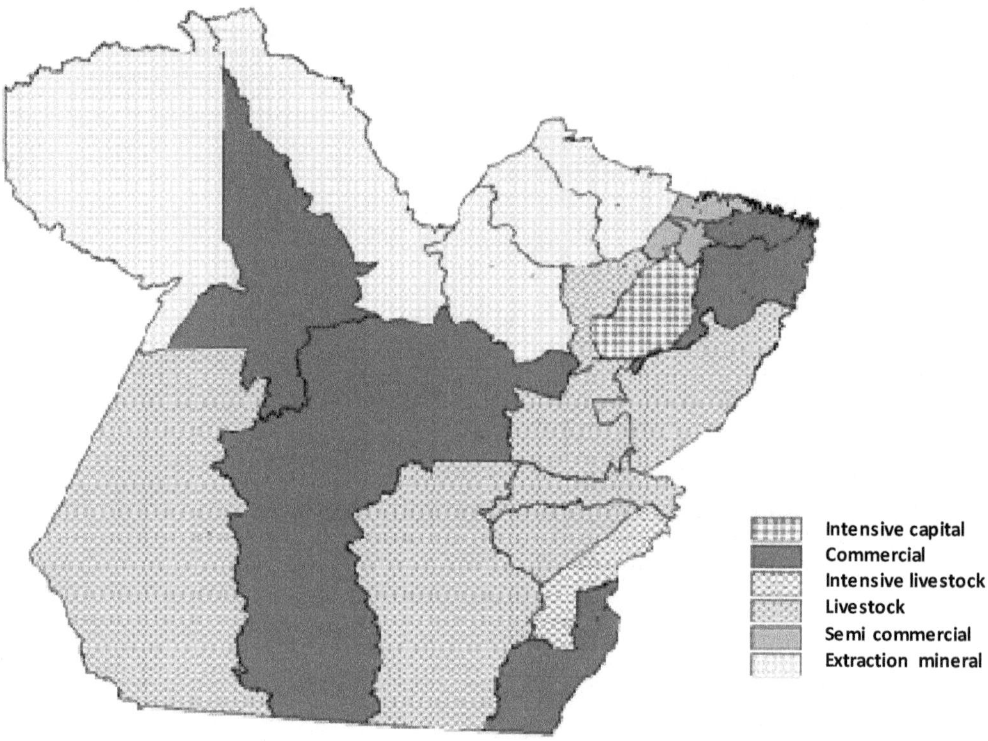

Figure 2. Regionalization of the technological modernization of agriculture in the State of Pará.

the installation of refrigerators, with the positive balance growth of the herd during the past ten years. It is important to mention that these last two groups of MRHs, unlike the first, are very well served by an extensive infrastructure of transport, energy and services.

Conclusions

The model of agricultural growth in the state of Para, centered mostly in the extractive exploitation of natural resources, shows very significant differences regarding the spatial level of technology used in productive activities. In addition, six clusters identified by statistical methods of analysis of factor and cluster find a spatial organization resulting from the various historical processes of land occupation in the state. The inter-regional differences are determined in three economic activities: commercial agriculture, livestock, farming and mining. The technological modernization of agriculture in the State of Para provides to the set of MRHs in question a well defined spatial segmentation. The process occurred very differently by incorporating spaces in which the tone of the investment, financing and technical management in agriculture was given in different business manners. Therefore, the structural changes occurring within the state agriculture during the last two decades, increases dramatically at certain MRHs, the regional income by guiding most of the production for the export market. But, it leads at the same time, to a significant reorganization of the agricultural space in other MRHs, where prevailing high rates of poverty and unemployment, deepening regional inequalities.

REFERENCES

Aaker DA, Day GS (1983). Marketing research. New York: Wiley & Sons.

Amin MM (2002). Situação da pecuária na Região Norte: tendências e perspectivas. IN: O Fundo Constitucional de Financiamento do Norte e o Desenvolvimento da Amazônia, Santana AC de. Belém: M&S Editora.

Bahiigwa G, Rigby D, Woodhouse P (2005). Right Target, Wrong Mechanism? Agricultural Modernization and Poverty Reduction. World Dev. 33:481-496.

Boyd HW, Westfall R, Stasch SF (1981). Marketing research: texts and cases. Ontario: Irwin Inc.

Churchill GA (1982). Marketing research: Methodological foundations. Illinois: The Dryden Press.

David B, Dirven M, Vogelgesang F (2000). The impact of the new economic model on latin americas agriculture. World Dev. 28:1673-1688.

Diederen P (2003). Modernisation in agriculture: what makes a farmer adopt an innovation? Int. J. Agric. Resour. 2:328-342.

Figueiredo NMS, Hoffmann RA (1998). Dinâmica da modernização da agricultura em 299 microrregiões homogêneas do Brasil: 1975, 1980 e 1985. In: XXXVI Congresso Brasileiro de Economia e Sociologia Rural. Poço de Caldas, Anais. pp. 439-450.

Gardner BL (2001). Agriculture, Economics of International Encyclopedia of the Social and Behavioral Sciences. 32:337-344.

Gollin D (2010). Agricultural Productivity and Economic Growth. Handb. Agric. Econ. 4:2825-2866.

Hair JF, Tatham RL, Anderson RE, Black W (1998). Multivariate Data Analysis. 5th Ed. New York: Prentice Hall.
Hartman HH (1976). Modern factor analysis. Chicago: The University of Chicago Press, 3rd Ed.
Hoffmann R (1992). A dinâmica da modernização da agricultura de em 157 microrregiões homogêneas do Brasil, Revista de Economia e Sociologia Rural 30(4):271-290.
Instituto Brasileiro de Geografia e Estatística – IBGE (2006). Censo Agropecuário de 2006 – Pará. Rio de Janeiro.
Instituto Brasileiro de Geografia e Estatística – IBGE (2008). Produção Agrícola Municipal. Rio de Janeiro.
Jin M, Yun H, Jeong CS, Heo JH (2011). Factor analysis and multiple regression between topography and precipitation on Jeju Island, Korea. J. de Hydrol. 410:189-203.
Johnston BF, Mellor JW (1972). El papel de la agricultura en el desarrollo económico. IN: Desarrollo Agricola, Flores, E. Mexico: Fundo de Cultura.
Kageyama A, Buainain A, Reydon B, Graziano J, Silveira JM, Fonseca M, Ramos P, Fonseca R, Belik W (1990). O novo padrão agrícola brasileiro: do complexo rural aos complexos agroindustriais. IN: Agricultura e Políticas Públicas. Delagado G, Gasques JG, Villaverde CM (Coord.). Brasília DF: IPEA, Série IPEA, P. 127.
Mardia K, Kent J, Ibby J (1980). Multivariate Analysis. New York: Academic Press.
Pastore J (1973). Agricultura e desenvolvimento. Rio de Janeiro: APEC/ABCAR.
Nurske R (1957). Problemas de formação de capital em países subdesenvolvidos. Rio de Janeiro: Editora Civilização Brasileira.
Pena HWA, Costa R, Bahia PQA (2008). Dinamic of exportation in Para state between 2004 and 2005. Facientifica 1:126-142.
Pison G, Rousesseeeuw PJ, Filzmoser P, Croux C (2003). Robust facto analysis. J. Multivar. Anal. 84:145-172.
Rodrigues A, Paulo E (2009). Introduction to multivariate analysis. In: CORRAR, L. J; PAULO, E; FILHO, J. M. D (Eds.). Multivariate analysis to administration and economy courses. São Paulo: Atlas, pp. 1-72.
Santana AC de (2002). Impactos socioeconômicos do FNO nas atividades produtivas da Região Norte: análise de insumo-produto. IN: O Fundo Constitucional de Financiamento do Norte e o Desenvolvimento da Amazônia, Santana AC de. Belém: M&S Editora.
Souza RF, Khan AS (2001). Modernização da agricultura e hierarquização dos municípios maranhenses. Revista de Economia e Sociologia Rural, Brasília 39(2):81-104.

15

Analysis of economic capacity utilization in the Nigerian sugar industry (1970 - 2010)

Sunday B. Akpan[1], Inimfon V. Patrick[1], Daniel E. John[2] and Samuel J. Udoka[1]

[1]Department of Agricultural Economics and Resources Management, Akwa Ibom State University, Ikot Akpaden, Mkpat Enin, Akwa Ibom State, Nigeria.
[2]Department of Agricultural Economics and Extension, University of Uyo, Akwa Ibom State, Nigeria.

The study analyzed economic capacity utilization rates in the sugar industry for the period 1970 to 2010 in Nigeria. Secondary data obtained from sugar firms; Central Bank of Nigeria; National Bureau of Statistics and the Federal Ministry of Finance were used in the study. Stochastic Cobb-Douglas cost functions for the sugar industry were estimated from which indices of economic capacity utilization rates were obtained. Trend in the economic capacity utilization rate showed undulated pattern with an average index of 60.30% and excess economic capacity of 39.70%. Multiple-regression of various forms based on the ordinary least squares technique was used to determine factors that influence the performance indicators in the industry. Empirical results revealed that economic capacity utilization rates in the sugar industry was influenced by the inflation rate, per capita real GDP, energy consumption of the industry, federal government expenditure on the sugar industry and the period of liberalization. The result of the regression and descriptive analyses revealed that the sugar industry in Nigeria was constrained by insufficient production inputs. Policy measures aimed at reduction or maintaining a steady or less fluctuated inflation rate in the country, expansionary aggregate demand, increase funding to agencies that have direct dealings with the sugar production and adequate provision of electricity to the industry as well as the adoption of the liberalization industrial policy on sugar industry were recommended.

Key word: Sugar, industry, capacity, utilization, economic, Nigeria.

INTRODUCTION

The sugar sub-sector has contributed to the development of the Nigeria's economy (NSDC, 2010). The importance of the sub-sector is derived from its contribution to the employment, development of other subsidiary industries and food self sufficiency as well as its significant impact on the rural economy (Nwaobi, 2005; ADB, 2000; ADF, 2000). In Nigeria, the demand for household sugar consumption remains firm, the soft drink production alone accounts for about half of the total industrial sugar usage in Nigeria (Michael, 2010). The domestic consumption of sugar in Nigeria is in excess of one million tons per annum. Domestic production of sugar in the country had suffered considerably in the past years. For instance, the domestic output declined from 51,080 tonnes in the period 1988 to 1990 to 5,597 tonnes in the period 2001 to 2003 (Wada et al., 2001; Olomola, 2007 and SSC, 2010 (Table 1). Currently, domestic production of sugar is slightly less than 5% of the country's annual requirement (CBN, 2008; NSDC, 2012).

Data in Table 1 reveals that from 2000 to 2003, the domestic sugar production declined significantly reaching all time low value of less than 1.00% of domestic sugar consumption in the country. The dismal performance of the sub-sector had been attributed to diverse factors

Table 1. Sugar supply and import price of sugar in Nigeria (1970 to 2008).

Year	Average domestic output (tons)	Average import (tons)	Average total supply (tons)	Average Import Price ₦/ton	Share of domestic to total output (%)	Share of import to total output (%)
1970-1972	38141	114158	152299	144.4	33.41	66.59
1973-1975	42594	99335	141929	424.6	30.01	69.99
1976-1978	34074	327382	361458	332.6	9.43	90.57
1979-1981	36296	632379	668675	349.8	5.43	94.57
1982-1984	37778	571562	609340	293.7	6.20	93.80
1985-1987	51872	450130	502002	465.2	10.33	89.67
1988-1990	51080	292766	343846	1878.5	14.86	85.14
1991-1993	40735	485540	526275	6681.5	7.74	92.26
1994-1996	45577	390718	436295	7696.6	10.45	89.55
1997-2000	13654	729870	743524	10980	1.84	98.16
2001-2003	5597	903066	908663	25229	0.62	99.38
2004-2008	11194	350113	361307	42625	3.20	96.80

Sources: National Sugar Company Document (NISUCO) (1999, 2000); Savannah Sugar Company limited (SSC) (2000), and FAO database (2011).

including economic, environmental, social, technology and factory based hindrances (Lafiagi, 1984; Wada et al., 2001; Akpan et al., 2011, 2012a, 2012b).

The two major integrated sugar plants at Bacita in Kwara state (Nigerian Sugar Company) and Numan in Adamawa state (Savannah Sugar Company) were established in 1961 and 1977, respectively, following the adoption of import substitution industrialization policy in the country (ISYB, 1978). The aims were to encourage technological development, reduce the volume of imports and encourage foreign exchange savings by producing locally some of the imported consumer goods (Ayeni, 1981; Ekeocha, 2009). The two sugar plants had a combined installed capacity of 105,000 tonnes per annum or less than 10% of the country's annual requirement (FMI, 2003). Due to some rather complex factors, the major existing sugar companies at Bacita and Numan whose combined installed capacity was expected to climb to 165,000 tons per annum after their expansion programme initiated by the federal government in collaboration with the African Development Bank and African Development Fund in 1989 and 1991, respectively could not fulfill the sub-sector expectations. Besides, the two major sugar producing companies, the other 2 mini sugar firms at Sunti (Niger State) and Lafiagi (Kwara State) were producing relatively small quantities of Sugar (i.e. less than 1,000 tons per annum each) (Wada et al., 2001; Nwaobi 2005; NSDC, 2006).

The general performance of the sugar sub-sector was fair in the early 1970s as indicated in Table 2. During this period, the sugar industry had fully integrated its operations backward through its direct involvement in sugarcane farming and sourcing of other raw materials locally (Ayayi, 1988; Akpan et al., 2012a). Towards the middle of 1980s and the late 1990s, the performance of the sub sector started to decline. The index of sugar production declined from 117.8% in the period 1986 to 1990 to 47.7% in 2001 to 2005 periods. By the middle of 1980's, the country's foreign exchange earnings declined significantly arising from the oil glut. The high import dependence manufacturing sector in the country became a serious liability on the economy (Isola, 2006).

The prevailing economic environment and the industrial policies during 1970s to early 1980s favored an average capacity utilization rate of above 50% in the sugar cocoa confectionery sub-sector (CBN, 2006; MAN, 2006). As revealed in Table 3, the agro based industries generally witnessed decline in productivity during the structural adjustment programme (SAP) period. The average technology based capacity utilization rate of the sugar cocoa confectionery industry stood at 40% in 1986 and declined to 36% during the early structural adjustment programme (SAP) period. The instability in some macro-economic variables in the Nigerian economy and agro-based firm related constraints during the SAP period probably contributed to the decline in the productivity of the sugar industry. The decline in the sub sector productivity might have manifested through the influence of rising inflation rate, low external reserves which constrained importation of sugar industry's equipment, deteriorating value of naira as well as demand and other production constraints imposed by low real GDP per capita during the period ((Isola, 2006; Akpan et al., 2012b).

In the early 1990s, the Nigerian sugar sub-sector was still largely underdeveloped with untapped resources and potentialities. The 4 existing companies were completely government owned and were characterized by low productivity occasioned by managerial, financial and infrastructural and technological constraints. The awfully low production by the existing sugar companies could only satisfy about 5% of the nation's requirement and the

Table 2. Average index of agro-based manufacturing sub-sectors in Nigeria (1985 = 100)

Agro-based Industry	1970 - 1975	1976 - 1980	1981 - 1985	1986 - 1990	1991 - 1995	1996 - 2000	2001 - 2005
Sugar	420.3	534.3	149.6	117.8	101.2	56.4	47.7
Textiles	110.6	171.4	100.9	100.8	117.2	97.4	94.2
Foot work	253.6	255.9	111.1	62.1	73.7	47.7	45.0
Total GR	24.8	80.5	-169.7	-65.5	20.9	-58.9	-24.4
Ave.GR. (%)	8.3	26.8	-56.6	-21.8	7.0	-19.6	-8.1

Source: Computed by using data from CBN statistical bulletin (2006).

Table 3. Economic and Industrial Sector Performance in Industrial Policy Regimes in Nigeria.

Indicator	Import Substitution Era			Liberalization Era				
	1970-1975	1976-1980	1981-1985	1986-1990	1991-1995	1996-2000	2001-2005	2006-2010
Mean macroeconomic variable indicator								
Inflation rate (%)	14.3	13.0	19.4	20.5	48.9	12.3	15.7	9.65
Official exchange rate (₦/$)	0.66	0.69	0.77	5.90	19.16	54.78	127.84	134.08
External reserve (₦b)	1.19	3.06	1.40	11.98	39.2	361.6	1869. 7	6366.194
Index of energy consumption (1985=100)	26.3	64.6	122.8	95.6	95.7	83.3	167.4	NA
Real GDP per capita (₦ / person) (1985=100)	177.2	735	3.25	962	674	547	819	4562.63
Real foreign private investment in Manu and processing (₦b)	2.79	3.17	2.74	2.20	1.53	1.09	1.26	NA
Index of agricultural production (1990=100)	65.62	56.44	58.74	79.20	121.6	141.5	158.2	NA
Index of manufacturing production (1985=100)	32.85	75.42	105.7	135.4	154.7	137.2	145.6	NA
Mean agro based industry capacity utilization rate (%)								
Meat and dairy product	NA	NA	55.85	30.07	30.34	36.46	63.12	461.01
Vegetable and grains mill	NA	92.37	48.97	33.40	35.54	18.34	51.18	48.74
Bakery product	66.4	71.48	58.64	37.90	37.90	21.45	58.23	62.11
Sugar cocoa confectionery	NA	59.59	53.12	42.20	33.12	31.42	31.32	59.00
Textile	79.7	80.72	59.84	51.20	48.34	34.30	42.61	59.66
Leather products	77.2	75.08	54.50	45.00	42.38	37.78	45.38	50.50
Tyres and tube	NA	NA	42.58	53.04	41.70	30.80	42.62	49.75
Leather foot wear	NA	47.5	76.69	55.56	41.18	24.57	26.0	74.99
Average agro industry capacity utilization rate (%)	75.10	73.72	56.12	43.10	38.46	28.01	47.4	51.70
Average industry capacity utilization rate (%)	76.6	71.12	53.58	41.14	35.40	33.59	46.18	41.68

Source: Computed by using data from Central Bank Nigeria Statistical Bulletin (2006) and World Bank Reports (20021). NA means data not available.

wide gap between sugar demand and supply was filled through importation with huge amount of foreign exchange requirement. With the dwindling fortune of the federal government resources, the existing sugar companies were wallowed in low productivity due to inadequate finance for both recurrent and capital expenditure (FMI, 2003). This situation further deepened the fortune of the local sugar production since all sugar companies were government owned. In an attempt to accelerate the domestic sugar production, the National Sugar Development Council (NSDC) was established by decree 88 of 1993. The NSDC was mandated to develop strategies that would promote the local production of sugar such that 70% of the country's sugar requirement would be met by domestic production (Busari et al., 1996 and FMI, 2003). Based on the government policy of direct participation and investment in the sugar industry, NSDC strategies were the expansion and rehabilitation of the 4 government owned sugar industries, establishment of 5 medium scale and many mini sugar plants in the country as well as the establishment of the sugarcane Research Development and Training Center. The Council however recorded some successes in implementing some of its strategies but could not still upsurge local production of sugar in the country (FMI, 2003).

Following the government reform programme on privatization and commercialization between 2001 and till date; the 2 integrated sugar companies and 2 mini sugar plants in Nigeria were partially privatized. The aims were to promote efficiency in resource utilization, increase productive capacity and increase the role of the private sector in the sugar industry (Zayyad, 2007). Despite this lofty attempt by the government to strengthen the productive capacity of the sugar sub-sector, the productivity of the sub-sector continued to decline. The average index of production in the sugar industry was -17.9% in the period 1970 to 2005 (CBN, 2006). During the post Structural Adjustment Programme era, growth in the sugar industry in Nigeria was hindered due to increase in manufacturing cost (Ogunbayo, 2009). The average capacity utilization for the sugar cocoa confectionery sub sector during post SAP period as published by official sources in Nigeria was below 30% (MAN, 2009; CBN, 2009).

Problem statement, objectives and justification of the study

Capacity utilization has an important bearing on the financial performance of any firm and the entire industry. It is widely used in business cycle analysis to characterize the situation of individual industry or the whole economy and to assess the appropriateness of the economic policy (Danish, 2003). The Nigerian monetary policy, among other things, aims at achieving full employment of resources without inflation (Anyanwu et al., 1997). Consequently, over the years the government has employed a number of monetary policy measures to increase capacity utilization in the economy and at the same time curb inflation. Government policy measures have varied from the pre-structural Adjustment Programme (Pre-SAP) period (1970 to 1985) to the SAP (1986 to 1993) and post-SAP (1994 to date) periods. Direct monetary control techniques were employed during the pre-SAP period. The major tools of monetary policy were administered interest and exchange rates, special deposits by banks, prescription of cash reserve requirements, selective credit controls and credit ceilings (Anyanwu et al., 1997).

In the SAP period (1986 to 1993), indirect measures were used to control the ability of banks to extend new credits alongside credit ceilings. The measures included the deregulation of interest rates, increase in commercial banks cash reserve requirements and its extension beyond demand deposits to include time and savings deposits. Other measures were the mopping of banks' excess liquidity through the issuance of stabilization securities and the transfer of public sector accounts from banks to the Central Bank of Nigeria (CBN). On June 30, 1993, the CBN formally introduced its open market operations (OMO). In the post-SAP period (1994 to date), administratively controlled measures were first adopted in1994 and were abandoned in 1995 for policy of guided deregulation. Banks became directly involved in equity funding and management of small-scale enterprises. Apart from monetary policies, the government also employed some fiscal policy measures to ensure full employment of resources. These measures include tax holidays, tariff protection, import duty relief, bans on imports and the provision of credit facilities.

In the sugar industry, some specific policies employed over the years to boost sugar production in the country included 50% tariff on the importation of white sugar, 5% levy on imported raw sugar, free excise duties on sugar production, reduction of import duties on sugar industry machineries, 5–year tax holiday to sugar refineries and privatization of the major sugar firms in the country, as well as, the sugar expansion programme in collaboration with the African Development Bank (ADB) and African Development Fund (ADF), 1989 and 1991 respectively. These measures were meant to stimulate the local production and hence increase the productivity and capacity utilization in the sub-sector. In spite of these measures, Nigeria still imports more than 90% of its sugar. Nigeria is the largest consumer of sugar in the West African sub-region and second in Africa (ADB and ADF, 2000). The country also has a large area of cultivable land, suitable for the growing of industrial sugarcane (Busari et al., 1996). Despite the favorable agro-climatic and edaphic conditions for the production of sugar-cane in addition to the long period of existence of sugar mills; sugar requirements of the country remain largely unmet from domestic sources (Olomola, 2007).

The domestic sugar production has oscillated between 7,000 tonnes and 55,000 tonnes per annum from1969 to 2010 (Wada et al., 2001; SSC, 2010).

In Nigeria, the issue on industry's economic capacity utilization rate is relatively new compared to other methods of capacity utilization especially in the agro-based industries. Several authors in Nigeria have worked on technology based capacity utilization in many industries (Fabayo, 1981; Ukoha, 2000; Soderbom et al., 2002; Salimonu et al., 2006; Adeel et al., 2006; Raimi et al., 2009; Akpan et al., 2011, 2012b). Most studies on the concept were based on survey opinions of firms on capacity utilization rate rather than empirical estimation in the individual industry through optimization of the industry's resource endowment (Soderbom et al., 2002; Adeel et al., 2006). Other studies based their analyses on the data published by the official sources such as Central Bank of Nigeria (CBN) and Manufacturing Association of Nigeria (MAN) with no consideration on capacity utilization estimation procedures (Fabayo, 1981; Ukoha, 2000; Salimonu *et al.,* 2006 and Raimi et al., 2009). Data on capacity utilization rates in the sugar-cocoa confectionary sub-sector, as reported by independent official sources showed that capacity utilization rates declined from an all time high value of 85% in 1975 to 50% in 1983, and remained consistently below 50% from 1983. This study investigated whether the sugar industry is really operating at half capacity in the sense that firms could double output without experiencing a rise in average costs.

Therefore, this study differs from the previous ones conducted in Nigeria as it estimated the economic capacity utilization rates for the sugar industry using the sugar industry production and cost data. In addition, it analyzed the factors which influence economic capacity utilization in the industry. Hence, the result of the study is a reliable quantitative fact and source of reference to policy makers to efficiently make relevant policies that can promote the sugar industry's performance in Nigeria. In addition, the results would serve as a useful screen board for future analysis of capacity utilization in any sector of the economy. Furthermore, this study provided a frame of reference for agricultural economists, economists, manufacturers, planners and students who might be carrying out studies on capacity utilization.

The concept of economic capacity

The earliest work on the economic concept of firm's capacity is that of Cassels (1937). The economic capacity takes explicit account of economic factors like cost, price, revenue and profit. It is defined as the optimum output of a firm from economic point of view. This approach considers capital as a quasi-fixed input, and allows for distinction between short and long-run cost curves. In the long-run, capacity can be adjusted in order to achieve optimal output (cost-minimizing, profit-maximizing) level. In the short-run, capital is fixed and only the variable inputs can be varied. Hickman (1964) defined economic capacity of a firm as that output level at which the short run average total cost curve is at its minimum; while Klein (1960) and Friedman (1963) defined economic capacity as the output level at which the long-run and short-run average total cost curves are tangent.

The relationship between the 2 notions of economic capacity measures depends upon the degree of scale of economies of a firm. Berndt and Hesse (1986) advocated that, under the assumption of prevailing constant returns to scale in the long-run, the tangency point between the long run and short–run average total cost curves will coincide at a point where the long-run and the short run average total cost curves reach their minimum. Hence, the two economic measures of capacity would be equivalent. The first two measures of economic capacity are termed primal economic capacity because they are directly measured in physical output and expressed in physical unit. The third measure of economic capacity proposed by Berndt and Morrison (1981), and Morrison (1985) is considered a dual-based concept and thus defines economic capacity in terms of firm's cost. They define economic capacity as corresponding to the shadow total cost of a firm. The shadow total cost is defined as the cost of the variable inputs plus the shadow cost of quasi-fixed inputs.

LITERATURE REVIEW

Optimizing economic variables of firms to derive economic capacity and utilization is relatively new in the literature especially in the sub-Saharan African countries. Berndt and Morrison (1981) used quadratic cost function to estimate capacity utilization rates for the U.S. manufacturing sector over the period 1958 to 1977. The model consisted of 3 variable factors, energy, materials and production workers as well as 2 quasi-fixed factors, capital and non production workers. He discovered capacity utilization rates greater than unity for the entire period. He also applied the same methodology to the U.S.A. automobile industry and obtained capacity utilization rates that exceeded unity for the years considered. While studying the performance of the Irish manufacturing sector, Kenny (1996) applied the translog Cost Function Approach on data from the manufacturing sector and under the assumption of long-run constant return to scale to estimate economic capacity utilization rates for the sector. In recognition of the dichotomy that characterized Irish industry, the model was fitted to 2 individual sub-sector classification; the hi-technology and traditional manufacturing sub sectors. The results revealed that substantial degree of excess capacity existed in the sector during the period 1970 to 1990. The traditional industrial sub-sector had a significant excess capacity compared to the hi-technology sub-sector. World Barik (2000) estimated the economic rate of capacity for the Indian paper industry for the period 1973 to 1974 and

1997 to 1998 using the translog Variable Cost Function. He found that under utilization of economic capacity prevailed in the industry, and also a decline in the rates of capacity utilization over time. Prior and Nelda (2001) in their study, estimated capacity utilization rate and cost efficiency in the chemical industry in Romania between period 1996 and 1997. They employed Cost Efficiency Data Envelopment Analysis (DEA) methodology. The result obtained showed high inefficiency and low capacity utilization rates among the industries. Hashim (2003) studied the trend in economic capacity utilization rates in Indian airlines for the period 1964 to 1990. He used the translog cost function to estimate the economic capacity utilization rates across the periods. The estimation of economic capacity utilization rates was based on 2 alternative measures of economic output. The first was where the short-run average cost was minimum, and the second was where the short-run and long-run average cost curves were equal. The results reveal an average estimated economic capacity utilization rate of 0.32 and 0.37 for the two methods respectively.

The results further showed that economic capacity utilization rates for the Indian airlines was generally poor, showing downward trend pattern in the study period. Lecraw (1978) analyzed factors that influence economic capacity utilization of 200 manufacturing firms in Thailand during the period 1962 to1974. He estimated profit-maximizing capacity utilization rate for each firm by using the projected balance sheets and income statements that the firms had prepared at the time of their initial investment. Their 'optimal' rate was roughly twice the rates chosen by the firms. The extent of non-optimal capacity underutilization of a firm was determined by the nationality of the firm's owner, entry date, number of firms in the industry, projected profits, and the manager's perceived risk of multi-shift operations. Dunlevy (1980) and Hayes and stone (1983) in their independent investigations in U.S.A found a significant positive relationship between capacity utilization rates in the industrial sector and the country's exports.

McElthattan (1985) investigated the relationship between capacity utilization rate in the industrial sector and inflation rate in U.S.A. She obtained a significant and positive relationship between the two variables. She however inferred from her regression results that for each percentage point, all industries capacity utilization rate exceeded 82%, inflation rate would accelerate by about 0.15% points. Earlier, Franz and Gordon (1993) discovered that capacity utilization rate depends more on inflation than on unemployment in both U.S.A and Germany economies. They also confirmed the non-accelerating inflation rate at capacity utilization rate for the U.S.A of about 82%, using Federal Reserve Bank Measures. Similar results were obtained by Garner (1994) and Yoo (1995). Gokcekus (1997) tested the hypothesis that trade liberalization increases economic capacity utilization in Turkish rubber industry. Using Generalized Leontief cost function, he established a positive relationship between the 2 variables. Seth (1998) established a positive link between industrial capacity utilization rates in India and public investment in infrastructures, capital, intermediary import and adoption of liberal policy. Kim (1999) analyzed the determinants of economic capacity utilization in U.S.A manufacturing sector. Evidence showed that capital stock, price of materials, capital price have significant negative relationship with the economic capacity utilization rates; while energy price, labour price and output have significant positive influence on the economic capacity utilization rate in the manufacturing sector. In India, Azeez (2002) investigated the impact of Indian industrial reform policies on the economic capacity utilization rate of the industrial sector. He discovered three distinct phases relating to economic capacity utilization rate movement during the policy era. Phase 1 (1974 to 1984) was characterized by relatively wide fluctuations; phase 2 (1985 to 1990) witnessed relatively stable fluctuation, while phase 3 (1991-1998) exhibited the characteristics of the phase 1. According to him the impact of the industrial reform on the industries economic capacity utilization rate was not remarkable. Kumar et al., (2009) used time series data from the period 1974 to 2005 to analyze the trends in the capacity utilization rates in the sugar industry in India. The result revealed that, the industry was operating with an excess capacity of 13% in each of the study year. The result also showed that, capacity utilization declined during the post reform years, and that the availability of raw materials was the most significant variable explaining the variation in the capacity utilization rate in India's sugar industry.

In Nigeria, Adeel et al. (2006) employed the survey and expert opinion approach to estimate capacity utilization rate among Nigerian firms. They discovered that firm's capacity utilization rates were affected by erratic power supply, variations in demand, insufficient capital and insufficient imports and domestic raw materials. Ukoha (2000) studied the determinants of the manufacturing capacity utilization rate in Nigeria in the period 1970 to 1988. He employed OLS method on secondary data published by the Central Bank of Nigeria. The result revealed that, the real exchange rate, federal government capital expenditure on the manufacturing sector and the per capita real income had positive effects on the manufacturing capacity utilization rate. On the other hand, the inflation rate and the real loans and advances to the manufacturing sector had negative effects on the capacity utilization rate of the sector. Akpan et al. (2011) investigated the influence of firm related factors and industrial policy regime on technology based capacity utilization in sugar industry in Nigeria. The empirical result reveals that sugar cane price and sugar industry's real energy consumption have significant negative relationship with the technology based capacity utilization in the sugar industry in Nigeria. On the other hand, the wage rate of skill workers, industry's, real research expenditure, human capital and period of import substitution have significant positive influenced on the technology based capacity utilization rate in the industry Akpan et al. (2012a) established the empirical relationship among technical efficiency, macroeconomic variables and industrial policy regimes in the Nigeria's sugar industry. Their result revealed that technical efficiency was influenced by the industrial sales growth, capital-labour ratio, official tariff rate on sugar import, real exchange rate and the content of the liberalization policy period. Akpan et al. (2012b) also analyzed the impact of macro-economic variable fluctuation on technology based capacity utilization in the sugar industry in Nigeria. The empirical results showed that the real sugar import, exchange rate, import price of sugar, parallel market exchange rate premium and tariff rate on sugar import were significant variables that influenced technology based capacity utilization rate in the industry.

THEORETICAL FRAMEWORK

Measuring economic capacity and capacity utilization rates using stochastic cost efficiency frontier (ECUR)

Following the fundamental assumption that capacity utilization rate is a short-run concept, and that variable inputs are efficiently utilized given the constraints imposed by quasi-fixed factors; the original efficiency stochastic cost frontier (SCF) was modified to represent economic efficiency capacity utilization rate (EECUR) by incorporating only the quasi-fixed factors into the firm's cost function (FAO, 2006). This implies that frontier cost is determined by efficient use of quasi-fixed input prices.

$$EECUR_t = \frac{C_j^*}{C_j} = \frac{f(X^*;\beta)\exp(V_j)}{f(X^*;\beta)\ \exp(V_j - U_j)} = \frac{1}{\exp(-U_j)} \tag{1}$$

Where C^*_j is the minimum or the frontier cost and C_j is the actual cost of production. X_j^* is the price of quasi-fixed inputs of j's firm. Economic efficiency capacity utilization rate (EECUR) is a biased index because it incorporates both capacity utilization and economic efficiency of fixed inputs. Unbiased economic capacity utilization rate (ECUR) was derived by dividing the index of economic efficiency capacity utilization rate (EECUR) by the cost efficiency score estimated in the traditional manner, such that;

$$ECUR_t = \frac{EECUR_t}{EE_t} \tag{2}$$

Where EE is the cost efficiency score computed for all factors of production. Firm's economic capacity is estimated as thus;

$$Economic\ Capacity = \frac{Actual\ TVC_t}{ECUR_t} \tag{3}$$

Hence, a firm that has full economic capacity utilization rate operates on the cost efficiency frontier (ECUR = 1), while those with economic capacity utilization rate less than unity (that is, ECUR < 1) operate below the cost efficiency frontier. This implies that the firm's fixed inputs are under-utilized and investment disincentive exists on the fixed factors of such firm. When economic capacity utilization rate is greater than unity (that is, ECUR >1), it implies that fixed inputs are over utilized and there is high tendency of investment incentive on the fixed factors of production (Morrison, 1985).

Klein capital utilization model

The relationship between capacity utilization rate of a firm and exogenous factors is found in Klein and Preston (1967) capital utilization model. In the model, they assume that;

$$\left\{\frac{K_t^f}{K_t}\right\} = \left\{\frac{L_t^f}{L_t}\right\} \tag{4}$$

Where, K_t^f and L_t^f are desired capital and manpower levels, while K_t and L_t are actual level of capital and manpower respectively. They also relate firm's output gap to manpower change as thus;

$$\left\{\frac{Y_t}{Y_t^f}\right\} = \left\{\frac{L_t^f}{L_t}\right\} \tag{5}$$

Where Y_t is the actual output and Y^f_t is the full employment level of output.
Combining Equation (4) and (5)

$$\left\{\frac{Y_t}{Y_t^f}\right\} = f\left[\left(\frac{K_t^f}{K_t}\right), \left(\frac{L_t^f}{L_t}\right)\right] \tag{6}$$

Where the firm output gap, $\left\{\frac{Y_t}{Y_t^f}\right\}$ represents the capacity utilization rate at period "t" (Klein and Preston, 1967; Johansen, 1968). Attaching log to both side of the equations and assuming Cobb- Douglas production function;

$$LnY_t - LnY_t^f = LnA_0 + Ln\left(K_t^f / K_t\right)^{\alpha} + Ln\left(L_t^f / L_t\right)^{\beta} + LnZ_t \tag{7}$$

Following the flexible investment function;

$$I_t = \delta(K^* - K_{t-1}$$

Where K* is the desired capital stock. Then

$$K^* = K_t^f = \delta^{-1}(I_t - K_{t-1}) \quad (8)$$

Substituting Equation (8) into (7) will produce;

$$LnY_t - LnY_t^f = LnA_0 + Ln[\delta^{-1}(I_t - k_{t-1})K_t^{-1}]^\alpha + Ln(L_t^f/L_t)^\beta + LnZ_t \quad (9)$$

Also, firm's demand for labour depends on the real wage rate in the economy. Hence at full employment level, wage rate corresponds to (W/P_0), while (W/P_1) corresponds to wage rates below equilibrium level. Thus;

$$\left\{\frac{L_t^f}{L_t}\right\} = \frac{(W/P_0)}{(w/P_1)} \quad (10)$$

Where "W" is the labour wage and "P" is the general price level in the economy. Substituting Equation (10) into (9) produces;

$$LnY_t - LnY_t^f = LnA_0 + Ln[\delta^{-1}(I_t - K_{t-1})K_t^{-1}]^\alpha + Ln[(W/P_0)/W/P_1]^\beta + LnZ_t \quad (11)$$

$$LnY_t - LnY_t^f = LnA_0 + \alpha[Ln(\delta^{-1}(I_t - K_{t-1})K_t^{-1}] + \beta Ln[(W/P_0)/(W/P_1)] + LnZ_t \quad (12)$$

In this framework, output gap defined as capacity utilization rate occurs as a result of the current investment level of a firm, the previous accumulated capital stock and the real wage rate influenced by the general price level in the economy. Their impact on firm output gap or capacity utilization rate is transmitted through factors specific elasticities. This framework assumes that, the output observed in any time period is the equilibrium level for observed rate of utilization of the inputs (Klein and Preston, 1967). Hence, other exogenous variables that affect capacity utilization can also be conceptualized in the same manner.

MATERIALS AND METHODS

Study area

The study was conducted in Nigeria; the country is situated on the Gulf of Guinea in sub Saharan Africa. It lies between 4° and 14° north of the equator and between longitude 3° and 15° east of the Greenwich meridian. Nigeria has a total land area of 923,768.622 km^2 or about 98.3 million hectares and a population of over 140 million (NPC, 2006). Industrial sugarcane is cultivated in commercial quantities in the northern part of Nigeria, and is mostly cultivated in irrigated lands or swampy areas. Niger state, Kwara state, and Adamawa state are the major industrial sugarcane producers in the country (Lafiagi, 1984). There are 4 major sugar producing firms and two sugar refineries in Nigeria.

These are: Nigeria Sugar Company at Bacita, Kwara State established in 1964 with initial installed capacity of 40,000 tons/annum; Savannah Sugar Company Limited at Numan, Adamawa State established in 1980 with initial installed capacity of 65,000 tons/annum; Lafiaji Sugar Company in Kwara State and Sunti Sugar Company in Niger State. The last 2 are mini sugar plants.

The refineries are BUA and Dangote located in Lagos state. The refineries are not involved in direct production, but refine imported semi processed sugar from Brazil and other sugar producing countries (NSDC, 2010).

Data source

Data used in the study were purposely collected from the two sugar producing firms in Nigeria. These firms depend fully on the domestic industrial sugarcane for the production of sugar and produced more than 95% of domestic produced sugar in the country (NSDC, 2010). Also, macro economic data published by the Central Bank of Nigeria (CBN), National Bureau of Statistics (NBS), Federal Ministries of Finance, Agriculture and Rural Development as well as labour and Productivity were used in the analysis. The sugar firms selected were: Bacita Sugar Company in Kwara state and Savanna Sugar Company in Adamawa state. The data collected covered the period of 1970 to 2010.

Analytical techniques

Estimation of economic efficiency (EE), economic efficiency capacity utilization rate (EECUR) and economic capacity utilization rate (ECUR)

Economic efficiency (EE) was estimated from Equation 13

$$EE = \frac{TVC_j^*}{TVC_j} = \frac{(P_j, Q_j, Z_j, \beta)exp(V_j)}{(P_j, Q_j, Z_j, \beta)exp(V_j - U_j)} = \frac{1}{exp(-U_j)^*} \quad (13)$$

Where, TVC_j is the actual total variable cost of production, TVC^*_j is the frontier total variable cost, P_j represents the prices of all inputs of ith firm, Q_i is the output level, and Z_j represents other variables. The economic efficiency (EE) of the sugar industry presented in Equation (13) was estimated using Equation (14) specify in log-linear as follows:

$$LnTVC_t = \delta_0 + \delta_1 LnWNP_t + \delta_2 LnPLP_t + \delta_3 LnRPK_t + \delta_4 LnWPW_t + \delta_5 LnPS_t + \delta_6 LnEC_t + \delta_7 LnPOI + \delta_8 LnSOI_t + \delta_9 LnTEP_t + V_j - U_j \quad (14)$$

Where; TVC_t = annual total variable cost of the sugar industry (₦) WNP_t = average wage rate of non-production workers (₦) $(\delta TVC_t/\delta WNP_t < 0)$; PLP_t = price of land used for sugarcane production (₦) $(\delta TVC_t/\delta PLP_t > 0)$; RPK_t = real depreciation cost of capital as a proxy of the rental price of capital stock (₦) $(\delta TVC_t/\delta RPK_t > 0)$; WPW_t = average annual wage rate of production worker (₦) $(\delta TVC_t/\delta WPW_t > 0)$; PS_t = average annual price of sugarcane (₦/tonne) $(\delta TVC_t/\delta PS_t > 0)$; EC_t = real energy consumption, proxies by annual expenditure on energy (₦) $(\delta TVC_t/\delta EC_t > 0)$; POI_t = average price of other inputs (₦) $(\delta TVC_t/\delta POI_t > 0)$; SO_t = sugar output (tonnes) $(\delta TVC_t/\delta SO_t > 0)$; TEP_t = technological progress captured by time trend $(\delta TVC_t/\delta TEP_t > 0)$

The economic efficiency capacity utilization (EECU) was measured by using only quasi-fixed inputs in the cost functions of Equation (15) as shown in Equation (16).

Table 4. Maximum estimates of Cobb-Douglas stochastic cost function for the sugar industry in Nigeria as defined in equation (14) and (16).

Variable	Equation for EE indices (Equation 14)	Equation for EECUR indices (Equation 16)
Constant	-0.1882 (-0.0932)	0.5235 (-1.7623)*
Wage rate of non prod. workers (WNP_t)	-0.2685 (-0.3839)	0.2612 (0.4174)
Land price (PLP_t)	-0.1278 (-1.8235)*	-0.2586 (-2.1612)**
Depreciation cost (RPK_t)	0.5039 (3.6650)***	0.4902 (3.2466)***
Sugar output (SO_t)	0.9259 (0.7854)	-
Wage rate of prod. worker (WPW_t)	0.7709 (2.0479)**	-
Price of sugarcane (PS_t)	0.9154 (0.2374)	-
Expenditure on electricity (EC_t)	-0.3379 (-0.5446)	-
Price of other inputs (POI_t)	0.1614 (2.1173)**	-
Technology progress (TEP_t)	0.4417 (0.5260)	0.5123 (0.4565)
Sigma square gamma (δ^2)	0.5351 (2.4249)**	0.8435 (4.5056)***
Gamma (λ)	0.4171 (1.4150)	0.6529 (3.7246)***
Log-Likelihood	-0.4425	-0.4826
LR Test	15.4047	25.4047

*, ** and ***, 5, 10 and 1% significant levels respectively. Figures in brackets are t-values, and variables are as defined in Equations (14) and (16).

$$EECUR = \frac{TVC_j^*}{TVC_j} = \frac{f(X_j^*;\beta)exp(V_j)}{f(X_j^*;\beta)exp(V_j - U_j)} = \frac{1}{exp(-U)^{**}} \quad (15)$$

To estimate economic efficiency capacity utilization (EECUR) presented in Equation (15), we specify Equation (16) in log-linear form as follows:

$$LnTVC_t = a_0 + \alpha_1 LnWNP_t + \alpha_2 LnPLP_t + \alpha_3 LnTEP_t + V_j + U_j \quad (16)$$

The variables have the same meaning as in Equations (13). An unbiased estimate of economic capacity rate (ECUR) was estimated using the results of Equations (14) and (16), as follows:

$$ECUR = \frac{EECUR}{EE} \quad (17)$$

Where;

$$CUR_t = \gamma_0 + \gamma_1 INFL_t + \gamma_2 PDSC_t + \gamma_3 LAP_t + \gamma_4 RER_t + \gamma_5 PGDP_t + \gamma_6 EC_t + \gamma_7 LOA_t + \gamma_8 SIM_t + \gamma_9 FCA_t + \gamma_{10} DSC_t + \gamma_{11} D + U_t \quad (18)$$

CUR_E = Economic capacity utilization rate (ECUR) for the sugar industry in Nigeria; $INFL_t$= inflation rate at period t (%); $PDSC_t$ = average annual price of domestic sugar cane (₦/tonne); LAP_t = labour productivity in the sugar industry [defined as total domestic output divided by total number of workers in the sugar industry (tonnes/person)]; RER_t = real exchange rate (₦/$); $PGDP_t$ = per capita real GDP (2003 =100) (₦); EC_t = energy consumption proxy by annual expenditure on energy (₦); LOA_t = real loans and advances to sugar industry (₦); $SIMPt/GDP_t$ = ratio of sugar import to the GDP (%); FCA_t = share of Federal Government Capital expenditure on the sugar industry in the GDP; DSC_t = domestic produced sugarcane used as input in the industry (tons); D = dummy variable which takes the value of 1 for the liberalization period (1986-2010), and 0 for otherwise (1970- 1985); U_t = stochastic error term.

RESULTS AND DISCUSSION

Estimation of economic capacity utilization (ECUR) equations for the sugar industry in Nigeria

Maximum likelihood estimates of the Cobb-Douglas stochastic cost function for the sugar industry are presented in Table 4. The Cobb-Douglas stochastic cost function in Equation (14) was defined for all the factors of production and was used to generate indices of cost or economic efficiency. Equation (16) was defined for only input considered quasi-fixed inputs and was used to generate economic efficiency capacity utilization indices for the sugar industry. The result revealed a signific sigma squared coefficients of 0.5351 at 5% level of probability for Equation (14) and 0.8435 at 1% level of probability for Equation (16). These indicate good fit and correctness of the specified distribution assumption of the composite error term for the models. The variance ratios (λ) in Equation (14) *and* (16) indicate the proportion of variations in the total variable cost in the sugar industry in Nigeria that is due to deviation from full economic efficiencies and economic capacity utilization respectively. In Equation (16), a gamma ratio of 0.4171 suggests that the presence of economic or cost inefficiency in the sugar industry in Nigeria explained about 41.71% variations in the total variable cost of the industry. In the same way, the gamma ratio of 0.6529 for Equation (16) implies that about 65.29% of variations in the total variable cost are

Figure 1. Trend in economic capacity utilization rates in the sugar industry in Nigeria.

attributed to cost inefficiency and unutilized cost capacity. The generalized likelihood ratio tests for the equations are highly significant and this confirms the presence of one - sided error component in the composite error terms. Therefore, the result of the diagnostic test confirmed the relevance of the stochastic parametric cost function and maximum likelihood estimation technique.

The empirical results reveal that land price (PLP_t), depreciation cost (RPK_t), wage rate of production workers (WPW_t), and price of other inputs (POI_t) were significant production variables that affect the value of total variable cost in the sugar industry in Nigeria.

Trend in the estimated economic capacity utilization rates in the Nigerian sugar industry

The trend in the estimated economic capacity utilization in the sugar industry is shown in Figure 1. The trend displayed highly undulated pattern throughout the considered years. In all observations, economic capacity utilization rate (ECUR) was less than unity with an average value of 60.30%; implying that the industry had an excess economic capacity utilization of about 39.70%. This means that the industry needed about 39.70% economic capacity gain to reach the economic or optimum capacity frontier. The present of the excess economic capacity utilization in the industry implies that the industry suffered from insufficient fund needed to cover the cost of production. This means that the industry was constrained by insufficient financial resources which prevented the attainment of the optimum economic capacity utilization level.

Determinants of economic capacity utilization rates of the Nigerian sugar industry

Table 5 reports the result of estimation of the economic capacity utilization equations in sugar industry in Nigeria. The linear form of the specified equations was picked as the lead equation following the result of the diagnostic tests and the number of significant independent variables. For the lead equation, R^2 is 0.848 denoting that about 84.80% of variations in economic capacity utilization rate were explained by the specified independent variables. The F-statistic of 3.088 is significant at 1% probability level, implying that the R^2 is significant and the model has goodness of fit. Durbin- Watson statistic of 2.74 indicates that auto-correlation might pose a minor problem. The empirical result revealed that the coefficient of inflation rate ($INFL_t$) is statistically significant (at 10% level) and negatively related to the economic capacity utilization rates in the sugar industry. This relationship indicates that an increase in the inflation rate will lead to a decrease in economic capacity utilization rates in the sugar industry. For instance, 10% increase in inflation rate will result in 0.01% decrease in economic capacity utilization rate. A similar result has been reported for the manufacturing sector in Nigeria by Ukoha (2000). However, the result is contrary to the findings of McElthattan (1985), Franz and Gordon (1993), Garner (1994) and Yoo (1995) in the United State of America.

The result also revealed a significant positive (at 5% level) relationship between the per capita real GDP ($PGDP_t$) and economic capacity utilization rate in sugar industry in Nigeria. The result is in line with the economic theory, because an increase in per capita real GDP

Table 5. Economic capacity utilization rate equations in sugar industry in Nigeria

Variable	Linear (L)	Exponential	Semi-log	Double- log
Constant	0.703 (5.60)***	-0.321 (-1.61)	0.515 (1.37)	-0.639 (-1.05)
$INFL_t$	-0.001 (-1.98)*	-0.002 (-2.16)**	-0.003(-0.23)	-0.009 (-0.35)
$PDSC_t$	-9.20e-008 (-0.03)	-3.69e-007 (-0.09)	-0.032 (-0.79)	-0.063 (-0.95)
LAP_t	-0.005(-0.69)	-0.009 (-0.83)	-0.032 (-0.18)	-0.068 (-0.24)
RER_t	-1.73e-005 (-0.02)	-5.92e-005 (-0.05)	0.0006 (0.02)	0.003 (0.07)
$PGDP_t$	6.85e-005(2.64)**	0.0001 (2.79)**	0.009 (0.32)	0.008 (0.15)
EC_t		-0.0002 (-1.67)*	-0.020 (-1.56)	-0.040 (-1.75)*
LOA_t	-0.0001 (-1.89)*	5.19e-006 (1.56)	0.003 (0.88)	0.005 (0.91)
$SIMP_t/GDP_t$	3.07e-006 (1.46)	-0.161 (-0.89)	-0.010 (-0.67)	-0.016 (-0.66)
FCA_t	-0.086 (-0.76)	8.77e-005 (2.21)**	-0.011 (2.33)**	0.017 (2.27)**
DSC_t	5.38e-005 (2.14)**		0.031 (0.75)	0.056 (0.83)
Policy dummy	1.81e-007 (1.03)	3.21e-007 (1.15)	0.179 (2.71)**	0.302 (2.80)**
	0.213 (2.93)***	0.353 (3.07)***		
R^2	0.848	0.819	0.417	0.442
F-Statistic	3.088***	3.647***	1.822*	2.0215*
DW-test	2.74	2.73	2.25	2.17
Normality test	7.957 (0.0187)**	4.297(0.1166)	7.872(0.0195)*	5.321(0.0699)*
Hetero-test	0.336 (0.9709)	0.369 (0.9586)	0.102 (1.0000)	0.112 (0.9999)
RESET –test	7.836 (0.001)***	6.383 (0.004)****	2.462(0.1280)	5.027 (0.0334)**
Schwarz Criterion	-89.04		-78.56	-40.14
Akaike Criterion	-109.98	-52.38	-99.13	-60.41
Hannan- Quinn C.	-101.98	-72.64	-91.80	-53.08
loglikelihood	66.65	-65.32	61.57	42.21
		48.32		

*,** and ***, 10, 5 and 1% significance levels respectively. Figures in bracket are t-values and variables are as defined in equation *(18)*. L means lead equation.

raises the demand level of consumers in the economy. Increased in the per capita real GDP would exert positive influence on the industry's net returns. This has a tendency to boost the economic capacity of the industry through increase ability to procure more production inputs. The result implies that the sugar industry in Nigeria was not demand constrained. Ukoha (2000) obtained a similar result for the Nigeria's manufacturing sector. The expenditures on energy in the sugar industry in Nigeria had a significant negative effect (at 10% level) on the economic capacity utilization rate of the industry. The result implies that increase in energy consumption decreases economic capacity utilization rate of the industry. The finding agrees to the *a priori* expectation as increase in energy consumption will tend to increase the total variable cost of the industry. This has an adverse effect on the net returns of the industry and the economic capacity utilization rates.

The coefficient of the federal government capital expenditure on sugar industry (FCA_t) was significant (at 5% level) and was a positive determinant of economic capacity utilization rate in the sugar industry. The result implies that, as the federal government subventions to the sugar industry increase the economic capacity utilization rates of the industry also increase. For instance, one million naira increase in the federal government subvention to the sub-sector will result in 0.000054% gains in economic capacity utilization rate in the industry. The reason for the result might be attributed to the fact that the sub-sector was completely owned by the federal government before it was privatized.

The liberalization policy period (D) had a significant positive influence on the economic capacity utilization in the sugar industry. This implies that the industrial policies embedded in the liberalization period had significant positive influence on the economic capacity utilization rate in sugar industry. The result agrees with the findings of Gokcekus, (1997) in Turkey, Earl and Amos (2002) in Romania, Phillipe and Robin (2003) in UK and Akpan et al., (2012a) in Nigeria.

CONCLUSION AND RECOMMENDATIONS

The study used sugar industry based data and macro-economic data from 1970 to 2010 to analyze economic capacity utilization rates in the sugar industry in Nigeria. Sugar industry based data collected were production and cost data. The macro-economic data used were inflation rate, exchange rate, GDP, tariff rates on sugar import, consumer price index, parallel and official exchange rates among others. Unit root tests were conducted on the specified variables in economic capacity utilization equation and their stationary determined. Similarly, the specified cost function variables for the industry were used at their levels to estimate the economic capacity utilization indices and economic efficiency indices from which the unbiased economic capacity utilization rates in the sugar industry were calculated. Multiple-regression equation of various forms was estimated based on the ordinary least squares method and used to determine factors that influence the economic capacity utilization rate.

Also, the patterns of fluctuations in the estimated economic capacity utilization rate in the industry showed undulated trend throughout the study period with an average value of 60.30% and excess capacity of about 39.70%. The finding also revealed that the economic capacity utilization rate in the industry had significant positive association with per capita real GDP, share of federal government expenditure on sugar industry in the GDP and liberalization policy period. The inflation rate and energy consumption had significant negative relationship with the industry economic capacity utilization rates.

To improve economic capacity utilization rate in the sugar industry in Nigeria, the study advocated for a policy package that either reduces or maintains a steady inflation rate in the country as this will enhance increase capacity utilization in the sugar industry in Nigeria. It is also recommended that an appropriate policy measure that aim at expansionary aggregate demand as a means of promoting capacity utilization in the sugar industry should be introduced. Such policy measure should be designed to avoid inflationary tendencies. Government should strengthen the power sector to provide constant electricity to sugar industry in Nigeria. This will help to lower the total variable cost of the industry and increase the net returns as well as the capacity utilization of the industry. Furthermore, the industrial policy package during liberalization era will promote economic capacity in the industry.

REFERENCES

Adeel M, Teal F, Simon B (2004). The Performance of Nigerian Manufacturing Firms. Report on the Nigeria Manufacturing Enterprise Survey.

African Development Bank (ADB) (2000). Operations Evaluation documents.

African Development Fund (ADF) (2000) Operations Evaluation documents.

Akpan SB, Udo UJ, Essien UA, (2011). Influence of Firm related factors and Industrial Policy Regimes on Technology Based Capacity Utilization Rate in Sugar Industry in Nigeria. Agric on-line Papers in Economics and Informatics. III, 3:15-22.

Akpan SB, Obasi O, Ukoha CE. Onyenweaku-Daniel EJ (2012a). Empirical Relationship among Technical Efficiency, Macroeconomic Variables and Industrial policy Regimes: A case study of Sugar Industry in Nigeria. Int. J. Econ. Manag. Sci. 1(10):01-11.

Akpan SB, Ini-mfon VP, Daniel EJ (2012b). Impact of macro-economic variable fluctuations on technology based capacity utilization in sugar industry in Nigeria. Int. J. Econ. Manag. Sci. 1(11):26-34.

Anyanwu JC, Oyefusi A, Oaikhenan H, Dimowo FA (1997). The Structure of the Nigerian Economy (1960-1997). Onitsha Joanee Educational Publisher.

Ayayi DD (1988). Spatial Patterns of Production Sub-contracting in Nigeria. Nig. J. Econ. Soc. Stud. (NSESS), 42(1):95-111.

Ayeni B (1981). The Spatial Distribution of Manufacturing Industries in Nigeria. Technical Report No. 2 Department of Geography, University of Ibadan.

Azeez EA (2002). Economic Reforms and Industrial Performance'. An Analysis of Capacity Utilization in Indian manufacturing Sector. A Technical Working paper at the Institute of Economic Growth, Delhi India.

Barik K (2000). Capacity utilization in Indian paper industry, Publication of the School of Social Science, India Gandhi National Open University, New Delhi.

Berndt E, Hesse D (1986). Measuring and Assessing Capacity Utilization in the Manufacturing Sectors of Nine OECD Countries. Europ. Econ. Rev. 30:961-989.

Berndt ER, Morrison CJ (1981). Capacity Utilization: Underlying Economic Theory and Alternative Approach. Am Econ. Rev. 71:48-52.

Busari LD, Misari SM (1996). Sugar and sugarcane in Nigeria. A Publication of NCRI in Nigeria.

Cassels JM (1937). Excess Capacity and Monopolistic Competition. Q. J. Econ. 51:424-443.

Central Bank of Nigeria (CBN), Statistical Bulletin, (2006, 2008, 2009).

Danish AH (2003). Capacity Utilization in Indian Airlines. A Postdoctoral Fellowship Study at the Institute of Economic Growth, Delhi, India.

Dunlevy J (1980). Aggregate Stocks and the Relationship between United State's Business Cycle Fluctuation and Export Performance. J. Econ. Integr. 13(1):163-198.

Ekeocha PC (2009). Evaluation of the effectiveness of industrial policy reforms in Nigeria 1999-2003: A computable general equilibrium simulation. Analysis presented at the 50th annual conference of the Nigerian Economic Society (NES) in Abuja September, 22-24th, 2009.

Fabayo A (1981). Production Function Measure of Capacity Output and Rate of Capacity Utilization in Selected Nigerian Industries: 1974-1975. Ife Social Sci. Rev. 2 (2):0-27.

FAO Corporate Document on Capacity Utilization in Fishery Sub-sector (2006).

Federal Ministry of Industry Abuja (FMI), Nigeria (2003). "National Sugar Policy". A Publication of National Sugar Development Council of Nigeria.

Franz W, Gordon RJ (1993). German and American Wage and Price Dynamics: Europ. Econ. Rev. May, pp. 719-762.

Friedman M (1963). More on Archibald versus Chicago Review of Economic Studies, 30:65-67.

Garner C, Alan (1994). Capacity Utilization and USA Inflation. Federal Reserve Bank of Kanakas City, Econ. Rev. 79(4):1-21.

Gokcekus O (1997). Trade liberalization and capacity utilization; New evidence from Turkish Rubber industry. Empir. Econ. 23;561-571.

Hashim DA. (2003). Capacity utilization in Indian airlines: A paper presented to Institute of Economic Growth., Delhi, India.

Hayes S, Stone J (1983). Specification of Supply Behaviors. Rev. Econ. Stat. 64:271-281.

Hickman BG (1964). On a new Method for Capacity Estimation. J. Am. Stat. Assoc. 59:529-549.

Isola WA (2006). Market reform and de-industrialization in Nigeria: 1986 -2003. Essays on Economic Development. MPRA paper no. 1025, 2006. http:// mpra.ub.uni-muenchen.de/ 1025/

ISYB (1978). International Sugar Year Book (1978), Publication of the Federal Ministry of Industry.

Kenny G (1996). Capacity Utilization in Irish Manufacturing Sector.

Technical Paper Presented to Central Bank of Ireland.
Kim HY (1999). Economic capacity utilization and its determinants: theory and evidence. Rev. Indust. Organ. 15;321-339.
Kim YC, Kwon JK (1977). The Utilization of Capital and the Growth of Output in a Developing Economy: the Case of South Korean Manufacturing. J. Dev. Econ. 4:265-78.
Klein LR, Preston RS (1967). Some new results in the measurement of capacity utilization. Am. Econ. Rev. 57:34-58.
Klein LR (1960). Some Theoretical Issue in the Measurement of Capacity. Econometrica 28:272-286.
Kumar S, Nitin A. (2009). Analyzing regional variations in capacity utilization of Indian sugar industry using non- parametric frontier technique. Eurasian J. Bus. Econ. 2(4):1-26.
Lafiagi MS (1984). Towards Self-Sufficiency in Sugar Production in Nigeria. Memo Submitted to Federal Ministry of Commerce and Industry pp. 1-46.
Lecraw DJ (1978). Determinants of capacity utilization by firms in less developed countries. J. Dev. Econ. 2(2):139-153.
Manufacturing Association of Nigeria Publication (2006, 2009).
McElthattan R (1985). Inflation, Supply Stock and the Stable Inflation Rate of capacity Utilization. Federal Reserve Bank of San Francisco, Econ. Rev. Winter 1:45-63.
Michael D (2010). USDA foreign Agricultural service on sugar imports in Nigeria. A comprehensive report on sugar imports in Nigeria. http://www.nigerianseminarsandtrainings.com. Retrieved 10-12-2010.
Morrison CJ (1985). Primal and Dual Capacity Utilization, An application to Productivity Measurement in the USA Automobile Industry. J. Bus. Econ. Statis. 3:312-324.
National Population Commission (NPC) (2006). National Census Data. www.population.gov.ng/. Retrieved 10-12-2010.
National Sugar Company Document (NISUCO) (1999). A Publication of National Sugar Development Council, Federal Ministry of Industry, Abuja, Nigeria.
National Sugar Development Council Document (NSDC) (2006, 2010 and 2012). A Publication of National Sugar Development Council, Federal Ministry of Industry, Abuja, Nigeria. http://www.nsdc.gov.ng/
Nigerian Sugar Company Ltd., Bacita Estate (NISUCO) (2000). Production Statistic for each Agricultural Year from 1965 to 2000.
Nwaobi GC (2005). The Nigerian sugar industry (1960-1987). 0501001, EconsWPA.
Ogunbayo M (2009). The fall of the Giant Unpublished Paper. Extracted from internet, (http://www.newswatch.ngr.com, 10-01-2011)
Olomola AS (2007). Competitive commercial Agriculture in Africa study (CCAA), Nigeria case study. A final report submitted to the Canadian International Development agency (CIDA) and the World Bank.
Prior D, Nela F (2001). On the Measurement of Capacity Utilization and Cost Efficiency: A Non- Parametric Approach at Firm Level. Pesquisa Operational 22(2):247-263.
Raimi L, Adeleke I (2009). A Review and Measurement of Capacity Utilization in the Nigerian Economy. Humanomics 25(4):274-284.
Salimonu KK, Gafar OA, Akintola JO (2006). Trend Analysis of Capacity Utilization of Agro-based Industries in Nigeria. Agric. J. 1(4):311-314.
Savannah Sugar Company limited (SSC) (2006) 2010. Production Data for Savannah Sugar Company Ltd., Numan, 1983-1999. The Punch Newspaper, Thursday Sept. 14, 2000, P. 23.
Seth K (1998). Economic Reforms and Capacity Utilization in Indian Industry", Productivity 39(3):391-98.
Soderbom M, Teal F (2002). The Performance of Nigerian Manufacturing Firms. Report on the Nigerian Manufacturing Enterprise Survey 2001.
World Bank (2001), Nigeria: poverty in the Midst of Plenty, The Challenge of Growth with Inclusion.
Ukoha OO (2000). Determinants of Manufacturing Capacity Utilization in Nigeria, 1970-1988. Nig. J. Econ. Soc. Stud. 42(2):121-129.
Wada AC, Agidi G, Ishaq MN, Busari LD (2001). Current Status of Sugar Research and Development in Nigeria. Sugar Technol. *3(1):4-12.*
Yoo PS (1995). Capacity Utilization and Price within Industries, Federal Reserve Bank of St. Louis Review 77(5). Sept. - Oct. 15-25.
Zayyad HR (2007). Privatization and Commercialization in Nigeria. Unpublished Work in the Internet (http://unpanl.un.org/intradoc/groups/public/docume. 15-10-2010).

Pomegranate marketing margins in the Neyriz, Iran

Mahdi Babaei[1*], Elahe Asemani[2], Saeid Gazmeh[3], Soheyla Rezazadeh Shamkhal[4] and Ebrahim Shirmohammadi[5]

[1]Young Researchers Club, Neyriz Branch, Islamic Azad University, Neyriz, Iran.
[2]Department of Agriculture Economic, University of Zabol, Iran.
[3]Department of Computer and Information Technology (IT), Science and Research Branch, Islamic Azad University, Fars, Iran.

Marketing agricultural products while establishing a balance between supply and demand in various stages of production to final consumption stabilize farm income and satisfy consumer demand. Therefore, agricultural marketing study seems necessary. In this study, pomegranate marketing in Neyriz city of Fars province was investigated and its various marketing channels were also evaluated. The results showed that, the share of pomegranate growers and marketing margin are almost good. Lack of suitable storage and fridge was one of the major problems of pomegranate growers. For improvement of the current situation, the formation of a cooperative of pomegranate growers in the region is recommended, for decreasing production costs and making storages and fridges enough for regulation of demand and supply.

Key words: Marketing, marketing margin, pomegranate, Neyriz city.

INTRODUCTION

Marketing refers to all attempts made by economical job creators for planning, producing, distributing, selling products and giving service to end-users and meeting their needs. Marketing agricultural products, in addition to forcing a balance between demand and supply in various stages from production to end consumption, makes stabilization of the agriculture income and meeting consumer needs possible (Karbasi, 2009).

Iran is birthplace of pomegranate and has the first rank of variability, quality, under cultivation area, production and export. Cultivation and culture of pomegranate has been common in Iran since the ancient era. Pomegranate fruit has been considered as a saint fruit by Muslims and the advantages of it are cited in 142 to 99th paradigm of Sura Anaam. Presently, 70 acre of farms are under cultivation in our country which is intended to add 15 more acres till the end of the 5th program (Karimi, 2007).

Fars has the first status in pomegranate cultivation among all provinces. The cities which have the most production cultivation area are Neyriz, Arsenjan, Kazeroon, Estahban, Jahrom, Marvdasht, Firooz abad, and Mamasani. Known species of pomegranate in the province are, Robab, Farough, Beritt, acidulent, Atabaki, shirin Shahsavaar keder, and green opaque sour. Export species of pomegranate are the best with view to quality, taste and market attention. Red grain of acidulent Robab species has been favoured in many countries especially in middle Asia. So, pomegranate can play a significant role in the economy due to its alimentary needs, economical value and export possibility. Pomegranate is pre-eminent among all horticultural products due to its potential for producing various kinds of alimentary and industrial products (Behzadi, 2000).

Neyriz city has 10540 km^2 of which 90000 acres of it, is fallow. 20000 people of this city are farmers which account for 18.6% of its total population. Average lands for each beneficiary is 3 acres. Its gardens account for 53.3% of total lands of this city which is 48000 acres. The proportion of considered pomegranate gardens is 4000 acres which makes this city a center of pomegranate

*Corresponding author. E-mail: mbabaei65@yahoo.com.

production in our country. This county is a resource of pistachio, pomegranate, fig, almond, corn, wheat and malt due to its high potential for farming. With regard to the statistics of the year 1387, pomegranate had the most production harvest with 44% productions which was equal to 6000 ton, and ranked first in the production of other specialties of pomegranate. The weather of this city is semi-arid whose maximum temperature for a 20-day period in June to July is 43°C and the minimum for a week in December to January is -10°C (Jehad Agriculture Office promotion, 2009).

The result obtained from evaluating the marketing margin of grape in Iran have indicated that during the considered period, average retailing margin of both products is more than the wholesale margin (Karbasi, 2009). In investigating and analyzing efficiency and barberry marketing in Khorasan, marketing margin, retailing, wholesale and marketing efficiency have been evaluated and determined. Findings have indicated that increasing technical efficiency of barberry producers is possible by reducing distance between them. Marketing margin of retailing came from more wholesale (Dehghanian et al., 2007). Findings obtained from this study have indicated that indices of production, export and transportation costs are the most factors effecting marketing margin of potato. The main obstacles of marketing this product are small size, potato spoiling, seasonality of this product, expanding active marketers with different monetary capacity, price instability and fluctuation, lack of quality standards, grading and packaging etc. (Beik Zadeh, 2008). Review of marketing margin investigations in Iran have shown lack of evaluating one of the most important products in the production inventory and exporting non-oil products.

Kizilaslan and Elmali (2012) investigated the effects of marketing structure of grapes on the consumers; the marketing margin was calculated based on the producer sale price and the market price (the consumer price) of grapes. The results show that the production of grape is a profitable activity, but it has been determined that the mediators between producers and consumers provide a higher income when compared to producers. The margin mediator is found to be 77.05%.

Preliminary data and information were obtained from gardeners and pomegranate producers and interview with the manager of herbaceous production institute of Neyriz, and were achieved in form of questionnaire from mentioned reference for the year 2010. This research is aimed at recognizing pomegranate of the city and finding marketing troubles and giving comments to solve those problems.

MATERIALS AND METHODS

Marketing margin

Information about each of these variables to complete the questionnaire and survey methods designed with the visiting farmers Neyriz city, were collected in 2009. The sample consisted of 248 farmers which were surveyed and the sampling method used is simple random sampling. Data processing and analysis were performed using Excel software. Marketing margin is described as divergence difference between purchase prices of a product in a competitive market which is determined by the producer. On the other hand, marketing margin include price of all activities and operations done by marketing institutes and dealers. So, the main scale for determining marketing margin is the difference between prices received by the end producers. Marketing margin is a function of distance, packaging and conversion, paid tax and duties, competition in market, volume mass of product, spoilage and calculated as follows:

$$Mr = Pr - Pw$$

$$Mw = Pw - Pf$$

$$Mm = Mr + Mw = Pr - Pf$$

Mr = retailing margin, Mw = wholesale margin, Mm = total market margin, Pr = retailing price, Pw = wholesale price, Pf = on farm price (Karbasi, 2009).

RESULTS AND DISCUSSION

In this study, pomegranate marketing in Neyriz city was investigated. Pomegranate production of this city has been 20 ton per acre on average for 4 successive years. Neyriz has achieved first place of pomegranate production in our country and the most harvest in this city was 116 ton per acre.

With regard to national pomegranate congress, pomegranate of this city is known for pomegranate traders. Irrigation methods used in pomegranate gardens are 80% flood irrigation with 15 days period irrigation on average and little percentage is drip irrigation. Average income of each pomegranate beneficiary has been 50 million Rials. Second and third grade pomegranate were delivered to concentrate factory for processing because price of first grade pomegranate is 800 and prices of second and third grades are 5000 and 2500 Rials, respectively. After processing, pomegranate sauce or juice is acquired per one kilogram, second and third grade pomegranate and will be sold 1000 Toman. The amount of export pomegranate in 1388 was 5000 ton whose canal is drawn formally and informally (Table 1). Pomegranate was exported to East Asia, Japan, Korea, Persian Gulf margin, Ukraine and some European countries at price of 15000 Rials. Farmer income per ton is summarized in Table 1. Harvest cost is 3000 thousands Rials per each acre in year. Retailer expenditure by considering transportation and packaging cost in markets and self-consumption in addition to being far from market is 1000000 Rials per ton on average. The price of first grade pomegranate for market consumption is 8000000 Rials per ton which is given in retailer expenditure table. Store rent, operation and other costs are not available because the amount of daily retailer

Table 1. Farmer income per each ton.

First grade pomegranate	Second grade pomegranate	Third grade pomegranate	Total (Rials)
20125000	1000000	875000	22000000

Source: Study data.

sales is not accessible. It is considered that the price of the retailer sale is 12000 Rials, and then its income is 12000000 Rials per each ton. If other expenditure is eliminated, its income will be 3900000 Rials.

The mentioned costs are true for all selling and buying canals. Packaging cost paid by wholesaler is 2000000 Rials per each ton which is for buying cartons and packaging cost is 1000000 Rials per ton. The most important expenditure is transportation cost which equals 2500000 Rials considering its delivery to importers in Bandar Abbas. 8000000 Rials per ton were paid to farmers by wholesalers in 1388, total costs were 11500000 Rials and product sales to exporters were 15000 Rials which is the only exporter income and equals to 15000000 Rials per ton. In this study, it is assumed that the buying is performed by a factory but practically it is done by native dealers and 100000 Rials paid for dealing, transportation cost is 90000 Rials and buying price is 3000000 Rials per ton; the other costs of concentrate factory are not available, therefore, explicit statistics cannot be reported in this stage. Processed pomegranate was 2000 ton which was shipped to concentrate factory for processing by native dealers with the price of 2500 than 5000 Rials.

Wholesale, retailing and all various sorts of pomegranate margins are cited for market delivery separately. Among the three sorts of pomegranate (first, second and third grades) 12500 kg of first grade pomegranate per ton are delivered for export marketing and 125 kg of the mentioned sort are kept for self-consumption in the city. 250 kg of second grade and 250 kg of third grade pomegranate are used for processing. Wholesale margin for the first path (export) is as follows:

Considering the equation below and export price of pomegranate (which is 13000 Rials) and price of buying first grade pomegranate (which is 8000 Rials), the marketing margin for export is 5000 Rials:

$$Mw = 13000 - 8000 = 5000$$

Total market margin is calculated from difference between retailing price and on-farm price considering obtained data, retailing price in export is 15000 Rials, on-farm price is 8000 Rials then total market margin is 7000 Rials which is estimated by the stated equation:

$$Mm = 15000 - 8000 = 7000$$

Retailing margin is calculated from the difference between retailing price and wholesale price considering obtained data, retailing price in export is 15000 Rials, wholesale price is 13000 Rials then retailing margin is 2000 Rials which is estimated by the stated equation:

$$Mr = 15000 - 13000 = 2000$$

Wholesale margin for self-consumption path is calculated from difference between wholesale price and on-farm price considering obtained data, wholesale price is 13000 Rials, on-farm price is 8000 Rials then wholesale margin is 5000 Rials which is estimated by the mentioned equation:

$$Mw = 13000 - 8000 = 5000$$

Total market margin for self-consumption path is calculated from the difference between retailing price and on-farm price or sum of retailing and wholesale margins considering obtained data, retailing price in export is 15000 Rials, on-farm price is 8000 Rials then total market margin is 7000 Rials which is estimated by the mentioned equation:

$$Mm = 15000 - 8000 = 7000$$

Retailing margin in self-consumption path is calculated from the difference between retailing price and wholesale price considering obtained data, retailing price in export is 15000 Rials, wholesale price is 13000 Rials, then retailing margin is 2000 Rials which is estimated by the mentioned equation:

$$Mr = 15000 - 13000 = 2000$$

With regard to obtained statistics, marketing margins in both paths are identical; meanwhile the only privilege of the first market is its great volume.

CONCLUSION AND RECOMMENDATIONS

One of the features of agricultural products is that they are produced in a short period under natural conditions but consumed in a longer period. For example, wheat which is the main food in household basket of Iranian consumers is produced in a short period and consumed in the same period of that year. One of the problems related to seasonality is that when the need for transportation vehicles increased in the harvest season, transportation cost increased as well. Additionally, if the product is not been delivered to store on time, more spoilage would occur. Storing problems and fridge loss result in severe spoil.

Presently, almost 20% of agriculture products are spoiled. From farm to consumption place with lack of marketing operation (transportation, storing and conversion), spoilage can be reduced. Spoilage reduction leads to supply increment. By increasing the capacity of concentrate factory and making fridge available, this problem can be resolved. As proposed by the manager of the Agriculture Institute and Pomegranate Producers Union with respect to the main problems of pomegranate in the year 2009 (storing problem and fridge loss to store product and sell it), government policies and decisions in the political-economical sector play an important role in every market.

REFERENCES

Behzadi SH (2000). Disperse and variability of pomegranate species in Iran. J. Agric. Econ. 1(3):31-37

Beik zadeh S, Chizari H (2008). Investigating marketing and factors effecting on marketing margin of potato. Develop. Agric. Econ. 15(57):113-200.

Dehghanian S, Shahnooshi N, Azin far Y (2007). Investigating, analyzing and marketing barberry producers in Khorasan. Agriculture and natural sources science. FAO, Uhttp://apps.fao.

Karbasi A (2009). Evaluating marketing margin of grape and raisin in Iran. J. Econ. Agric. Dev. (Agricultural Science and Technology), 23(2):95-102

karimi N (2007). Abstract of articles of national pomegranate congress in Neyriz city. J. Agric. Econ. 4(1):77-92.

Kizilaslan H, Elmali O (2012). Grape production costs and marketing margins in turkey. J. New Results. Sci. 1:26-32

17

Effects of planting depth on agronomic performance of two potato varieties grown in the Sahel region of Tunisia

Sayed CHEHAIBI, Wissem HAMDI and Khaoula ABROUGUI

High Institute of Agronomy, Chott Meriem 4042, Sousse, Tunisia.

Possibilities to enhance potato productivity in the Sahel region of Tunisia have been the domain of active investigation during the last several years. Many field experiments were carried out on the potential of using different plating depths to grow profitable potato crops. In the research station of the higher institute of agronomy of Chott-Mariem in the Sahel region of Tunisia, two varieties of potato: Alaska and Safrane, were mechanically planted at two different depths: 10 and 15 cm in a sandy loam soil. After three months of growth, the effect of planting depth was determined using fresh and dry weights of the above and below ground biomass as well as tuber yield. The results showed that in a light textured soil, plants of two potato varieties planted at 15 cm depth outperformed those planted at a 10 cm depth. In addition, for tuber yield, the Alaska variety was more productive than the Safrane variety with increasing yield of tubers than 2% for depths of 15 cm and 11, 9% for depths of 10 cm.

Key words: Potato, planting depth, light soil, agronomic parameters, Tunisia.

INTRODUCTION

The potato (*Solanum tuberosum* L.) belongs to the *Solanaceae* family, genus *Solanum* and is one of the most important food crops in Tunisia, as well as many other countries. It is a perennial that spreads by vegetative propagation and is grown as an annual species (Graham et al., 2001). However, tuber yield can be improved by choosing appropriate varieties and more rational cultivation techniques. Indeed, to obtain a homogeneous culture, the tubers should be distributed evenly with a specific spacing between rows and a uniform planting depth (Baarveld et al., 2002). Shallow planting is preferred in wet and heavy soils because in such soils deep planting of the tubers may lead to exhaustion of stored food before the sprouts emerge above the soil. Conversely, in light textured soils, where there is a risk of dehydration due to moisture stress, deep planting is essential. Deep planting has also advantage over shallow planting where temperatures are high.

In addition, Lambion et al. (2006) advocated that deeper planting may also limit the damage to tubers by certain pests. In addition to planting depth, crop yield is also influenced by variety, soil, temperature (Gopal et al., 1998), photoperiod (Pruski et al., 2001), light intensity (Gopal et al., 1998) nitrogen nutrition (Etemad and Sarajuoghi 2012) potassium (Naik and Sarkar, 1998) planting density, plant height, the number of stems on the surface, number of tubers formed by size. That depth of planting can be achieved through well-regulated machinery taking out a regular spacing of rows, a regular distribution of tubers in a planting depth and uniform coverage of land by ridging was reported by Haider et al. (2012). In the context of studies of the interaction of soil plant machine, we have found it useful to study the effects of mechanical planting depth on agronomic performance of two potato varieties grown in a sandy loam soil in the Chott Mariem region.

MATERIALS AND METHODS

Experimental site

The experiments were conducted from February to May in the research station of the agronomic height institute of Chott-Mariem (35° 54' N 10° 36' E), Sousse, Tunisia. The climate is Mediterranean, and is characterized by hot dry summers and moderate wet winters, an average annual precipitation of 230 mm and an annual mean temperature of 18.5°C. The area is located in an area under drip-irrigation, characterized by a sandy loam soil (52% silt, 29% sand and 19% clay). Previously, the soil was plowed to 28 cm depth with a moldboard plow and followed by plowing with a disc harrow and cultivar teeth.

Treatments and experimental design

The treatments consisted of two varieties (Safrane and Alaska) and two planting depth [T1 (10 cm) and T2 (15 cm)]. The experiment was laid out as a randomized complete block. The plot size was (30 x 50) m^2. Each plot consisted of 36 rows. Each row accommodated 178 plants. Plants in the central rows were used for determination of agronomic performance, leaving aside those in the two border rows as well as those at both ends of each row to avoid edge effects. Tubers of the two different potato varieties were planted at the spacing of 80 cm between rows and 28 between plants. For each depth the plants samples were allocated randomly to the treatments and the analysis was repeated three times for each sample. Therefore each depth was represented by 27 plants at 9 plants per elementary for each plot. For each variety, the effect of planting depth on agronomic parameters of the plant was considered at three growth stages (Stages 1, 2 and 3) corresponding, respectively to 60, 90 and 110 days.

Statistic

The statistical analyses were implementing with XLSTAT (2012). The analysis of variance ANOVA and post-hoc multiple comparisons Turkey's test ($P < 0.05$) was used to examine the effects of depth and variety on the variation of agronomic parameters.

Fertilizer and pesticide treatments

Before planting each plot received 12.5 t/ha manure, 50 kg/ha triple superphosphate and 150 kg/ha potassium sulfate. After planting (one month later), an additional 150 kg/ha of ammonium nitrate was added. In addition, maintenance operations were conducted on plants at the Unilax (250 ha), at Gelben (250 g/hl) and C-Tattoo (150 cc/hl). Finally, the potatoes are manually picked from the field.

Measured parameters

Measurements on plants were recorded for above-ground (stem + leaf) and below-ground (root + stolon) fresh and dry weights. The weights were determined for three representative plants in each plot of each treatment. This was done three times during the growing cycle; at 60, 90 and 110 days after planting.

After determining the fresh weight of above- and below-ground portions, they plants samples were placed in an oven at 80°C for 24 h to determine dry weight. Tuber yield was determined at harvest. This measurement was made for the different plots by determining the average weight of 27 potato plants of each treatment at a rate of 9 plants per repetition.

RESULTS AND DISCUSSION

Fresh weight of aerial parts

From Table 1 it can be revealed that in sandy loam soil, the fresh above ground biomass increased between 60 and 90 days, and then decreased at 110 days after planting. Indeed, until the 90th day after planting, the plant formed its vegetation, and this fresh weight increased with time. In addition, after approximately 90 days, the photosynthetic products apparently began partitioning from the aerial portion to the tubers, which may explain the drop in fresh above-ground biomass between 90 and 110 days in both varieties.

It can be noted that, for both varieties, the best performance was obtained with plants planted at 15 cm depth. Indeed, after 90 days, fresh above ground biomass for potatoes planted at 15 cm depth was 29.9 and 10.7% greater than those planted at 10 cm depth for the Alaska and Safrane varieties, respectively. It seems that in light textured soil, planting deeper allows the crop to get more food reserves and water (McEwan, 2012) encouraging development and increasing fresh weight.

Dry weight of aerial parts

From Table 1, it can be seen that the dry above ground biomass for both varieties reached maximum approximately 90 days after planting and then began to decrease. This reduction in dry weight may be largely due to the effect of the state of ripening of the plant where a reduction in leaf area directly affects photosynthesis and, subsequently, carbon accumulation.

The latter in turn affects the rate of accumulation of dry matter in the various aerial organs. Moreover, it can be deduce that the dry weight of aerial parts evolved in the same direction as that of fresh weight with a better yield obtained with plants grown at 15 cm depth. Indeed, 90 days after planting, the dry weight recorded for plants grown at the depth of 15 cm depth was 50.6 and 35% greater than those grown at the depth of 10 cm for the Alaska and Safrane varieties, respectively.

Fresh below-ground biomass

The below ground portion is the most important part of the plant since it is the tubers that provide the nutritional value of potato (Le et al., 2002). It also bears many thin and fibrous roots that can penetrate deep into the soil if it

Table 1. Measured parameters.

Variable	Depth (cm)	Safrane variety			Alaska variety		
		Days after planting			Days after planting		
		60	90	110	60	90	110
Fresh weight of aerial parts (g/plant)	15	88.7 ± 2.8	215.8 ± 6	101.8 ± 7.6	202.8 ± 8.1	232.7 ± 9.8	98.7 ± 4.6
	10	84.3 ± 2.5	192.6 ± 4	79.6 ± 5.6	150.8 ± 7	162.9 ± 5.9	83 ± 5.5
Dry weight of aerial parts (g/plant)	15	13.9 ± 1.1	64 ± 3.9	39.4 ± 3.2	16.6 ± 2.6	66.3 ± 5.6	19 ± 2
	10	12.83 ± 0.9	41.6 ± 4	23.1 ± 1.9	12.8 ± 1.3	32.7 ± 4.1	15 ± 1.4
Fresh below-ground biomass (g/plant)	15	12.5 ± 1.7	19.1 ± 3	12 ± 1.8	17.7 ± 2.4	24 ± 3.6	18.5 ± 1.7
	10	11.22 ± 1.5	15 ± 2.4	11 ± 1.7	14.9 ± 1.8	17.5 ± 2.4	15.7 ± 1.5
Dry weight of under-ground parts (g/plant)	15	1.85 ± 0.04	3.16 ± 0.08	1.8 ± 0.03	2.92 ± 0.08	3.25 ± 0.86	1.9 ± 0.56
	10	1.76 ± 0.07	2.6 ± 0.05	1.7 ± 0.05	2.9 ± 0.1	2.82 ± 0.53	1.7 ± 0.78

is sufficiently soft (Soltner, 2005). From Table 1, the fresh below-ground (roots and tubers) biomass increase until 90th days after planting in both varieties.

It is similar to the aerial part, which allows us to deduce a similar correlation between the increases in fresh weight of aerial part and the underground part. This can be explained by the importance of metabolic processes at the plant and the immigration of the aerial part of assimilates to the underground.

In addition, the results of Table 1 show a steady increase in fresh weight for both depths of planting. However, plants grown at 15 cm depth are characterized by the greatest fresh weight for all stages. Indeed, for the variety Alaska, 90 days after planting, the below ground biomass grown at 15 cm depth had an average weight of 24 g/plant *vs.* 17.5 g /plant for crops at established at 10 cm depth, an increase of more than 33%. The fresh weight was lower for the Safrane variety where mean below ground biomass was 19.1 g/plant and 15 g/plant for 15 and 10 cm depths, respectively. It seems that in this light textured soil, planting depth was important to allow plants to get more food reserves and to promote their development and therefore yield higher fresh weights. Also, on the conditions of root growth which depend essentially on the structural state of topsoil and the presence of zones of compaction due to tillage in wet weather.

Dry weight of underground parts

From the results of Table 1, it can be deduced that the average weight of the underground part significantly ($P < 0.05$) increased during the crop cycle. In the early stages of development, this growth was rapid, peaked at about the 90th day after planting and then decreased. This was for the two planting depths tested. Indeed, for the Safrane variety for planting at 10 cm depth the lowest fresh weight was 2.6 and 1.7 g/plant, respectively 90 and 110 days after planting. A 15 cm depth, the weights were only 3.16 and 1.7 g/plant, respectively 90 and 110 days after planting.

For the Alaska variety, and for the same dates (90 and 110 days after planting) dry weights corresponded to 2.82 g/plant and 1.7 g/plant for the 10 cm depth and 3.25 g/plant and 1.9 g/plant for the 15 cm depths. This may be due to the change in reserve accumulation that starts fast but gradually decreases as the plant grows (Muhinyuza et al., 2008). Consistent with the results of this study, Marofi et al. (2013) and Snapp and Kravchenko (2010) reported that relative decreases in dry weights of underground parts are common after senescence of the plants. As measured by the variance at the 5%, we determined a significant difference in growth between the two varieties of potato. This difference recorded between the two varieties may be due to genetic effect as explained by the number of stems per unit area that emerged, and subsequently, increasing the size of plants or that of planting density, which generally leads to an increase in the average weight (Memari et al., 2011).

Tuber yield

According to Figure 1, both varieties produced the highest tuber yields when planted at 15 cm depth, 17.7 and 26% greater than those planted at 10 cm depth for the Alaska and Safran varieties, respectively. It appears that there is considerable interaction between the level of release of tubers in the soil and yield of the plant. Indeed, a major planting deepest germinates more slowly but their roots grow best laterally and at depth and make the most of available water and nutrients. By contrast, at shallower planting depths, plant biomass was relatively

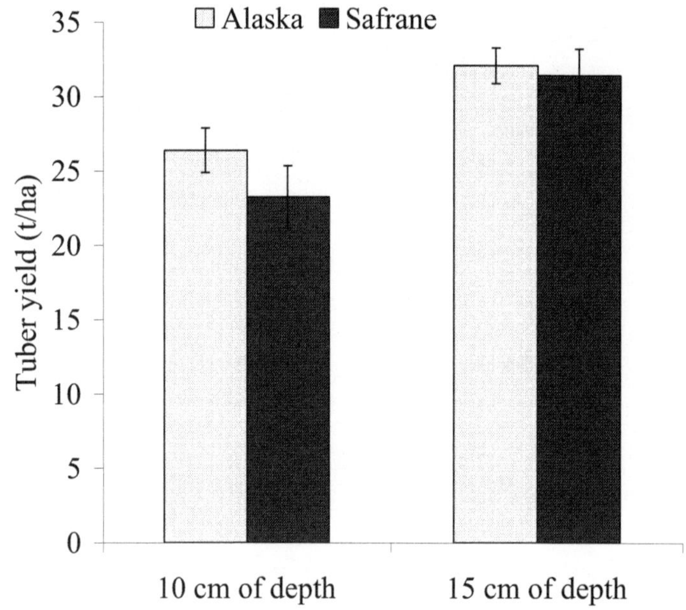

Figure 1. Tuber yield (t/ha).

less developed, leading to a lower tuber yield. Tuber weight depends primarily on the operating time of the leaf canopy (Snapp and Kravchenko, 2010), but also the conditions of operation and conditions of root growth. Studies of Abdulla et al. (1993) and Bohl and Love (2005) showed that potato yields increased when planting depth increased, but with a small tuber greening. Pavek and Thomotonr (2009) showed, in a study on the effects of planting depth, the performance of two commercial varieties of potatoes were lower for the reduced depth.

Similarly, Tamia et al. (1999) have indicated that in very compact horizons, root density can be higher than in the packed horizons, which directly affects performances. However, the complexity of the mechanisms of nutrition and plant growth and structural changes under the influence of climatic factors, biological and mechanical parameters are also of great influence.

Conclusion

At the end of this experimental work aimed to study the effects of planting depth on the mechanical performance of two varieties of potato, it appears that in the light soil, planting potatoes at 15 cm depth promoted better development of both below and above ground plant components as well as higher yields compared with potatoes planted at 10 cm depth. Indeed, plants grown at the shallower depth provided returns of 26.4 and 23.24 t/ha, respectively for the Alaska and Safrane varieties. By contrast, those grown at the greater depths resulted in yields of 32.09 t/ha for the Alaska variety and 31.42 t/ha for the Safrane variety.

From this, it appears that in light textured soil, planting potatoes at 15cm depth promotes better development of both above and below-ground organs of the plant. This may be due to the fact that in a light-textured soil with limited compaction, moisture conditions are typically better, which reduces the risk of dehydration and promotes normal development of roots, especially when the crop cycle coincides with the relativity warm months.

ACKNOWLEDGEMENT

We would like to thank David Pelster, researcher in Soils and Crops Research and Development Centre, Agriculture Canada, for his collaboration to finish this work.

REFERENCES

Abdulla A, Wahdan HM, Wahby MF (1993). Yield and Physical Properties of Potato Tuber as Influenced by Planting depth. J. King Saud Univ. 5:2.

Bohl WH, Love SL (2005). Effect of planting depth and hilling practices on Total, U.S. No. 1, and field greening tuber yields. Am. J. Potato Res. 82(6):441-450.

Etemad B, Sarajuoghi M (2012). Study of the Effect of Different Levels and Application Timing of Nitrogen Fertilizer on Yield and Number of Potato Tuber of Agria in Ghorveh, Iran. Ann. Biol. Res. 3(3):1385.

Gopal J, Minocha JL, Dhaliwal HS (1998). Microtuberization in potato (*Solanum tuberosum* L.). Plant Cell Rep. 17:794-798.

Graham T, Van de Fliert E, Campilan D (2001). What happened to participatory research at the International Potato Center. Agric. Hum. Values 18:429-446.

Haider M, Wasim A, Chaudhary M, Pervez MA, Asad HU, Raza SA, Ashraf I (2012). Impact of foliar application of seaweed extract on growth, yield and quality of potato (*Solanum tuberosum* L.). Soil Environ. 31(2):157.

Lambion J, Toulet A, Traente M (2006). Plant protection cultivation of organic potato, Fact 2: The fight against pests. Tech. Inst. Org. Agric. Paris-France P. 4.

Le CL, Thomas D, Nowbuth L (2002). Conservation of potato in vitro and characterization of cultivated varieties in Switzerland. Switzerland Agric. 34(3):133-136.

Marofi S, Parsafar N, Rahim G, Dashti F, Marofi H (2013). The effects of wastewater reuse on potato growth properties under greenhouse lysimeteric condition. Int. J. Environ. Sci. Technol. 10(1):133.

McEwan G (2012). Flooding means losses for many potato growers. Hortic. Week P. 36.

Memari T, Elnaz F, Yarnia, F, Noshin AV (2011). Effect of different irrigation level on yield potato 2011. Ann. Biol. 2(6):269.

Muhinyuza JB, Nyiransengiyumva S, Nshimiyimana JC, Kirk W (2008). The effect of the application frequency and dose of mancozeb on the management of potato late blight in Rwanda." J. Appl. Biosci. 3:76-81.

Naik PS, Sarkar D (1998). Effect of potassium on the *in vitro* potato microtuber production. Biol. Plantarum 41(1):121-125.

Pruski KP, Duplessis T, Lewis T, Astatkie J, Nowak PC (2001). Jasmonate effect on *in vitro* tuberization of Potato (*Solanum tuberosum* L.) cultivars under light and dark conditions. J. Potato Res. 44:315-325.

Pavek MJ, Thorntonr RE (2009). Planting Depth Influences Potato Plant Morphology and Economic Value. American journal potato research, ISSN 1099209X CODEN AJPRFQ. 86(1):56-67.

Snapp E, Kravchenko A (2010). Potato Yield Variability across the Landscape. Agron. J. 102(3):885.

Tamia A, Moreau R, Fortier M, Yoro G (1999). Influence of tillage on the physical evolution of a ferraltic forest soil after motor clearing and Soil Management. 6(1):27-39.

XLSTAT (2012). Running a partial least squares (PLS) discriminant analysis with XLSTAT-PLS. New York, NY. Addinsoft c1995-2012. Available at http://www.xlstat.com/en/learning-center/tutorials/running-a-partial-least-square-pls-discriminant-analysis-with-xlstat-pls.html# (accessed 12 June 2012).

18

The implications of refuge requirements for Bt cotton in India on world cotton markets

Rohit Singla[1] , Phillip Johnson[2] and Sukant Misra[2]

[1]Department of Agricultural Economics, McGill University, Montreal, QC, Canada.
[2]Department of Agricultural and Applied Economics, Texas Tech University, Lubbock, TX, U.S.A

The study evaluated the potential impacts of refuge requirements for Bt cotton in India on world cotton markets. The objective was accomplished by estimating regional cotton yield functions in the Indian fiber model, and connecting it with rest of the world fiber model. Results revealed that the refuge requirements have potential to impact world cotton markets; magnitudes of impacts are smaller, however. The world cotton trade would be lower, and the cotton prices would be higher under higher refuge requirements, and vice versa. As the refuge requirements for an insecticidal Bt technology depend on its replacement rate, the time spent in R&D and regulation of the technology has implications for Bt crops refuge requirements and world commodity markets.

Key words: Bt cotton, India, refuge requirements, world cotton markets, world fiber model.

INTRODUCTION

A large number of studies reported that insect-resistant Bt crops (such as Bt cotton and Bt corn) have led to significant productivity gains, reductions in insecticide use, or both throughout the world. Some examples of these studies describing the farm-level impacts of Bt crops are: James (2009), Bennett et al. (2004, 2006), Purcell and Perlak (2004), Huang et al. (2002), Qaim (2003), Yorobe and Quicoy (2006), and Brookes and Barfoot (2007). Other than these farm-level studies, there are studies (Pekaric-Falak et al., 2001; Brookes et al., 2010; Frisvold and Reeves, 2007; Frisvold et al., 2006; Elbehri and Macdonald, 2004; Anderson et al., 2008; Falck Zepeda et al., 2000) that examined the trade, price and welfare effects of Bt crops adoption on world markets. These studies reported an increase in agricultural trade, a reduction in prices, and an increase in welfare of people all over the world after commercialization of Bt crops.

Despite the aforementioned positive impacts of Bt crops, one of the primary concerns in adopting Bt crops is the potential resistance by insects to the Bt toxin present in the Bt crop (Shelton et al., 2000). To address this concern, Environmental Protection Agency (EPA) established a mandate requiring Bt growers to grow a proportion of non-Bt refuge[1] along with Bt crop. The mandate provides farmers choice of a sprayed refuge option and an unsprayed refuge option. With sprayed refuge option, growers may plant up to 80% of their total acres to Bt varieties and at least 20% to non-Bt varieties and they are allowed to use conventional pesticide throughout. With the non-sprayed option, producers may plant 95% of their acres to Bt varieties, and spray Bt acres as needed with conventional insecticides; however, no insecticide may be used on 5% of refuge acres.

There are some studies (Livingston et al., 2004; Qiao et al., 2009, 2010; Singla et al., 2012; Hurley et al., 2001) that challenged the efficiency of EPA's universal mandate

[1]Refuges allow susceptible pests to thrive so they can mate with resistant pests that survive in the Bt crop fields thus extending the efficacy of the insect-resistant varieties.

on refuge requirements for Bt cotton and Bt corn. These studies provided some evidences of sustainability of productivity effects of Bt crops under a scenario of potential resistance development by various pests to the Bt toxin. Livingston et al. (2004), Qiao et al. (2009, 2010), and Singla et al. (2012) examined the refuge requirements for Bt cotton in the U.S., China, and India, respectively, in a bio-economic modeling framework. Livingston et al. (2004) found optimal structured refuges of 16%[2] for eleven year planning horizon for the U.S. cotton. Qiao et al. (2009, 2010) findings supported a 'zero refuge' policy for Bt cotton in China. Singla et al. (2012) found optimal refuge requirements of 42, 19, and 0% for North, Central, and South India, respectively, for a 15-year time horizon. Hurley et al. (2001) also examined Bt corn refuge requirements in the U.S. by employing a bio-economic model; they recommended optimal refuge requirements between 20 and 40%. All the refuge requirements discussed previously, however, were found sensitive to some biological parameters used in the model.

Frisvold and Reeves (2008) examined that any mandates on refuge requirements have potential to decrease the production and profitability of cotton in the short run because of lower yields of cotton planted in refuge area. Their study, however, was at micro level and did not estimate how a change in refuge requirements could potentially affect the global trade and prices. The current study contributes to the literature by estimating the impact of various refuge requirements for Bt cotton in India on world trade and prices. It is important to examine the impact of refuge requirements In India as it is one of the largest[3] producers of cotton in the world, and changes in proportion of area under Bt and non-Bt cotton has potential to alter world cotton trade flows and prices. We are considering only the impact of change in refuge requirements in India because the governments of two other major cotton producing countries, that is, the U.S. and China have already announced zero refuge policies for Bt cotton. The zero-refuge requirement for Bt cotton China is based on the idea that the abundant non-Bt host plants of the target pest provide sufficient natural refuges to delay resistance in the pest (Qiao et al., 2010). In case of the U.S., the stacked Bt varieties have replaced a significant portion of single-gene Bt varieties (USDA, 2012), and there is zero refuge requirements for stacked varieties (EPA, 2012). There is relatively a small portion of area left under single-gene Bt varieties, and it is likely to be replaced by stacked varieties in the near future. So, virtually, there are no refuge requirements for Bt cotton in China and the U.S. This article, therefore, evaluates trade and price impact of Bt cotton refuge requirements in India only.

There is no study that examined the potential impact of refuge requirements for Bt crops on world markets. However, there are many studies[4] examining the trade, price and welfare effects of Bt crops adoption on world markets. The studies by (Falck Zepeda et al., 2000; Frisvold et al., 2006; Pekaric-Falak et al., 2001) examined the trade, price and welfare effects of Bt cotton using partial equilibrium model. Brookes et al.. (2010) used similar methods to examine the production and price impact of biotech corn, canola, and soybean crops. The studies conducted by Pekaric-Falak et al., (2001) and Frisvold and Reeves (2007), however, employed general equilibrium modeling framework to estimate the economy-wide impact of Bt cotton on trade, prices and welfare. Elbehri and Macdonald (2004) also used general equilibrium framework to examine the trade and price impact of Bt corn.

The current study examines the potential impact of various refuge requirements for Bt cotton in India on world cotton markets by estimating the regional cotton yield models in the Indian fiber model, and then connecting the Indian model to rest of the world fiber model. The specific objectives of this study are (1) to estimate the cotton yield models for the three cotton growing regions in India (2) to estimate and compare the trade and price impacts under status quo; under efficient Bt cotton refuge requirements in India for 10- and 15-years planning horizons; and the refuge requirements mandated by EPA.

CONCEPTUAL MODEL

The conceptual analysis presented here provides the expected directional change in the world fiber market, with a change in refuge requirements in India. It can be hypothesized that increased refuge requirements for Bt cotton varieties in India would decrease the world cotton supply because of lower yield of cotton planted in refuges. A decrease in world cotton supply could potentially increase world cotton prices, ceteris paribus. Given that the demand for cotton is rising rapidly in India after the elimination of import quotas under the Multi-

[2]Optimal/efficient refuge requirements vary with the length of planning horizon, that is, number of years it would take for a new technology to replace an existing technology. Here, a refuge requirement of 16% is based on the assumption that a new technology (such as stacked varieties) would replace the existing single-gene Bt technology in 11 years. Lower refuge requirements will be required if a technology get replaced earlier, and vice-versa. EPA has announced zero structured refuge requirements for stacked Bt cotton varieties in the U.S. in 2007. Since then the adoption of stacked varieties increased. They cover 63% of total cotton area in 2012 (USDA, 2012).

[3]India accounted for about 23% of world cotton production from one-third of world's cotton acreage it possesses (USDA, 2009). Once upon a time a net importer of cotton, India is now the second largest exporter of cotton in the world after the U.S. (National Cotton Council, 2009; USDA, 2009). The introduction of Bt cotton in India in 2002-03 is considered as the primary reason of India's transition from a net importer to a leading exporter of cotton (James, 2009; Choudhary and Gaur, 2010).

[4]Bt adoption and refuges go hand in hand because of a presence of complementarity between them. An increase in area under Bt crops decreases refuge area and vice-versa. Methods used to examine the trade and price impact of Bt crops adoption can be used to examine the impact of planting refuges.

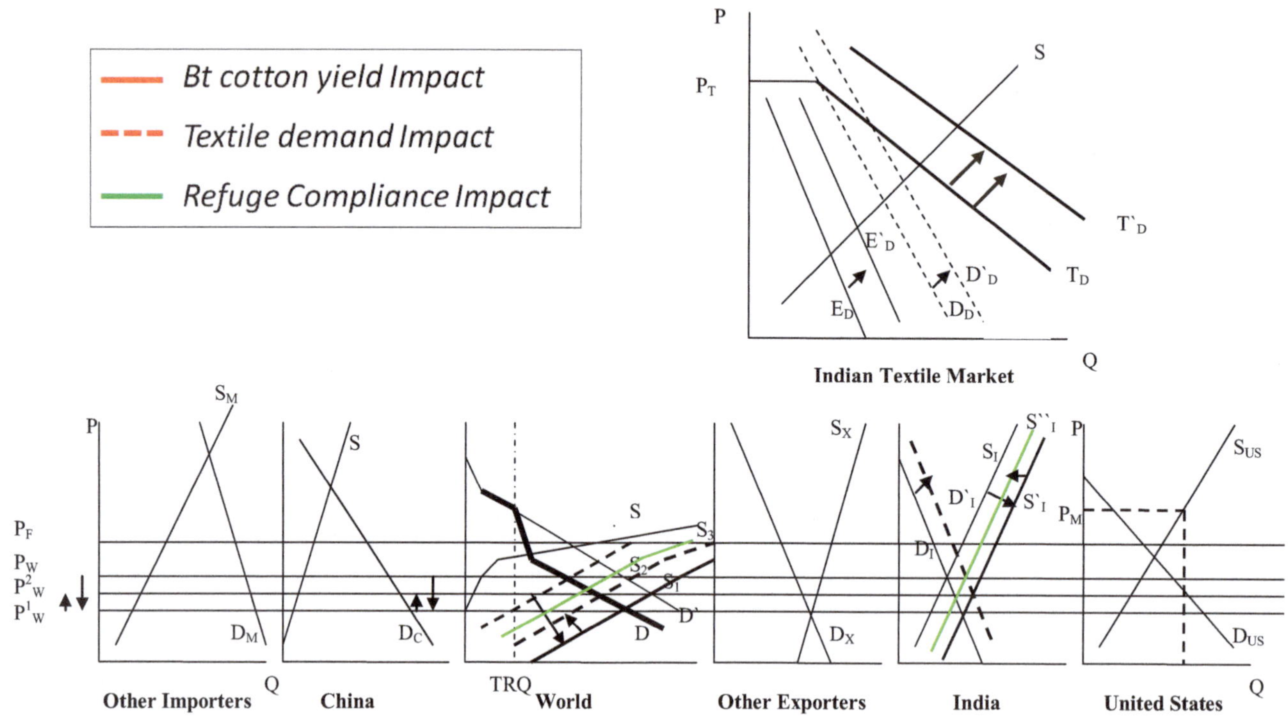

Figure 1. Impacts of refuge requirements on the U.S., India and world cotton markets

Fiber Arrangement (MFA)[5], a decrease in supply could have implications for the future trade flows of cotton.

A partial equilibrium analysis of a hypothesized cotton trade scenario including India, the U.S., China and the rest of the world (ROW) cotton importing/exporting countries is presented in Figure 1. India and the U.S. are presented as net cotton-exporting countries, implying that domestic supply is greater than the domestic demand for cotton. China is assumed to be a net cotton-importing country.

The conceptual analysis shows that the world price is P_W after considering the Chinese Tariff Rate Quota (TRQ) and the U.S. marketing loan program. The free trade price is shown as P_F. The conceptual model suggests that an increase in the supply of raw cotton in India (as a result of increased adoption of Bt cotton) would shift the supply curve from S_I to $S`_I$, which would shift the excess supply curve upward in the world cotton market from S to S_1. This should result in a decrease in world price from P_W to P^1_W and an increase in the quantity traded. It can be inferred that an increase in the world supply of cotton does not necessarily translate into sustained higher revenues/profits for adopters of Bt cotton as prices for cotton could fall worldwide (Bennett et al., 2004; Huang et al., 2002), provided there is not a concurrent increase in demand.

The rising domestic demand for textiles in India because of an increased standard of living in recent years, coupled with increased exports of cotton-based textiles associated with the elimination of import quotas under the multi-fiber arrangement (MFA), could increase demand for domestic and imported cotton in India. This is represented by the total demand for textiles increasing from T_D to $T`_D$ in Figure 1. Due to this increase in demand for textiles, the derived demand for cotton in India is expected to increase from $D_I$ to $D`_I$. This would result in a decrease of the excess supply in the world market from *S1* to *S2*, and an increase in the price from P^1_W to P^2_W, and a decrease in the quantity traded.

With an increase in refuge requirement, the supply of cotton is expected to decrease because of lower yield of cotton planted in refuges. In Figure 1, this is represented by a shift in the supply of cotton in the Indian market from $S`_I$ to $S``_I$. A decrease in supply would shift the excess supply curve to the left to S_3, resulting in a world price between P_W and P^2_W. Nevertheless, the net change in world price and trade is an empirical question and can only be determined by the various elasticities of demand and supply involved (Landes et al., 2005).

METHODS

The conceptual analysis suggested that the expected impact of increase in refuge requirements would be to alter cotton trade flows and increase world prices of cotton. The empirical model allows testing this hypothesis as well as the estimation of the magnitude of the change in price and trade flows. This is achieved by estimating regional cotton yield response models in the Indian fiber model

[5]MFA governed the world trade in textiles and garments from 1974 to 2004. MFA imposed quotas on the amount of textiles and garments developing countries could export to developed countries. It expired on 1 January 2005.

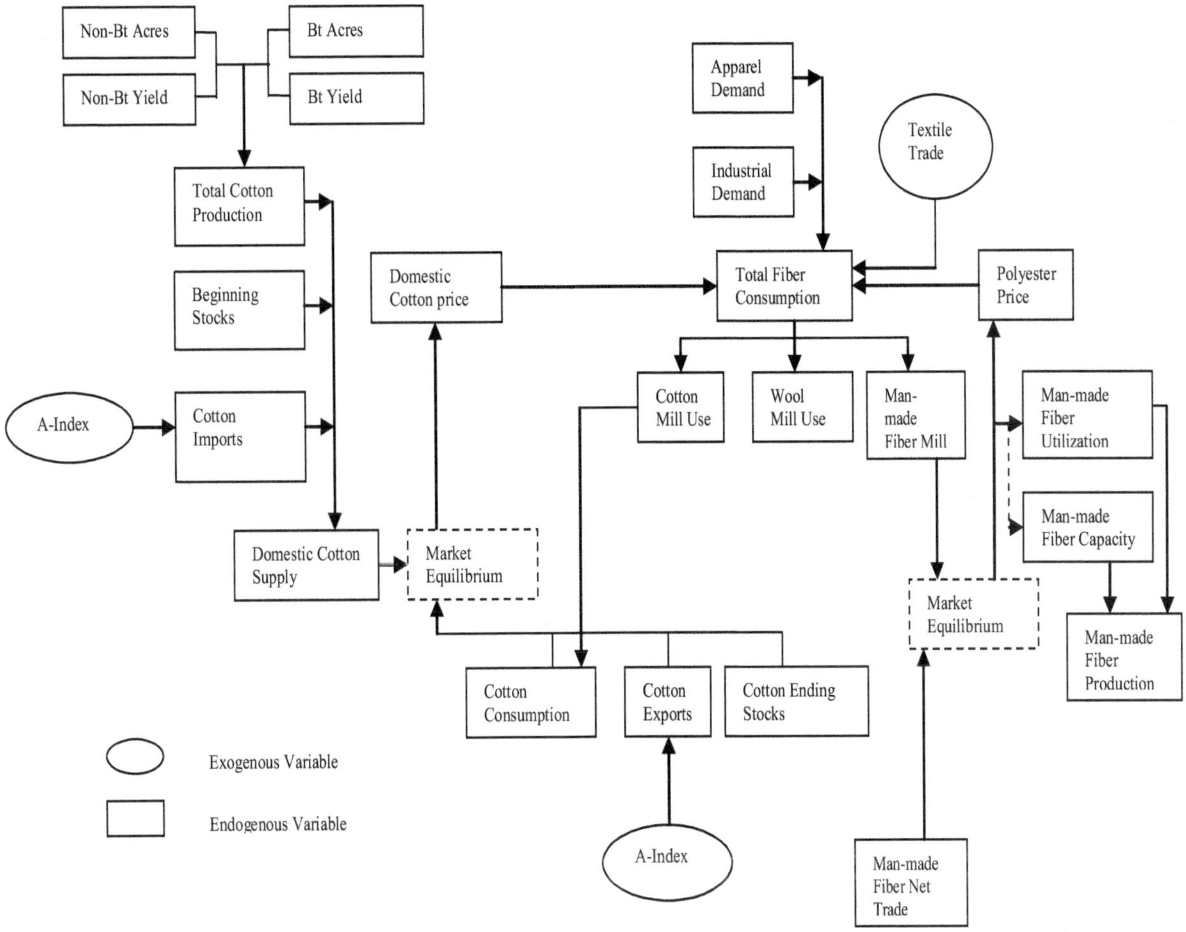

Figure 2. Indian fiber model.

component of the WFM[6], where proportion of area under Bt varieties is considered among one of the exogenous factors. The calibrated Indian Fiber model is then connected to the WFM to simulate the potential impact of various refuge requirements on world cotton markets.

To estimate the cotton yield models, cotton area in India is first divided among three regions based on differences in agro-climatic conditions across regions (Chaudhary et al., 2008; Chaudhary, 2005). Then cotton yield models for the three regions in India were estimated as a function of proportion of Bt area after controlling for other regional factors. The cotton yield model for j^{th} region can be specified as follows:

$$Y_{j,t} = f\,(Y_{j,t-1},\, q_{j,t},\, FU_{j,t},\, Irri_{j,t},\, t)$$ [7]

where $Y_{j,t}$ represents cotton yield in the jth region at time t; $Y_{j,t-1}$ is lagged cotton yield in the jth region; $q_{j,t}$ is the proportion of refuge a farmer is growing in the jth region at time t; $Irri_{j,t}$ is the area under irrigation for the jth region at time t, $FU_{j,t}$ is the fertilizer use in the jth region at time t; and t is the time trend. The cotton yield models were used to estimate the Bt and non-Bt cotton yields that along with the respective acreages under Bt and non-Bt cotton determine cotton supply in the Indian fiber model as shown in Figure 2. As shown in the Figure, Bt and non-Bt cotton areas and yields contribute to total cotton production in India after accounting for beginning stocks and imports. The model also takes into account competition among fibers such as cotton, man-made fibers, and wool at the mill level.

The baseline projections for supply, demand, and prices of cotton, man-made fibers, and textiles in the Indian fiber market were generated under status quo conditions where the acreage under Bt cotton is expected to rise. The baseline projections assume the continuation of current policies such as China's TRQ and U.S. marketing loan program for cotton. The policy scenario projections were made by shocking the Indian Fiber model with efficient refuge requirements for Bt and non-Bt cotton for 15- and 10-years planning horizons as examined by Singla et al. (2012), and the refuge requirements mandated by EPA. The refuge requirements for a 15-year planning horizon were 42%, 19% and 0% for North, Central and South India, respectively. The refuge requirements for a 10-year planning horizon, however, were 29%, 4% and 0% for North, Central and South India, respectively. The EPA refuge requirements used in the model were for those mandated for sprayed refuges, which is 20% for all the three cotton growing regions. Both baseline and policy scenario projections were developed for a 14-year time period beginning in 2012-2013 and ending in 2025-2026. The Indian Fiber model was connected

[6] WFM is a partial equilibrium structural econometric model developed by Pan and Mohanty (2004). A brief description of the model along with data used is given in Appendix A. Two major applications of world fiber model are presented in Chaudhary et al (2008) and Pan et al (2007).

[7] The regional time-series data (on cotton yield and other factors) used to estimate this model were obtained from Indiastat.com.

Table 1. Regression estimates of regional cotton yield models in India.

Independent variable	North	Central	South
Intercept	144.05 (90.01)	2212.54 (1059.31)*	268.99 (19.87)***
$q_{j,t}$	249.67 (246.69)	243.05 (64.42)***	204.55 (55.82)**
$Y_{j,t-1}$	0.62 (0.24)**	-	-
$FU_{j,t}$	-	-6.52 (3.11)*	-
$Irri_{j,t}$	-	-11439 (5986.81)*	-
$FU_{j,t}$* $Irri_{j,t}$	-	38.37 (17.25)**	-
t	-	-	9.27 (2.21)***
R^2	0.54	0.89	0.86
DW statistic	1.80	1.92	2.02
Number of observations	18	18	18

Figures in the parentheses are the standard errors. *, ** and *** indicate significance at 10, 5 and 1% levels, respectively.

to the World Fiber model to estimate the impacts on the world and the U.S. cotton markets. The policy effects were measured by comparing the differences between the policy scenario and baseline projections.

RESULTS AND DISCUSSION

The estimated coefficients of cotton yield equations for the three regions are presented in Table 1. The coefficient of proportion of area under Bt cotton ($q_{j,t}$) represents the yield difference between Bt and non-Bt cotton; and it was found to be statistically significant in Central and South India after controlling for several factors such as fertilizers, irrigation, time trend and lagged yield. The $q_{j,t}$ was not statistically significant for North India due to lack of a sufficient number of observations for Bt cotton because of its late adoption. The coefficients of determination (R^2) for the North, Central, and South regions were 0.54, 0.89 and 0.86, respectively. A low value of R^2 in the North region may be due to erratic monsoon rainfall and high weather variability, which was not being captured by the model.

The estimated regression coefficients were used in estimating cotton production and supply in the Indian fiber model, which was further connected to WFM to develop baseline projections of the potential impacts on world cotton trade and prices assuming current trend in area under Bt cotton (and refuge). Also the baseline projections assumed a continuation of current policies including MFA quotas elimination and China's TRQ, which are based on World Trade Organization (WTO) commitments. The projections for fiber demand, supply and prices were developed for 14-year period under a set of assumptions for exogenous variables. After developing the baseline, three alternate scenario projections were developed for three different levels of refuge requirements. These include regional refuge requirements under 15-year planning period (Scenario 1), under 10-year[8] planning period (Scenario 2), and those under EPA mandate (Scenario 3). Baseline and scenarios projections were made for India, the U.S., and world cotton markets as shown in Tables 2, 3 and 4, respectively. Average trade and price effects under the three scenarios are compared in Figures 3 and 4.

As shown in Table 2, average cotton yield in India would decline with an increase in refuge requirements for Bt cotton in India. Yield would decrease by an average of 1.89, 0.86 and 1.04%, for refuge requirements under 15-year planning period, under 10-year planning period, and under EPA mandate, respectively. Average reduction in cotton yield under 15-year refuge requirements was the highest followed by EPA and 10-year refuge requirements. Cotton yield reductions were found to be larger than cotton area expansions thus resulting into a decreased cotton production under the three refuge requirements scenarios. A decreased cotton production led to lower mill use, lower exports and lower ending stock on average. Negative impacts on mill use, exports and ending stock were highest under 15-year refuge requirements followed by EPA and 10-year refuge requirements. Impacts of various refuge requirements on the U.S. cotton market are reported in Table 3. Farm price of cotton in the U.S. would increase on average by 0.48% under 15 year refuge requirements, 0.18% under 10-year refuge requirements, and 0.29% under refuge requirements mandated by EPA. Cotton production would also increase in the U.S. because of expectations of expansion and yield increments. Average mill use was almost unaffected under the three refuge requirements scenarios. Net cotton exports from the U.S. would rise on average because of a decrease in exports from India.

[8]A long term planning horizon assumes a longer period for a new technology to replace existing technology. A longer time in technology replacement corresponds to higher refuge requirements.

Table 2. Estimated effects of efficient Bt cotton refuge policy compliance in India on Indian cotton market.

Parameter	2012/2013	2013/2014	2014/2015	2015/2016	2016/2017	2017/2018	2018/2019	2019/2020	2020/2021	2021/2022	2022/2023	2023/2024	2024/2025	2025/2026	Avg.
Area								**000 Acres**							
Baseline	29776.03	28559.67	28037.13	28836.89	29923.74	31544.69	31679.84	31836.54	31944.74	32266.15	32413.36	33769.29	33776.36	33907.84	31305.16
Scenario 1 (%)	0.00	0.26	0.46	0.73	1.14	1.45	1.78	1.47	1.15	0.87	0.85	0.38	0.37	-0.14	0.77
Scenario 2 (%)	0.00	0.10	0.19	0.35	0.59	0.79	0.99	0.78	0.55	0.38	0.35	0.05	0.04	-0.28	0.35
Scenario 3 (%)	0.00	0.16	0.27	0.40	0.58	0.71	0.85	0.73	0.60	0.48	0.48	0.29	0.28	0.07	0.42
Yield								**Bales/Acres**							
Baseline	0.94	0.99	1.01	1.02	1.02	1.02	1.03	1.05	1.07	1.11	1.11	1.13	1.14	1.15	1.06
Scenario 1 (%)	-1.44	-1.71	-2.27	-3.10	-3.58	-3.89	-3.08	-2.25	-1.68	-1.60	-1.01	-0.76	-0.20	0.12	-1.89
Scenario 2 (%)	-0.55	-0.72	-1.10	-1.64	-1.96	-2.17	-1.64	-1.08	-0.72	-0.66	-0.27	-0.13	0.23	0.43	-0.86
Scenario 3 (%)	-0.88	-0.99	-1.22	-1.56	-1.75	-1.88	-1.53	-1.19	-0.95	-0.90	-0.67	-0.56	-0.32	-0.19	-1.04
Production								**000 Bales**							
Baseline	28005.47	28344.79	28190.04	29305.40	30608.35	32159.23	32504.24	33569.34	34077.29	35846.54	36101.56	38069.63	38534.12	38854.32	33155.02
Scenario 1 (%)	-1.44	-1.45	-1.82	-2.39	-2.48	-2.50	-1.36	-0.81	-0.55	-0.74	-0.16	-0.38	0.17	-0.02	-1.14
Scenario 2 (%)	-0.55	-0.63	-0.91	-1.30	-1.38	-1.40	-0.67	-0.31	-0.17	-0.29	0.08	-0.08	0.27	0.15	-0.51
Scenario 3 (%)	-0.88	-0.83	-0.95	-1.17	-1.19	-1.18	-0.70	-0.48	-0.35	-0.42	-0.19	-0.27	-0.04	-0.12	-0.63
Mill Use								**000 Bales**							
Baseline	20835.15	21977.75	24313.19	25061.34	25276.15	25520.99	25556.29	26095.42	27046.58	28216.76	29422.71	30958.35	31644.92	32630.71	26754.02
Scenario 1 (%)	-0.14	-0.12	-0.14	-0.19	-0.19	-0.18	-0.04	0.01	0.02	-0.02	0.06	0.01	0.10	0.04	-0.06
Scenario 2 (%)	-0.05	-0.05	-0.07	-0.11	-0.11	-0.11	-0.01	0.02	0.02	0.00	0.05	0.01	0.07	0.03	-0.02
Scenario 3 (%)	-0.08	-0.07	-0.07	-0.09	-0.09	-0.08	-0.02	-0.01	0.00	-0.01	0.02	0.00	0.04	0.01	-0.03
Net Export								**000 Bales**							
Baseline	7217.60	6355.20	6326.55	6025.92	6687.38	7718.51	7844.53	8245.86	7724.98	8253.13	7256.56	7651.99	7414.95	6735.40	7247.04
Scenario 1 (%)	-4.58	-5.72	-7.52	-10.73	-10.70	-9.90	-5.85	-3.54	-2.63	-3.17	-1.22	-1.92	0.37	-0.28	-4.81
Scenario 2 (%)	-1.76	-2.44	-3.69	-5.76	-5.90	-5.54	-2.95	-1.44	-0.89	-1.25	0.10	-0.43	1.06	0.72	-2.15
Scenario 3 (%)	-2.81	-3.30	-3.98	-5.30	-5.15	-4.69	-2.98	-2.02	-1.61	-1.82	-1.10	-1.33	-0.40	-0.74	-2.66
End Stock								**000 Bales**							
Baseline	8504.68	8971.78	6978.06	5651.18	4749.24	4120.11	3672.39	3346.85	3095.87	2912.94	2772.52	2665.89	2570.14	2484.45	4464.01
Scenario 1 (%)	-0.53	-0.75	-1.01	-1.39	-1.60	-1.73	-1.25	-0.85	-0.56	-0.50	-0.13	-0.12	0.24	0.24	-0.71
Scenario 2 (%)	-0.20	-0.32	-0.48	-0.73	-0.87	-0.97	-0.67	-0.41	-0.23	-0.20	0.04	0.03	0.25	0.24	-0.32
Scenario 3 (%)	-0.32	-0.44	-0.55	-0.71	-0.79	-0.83	-0.62	-0.46	-0.32	-0.29	-0.13	-0.12	0.03	0.04	-0.39

Table 3. Estimated effects of efficient Bt cotton refuge policy compliance in India on the U.S. cotton market.

Parameter	2012/2013	2013/2014	2014/2015	2015/2016	2016/2017	2017/2018	2018/2019	2019/2020	2020/2021	2021/2022	2022/2023	2023/2024	2024/2025	2025/2026	Avg.
Farm price								**$/Bale**							
Baseline	75.23	74.53	75.89	77.20	78.02	78.61	78.80	79.09	79.93	79.92	81.01	82.67	85.08	85.63	79.40
Scenario 1 (%)	2.34	1.50	1.77	2.49	2.07	1.78	-0.85	-0.94	-0.70	0.14	-1.29	0.02	-1.44	-0.17	0.48
Scenario 2 (%)	0.90	0.69	1.00	1.47	1.24	1.08	-0.62	-0.73	-0.49	0.01	-0.90	-0.03	-0.96	-0.13	0.18
Scenario 3 (%)	1.43	0.83	0.84	1.12	0.91	0.77	-0.32	-0.33	-0.28	0.09	-0.48	0.04	-0.57	-0.05	0.29
Area								**000 Acres**							
Baseline	9725.79	9550.88	9686.34	9810.76	9886.45	9921.40	9987.91	10201.97	10240.74	10320.14	10384.52	10395.27	10480.64	10549.21	10081.57
Scenario 1 (%)	0.00	0.31	0.20	0.24	0.34	0.29	0.25	-0.10	-0.12	-0.09	0.02	-0.16	0.00	-0.19	0.07
Scenario 2 (%)	0.00	0.12	0.09	0.13	0.20	0.17	0.15	-0.08	-0.10	-0.07	0.00	-0.11	0.00	-0.13	0.03
Scenario 3 (%)	0.00	0.19	0.11	0.11	0.15	0.13	0.11	-0.04	-0.04	-0.04	0.02	-0.06	0.01	-0.08	0.04
Yield								**Bales/Acres**							
Baseline	1.60	1.63	1.63	1.63	1.65	1.66	1.68	1.72	1.72	1.72	1.72	1.73	1.73	1.74	1.68
Scenario 1 (%)	0.00	-0.02	0.14	0.09	0.10	0.15	0.13	0.13	-0.04	-0.05	-0.04	0.02	-0.08	0.01	0.04
Scenario 2 (%)	0.00	-0.01	0.05	0.04	0.06	0.09	0.08	0.08	-0.03	-0.04	-0.03	0.00	-0.06	0.00	0.02
Scenario 3 (%)	0.00	-0.01	0.09	0.05	0.05	0.07	0.06	0.06	-0.01	-0.02	-0.02	0.01	-0.03	0.00	0.02
Production								**000 Bales**							
Baseline	15602.13	15576.02	15813.50	16009.82	16346.16	16471.85	16815.58	17497.11	17568.33	17732.89	17886.74	17954.09	18143.37	18313.88	16980.82
Scenario 1 (%)	0.00	0.29	0.34	0.33	0.44	0.44	0.38	0.02	-0.16	-0.14	-0.02	-0.14	-0.08	-0.19	0.11
Scenario 2 (%)	0.00	0.11	0.15	0.17	0.26	0.26	0.23	0.00	-0.13	-0.11	-0.03	-0.11	-0.06	-0.13	0.04
Scenario 3 (%)	0.00	0.18	0.20	0.16	0.20	0.20	0.17	0.02	-0.05	-0.05	0.00	-0.05	-0.02	-0.07	0.06
Mill Use								**000 Bales**							
Baseline	3640.43	3559.65	3486.49	3472.98	3386.03	3366.43	3284.86	3275.69	3048.80	2918.03	2812.47	2798.62	2763.08	2523.93	3166.96
Scenario 1 (%)	0.01	0.02	0.02	0.03	0.02	-0.04	-0.05	-0.05	-0.07	-0.05	0.00	0.05	0.07	0.14	0.01
Scenario 2 (%)	0.00	0.01	0.01	0.02	0.02	-0.01	-0.02	-0.03	-0.05	-0.04	-0.01	0.02	0.03	0.06	0.00
Scenario 3 (%)	0.00	0.01	0.01	0.01	0.00	-0.03	-0.03	-0.02	-0.02	-0.01	0.01	0.03	0.04	0.08	0.01
Net Export								**000 Bales**							
Baseline	11569.71	12012.94	12308.14	12514.99	12874.97	13158.22	13588.23	14152.76	14582.30	14871.81	15155.91	15251.71	15492.86	15953.27	13820.56
Scenario 1 (%)	0.43	0.26	0.47	0.53	0.49	0.51	0.10	0.00	-0.15	-0.06	-0.18	-0.04	-0.26	-0.13	0.14
Scenario 2 (%)	0.17	0.12	0.23	0.30	0.29	0.30	0.04	-0.03	-0.11	-0.06	-0.13	-0.04	-0.18	-0.09	0.06
Scenario 3 (%)	0.27	0.14	0.24	0.25	0.23	0.23	0.06	0.01	-0.06	-0.01	-0.06	-0.01	-0.10	-0.05	0.08
End Stock								**000 Bales**							
Baseline	4092.00	4095.43	4114.31	4136.17	4221.34	4168.55	4111.05	4179.71	4116.94	4059.99	3978.34	3882.10	3769.52	3606.20	4037.98
Scenario 1 (%)	-1.23	-0.88	-0.98	-1.33	-1.11	-0.95	0.30	0.45	0.36	-0.03	0.57	0.05	0.68	0.24	-0.27
Scenario 2 (%)	-0.47	-0.40	-0.54	-0.78	-0.67	-0.58	0.23	0.35	0.25	0.02	0.40	0.05	0.46	0.17	-0.11
Scenario 3 (%)	-0.75	-0.49	-0.47	-0.60	-0.49	-0.42	0.11	0.16	0.14	-0.03	0.21	0.00	0.27	0.07	-0.16

Table 4. Estimated effects of efficient Bt cotton refuge policy compliance in India on the world cotton markets.

Parameter	2012/2013	2013/2014	2014/2015	2015/2016	2016/2017	2017/2018	2018/2019	2019/2020	2020/2021	2021/2022	2022/2023	2023/2024	2024/2025	2025/2026	Avg.
A-Index								$/Bale							
Baseline	93.26	90.95	90.84	93.08	93.80	94.12	94.19	94.45	96.48	96.29	97.43	98.29	102.16	104.89	95.73
Scenario 1 (%)	2.71	1.57	2.01	2.80	2.28	2.01	-0.98	-0.97	-0.77	0.17	-1.47	0.06	-1.68	-0.20	0.54
Scenario 2 (%)	1.04	0.74	1.13	1.66	1.37	1.21	-0.72	-0.77	-0.53	0.02	-1.03	0.00	-1.12	-0.15	0.20
Scenario 3 (%)	1.66	0.86	0.95	1.26	1.01	0.88	-0.36	-0.33	-0.30	0.11	-0.55	0.06	-0.67	-0.06	0.32
Area								000Acres							
Baseline	85300.79	83472.69	83104.56	84474.07	85877.64	87826.65	88593.82	89388.31	90017.00	90935.60	91634.47	93587.23	94188.46	94788.31	88799.26
Scenario 1 (%)	0.00	0.24	0.29	0.42	0.62	0.74	0.85	0.58	0.41	0.29	0.32	0.07	0.11	-0.16	0.34
Scenario 2 (%)	0.00	0.09	0.12	0.20	0.33	0.41	0.48	0.30	0.18	0.11	0.12	-0.04	-0.01	-0.18	0.15
Scenario 3 (%)	0.00	0.15	0.17	0.22	0.31	0.36	0.40	0.29	0.22	0.17	0.18	0.08	0.10	-0.01	0.19
Production								000Bales							
Baseline	120605.30	120927.88	122311.45	125578.19	127996.65	130512.48	133042.25	135854.90	137922.23	141335.58	143190.21	146780.31	148790.68	150282.66	134652.20
Scenario 1 (%)	-0.32	-0.14	-0.22	-0.32	-0.27	-0.28	-0.01	-0.08	-0.08	-0.14	0.03	-0.12	0.07	-0.09	-0.14
Scenario 2 (%)	-0.12	-0.07	-0.12	-0.18	-0.15	-0.16	0.02	-0.02	-0.03	-0.07	0.05	-0.06	0.07	-0.03	-0.06
Scenario 3 (%)	-0.20	-0.07	-0.10	-0.15	-0.13	-0.13	-0.02	-0.05	-0.05	-0.08	0.00	-0.06	0.01	-0.05	-0.08
Mill Use								000Bales							
Baseline	115928.42	118949.01	123157.17	125731.44	127682.14	129530.85	132202.64	134752.82	137822.67	140858.31	142891.53	146528.35	148789.98	150592.46	133958.41
Scenario 1 (%)	-0.16	-0.16	-0.20	-0.27	-0.29	-0.31	-0.16	-0.11	-0.08	-0.11	-0.03	-0.07	0.01	-0.04	-0.14
Scenario 2 (%)	-0.06	-0.07	-0.10	-0.15	-0.16	-0.17	-0.08	-0.04	-0.03	-0.05	0.01	-0.02	0.03	0.00	-0.06
Scenario 3 (%)	-0.10	-0.09	-0.11	-0.14	-0.14	-0.15	-0.09	-0.06	-0.05	-0.06	-0.03	-0.05	-0.01	-0.03	-0.08
Net Trade								000Bales							
Baseline	39264.63	39439.31	40723.83	41353.28	43015.40	45038.72	46203.04	47837.11	47995.14	49374.03	48869.12	50095.06	50603.24	51190.85	45785.91
Scenario 1 (%)	-0.48	-0.61	-0.74	-0.99	-1.08	-1.10	-0.70	-0.40	-0.25	-0.27	-0.06	-0.05	0.14	0.16	-0.46
Scenario 2 (%)	-0.19	-0.26	-0.36	-0.53	-0.59	-0.61	-0.37	-0.17	-0.08	-0.10	0.04	0.03	0.15	0.16	-0.21
Scenario 3 (%)	-0.30	-0.35	-0.39	-0.49	-0.52	-0.52	-0.35	-0.22	-0.16	-0.16	-0.07	-0.06	0.02	0.03	-0.25

The estimated effects of refuge requirements on the world cotton market under baseline, and under three alternate policy scenarios are presented and compared in Table 4. The top set of numbers represents the world cotton price (A-Index) under the baseline scenario, as well as the projected world price under the policy scenarios. The world cotton price (A index) is expected to increase by 0.54, 0.20 and 0.32% for refuge requirements under 15-year, under 10-year, and under EPA mandate, respectively. Higher refuge requirements would decrease the world cotton production, which would push the cotton prices up. Although the area under cotton cultivation would increase in the future, the total production would decline because of relatively lower overall cotton yield resulted from planting refuges. A lower cotton production would pull down mill use and net cotton trade in the world.

Thus higher refuge requirements would decrease world cotton production and trade, and would increase world cotton prices. In the U.S., however, the net export of cotton would increase because of relatively lower export competition from India,

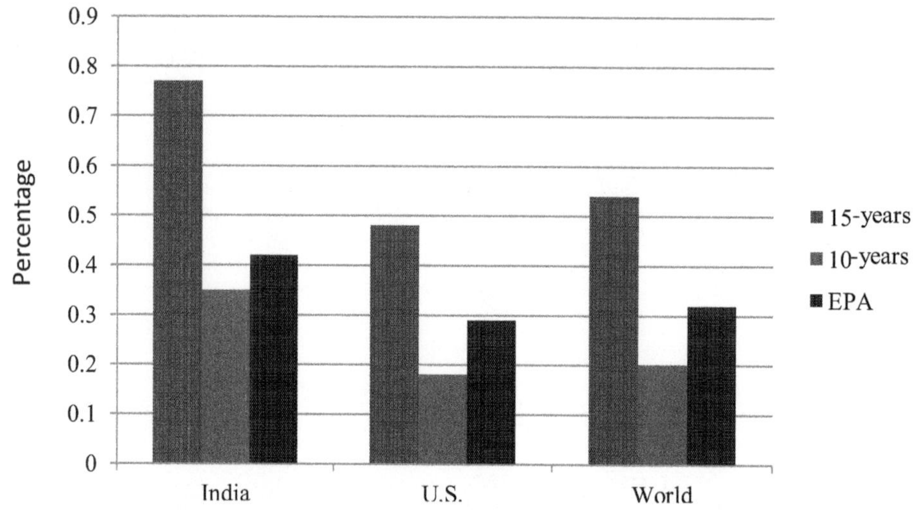

Figure 3. Average price effects of refuge requirements on world cotton markets.

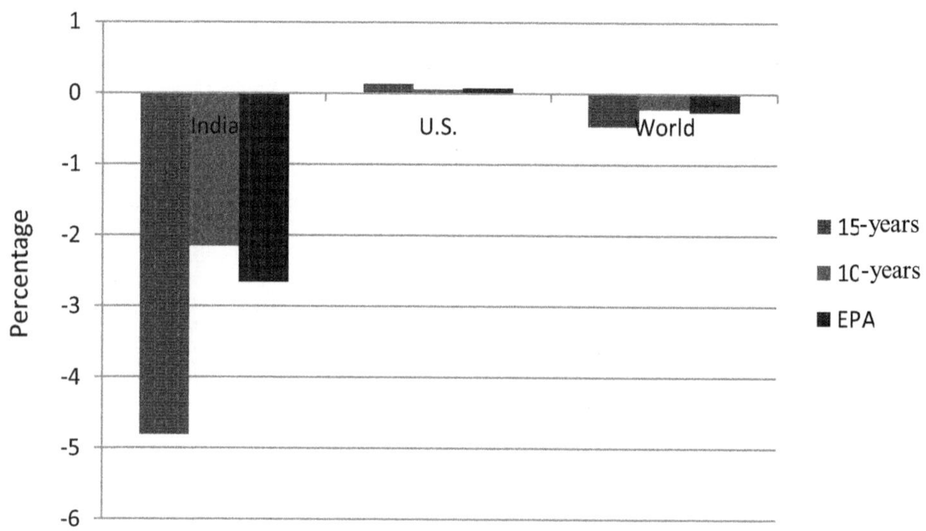

Figure 4. Average trade effects of refuge requirements on world cotton markets.

which stems from a decrease in cotton supply in India resulted from planting refuges. Although the magnitudes of trade and price effects on world markets look dwarf, the impact could sum up to a significant number at an aggregate level given the fact that India occupies 33% of total cotton area and contributes 23% to total cotton production in the world.

SUMMARY AND CONCLUSIONS

The study evaluates the potential impacts of refuge requirements for Bt cotton in India on world cotton markets by using a partial equilibrium world fiber model. The regional cotton yield models in the Indian fiber model were estimated using proportionate area under Bt cotton as an exogenous factor in the model. The Indian fiber model is then connected to an existing WFM to conduct baseline and scenarios projections. Three scenarios were considered, which includes optimal refuge requirements under 10-year and 15-year planning period; and refuge requirements under EPA mandate.

Simulation results reveal that cotton refuge requirements in India have potential to affect the domestic market as well as world cotton markets; the magnitude of impact is lower, however. World cotton trade is expected to decrease, and world prices of cotton are expected to rise under higher refuge requirements. In the U.S., however, the net export of cotton is likely to increase because of a decrease in cotton production and exports in India resulted from planting refuges. Although the percentage trade and price effects on world cotton markets look smaller, the impact could sum to be a larger amount at an aggregate level given the fact that Indian farmers cultivate about one-third of total cotton area and contributes about one-fourth to total cotton production in the world.

Comparisons of impacts under the three scenarios reveal that magnitudes of impacts depend on length of planning horizon or, in other words, replacement rate of insecticidal Bt technology. If an existing insecticidal technology is expected to be replaced by a new insecticidal technology early then there are lower refuge requirements and lower negative impacts, and vice versa. Technology replacement rate further depends on time spent in R&D and regulation to commercialize a new insecticidal technology; a significant time is spent on regulatory approval, however. Thus the time spent in R&D and regulations of an insecticidal crop technology is an important factor contributing indirectly to determine refuge requirements and their impacts on world commodity markets. A decrease in time in R&D and regulatory affairs of a new insecticidal agricultural technology has implications for refuge requirements and the world commodity markets.

ACKNOWLEDGEMENT

The authors thank Dr. Anwar Naseem (McGill University) for providing useful comments on the manuscript.

Abbreviations: EPA, Environmental Protection Agency; **USDA,** United States Department of Agriculture; **WFA,** World fiber model; **MFA,** multi-fiber arrangement; **ROW,** rest of the World; **TRQ,** tariff rate quota; **WTO,** World Trade Organization.

REFERENCES

Anderson KE, Valenzuela E, Jackson LA (2008). Recent and prospective adoption of genetically modified cotton: A global computable general equilibrium analysis of economic impacts. Econ. Dev. Cultural Change 56:265-296.

Bennett RM, Ismael Y, Kambhampati U, Morse S (2004). Economic Impact of genetically modified cotton in India. Agric. Biol. Forum 7:96-100.

Bennett RM, Kambhampati U, Morse S, Ismael Y (2006). Farm-Level economic performance of genetically modified cotton in Maharashtra, India. Rev. Agric. Econ. 28:59-71.

Brookes G, Barfoot P (2007). Global impact of biotech crops: socio-economic and environmental effects, 1996-2006. Agric. Biol. Forum 11:21-38.

Brookes G, Yu TH, Tokgoz S, Elobeid A (2010). The Production and price impact of biotech corn, canola, and soybean crops. Agric. Biol. Forum 13:25-52.

Choudhary B, Gaur K (2010). Bt Cotton in India: A country profile. ISAAA Series of Biotech Crop Profiles. ISAAA: Ithaca, New York.

Chaudhary J (2005). The future of Indian cotton supply and demand: Implications for the U.S. cotton industry. Ph.D. dissertation, Texas Tech University, Lubbock,TX.

Chaudhary J, Mohanty S, Misra S, Pan S (2008). The effects of MFA quota elimination on Indian fiber markets. Appl. Econ. 40:1083-1099.

Elbehri A, Macdonald S (2004). Estimating the impact of transgenic Bt cotton on West and Central Africa: A general equilibrium approach. World Dev. 32:2049-2064.

Environmental Protection Agency (EPA), Internet site: http://www.epa.gov/oppfead1/cb/csb_page/updates/2007/bollgard-cotton.htm (Accessed June, 2010).

Falck Zepeda JB, Traxler G, Nelson RG (2000). Rent creation and distribution from biotechnology innovations: The case of Bt cotton and herbicide tolerant soybeans in 1997. Agribusiness 16:21-32.

Frisvold GB, Reeves JM (2007). Economy-wide impacts of Bt cotton. Paper presented at Beltwide Cotton Conferences, New Orleans, Lousiana.

Frisvold GB, Reeves JM, Tronstad R (2006). Bt Cotton adoption in the United States and China: International trade and welfare effects. Agric. Biol. Forum 9:69-78.

Frisvold GB, Reeves JM (2008). The costs and benefits of refuge requirements: The case of Bt cotton. Ecol. Econ. 65:87-97.

Huang J, Hu R, Fan C, Pray CE, Rozelle S (2002). Bt cotton benefits, costs, and impacts in China. Agric. Biol. Forum 5:153-166.

Hurley TM, Babcock BA, Hellmich RL (2001). Bt corn and insect resistance: an economic assessment of refuges. J. Agric. Resour. Econ. 26:176-194.

Indiastat.com, Revealing India Statistically. Internet site: www.indiastat.com (Accessed August, 2008)

James C (2009). Global status of commercialized biotech/GM crops: 2009. ISAAA Brief No.41. ISAAA: Ithaca, New York.

Landes M, MacDonald S, Singh SK, Vollrath T (2005). Growth prospects for India's cotton and textile industries. Outlook Report CWS-05D-01, United States Department of Agriculture-Economic Research Service, Washington, DC.

Livingston MJ, Carlson GA, Fackler PL (2004). Managing resistance evolution in two pests to two toxins with refuge. Am. J. Agric. Econ. 86:1-13.

National Cotton Council, Internet site: http://www.cotton.org (Accessed June, 2009).

Pan S, Mohanty S, Welch M, Ethridge D, Fadiga M (2007). Effects of Chinese currency revaluation on world fiber markets. Contemporary Econ. Policy 25:185-205.

Pan S, Mohanty S (2004). Technical documentation of world fiber model. Cotton Economics Research Institute, Texas Tech University, Lubbock, TX.

Pekaric-Falak I, Meilke K, Huff K (2001). The trade effects of Bt Corn. Canadian Agri-Food Trade Research Network (CATRN) paper.

Purcell JP, Perlak FJ (2004). Global impact of insect-resistant (Bt) cotton. Agric. Biol. Forum 7:27-30.

Qaim M (2003). Bt Cotton in India: Field Trial Results and Economic projections. World Dev. 31:2115-27.

Qiao F, Wilen J, Huang J, Rozelle S (2009). Dynamically optimal strategies for managing the joint resistance of pests to Bt toxin and conventional pesticides in a developing country. Eur. Rev. Agric. Econ. 36:253-279.

Qiao F, Huang J, Rozelle S, Wilen J (2010). Natural refuge crops, buildup of resistance, and zero-refuge strategy for Bt cotton in China. Sci. China Life Sci. 53:1227-1238.

Singla R, Johnson PN, Misra SK (2012). Examination of regional-level efficient refuge requirements for Bt cotton in India. Agric. Biol. Forum 15:303-314.

Shelton AM, Tang JD, Roush RT, Metz TD, Earle ED (2000). Field tests on managing resistance to Bt-engineered plants. Nat. Biotechnol. 18:339-342.

United States Department of Agriculture- Foreign Agricultural Service, Internet site: http://www.fas.usda.gov (Accessed June, 2009.

United States Department of Agriculture- National Agriculture Statistics Service, Internet site: http://usda01.library.cornell.edu/usda/nass/Acre//2010s/2012/Acre-06-29-2012.pdf (Accessed July, 2012)

Yorobe JM, Quicoy CB (2006). Economic impact of Bt corn in the Philippines. Philippine Agricultural Scientist 89:258-267

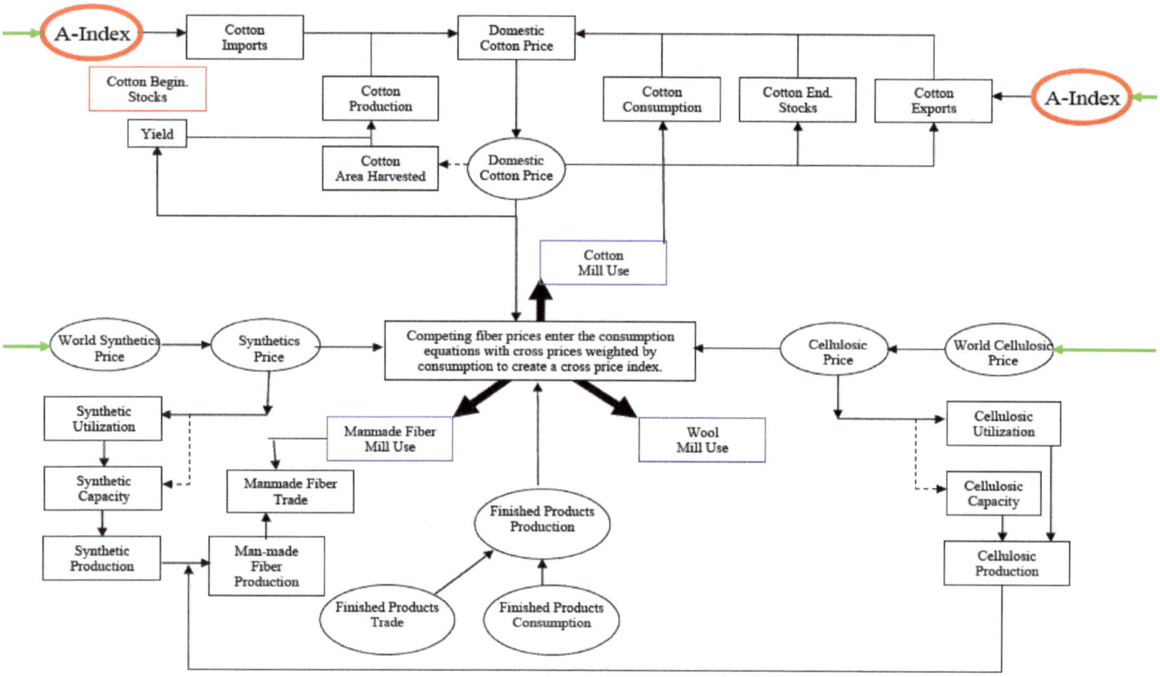

Figure A1. Representative country model.

APPENDIX A

World fiber model

The empirical world fiber model is a partial equilibrium structural econometric model developed by Pan and Mohanty (2004) at Cotton Economics Research Institute (CERI), Texas Tech University. The model includes supply, demand and market equilibrium for the cotton and man-made fibers for the U.S., India, China, and 21 other major cotton producing and consuming countries. A representative country model is presented in Figure A1.

Cotton production in each country and region defined in the model is derived from behavioural equations of area and yield. For geographically large cotton-producing nations such as the United States, China and India, cotton production is estimated in a regional framework to capture regional differences in climate, water availability, and other natural resources that influence crop planting decisions in different parts of each country. The manmade fiber production is derived from estimates of manmade fiber production capacity and utilization rates.

The textile sector in the model is used to determine the mill use of each fiber (cotton, manmade fiber, and wool). It is estimated using a two-step process. In the first step, total fiber demand (cotton, wool, and manmade) is calculated by summing fiber demand in apparel, home furnishing, floor covering, and other industrial sectors. In the second step, total fiber production is divided among cotton, manmade, and wool fibers based on the relative price of each as well as other non-price factors.

Two major applications of world fiber model are presented in Pan et al. (2007) and Chaudhary et al. (2008). While Pan et al. (2007) analyzed the effects of Chinese currency revaluation on world fiber markets, Chaudhary et al. (2008), on the other hand, examined the effects of MFA quota elimination on Indian and world fiber markets. Full explanation of the world fiber model is documented in Pan and Mohanty (2004).

The world fiber model uses data from various sources. Macroeconomic variables for India such as gross domestic product (GDP), population, exchange rate, GDP deflator and the average spot price of crude oil were obtained from International Financial Statistics published by the IMF. Cotton A-index price, US 1.5 denier polyester price, and US farm price sheer wool were collected from Cotton and Wool Yearbook of Economic Research Service, United States Department of Agriculture (ERS, USDA). Prices of polyester staple fibre and cotton fibre, and cotton tariff/duty in India were obtained from Foreign Agricultural Service, USDA and the Textile Commissioner's Office, Government of India (GOI). Minimum support price for cotton and competing crops were obtained from Ministry of Agriculture, India. The textile price index was gathered from the Handbook of Industrial Policy and Statistics 2001, India; at the same time, wholesale price index for food was obtained from the Handbook of Statistics on Indian Economy, 2001 on CD-ROM. Both indices were originally available on 1970/1971 and 1981/1982 base years, which were converted to1993/1994 base year for consistency. The textile price index for the year 1982 to 1984 was missing and had to be interpolated. The producer price for cotton and competing crops was obtained from the database of Food and Agricultural Organization (FAO). The consumer price index was gathered from the Ministry of Finance, GOI. Total fibre consumption, total cotton consumption and total man-made fibre consumption were obtained from the FAS/USDA, and the Textile Commissioner's Office, GOI. Wool and other fibre consumption were calculated by subtracting cotton and man-made from total fibre consumption. Similarly, man-made fibre capacity, utilization and man-made fibre production were also collected from the same sources. Data for cotton supply and demand were obtained from the FAS/USDA. The database consists of cotton area, yield, production, imports, exports, ending stocks and total domestic consumption.

Participation in land market and technical efficiency in Southern Ethiopia: A case study after 2005 land proclamation of Ethiopia

Tewodros Tefera[1] and D. V. Subaro[2]

[1]School of Environment, Gender and Deveopment Studies, Hawassa University, Hawassa, Ethiopia.
[2]Acharya RNGA Agricultural Universities, Hyderabad, India.

Agriculture is the mainstay of the Ethiopian economy contributing 45% to GDP and 80% to employment opportunity. Majority of the farmers in Ethiopia are smallholder farmers possessing less than a hectare of land. Land transfer mechanism in Ethiopia is limited to temporal land rental market and lease. Land sale and long term lease by farmers is outlawed by proclamation. The present study examines the technical efficiency of farmers who are operating under different tenurial structures and explains why some farmers (plots) are more efficient than the others. A stochastic frontier was used to estimate technical efficiency using data from 1786 parcel of land from 3 districts located in Oromia and Southern Regional Sates of Ethiopia The result of the analysis revealed that, the plot level technical efficiency ranged from 0.208 to 0.932 with mean value of 0.809. The study contested Marshalian conception of share tenancy as an inefficient institutional arrangement; it was found that, both share cropped in and out plots were more efficient than pure owner operated plots. The possible explanation for this finding is that, most of the share cropping arrangement was made between blood relatives that might evade the pervasive moral hazard problem in such tenurial arrangements. In addition most of the share cropping recipients was near landless and the productive use of the land is the only-option for them to meet their food security. The results of technical inefficiency model showed that, with the exception of slope other plot level characteristics which include; soil type and soil quality have significant positive effect on technical efficiency. While receiving land certificate, investment on soil conservation measures significantly reduces technical inefficiency, shallow soil depth has positive effect on technical inefficiency. The result accentuates that; the government should encourage temporal land transfer from less productive to efficient and from land surplus to land constrained households through land market.

Key words: Smallholder, stochastic frontier, technical efficiency, land tenure reform.

INTRODUCTION

The overwhelming population of Ethiopia are residing in rural area and eke out a living from farming. Arable land is becoming scarce and precious and the per capita landholding showing a consistent and declining trend. Over the last four decades the per capita land holding has shown a half cut (Jayne et al., 2002). Ethiopia is the only country in Sub Sahara Africa where its land policy remains static after the radical 1975 land reform which nationalized all rural land and made a state property. The 1975 land reform has brought mixed outcomes in the country. On the one hand, it abolished the exploitative tenant-landlord relationship and provided tenants' usufruct (non-free hold) right to land. In this regard, the reform was applauded for its egalitarian distribution of land and social justice.

However, its long term significance in improving the agricultural sector growth had fell short of the expectation mainly due to tenure insecurity and misguided socialist policy. Following the down fall of the Derg regime in 1991 a contested debate was opened on land policy among scholars, policy makers and donors. The debate has largely been carried out along two antagonistic arguments concerning property rights to land which include; privatization versus public land ownership (Samuel, 2006).

The Federal Democratic Republic of Ethiopia (FDRE) government has maintained the status quo of state landownership and overruled the privatization of land. The constitution which was erected in 1995 reaffirms the state ownership of land in Ethiopia (FDRE, 1995). It continues its predecessor regime land policy whereby only usufruct rights are bestowed upon landholders while the state enjoys eminent domain. The usufruct rights exclude the right to sell or mortgage the land. The government justifies its decision from the point of view of protecting farmers from losing their holding by distress selling and to avoid the possibility of resurgence of tenancy through land concentration on the hands of the wealthy.

Although the government position has attracted some support it has been attacked by advocates of a privatization of land. The latter argued that, state ownership of land prevents the development of a land market that facilitates the transfer of land to most efficient users, discourages farmers to invest on land, and thereby holds down land productivity as well as encourages unsustainable land use practices. At the heart of the land policy debate the government of Ethiopia introduces land proclamation at federal and regional level (FRLAUP, 2005; OR, 2007; SNNPR, 2007). The major departure in the new land proclamation ranges from decentralization of land administration to regional level to the introduction of land certificate to improve tenure security. It also allows land rental market and share cropping which is outlawed in the previous regime. The present study proposed to fill two gaps. First, it investigates the technical efficiency among the different landownership arrangements[1]. Secondly, it provides policy feedback for further refinment of the existing land policy.

METHODOLOGY OF THE STUDY

Background of the study area

The study was conducted in two regional states of Ethiopia which are; Southern Nation Nationalities and Peoples Region (SNNPR) and Oromiya Regional State (ORS). Three districts, one from the former and two from the later state were selected for the study. Shashemene and Arisi Negele districts were selected from Oromiya region for their importance as active trading centers along the main road to the capital city, Addis Ababa. The land pressure and conversion of land into non-agricultural purposes is likely to be higher and the land market is also assumed to be dynamic. The third district, Meskan, was selected from the southern region for comparison purpose, the area is also being known for land scarcity and land market was expected to be active in this woreda, and which may facilitate collection of valid and authentic information from the sample farmers.

Sample and sampling design

A multi stage random sampling techniques were employed in selecting the samples. In the first stage districts were selected purposefully. In the second stage eleven Peasant Associations[2] (PAs) were selected purposefully from the three districts. In the third stage a sampling frame was prepared comprising all households resides and cultivating farm land that is, own land or leased-in or share cropped-in land. A total of 394 households were selected using simple random sampling techniques.

Well structured and pre-tested questionnaires were used to elicit information from the selected sample households and their operational farm plots pertain 2007 to 2008 production year (mainly main season following June to August monsoon and whenever appropriate the small season which follow after the shower of March to April rain). A structured questionnaire which had three actions was used for the purpose. The first section was designed for collecting basic household socioeconomic information such as demographic, consumption, expenditure and marketing activities. The second section covered all the relevant information from individual farm plot such as input use and output levels, investment, land rental activity etc.

The plot level data also complemented with the information obtained from land certificate. The last one is partner schedule which was meant to collect information from land market partner to the main sample households. All information concerning the farm plots which were rented out by the main sample household to the land market during the 2007 to 2008 production year was collected by using partner questionnaire. A total of 1786 plots[2] (owned and rented in/share cropped in plots) were covered in the analysis. While rented in/share cropped in plots information were obtained from the main household, information regarding rented out/share cropped out plats were obtained from partner households (tenants). Hence, the partner households are identified after information was obtained on the status of the plot from the main household.

Analytical procedure

In this study, stochastic prduction function (SPF) was employed. The most important advantage of SPF approach is that, it allows for

[1]The federal land proclamation of 2005 states that, land redistribution may be used in relation to irrigation investment to ensure equitable distribution of irrigable land and land of deceased without hire will be distributed to landless. OR and SNNPR also adapted the two federal land proclamation cases of land redistribution in their 2007 land proclamations.

[2]Peasant Association refers to the smallest administration unit in Ethiopia. It was formed during the former socialist regime for the mobilization of rural community and to facilitate the trickle down of socialist ideology. Basically, peasant association has similar size. However, the population size is different from one peasant association to another as historically more people settled in fertile areas. Peasant association continued as the smallest administrative unit in the current regime too. [4]The price of agricultural commodity this year affects the supply (production) on the same commodity in the following year. Market guilt occurs when food aid or import coincides with harvest time affect pricing wrongly.

the introduction of statistical noise resulting from natural events which are outside the control of economic agents' such as the incidence of drought which is common in Ethiopia and other factors including market guilt[3] and luck. The SPF treats the disturbance term (ε) as being comprised of two components which are; standard independent statistical noise term (υ) and one sided non-negative random disturbance (μ), that is, $\varepsilon = \upsilon - \mu$. The white noise component, υ, that accounts for non idiosyncratic random effects, stands for a systematic error term assumed to be indenpendently and identically distributed (iid) as $N[0, \sigma^2_\upsilon)$. The second error term, μ, represents systematic effects that are not explained by the production fucntion and therefore are attributed to the agents' technical inefficiency. The inefficncy term μ is one sided since if $\mu = 0$, the agents would be lying on the production frontier, obtaining maximum production given the level of inputs. Where as, if $u_i > 0$, then, the agents would be operating at some level of technical inefficiency. The inefficiency effect term assumed to follow 'half normal[4]' being identically and independently distributed as $N|(0, \sigma^2_\mu)|$.

Following Farrell's (1957) techncial efficiency (TE) notation, a measure of TE for any given economic agent i would be given by the following ratio:

$$TE_i = \frac{E(Q_i \mid \mu_i, x_i)}{E(Q_i \mid \mu_i = 0, x_i)} \qquad (1)$$

Where, Q_i, x_i and μ_i are the vectors of output, input and ifnefficiency effect terms, respectively. Intern, the general stochstic frontier production function is usually defined by:

$$Q_i = F(x_i; \beta)\,\varepsilon^{(v_i - \mu_i)} \qquad (2)$$

Despite its well known limitation, a Cobb Douglas type of production function is used in the present study. Taylor et al. (1986) argued that, as long as interest rests on efficiency measurement and not on the analysis of the general structure of the production technology, the Cobb-Douglas production function provides an adequate representation of the production function. Moreover, in one of the very few studies examining the impact of functional form on efficiency, Kopp and Smith (1980) concluded that, "the functional specification has a discernible but rather small impact on estimated efficienc" (pp. 1058). That is why the Cobb-Douglas functional form has been widely used in farm efficiency analyses both for developing and developed countries (Battese, 1992; Coelli et al., 1998).

In the present study, technical efficieny analysis was computed at plot level. Since the status of the plot pricesely defined whether it was cultivated by owner operators or tenants, it enabled us to examine perfomence of technical efficency across various tenacy arrangements. The emperical model for plot level production function is specified as follows:

$$\ln Q_i = \beta_0 + \sum_{i=1}^{7} \beta_i \ln X_i + v_i - \mu_i \quad i=1,2,\ldots 7 \qquad (3)$$

Where, Q_i is the dependent variable in the production function showing total output value for the ith plot. ln represents the natural logarithm. Both the output value in ETB and inputs quantity are expressed in logarithms. Six input categories are defined as explantory vaiables in the production function. X_i is a vector of k inputs used in the production of ith crop and it is defined as follows:

x_1 = quantity of manure applied (kg/M^2), x_2 = draft animal power used in pair (oxen days), x_3 = fertilized applied (kg/M^2), x_4 = family labour (person day), x_5 = size of plot in ha, x_6 = value of seed (ETB[5]).

β_is'are unknown parameters to be estimated and V_i and U_i are random error term and non-negative random variables associated with technical inefficiency respectively. U_i is assumed to be independetly distributed such that, the technical inefficiency effect for the ith plot is obtained by truncation (at zero) of the normal distribution with mean μ_i and σ^2 such that:

$$u_i = \delta_0 + \delta_1 Z_1 + \delta_2 Z_2 + \delta_3 Z_3 + \delta_4 Z_4 + \delta_5 Z_5 + \delta_6 Z_6 + w_i \qquad (4)$$

Where, Z_1 = represent a dummy variable for plots registration in land certificate (1 = if it is registered in the land certificate, 0 otherwise), Z_2 = represent the soil type (1 = black, 2 = Dark brown, 3 = red, 4 = white, 5 = sandy), Z_3 = soil depth (1= shallow, 2 = medium 3 = deep), Z_4 = slope (1 = plain, 2 = foothill 3 = midhill 4 = steephill), Z_5 = plot qulaity (1 = poor 2 = medium 3 = good 4 = very good) Z_6 = dummy variable showing the presence or absence of soil conservation parcatice (1 if soil conservation structure is constructed, other wise 0).

The δ's are unknown parameters to be estimated. w_i is composed of u_i and u_i as defined earlier. It is assumed that, some farmers produce on the frontier and others do not. For this study the parameters of Equations 1 and 3 were estimated using the Maximum Likelihood (ML) method, following the likelihood function estimation by Battese and Corra (1977). Where, $\sigma^2_s = \sigma 2 + \sigma^2_v$ and $\gamma = \sigma^2 / \sigma^2_s$, and σ^2_u is the variance of u_i and σ^2_v is variance of u_i. And γ is defined as, the total variation of output from frontier which can be attributed to technical (in) efficiency.

RESULTS AND DISCUSSION

The prevailing tenancy structure

The sample households were found operating in eight types of tenancy structures. The majority (66%) was pure owner operator who cultivates their own farm holdings while the rest of 34% of households participated in the land rental market as a tenant or lessee or both. About 46% of the household participated in the share cropped type of tenancy while 40% of households took part in fixed rental land market. The remaining 14% was participated in both fixed and share cropped rental market. Examination of the tenancy structure shows that, of all the households that participated in the land rental market, 50% of the respondent household were situated at the demand side whereas, 46% participated as

er, the population size is different from one peasant association to another as historically more people settled in fertile areas. Peasant association continued as the smallest administrative unit in the current regime too. [4]The price of agricultural commodity this year affects the supply (production) on the same commodity in the following year. Market guilt occurs when food aid or import coincides with harvest time affect pricing wrongly.

[4]The price of agricultural commodity this year affects the supply (production) on the same commodity in the following year. Market guilt occurs when food aid or import coincides with harvest time affect pricing wrongly.

[5]Plot in this study refers to a specific parcel of land allocated for the production of one type of crop or intercepting. In the latter case most inputs are commonly used as the result the figures are divided equally for the two crops while yield data is collected separately

[6] ETB refers to Ethiopian Birr. Birr is the name of Ethiopia currency.

Table 1. Type and structures of tenancy (%).

Tenancy type	Household level (N = 394)	Plot levels (N = 1786)
Pure owner operator	65.99	83.76
Fixed rent tenants	5.84	4.03
Fixed rent lessors (land lord)	7.87	3.70
Share tenants	8.88	4.70
Share lessors	6.60	3.81
Fixed and share tenants	2.28	-
Mixed (lessors and tenant)	1.27	-
Fixed and share lessors	1.27	-
Total	100	100

supplier of land. Only 4% of the household were participated at both supply and demand side simultaneously. The plot level data reveals that out of the total 1786 plots, 290 plots (16%) were supplied to the land rental market. For the details of input-output information at plot level and the major crops grown in the study area refer Table 1A and 2A of the appendix (Table 1).

The land tenure structure most often indicates the level of land rental market participation and the direction of tenancy. Swamy (1988) and Chattopadhyay and Ghosh (1983) has shown that, the term and structures of tenancy prevailing in a given area influence the condition of demand and supply in the land lease market. The demand for land is a function of labour endowment that is, the extent of unemployed or underemployed family labour within the tenant household in relation to landholding size. The terms of tenancy such as arrangements of inputs allocation and output sharing in share cropping and also obligation of tenants and the duration of contract in the fixed rental market influence the demand for land.

The supply of land on the other hand depends upon the state of the art or methods of cultivation and the ground rent in relation to the marginal product of investment through direct cultivation. Under perfect condition tenancy equilibrium is attained, when the marginal product of capital equal to rent and when wage rate is equal to the excess of marginal product of land over rent while the former make the landlord indifference between self cultivation and renting out the latter makes the tenant indifference between renting in and working as labourers.

However, in most of the cases the land and labour market are less than been perfect and it prevents these conditions to be happened. The actual scenario is that, there are too many aspirant tenants at the demand side and few landlords are at the supply side. Hence, the land rental market is characterized by near monopoly at the supply side and near perfect competition at the demand side. However, when the landlord is poorly endowed with resources and rent out land under distress situation to respite this situation or due to lack of labour employment as in the case of the study areas, the supply and demand of land governed by not as such due to surplus or deficit of land at household level but due to the prevailing interlocked (such as social capital, resource transfer) and imperfect factors market (credit and labour).

Estimates of parameters for SFP function and inefficiency determinants

In the present study prior to proceeding to the analyses of technical efficiency and its determinants (Table 2); the presence of technical inefficiency was detected. The test was carried out by estimating stochastic frontier production function and conducting a likelihood ratio test assuming the null hypothesis of no technical inefficiency. The test statistics confirmed that the inefficiency component of the disturbance term (u) is significantly different from zero at 5% level suggesting that the null hypothesis of the technical inefficiency is rejected. Hence in the production input-output data for plot level inefficiency exist and it is indeed stochastic. The value of gamma (γ) (Table 2) further indicates that, there is 38% of variation in output is due to technical inefficiency. This means that, technical inefficiency is likely to have an important effect on in explaining output among the plots in the sample[6].

Following a one step approach of Coelli (1996), a stochastic frontier production function was estimated using Cobb-Douglas formulation where, the natural logarithm of output value per hectare is considered as dependent variable. In plot level crop production, technical efficiency is likely to be affected by a wide range of plot level characteristics, plot ownership (tenancy), tenure security and plot level investment in conservation measures.

The result as presented in Table 3 indicates that, the coefficients which denote the output elasticity of inputs

[7]Therefore, Maximum Likelihood Estimate (MLE) gives appropriate results rather than ordinary least square estimator (OLS).

Table 2. Detecting the presence of inefficiency.

Explanatory variable	Coefficients	Std. Err.	Z
Manure	0.006	0.009	0.66
Seed	0.114**	0.054	2.08
Fertilizer	0.050***	0.009	5.58
Draftanimal	0.425***	0.050	8.46
Plotsize	0.496***	0.047	10.54
Tlabout	-0.033***	0.013	-2.67
_constant	1.432***	0.203	7.06
Number of observation	1786		
Wald chi-square (χ^2) (6)	2196.13		
Prob.>chi2(6)	0.000		
σ^2_v	0.328	0.028	
σ^2_u	0.203	0.095	
$\sigma^2 = (\sigma^2_u + \sigma^2_v)$	0.532	0.058	
$\lambda = (\sigma^2_u / \sigma^2_v)$	0.787	0.121	
$\gamma = \frac{\lambda^2}{(1+\lambda^2)}$	0.382		
Likelihood-ratio test of sigma _u = 0: chibar2(01)	2.65**		

*** Significant at 1% level; ** Significant at 5% level; * Significant at 10% level.

Table 1. Plots level estimate of stochastic frontier for C-D type production function.

Variable	Coefficients	Standard error	t-ratio
Constant (β_0)	2.72	0.153	17.75***
In manure (β_1)	0.001	0.009	0.038
In draft animal (β_2)	0.514	0.039	13.18***
In fertilizer (β_3)	0.598	0.009	4.89***
In family labour (β_4)	0.044	0.049	12.27***
In seed (β_5)	-0.010	0.012	-0.877
In plot size (β_6)	0.277	0.027	2.84***
Technical inefficiency			
Constant (δ_0)	-5.260	3.526	-1.92**
In land certificate (δ_1)	-1.736	0.752	-2.31**
In soil type (δ_2)	-1.796	0.911	-1.97**
In soil depth (δ_3)	0.533	0.251	2.13**
In slope (δ_4)	-0.092	0.254	-0.364
In plot quality (δ_5)	2.632	1.120	2.35**
Lnadoption conservation (δ_6)	-0.850	0.402	-2.12**
$\sigma_u/\sigma_v = \lambda$	0.590		
$(\sigma^2_u + \sigma^2_v) = \sigma^2$	0.532	0.401	3.56***
σ^2_u	0.203		
σ^2_v	0.329		
Gamma (γ)	0.382	0.071	10.47***
Log likelihood	-1756.1		
N	1786		

*** Significant at 1% level; ** Significant at 5% level; * Significant at 10% level.

included stochastic frontier estimation were all positive except for seed. The negative sign for value of seed has indicated that farmers have over expended for seed. This is quite often the case as in mixed farming areas of

Table 4. Technical efficiency of plots under different tenancy arrangements (N %).

Technical efficiency	Technical efficiency				
	Owner operated plots	**Rented in plots**	**Rented out plots**	**Share in plots**	**Share out plots**
Below 75	238 (16)	2 (2.78)	7 (10.6)	10 (11.9)	5 (7.35)
75.1 to 90	1242 (83)	69 (95.83)	59 (89.4)	72 (85.7)	61(89.71)
Above 90	16 (1)	1 (1.34)	0	2 (2.4)	2 (2.94)
Mean	0.8071	0.8362	0.8063	0.8152	0.8238
Loss of output value due to inefficiency (ETB)	19.31	16.38	19.37	18.48	17.62
N	1496	72	66	84	68
F statistics			4.05***		

*** Significant at 1% level.

Ethiopia farmers exceed the normal seed rate for some cereals to thin out later to feed their animals as green fodder. The positive and significant value for other inputs implies that there is scope for increasing plot level productivity of crops in the districts under study. Fertilizer, draft animal and land come as the most important factors of production with elasticity of 0.598, 0.514, and 2.77, respectively. This implies that, ceteris paribus, an increase in the extent of fertilizer application, increase in oxen days and cultivable plots under crop production would significantly lead to increased output of crops. Similar results are reported by Barnes (2008) and Basnayake and Gunaratne (2002) among Scottish cereal producers and Sri Lanka tea smallholders respectively. The returns to scale (RTS) value, 1.424[7] obtained from the summation of the coefficients of the estimated coefficients of elasticity confirm that, plot level production in the study area is in Stage I of the production frontier. This stage is characterised by increasing return to variable inputs (Table 3).

Plot level technical efficiency

The model overall explanatory powers are good with significant log likelihood ratio test ($\chi^2 = 21.51$, $p < 0.05$). The null hypothesis which specifies the technical inefficiency effect is not present ($H_{0:}\ \gamma=0$) can be rejected as the gamma value 0.38 is significant at 1% level implying that inefficiency exist and is indeed stochastic. The estimated value of σ^2_u and σ^2_v were 0.204 and 0.329, respectively. The estimate of the total error variance sigma square (σ^2) value of 0.532 implying that, 53% of the difference between the observed and the maximum possible production for the plots considered are due to existing differences in the technical efficiency levels or management practice among the producers.

The value of gamma (γ) further indicates the presence of inefficiencies in the production of crops. In other words, about 38% of the difference between the observed and the frontier output was mainly due to the inefficient use of resources, which are under the control of farmers. The result corroborates with the findings of Rama Rao et al. (2003), Bhende and Kalirajan (2007) from India, Kariuki et al. (2008) from Kenya; Getu (1997), Ahmed et al. (2002), Gavian and Ehui (1999); Tesfaye et al. (2005); Kassie and Holden (2007); and Bamlaku et al. (2009) from Ethiopia, they reported the prescience of inefficiency in smallholder farming. Table 4 shows the frequency distribution of estimated technical efficiency and mean plots efficiency by tenancy types.

The results of technical inefficiency model shows that, except slope all plot level characteristics and other explanatory variables included in the model have significant effect on technical efficiency. Receiving land certificate, soil type and adoption of soil conservations have a significant positive effect on technical efficiency. On the other hand, shallow soil depth and poor soil quality have a significant negative effect on technical efficiency. Taking each of these technical inefficiency variables in turn we find that, receiving land certificate is significant (t-ratio = - 2.31), showing thatfarmers who received land certificate for their plots are more efficient than those who did not receive certificate. Soil type of the plot is also found to be significant (t =-1.970) which suggest that, black and dark brown soils contributes more efficient production than white and sandy soils. Soil depth is significant at 5% level (t = 2.13), supporting the argument that, shallow depth soil reduce technical efficiency (Tchale and Sauer, 2007). Likewise poor quality plot found to have a significant negative effect on technical efficiency in that owning poor quality plot lead to technical inefficiency.

As mentioned earlier, the technical inefficiency effect is significant; thus, the technical efficiencies of the sample

[8]The calculated value of return to scale (1.424) is tested for its statistical difference from constant return to scale using t-test approach. The result indicates that the calculated t value 2.92 is greater than the tabulated t value 2.447 at 5% level and 6 degree of freedom. Hence implying the hypothesis of constant return to scale is rejected at 5% level.

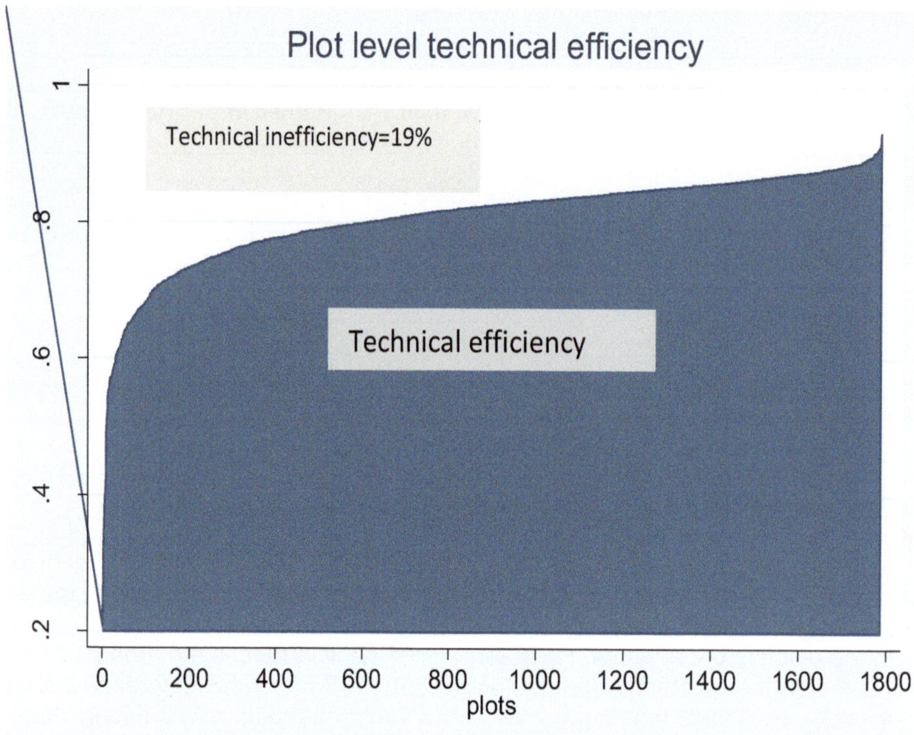

Figure 1. The cost of technical inefficiency.

households are equal to 1 when the plot level production was on the frontier or less than one. The cost accrued to the farmers due to the existence of technical inefficiencies is considerable, ranging from 7 to 79% in terms of loss of outputs value. The area bounded below the concave curve in Figure 1 indicates the technical efficiency, while the upper area represents technical inefficiency. The technical inefficiency area amounts about 19% in the output value on average due to technical inefficiency that can be bridged by efficient use of the existing resources under the prevailing.

Tenancy and plot level technical efficiency

In theory, technical efficiency level ranges between zero and one. The higher the technical efficiency value (close to one) the higher the efficiency of the farm (Coelli, 1994). The efficiency levels in this study ranged from 0.208 to 0.932[8] with a mean of 0.809. This implies that, if an average plot of land to achieve the efficiency of the most efficient counterpart, then the average operator would realize up to 13.2% more output from the same resources[9]. In terms of tenure structure the technical efficiency of the five types of tenancy were examined. The finding indicates that, rented in plots was found to have the highest technical efficiency level with a mean of 0.84. Contrary to the Marshalian conception of share tenancy as inefficient institutional arrangement we found both share in and share out plots were more efficient than pure owner operated with mean technical efficiency of 0.815 and 0.824, respectively. The possible explanation for the share tenant managed plots technical efficiency superiority over pure owner operated and rent out plots might be due to the tenancy arrangement was mostly done between blood relatives and in-laws and this might reduce moral hazard and associated disincentives. For instance Sadoulet et al. (1997), using data from a 1992 survey of three Philippine villages, test for efficiency differences across sharecrop contracts made among kin and impersonal sharecrop contracts and concluded that the technical efficiency of tenancy with kin was superior to that of non-kin. The absence of noticeable difference between share cropping and other tenancy arrangements was also reported by Ashok Rudra (1973) from India.

The observed technical efficiency difference among

[10]Getu (1997) measured the technical efficiency of farmer as well as plot level in Babile area of Ethiopia for 1993 and 1994 production seasons and he reported a technical efficiency ranged from 0.20 to 0.91 in 1993 and 0.30 to 1 in 1994 which is similar to the findings of this study.

[11] $\left(1-\frac{0.809}{0.932}\right)$

various tenancy structures was significant at 1% level indicating positive potential role of land rental market for efficiency (Table 4). It allocates land from less efficient owner operator to efficient tenants which are commendable policy implication from both equity and efficiency point of view. First, since both study regions dissociate from future land redistribution except for irrigated areas and with the community consent, the available best alternative is market oriented allocation of land and should be encouraged through appropriate policy intervention. Second, the existing land ownership inequality between the generation who benefited from last redistribution and the current generation who are near landless was overcome by land rental market. Despite the efficiency and equity merit of land rental market both Oromiyia and SNNP regions put some restrictions on the land market operation. In both regions landholder is allowed to lease out up to half of the land under his or her holding with the justification of protecting household food production. This means under the prevailing average land holding a farmer after leasing out is left out with 0.5 ha or less for self cultivation which is not viable and economical to use modern technologies on the face of chronic production risk (drought, infestation, market etc) and the prevailing methods of production.

In addition this restriction tied up farmers to a tiny parcel rather than contemplating alternative non-farm and migration livelihood options. This mainly because of the land policy is narrowly viewed food security from production perspective and this critically undermines the possibility of acquiring food security from non-agricultural livelihood options. In terms of duration of lease Oromiya region specified three years if the renter use traditional technology and 15 years if the renter uses modern technology. SNNP region relax the duration of lease out up to five years for users of traditional technology and up to 10 years if renter uses modern technology. From the point of view, crop production which enables to reap short term, benefit the lease period can be considered adequate but it can hampers long term investments in agriculture which normally requires longer gestation period.

Conclusion

The findings of this study showed that, there is efficiency gain as a result of land allocation through land rental market (both fixed and share cropping contracts). Contrary to the Marshalian conception of share tenancy as inefficient institutional arrangement, we found that, both share cropped in and out plots were more efficient than pure owner operated plots. However, the restriction imposed on the size and duration of land lease through the land proclamations in the study regions has constrained households from tapping the full potential benefit of land rental market.

The area of land that can be leased and the length of time the rental agreement lasts impede not only the land rental market efficiency but also labour mobility which consequently tied up households on farming sub-economic plots. Lifting the ceiling on the land area to be leased and relaxing the time limit on rental contract is an important step and can serve as a natural experiment to study the responses of households to such changes in terms of land transfer, its direction and the possible negative consequences before embarking to fundamental reforms like land privatisation. In addition removing the restriction on duration of lease will motivate tenants to invest on the land that can improve land management and overall efficiency.

REFERENCES

Ahmed MM, Gebremedhin B, Benin S, Ehui S (2002). Measurement and source of techncical efficiency of land tenure contracts in Ethiopia. J. Environ. Dev. Econ. 7(3):507-528.

Bamlaku A. Alemu E, Nupppenau A, Bolland H (2009). Technical efficiency across agro-ecological zones in Ethiopia: The Impact of poverty and asset endowment. Agric. J. 4(4):202-207.

Barnes A (2008). Technical Efficiency Estimates of Scottish Agriculture: A Note. J. Agric. Econ. 59(2):370-376.

Basnayake BMJK, Gunaratne LHP (2002). Estimation of Technical Efficiency and It's Determinants in the Tea Small Holding Sector in the Mid Country Wet Zone of Sri Lanka. Sri Lanka J. Agric. Econ. 4:137-150.

Battese GE, Corra GS (1977). "Estimation of a Production Frontier Model:With Application to the Pastoral Zone of Eastern Australia". Austr. J. Agric. Econ. 21:169-179.

Battese GE (1992). Frontier production functions and technical efficiencies: A survey of empirical applications in agricultural economics. Agric. Econ. 7:185-208.

Bhende MJ, Kalirajan KP (2007). Technical efficiency of major food and cash crops in Karnataka (India). Indian J. Agric. Econ. 62:2.

Chattopadhyay M, Ghosh SK (1983). Tenurial Contracts in a Peasant Movement Belt. Field Survey Data on Maxalbari, Kharibari and Phansidewa Regions: Economic and Political Weekly. Rev. Agric. P. 71.

Coelli TJ, Rao DSP, Battese GE (1998). An introduction to efficiency and productivity analysis. Kluwer-Nijhoff, Boston.

Coelli TJ (1994). A guide to Frontier Version 4.1: A computer Programme for Stochastic Frontier Production and Cost Function Estimation. Mimeo, Department of Economics, University of New England, Armidale.

Coelli TJ (1996). A Guide to frontier version 4.1: A computer programme for stochastic frontier production and cost function estimation, mimeo, Department of Econometrics, University of new England, Armidale.

Farrell MJ (1957). The measurement of productive efficiency. J. Royal Stat. Soc. Series A. 120:253-290.

FDRE (Federal Democratic Republic of Ethiopia) (1995). The Constitution of the Federal Democratic Republic of Ethiopia. 1st Year No. 1 Addis Ababa, 21st August, 1995.

FRLAUP (2005). Federal Rural Land Administration and Utilisation Proclamation. Federal Democratic Republic of Ethiopia. 456/2005.

Gavian S, Ehui S (1999). Measuring the production efficiency of alternative land tenure contracts in a mixed crop-livestock system in Ethiopia. Agric. Econ. 20:37-49.

Getu H (1997). Analysis of Performance of variation: the case of Smallholder Farms in Easter Hararghe Region. Unpublished M.Sc. Thesis, Alemaya University of Agriculture.

Jayne TS, Takashi Y, Michael TW, David T, Rui B, Antony C, Ballard Z, David N (2002). "Smallholder income and land distribution in Africa: Implications for poverty reduction strategies. P. 59.

http://www.aec.msu.edu/agecon/fs2/psynindx.htm (last accessed in April 28, 2010).
Kariuki DK, Ritho CN, Munei K (2008). Analysis of the effect of land tenure on technical efficiency in smallholder crop production Kenya. University of Hohenheim, October 7-9, 2008. Conference on International research on Food security, Nat. Resour. Manage. Rural Dev. P. 3.
Kopp RJ, Smith VK (1980). Frontier production function estimates for steam electric generation: A comparative analysis. S. Econ. J. 47:1049-1059.
Kassie M, Holden S (2007). "Sharecropping efficiency in Ethiopia: threats of eviction and kinship," Agric. Econ. Int. Ass. Agric. Econ. 37(2-3):179-188.
OR (Oromia Region) (2007). Proclamation to amend the proclamation No. 56/2002, 70/2003 and 103/2005 of Oromia Rural Land Use and Administration.
Rama Rao CA, Chowdry KR, Reddy YVR, Krishna Rao GV (2003). Measuring and Explaining technical efficiency in crop production in Andra Pradesh. Indian J. Agric. Econ. 58(4):768-780.
Sadoulet E, de Janvry A, Fukui S (1997). The Meaning of Kinship in Sharecropped Contracts. Am. J. Agric. Econ. 79(2):394-406.
Samuel G (2006). Land, Land Policy and Smallholder Agriculture in Ethiopia: Option and scenario. Paper presented at the Future Agriculture Consortium Meeting at the Institute of Development Studies 20-22 March, 2006.
SNNPR (2007). SNNPR Rural Land Administration and Utilization Proclamation.
Swamy DS (1988). Agricultural Tenacy in the 1970's. Indian J. Agric. Econ. 43(4)555-568.
Tchale H, Sauer J (2007). The efficiency of maize farming in Malawi. A bootstrapped translog frontier. Cahiers d'économie et sociologie rurales. pp. 82-83.
Tesfaye G, Ruben R, Pender J, Kuyvenhoven A (2005) Resource use efficiency on own and sharecropped plots in Northern Ethiopia: determinants and implications for sustainability.

Appendix

Table 1A. Input and output level at plot level.

Variable	Districts			
	Shashamene	Arsi Negele	Meskan	All
Landholding size (ha)	1.12 (0.070)	1.33 (0.075)	0.78 (0.045)	1.11(0.41)
Output per ha (kg)	1397 (126)	1361 (70.4)	1024 (70)	1286 (58)
Weighted index output value (ETB/ha)	3373.93 (250.83)	3726.40 (98.35)	4013.94 (149.37)	3692.70(102.36)
Seed Value (ETB)	533.64(35.86)	543.87(27.48)	185.67(13.89)	448.84(18.92)
Manure (kg /ha)	875.4 (186.4)	217.4(43.4)	440.2(363)	988.2(312.7)
Draft animal (oxen days per ha)	107.00(2.12)	101.38(1.63)	130.69 (5.02)	111.99(1.784)
Fertilizer (DAP+UREA) (kg/ha)	72.83(4.64)	92.67(8.06)	41.70(4.13)	45.78(0.95)
Male labour allocated (person days per ha)	44.25(1.80)	43.78(0.85)	49.96(2.20)	45.78(0.95)
Female labour allocated (person days per ha)	33.69(2.12)	21.10(0.66)	30.62(1.82)	28.19(0.94)
Hired labour used (person days per ha)	5.63(0.38)	8.30(0.52)	6.80(1.12)	6.95(0.41)
Number of plots (fragmentation)	3.02(0.09)	3.68(0.12)	5.54(0.26)	3.91(0.10)
Total gross income (ETB)	7497.53(801.21)	9283.70(805.05)	6531.64(337.86)	7904.79(433.82)
Per capita income (ETB)	1245.23(114.62)	1397.71(109.91)	1226.66(103.92)	1296.60(64.83)

Figures in the table are mean followed by standard error (SE).

Table 2A. Plots allocation for crop production by districts.

Crop type	Districts			
	Shashamene	Arsi Negele	Meskan	All
Total plot numbers	600 (33.60)	654 (36.62)	532 (29.79)	1786 (100)
Wheat	68 (11.33)	168 (25.70)	24 (4.51)	260 (14.56)
Barely	13 (2.15)	28 (4.30)	12 (2.26)	53 (2.97)
Tef[12]	98 (16.32)	84 (12.90)	144 (27.12)	326 (18.25)
Sorghum	13 (2.15)	15 (2.30)	68 (12.81)	96 (5.38)
Maize	147 (24.50)	208 (31.80)	201 (37.78)	566 (31.13)
Haricot bean	16 (2.65)	13 (2.00)	68 (12.81)	97 (5.43)
Horse bean	1 (0.17)	13 (2.00)	5 (0.94)	19 (1.06)
Chickpea	3 (0.5)	-	-	3 (0.17)
Sweet potato	4 (0.67)	1 (0.15)	-	5 (0.28)
Chilli pepper	-	-	7 (1.32)	7 (0.39)
Potato	220 (36.67)	100 (15.30)	-	320 (17.92)
Kale	12 (1.99)	10 (1.54)	1 (0.19)	23 (1.29)
Cabbage	5 (0.83)	2 (0.31)	-	7 (0.39)
Onion	-	12 (1.84)	2 (0.38)	14 (0.78)
Enset	73 (48.32) †	31 (21.38) †	-	104 (5.82) †
Chat	-	-	17(17) †	17(0.95) †

Figures in the table are number of plots allocated for each crop followed by percentage. † indicate number of households.

20

Viability of tobacco production under smallholder farming sector in Mount Darwin District, Zimbabwe

James Masvongo, Jackqeline Mutambara and Augustine Zvinavashe

Department of Agricultural Economics and Extension, Faculty of Agriculture, University of Zimbabwe, Box MP 167, Mount Pleasant, Harare, Zimbabwe.

Smallholder commercial agriculture in developing countries is hinged on cash crop production and in Zimbabwe tobacco is increasingly becoming an important smallholder cash crop. This study therefore, analyzed the viability of tobacco production by smallholder farmers in Zimbabwe. Cross sectional survey data was collected for the 2010 and 2011 production season from 60 smallholder households in the Mount Darwin District of Mashonaland Central Province in Zimbabwe. Data were analyzed using descriptive statistics, gross margin analysis, breakeven analysis and ordinary least square (OLS) criterion to determine the viability and determinants of income earnings from tobacco by farmers. The study revealed that, smallholder tobacco production was viable, with farmers achieving average yield of 2052 kg/ha, average price of US $2.45 per kilogram and earning, on average, about US $2352 per hectare as gross margin. Break-even analysis revealed a margin of safety of 50% with respect to both yield and prices, indicating that, small-scale tobacco production will remain lucrative even at much lower prices and yields. Regression analysis showed that, off farm employment was inversely related to revenue earnings from tobacco with coefficient (-0.058) and the relationship was significant at 5%. Price and yield were positively related to tobacco gross margins with coefficients 0.865 and 1.001 and the relationship was significant at 1%. It is therefore concluded that, tobacco production is viable and can improve incomes for smallholder farmers. To improve income earnings from tobacco, there is a need for farmers to focus on farm production for better yields and improve quality for better prices.

Key words: Break even, gross margins, income, off farm employment, yields, quality.

INTRODUCTION

In Zimbabwe, the 2000 fast track land reform has been characterised by radical reconfigurations of land, production, economy and livelihoods in the rural landscape (Mavedzenge et al., 2008) and brought about benefits and opportunities as well as costs, challenges and pitfalls (Scoones et al., 2010). What is clear is that, there is a move to commercialise small-scale production, and integrate more effectively black indigenous farmers into the national economy, hence the increased participation of smallholder farmers, who now command the majority of the land used for agricultural purposes in Zimbabwe (Shumba and Whingwiri, 2006).

"Smallholder farmers" is used more generally to describe rural producers, predominantly in developing countries, who farm using mainly family labour and for whom the farm provides the principal source of income (Ellis, 1988) and in Zimbabwe it is used loosely to define indigenous black farmers. In Zimbabwe smallholder agriculture has been traditionally based on a wide range of rain fed, seasonal food crops for balanced household nutrition and risk aversion. Small scale farming has been important for food security, contributing 42% of total maize production (the staple in Zimbabwe) in 1980 increasing to 60% in 1985 and then 70% in 1999 (Rukuni, 2006). Commercial agriculture in Zimbabwe is important for generating raw materials for the manufacturing industry, generation of foreign earnings, and ultimately economic growth. In addition, commercial production mainly focuses on year round production of high value cash and export crops, using production systems based on modifications of the natural environment and use of highly mechanised machinery.

The new agrarian structure in Zimbabwe entails commercialisation of smallholder agriculture, hence the need for smallholder diversification into high value crops like tobacco. Tobacco "the golden leaf" has been the single most important export commodity in Zimbabwe. Tobacco generates in excess of USD650 million in foreign revenue annually, by 1991 Zimbabwe had become the world leading exporter of flue cured tobacco, accounting for around 40% of its foreign currency earnings and contributing approximately 10% of country's GDP (Muir-Leresche, 2006). Tobacco generates employment more than any other crop in the country. It directly employs over a million people, and many more in the downstream industries. The area under flue cured tobacco increased from 61 180 ha in 1950 to 84 857 ha in 2000, the quantity of tobacco produced also increased from 47 294 tonnes in 1950 to a peak of 236 946 tonnes in 2000 (Cole and Cole, 2006).

The 2000 agrarian reforms also transformed the structure of the tobacco sub-sector. There were 15000 registered tobacco growers in 1998, currently there are more than 64 000 predominantly smallholder tobacco growers in Zimbabwe (Tobacco Industry and Marketing Board (TIMB), 2011). The increase in the number of producers has been accompanied by decline in the area grown per farmer and of concern is also the decline in productivity. Tobacco productivity declined from 2200 kg/ha in 1998 to about 700 kg/ha in 2001(Ministry of Lands and Agriculture, 2005). Tobacco output plummeted from a record level of 267 million kg in 2000 to 73 million kg in 2007 (Dawes et al., 2009). There was a marked decrease in the value of tobacco exports in 2001 from US $640 million to US $204 million in 2002 and US $396 in 2004 (FAO, 2004), weakening Zimbabwe's competitive position in the world market for flue cured tobacco in favour of China, India, and Brazil. Tobacco has been a highly profitable and lucrative crop for commercial farmers in Zimbabwe (Rukuni, 2006) however for smallholders there is need for research based inquiry to ascertain economic benefits of choosing tobacco as a cash crop enterprise.

In Zimbabwe as in many developing countries, smallholder agriculture was viewed as a failed sector, and smallholder farmers as severe degraders of the environment. This notion is however debatable. For example in Zimbabwe prior to the fast track land reform program, and because of colonial injustices, smallholders typically occupied communal areas in more fragile and marginal environments. In addition Zimbabwe has a well-documented post-independence smallholder productivity success story, though some argue that the drastic increase in output from the smallholder sector has been largely a result of the increase in areas planted while crop yields have remained almost static (Takavarasha, 1994). Smallholder farmers in Zimbabwe now have access of land in high potential areas and questions regarding potential of small scale commercial agriculture to successfully produce cash crops of economic importance and their potential in improving smallholder household incomes remain unanswered. This study was therefore designed to (i) investigate viability of tobacco as a smallholder cash crop in Zimbabwe and (ii) determine the determinants of viability.

MATERIALS AND METHODS

Study area

The study was carried out in Mount Darwin, a District in the Mashonaland Central province of Zimbabwe. Mashonaland Central province accounts for about 30% of Zimbabwe's tobacco output (TIMB, 2011) and Mount Darwin is in Natural region III, a semi-intensive farming region with moderate rainfall. The study area comprises mainly of small scale farmers who are into cash and food crop production.

Sampling and sample size

Using a two-stage-selection approach, a random sample was selected. The first stage identified all smallholder tobacco farmers in Mount Darwin using the TIMB database. The second stage identified, from a list of all smallholder tobacco farmers, all newly resettled farmers. Finally the identified farmers were stratified in to 3 main tobacco growing areas based on information provided by district extension workers from which a random sub-sample of 20 was drawn from each for an in-depth study.

Data collection

A structured questionnaire was used to interview the selected farmers. Secondary data for each household on yield, credit provided to the household, price, area, average weight of tobacco bales delivered and total sales revenue was obtained from TIMB. Secondary data was considered to be more reliable than the primary data obtained from interviewing the smallholders. Data from the structured questionnaire was therefore cross-referenced with secondary data and in cases of discrepancies, secondary data was given in preference.

Data analysis

Data analysis methods used were descriptive statistics, gross margin analysis, break even analysis and regression analysis. Details of these analytical techniques are given in this section.

Descriptive statistics

Descriptive statistics were used as a preliminary investigation procedure to gain an understanding of inherent significant socio-economic characteristics of the smallholder farmers.

Gross margin analysis

Johnson (1985) defines the gross margin as, the gross income from an enterprise less operating (variable costs) of production. Although a gross margin is not profit as it does not include fixed or overhead costs such as depreciation, interest payments, rates and permanent labour, which have to be met regardless of enterprise size. Gross margin serves as the unit of analysis in evaluating the economic performance of an enterprise and gives an indicator of the viability of an enterprise and its potential contributing to household income. Gross margins are generally quoted per unit of the most limiting resource, for example, land, hence crop gross margins are provided on a per hectare basis.

Gross margins were computed as follows:

$$GM = P_q Q + \sum_i PiXi$$

Where, GM is gross margin in US dollars per hectare for the tobacco crop enterprise; P_q is the price of tobacco per kg; Q is the quantity of crop output per hectare in kg; *Pi* is the price of the ith variable input used in tobacco production; and Xi is the quantity of the ith variable input per hectare.

Breakeven analysis

The breakeven price is the minimum price per unit required to cover all production costs at the anticipated yield and was computed as follows:

$$Breakeven\ Price = \frac{\text{Anticipated Total Production Costs}}{\text{Anticipated Yield}}$$

The breakeven yield is the minimum yield required to cover all costs at the anticipated price per unit and was computed as follows:

$$Breakeven\ Yield = \frac{\text{Anticipated Total Production Costs}}{\text{Anticipated Price}}$$

Ordinary least square (OLS) analysis

For the study the dependant variable was total tobacco gross income (US$/ha). Though profit maximisation is the objective in smallholder cash crop production and ideally, one would wish to model profits directly through the incorporation of variable costs, fixed costs and revenues pertaining to the input-output relationship. Profit is difficult to measure for a single crop in a smallholder setup due to unavailability of records on crop-specific inputs (Fulginiti and Perrin, 1998). In addition the dependence on family labour often makes it difficult to disaggregate and allocate it for specific intra-household cropping activities. The Cobb Douglas functional form was used for its simplicity, flexibility and also the empirical support it has received through wide applications on data for various industries and countries.

$$\text{Log}Y = \beta_0 + \beta_1 \text{Log}X_1 + \beta_2 \text{Log}X_2 + \beta_3 \text{Log}X_3 + \beta_4 \text{Log}X_4 + \beta_5 \text{Log}X_5 + \beta_6 \text{Log}X_6 + \beta_7 \text{Log}X_7 + \beta_8 \text{Log}X_8 + \beta_9 \text{Log}X_9 + \beta_{10} \text{Log}X_{10} + \beta_{11} \text{Log}X_{11} + \mu$$

Where, Y = Tobacco Gross Income/ha; X_1 = Tobacco area; X_2 = Variable costs/ha; X_3 = Labour; X_4 = Cattle; X_5 = Employment off farm; X_6 = Training in Agriculture; X_7 = Sex; X_8 = Age; X_9 = Price/kg; X_{10} = Yield/ha; X_{11} = Maize area; $\beta_0,\ldots,\beta_{11}$ = Parameters to be estimated; u = Random error.

Tobacco area

Tobacco area is the total area allocated for tobacco cultivation by the household in hectares. Increasing the area under cultivation can possibly bring about economies of scale associated with efficient use of fixed and highly specialised assets required by tobacco, for example curing barns and grading sheds. The expected effect on revenue is positive.

Variable costs

Variable costs in US dollars were computed by summation of expenditure on all variable inputs for the tobacco enterprise per hectare. A household spending more on fertilisers, crop protection chemicals, fuel and seeds can be expected to achieve high yields and better quality. Revenue from the tobacco enterprise is expected to thus increase with intensity of variable input use.

Labour

This is the number of hired workers engaged by the household specifically for tobacco production. Tobacco competes with maize for labour as the two crops are grown in the same season, and many key operations like planting, weeding and harvesting coincide. In smallholder agriculture family labour is usually prioritised for food security crops, therefore engaging hired labour thus, reduce loses due to weed competition, pre and post harvesting loses for cash crop enterprises. The expected effect on revenue is positive.

Cattle ownership

Cattle ownership was measured by the number of cattle belonging to the household. Addition of cattle as a variable in the model was justified for smallholders as cattle are a source of draft power and just like labour, cattle can be considered as direct inputs. A household endowed with more cattle can timely execute key operations such as, land preparation, weed control, harvesting and transportation, all of which require cattle draft power. The expected impact on revenue is positive

Off- farm employment

Some small-scale farmers are full time farmers while some have full time formal employment in urban areas, hence are part time farmers. The expectation was that full time farmers (X_5 = 1) could realise more revenue from their farming operations, as they are resident on their farms. Full time farmers allocate more time to their farming operations and accord more attention to their farming enterprises than those having off farm formal employment (X_5 = 2). Off farm employment was therefore anticipated to negatively impact tobacco revenue.

Training in agriculture

Training in agriculture was regarded as any form of training in agriculture by the household head. Training brings about better decision making, better use of production technology, adoption of appropriate technology and adherence to recommended production practices. Those trained in agriculture (X_6 = 2) were therefore expected to be more productive than those without any training in agriculture (X_6 = 1). Tobacco is a highly specialised crop whose yield and quality is very sensitive to quality of management and hence, training in agriculture was expected to impact tobacco revenue positively.

Sex

Tobacco has high labour and capital demands. Crop ownership and gender are therefore important in Zimbabwe, where men control all important household resources and make decisions on family labour and household resource allocation. If the tobacco crop belongs to men (X_7 = 2) it was anticipated that, productivity will be high and if the tobacco crop belongs women (X_7 = 1), productivity was expected to be low.

Age

Age was measured by the number of years for the household head. Even though accumulation of farming experience and production capital comes with age. In Zimbabwe older farmers have bigger household sizes and are more worried about food security. When food crop production takes precedence, efficiency in cash crop production can be compromised. Furthermore, in developing countries age is usually negatively correlated to education and literacy levels. The younger being likely more educated and literate, the expectation was age negatively impact productivity.

Price

In tobacco production, price varies with quality. Farmers achieving better quality attain better prices and ultimately high revenue. Agronomic practices as well as the curing process influence leaf quality. The expectation was that, price positively impact tobacco revenue.

Yield

Marketable yield in smallholder tobacco is dependent on agronomic practices and more important for smallholder farmers are post-harvest loses. Yield was expected to positively impact revenue.

Maize area

Maize is the staple in Zimbabwe and competes with tobacco for land, labour and other production inputs. The expectation was that, as the area under maize cultivation increases, tobacco revenue would be reduced.

RESULTS

General socio-economic characteristics

According to results in Table 1 the mean age of surveyed farmers was 40.58 years indicating that, most smallholder tobacco producers were fairly middle aged farmers. Farmers had an average 6.4 ha and most (5.1ha) of it was being cultivated. The average area under maize and tobacco were 1.41 and 1.3 ha, respectively. On average farmers had about 15.8 years of experience in general farming but had fewer years of farming tobacco (7.1 years). Employment of permanent workers was very minimal with an average of 0.8 per household; however farmers relied more on seasonal workers who averaged 6.22 per household per season. Ownership of equipment such as tractors and cars was very low with an average of 0.01 and 0.2, respectively. On average household ownership of cattle was 5.95 indicating that most farmers had cattle. The average price of tobacco per kilogram was US $ 2.45 and the average yield was 2052 kg/ha.

There was a lot of private sector support and involvement in tobacco production as most of the households received inputs on credit and hence, used very high rates of fertiliser, applying 321 kg/ha mean basal fertiliser, and on average 86 kg/ha of top dressing. All farmers reported using firewood for curing their tobacco and only 30% reported using firewood in combination with coal. Many of the small scale farmers were full time farmers and only 25% of the farmers had formal employment and farming was part time. More than half of the farmers had never been exposed to any form of training in agriculture and only 45% of the farmers had received training in agriculture. Tobacco production was dominated by males and in 82% of the households the tobacco crop belonged to a male household member.

According to results in Table 2 the mean total revenue from tobacco per hectare was US $5120 and the mean total variable costs per hectare were US$2768 resulting in average gross margins of US$ 2352 per hectare. This outcome shows that tobacco production by smallholder farmers was viable. From Table 3, the computed breakeven yield was 1100 kg/ha and the breakeven price was 1.21/kg. The margin of safety was 46 and 50% with respect to both yield and prices respectively (Table 3).

Determinants of income earnings from tobacco

Regression analysis showed that off farm employment was inversely related to revenue earnings from tobacco with coefficient (-0.058) and the relationship was significant at 5%. Price and yields were positively related to tobacco revenue with coefficients 0.865 and 1.001 and the relationship was significant at 1% (Table 4).

The coefficient of determination (R^2) of 90.7% indicated the strong explanatory power of the model with only 9.3% of the variation in the dependent variable (income earnings from tobacco) being explained by other factors nor specified in the model.

DISCUSSION

Viability of tobacco production

Tobacco production was observed to be an important

Table 1. Descriptive analysis of the general socio-economic characteristics.

Variable	Minimum	Maximum	Mean	Standard deviation
Age of household head	26	71	40.58	10.40
Total land area (ha)	2	13.5	6.4	6.9
Area under cultivation (ha)	1	9.5	5.1	3.4
Farming experience (years)	3	45	15.8	8.9
Tobacco experience (years)	2	25	7.1	5.8
Permanent workers	0	4	0.8	1.1
Tractors	0	1	0.01	0.2
Cars	0	2	0.2	0.4
Seasonal workers	0	15	6.22	3.69
Number of cattle	1	18	5.95	3.88
Area under maize (ha)	.5	4.0	1.41	0.62
Area under tobacco (ha)	1	3	1.3	0.51
Price/kg (USD)	1.18	3.63	2.45	0.50
Variable costs/ha (USD)	2365	3139	2768	206
Yield/ha (kg)	207	3938	2052	1
Total revenue/ha (USD)	370.2000	11037.26	5120.06	3
		Frequency	**%**	
Formal employment	Yes	45	75	
	No	15	25	
Training in agriculture	Yes	27	45	
	No	33	55	
Sex of household head	Male	49	82	
	Female	11	18	

Table 2. Smallholder tobacco gross margin analysis.

Variable	Minimum	Maximum	Mean	Standard deviation
Gross margin/ha	-2397	8269	2352	2.20
Total revenue/ha	370	11037	5120	2.89
Total variable costs/ha	2365	3139	2768	206

Table 3. Breakeven analysis.

Variable	Yield kg/ha	Price US $/kg
Achieved	2052	2.45
Breakeven	1100	1.21
Margin of safety (%)	46	50.6

smallholder enterprise, as farmers allocated approximately equal land areas to tobacco and maize (the staple in Zimbabwe). The primary objective in smallholder tobacco production is profit maximization. This is because tobacco smallholder is strictly a cash crop and about 98% of the tobacco produced in Zimbabwe is exported (TIMB, 2011). The mean gross margin was US $2352 per hectare, indicating that, tobacco is a viable smallholder cash crop and contributes significantly to rural household income. This finding has been acknowledged by a number of authors. Keyser (2002) noted that, tobacco generated direct income for commercial farmers and indirect (wage) income for smallholder farmers, though the current study showed a transition of smallholder tobacco income from being indirect to direct income.

Table 4. Regression results to determine factors affecting income earnings from tobacco production.

Variable	Beta	Standard error	Significance
Constant	-0.040	0.101	0.690
Sex	-0.013	0.026	0.613
Age	0.021	0.032	0.528
Training in agriculture	0.011	0.021	0.596
Labour	0.010	0.010	0.315
Cattle	0.010	0.012	0.412
Off farm employment	-0.058	0.026	0.027
Maize area	-0.026	0.018	0.164
Tobacco area	-0.006	0.024	0.802
Price	0.865	0.034	0.000
Expenditure on variable inputs	0.019	0.030	0.531
Yield kg/ha	1.001	0.011	0.000

R Square 0.907.

From break even analysis, the margin of safety in tobacco production was more than 50%. This implies that, tobacco will remain relatively profitable even at lower yields and prices. This notion has also been supported by Rukuni (2006), though on large scale farming. For large scale commercial farmers' tobacco was observed to offer good financial returns even after a large drop in yield and price and would continue to be attractive even under progressively difficult conditions.

Determinants income earnings from tobacco production

It was also revealed from the study that, the majority of the respondents (75%) were full-time farmers, while others engaged in other occupations apart from farming and there was a negative correlation between off-farm employment and tobacco income. The finding of the study is consistent with Deiniger and Olinte (2001) who, studying data from Colombia found that, specialisation in either farm or non-farm activities increased wealth and income levels. However, this is not consistent with general findings in Africa that households that engage in both agricultural and non-agricultural activities are richer, both in income flows and in endowments of assets (Piesse et al., 1999). Diversification caters for livelihoods security, by spreading risk and uncertainties of agriculture; it however compromises on time allocation on various farm enterprises especially for high value crops like tobacco.

Leaf quality and yield had a significant positive effect on tobacco returns. Total revenue from a cropping enterprise is a product of yield and price. Leaf quality for tobacco is reflected in the price per kg achieved. Smallholder farmers achieving higher prices and high yields therefore achieve better returns. Yield and quality depend on good agronomic practices, proper curing and minimisation of post harvesting losses and leaf spoilages.

Age of the household head had the expected effect (negative sign) on tobacco farm returns, implying that, younger tobacco farmers had more income than older farmers. Though not significant the notion of age being negatively correlated to performance is however supported in literature. Yaron and Dinar (1992) argued that, old aged farmers generally tend to resist change than young farmers who quickly adopts new appropriate technology. Training in agriculture as expected was positively correlated to tobacco revenue though not significant at 5% level. Similar conclusions were also observed by Mutandwa et al. (2008), who in an attempt to ascertain whether training in tobacco farming influences tobacco yields and returns among smallholder farmers concluded that, returns for trained farmers were greater than untrained farmers though statistically insignificant.

Conclusions

Gross margin analysis revealed that, tobacco production is a viable smallholder enterprise, and therefore has potential to contribute towards improving incomes and livelihoods of rural small scale farmers in Zimbabwe. From break even analysis, tobacco will remain relatively profitable even at lower yields and prices. The study therefore concluded that, farmers should consider tobacco production as a crop of choice as it is a viable crop and its returns remains favourable under diverse conditions.

All the farmers reported to be using firewood in tobacco curing. This brings about questions regarding long run sustainability of smallholder tobacco production. Deforestation implies switching over to coal and electricity for tobacco curing. In the long run, small scale tobacco production costs are therefore expected to rise and profitability will decline. There is therefore need for government intervention to raise awareness about the negative effects of deforestation. Apart from enforcing environmental protection laws, there is need to ensure each tobacco farmers establish replacement wood lots.

The observed negative significant relationship between off farm employment and revenue from tobacco reveals that, farmers in formal employment are not productive. There is therefore need for government policy reconsiderations in land allocations, and input support to prioritise productive full time farmers. To improve income earnings from tobacco, there is also a need for farmers, government and other interested stakeholder to focus on improving yield and quality of the crop for better prices. There is also a need to strengthen education, extension, research and development of appropriate technologies for smallholders given that previous research and development efforts were targeted on large scale farming.

REFERENCES

Cole L, Cole SJ (2006). Tobacco research and development. In: Rukuni M, Tawonezvi P and Eicher C. Zimbabwe's agricultural revolution revisited. University of Zimbabwe Publications. Harare. pp. 403-414.

Dawes M, Murota JR, Masara C, Sola P (2009). Inventory of Smallholder Contract Farming Practices in Zimbabwe. Support Capacity for Enhanced Market Access and Knowledge Management (SCAPEMA).

Ellis F (1988). Peasant Economics: Farm Households and Agrarian Development. Cambridge Univ Press. Cambridge, UK.

Deiniger K, Olinte P (2001). Rural Nonfarm Employment and Income Diversification in Colombia. World Dev. 29(3):455-465.

Food and Agriculture Organization (FAO) (2004). Statistics-Online Database, FAOSTAT, Http://www.fao.org/es/ess/toptrade.asp.

Fulginiti LE, Perrin RK (1998). Agricultural productivity in developing countries. Agric. Econ. Blackwell. 19(1-2):45-51.

Johnson DT (1985). The business of farming: a guide to farm business management in the tropics. London and Basingstoke: Macmillan Publishers Ltd.

Keyser JC (2002). The Costs and Profitability of Tobacco Compared to Other Crops in Zimbabwe. Economics of Tobacco Control Paper No 1. HNP Discussion Paper. World Bank.

Mavedzenge BZ, Mahenehene J, Murimbarimba F, Scoones I, Wolmer W (2008). 'The Dynamics of Real Markets: Cattle in Southern Zimbabwe following Land Reform'. Dev. Change 39(4):613-639.

Ministry of Lands and Agriculture (2005). Crop Statistics, Government of Zimbabwe Publications, Harare, Zimbabwe.

Muir-Leresche K (2006). Agriculture in Zimbabwe. In: Rukuni M, Tawonezvi P and Eicher C (eds). Zimbabwe's agricultural revolution revisited. Univ. Zimbabwe Publications pp. 99-114.

Mutandwa E, Gadzirayi TC, Foti R (2008). Measuring payoffs to agricultural traning among smallholder tobacco producers. Int. NGO J. 3(3):033-037.

Piesse J, Simister J, Thirtle C (1999). Leveller for Some? Non-Farm Income and Euqality in Zimbabwe. Mimeo, Birkbeck College, University of London.

Rukuni M (2006). Revisiting Zimbabwe's Agricultural Revolution. In: Rukuni M, Tawonezvi P, Eicher C. Zimbabwe's agricultural revolution revisited. University of Zimbabwe Publications, Harare. pp. 143-160.

Scoones I, Marongwe N, Mavedzenge B, Mahenehene J, Murimbarimba F, Sukume M (2010). Zimbabwe's Land Reform: Myths and Realities, London: James Currey.

Shumba EM, Whingwiri EE (2006). Commercialisation of Smallholder agriculture. In: Rukuni M, Tawonezvi P, Eicher C (eds). Zimbabwe's agricultural revolution revisited. University of Zimbabwe Publications, Harare. pp. 577-593.

Takavarasha T (1994). Agricultural price policy. In: Rukuni M and Eicher CK (eds), Zimbabwe's agricultural revolution, University of Zimbabwe Publications, Harare.

Tobacco Industry and Marketing Board (2011). Flue Cured annual statistical report, Harare. Online statistics.Online. Available: www.timb.co.zw.

Yaron D, Dinar A (1992). Innovation on family farms: The Nazareth Region in Israel. J. Agric. Econ. P. 74.

The role of small scale irrigation in poverty reduction

Adugna Eneyew[1], Ermias Alemu[2], Mekonnen Ayana[2] and Mihret Dananto[3]

[1]Department of Agricultural Economics and Extension, Jimma University, P. O. Box 307, Ethiopia.
[2]Department of Water Resource and Irrigation Engineering, Arba Minch University, Ethiopia.
[3]School of Bio-systems and Environmental Engineering, Hawasa University, Ethiopia.

Poverty is of multidimensional characteristics affecting nearly a billion world population. Especially, a third of sub Saharans fall under poverty. The emergence of climate change coupled with the incidence of drought, are worsening the situation. The only option to escape this challenge is through the development of water resource projects. In attempting to do so, Ethiopia has yet developed not more than 5% of the irrigation potential. Much of this is owned and poorly managed by small holder farmers. The purpose of this study is thus to investigate whether small scale irrigation schemes contribute to poverty reduction or not. Based on 313 sample households from the Rift Valley Lake Basins, it was observed that irrigation improved household income and contributed to poverty reduction. However, the enhanced poverty impact of irrigation was constrained due to unsatisfactory performance and imperfect market. Thus, enhancing the capacity of water user associations through provision of training, market linkage and finance are a necessary step to improve irrigation performance towards poverty reduction.

Key words: Ethiopia, lake basin, rift valley, binomial logit.

INTRODUCTION

Ethiopia is predominantly an agricultural country where agriculture accounts for about 45% of the country's Gross Domestic Product (GDP), 65% of the total exports and 85% of employment (MoFA, 2007). One of the features of the Ethiopian agriculture and the national economy at large is its inability to produce sufficient food to feed the population (Samuel, 2006). In history, Ethiopia is characterized by famine as a result of high population pressure, resource base depletion and drought that affects the rain-fed agriculture significantly (Berhanu, 2001; Bruce et al., 1994). It has been documented that low farm production and productivity resulting from use of backward technology and other productivity-enhancing modern inputs are the major reasons for rampant poverty and food insecurity in rural areas (FDRE, 2010; Samuel, 2006). Poverty reduction is the first millennium development goal. Poor countries like Ethiopia were expected to halve the number of people living below one dollar by the end of 2015 (MoFED, 2010). Since 1992, the Government of Ethiopia has been carrying out measures to reduce poverty in the context of a series of reform programmes in the political, economic and social spheres (FDRE, 2003, 2010). Thus, following government efforts, poverty has declined from 45.5% in 1995/1996 to 29.6% in 2010/2011 (MoFED, 2012). Consensus has been reached by the government and donors that any solution to further reduces rural poverty must focus on increasing the production and productivity of smallholder agriculture (FDRE, 2010). Creating access to fertilizer, improved seeds, agricultural credit and thereby bringing significant growth in crop production is the major concern of national strategy (Samuel, 2006).

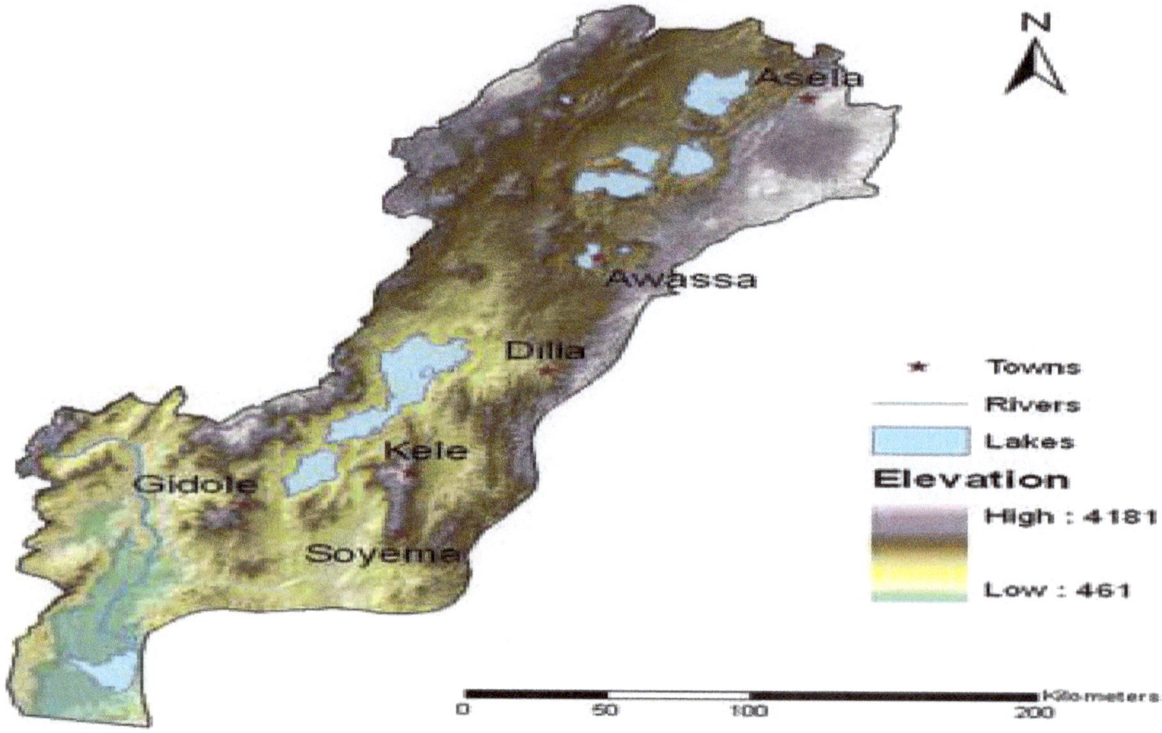

Figure 1. Ethiopian rift valley river basin (Bekele et al., 2007).

While technology is important, the issue of drought and rain fall variability is of paramount important. In order to address these challenges as a vital resource in agriculture, irrigation water contributes a lot in productive and livelihood activities of farmers.

Ethiopia is a water tower of Africa. A large number of rivers flowing on either side of the rift valley form a drainage network that covers most of the country. The government has focused to develop the sub-sector to fully tap its potentials (Mekuria, 2003; MoFED, 2006, 2010). Special attention is given to small scale irrigation development for their low capital requirement. In spite of this, the attention paid for this sector, the development of irrigation has not picked up. Even though some efforts have been underway to develop small scale irrigation (SSI) schemes; yet, Ethiopia has developed only 5% of the irrigable land (World Bank, 2006). Furthermore, it is noticed that the existing irrigation farms are operating at sub-optimal levels and many of the SSI projects have been operating below the required economic efficiency (Getaneh, 2011; Mekuria, 2003). Several studies have documented poverty-related benefits and costs of irrigation (Hussain, 2004). Most of them indicated irrigation can increase production and productivity. This, in turn, opens up new employment opportunities, both on-farm and off-farm, and can improve incomes, livelihoods and the quality of life in rural areas (Getaneh, 2011; Hussain, 2004; Oni et al., 2011). However, there are no available studies that assessed the poverty impact of SSI in the Ethiopian rift valley lake basins.

The purpose of this study is therefore to answer whether small scale irrigation schemes; while under performing, contributes to poverty reduction or not in the study sites. Since poverty reduction is the ultimate measure of development effectiveness; this study investigated the role SSI played in rural poverty reduction. The remainder of this study presents methodology, results, discussions and conclusions.

METHODOLOGY

The study site

The rift valley basin has an area of 52,739 km^2, covering parts of the Oromia, SNNPR regions. The total mean annual flow from the river basin is estimated at about 5.6 BMC. Large-scale irrigation potential is estimated at 45,700 ha with an estimated total irrigable area of 139,300 ha (Figure 1). The basin is endowed with a number of lakes of varying size with high environmental significance (Bekele et al., 2007).

Sampling and data

Multistage sampling procedure was followed to select respondents. In the first stage, four SSI schemes in the rift valley were selected purposively (Gedemso and Argeda from Oromiya; Ebala and Bedeneyalemtena from Southern Nations Nationalities and peoples

/SNNPR/ Regions). In the second stage, households from the head, middle and tail of the schemes were selected randomly, which comprises 145 users and 168 non users. Quantitative data on resource endowments and assets, average landholding size, livestock holding, incomes, expenditures and employment; demographic and social indicators like family size, dependency ratio and education level were collected from sample households through interview schedule. Qualitative data on the community perceptions about the benefit of irrigation and constraints were gathered from community representatives through focus group discussion.

Data analysis

Data was entered and analyzed using statistical package for social sciences (SPSS version 16). The descriptive analysis is based on means and standard deviations computed from the data. Independent sample t and chi square tests were used for assessing the difference between irrigation users and non-users in terms of socio-economic factors. The poverty line is measured based on cost of basic needs (CBNs) derived from the lowest income quartile and poverty indices were computed using Foster Greer and Thorbecke (FGT) formula. Foster et al. (1984) have suggested a useful general index for poverty measures. Their class of poverty indices takes the following form:

$$\rho = \frac{1}{N}\sum_{i=1}^{q}\left[(Z_p - Y_p)\right]^{\alpha}$$

Where Zp denotes the poverty line, Yi the expenditure or income of the i-th poor household (or individual), N the total number of households and q the number of households whose expenditures or incomes are below the poverty line. Thus, if α = 0, index Pα becomes: P_0 = q/N, which has been referred to as the head-count index; if α is 1, poverty gap index and if α is 2 poverty severity index.

A logistic regression model was used to analyze the impact of small scale irrigation schemes on household poverty status. Similar studies have used binomial logit model in irrigation impact analysis (Farah et al., 2001; Getaneh, 2011; Oni et al., 2011). Thus, poverty is the dependent variable, and is determined by independent variables such as irrigation use, household characteristics, asset holdings and access to services. The dependent variable is binary (1 if the household is poor and 0 if the household is non-poor). Following Gujarati (2003), the probability that the ith household is poor is given by:

$$Pi = E(Y = 1/Xi) = \frac{1}{1 + e^{-(\beta_0 + \beta_i \chi_i)}} \quad (1)$$

For ease of exposition, the probability that a given household is poor is expressed as:

$$\rho_i = \frac{1}{1 + e^{-(Z_i)}} \quad (2)$$

Probability for not poor is $1-P_i$. Thus,

$$\frac{P_i}{1-p_i} = \frac{1+e^{z_i}}{1+e^{-z_i}} \quad (3)$$

is the ratio of the probability that a household was poor to the probability of that it was non-poor. The natural log of Equation 3 is:

$$L_i = \ln\left[\frac{P_i}{1-P_i}\right] = Z_i = \beta_o + \beta_1\chi_1 + \beta_2\chi_2 + \ldots + \beta_n\chi_n. \quad (4)$$

Where Pi is a probability of being poor ranges from 0 to 1, Zi is a function of n explanatory variables (x) which is also expressed as:

$$Z_i = \beta_o + \beta_1\chi_1 + \beta_2\chi_2 + \ldots + \beta_n\chi_n \quad (5)$$

β_o is an intercept $\beta_1, \beta_2, \ldots\ldots\ldots\ldots\beta_n$ are the slopes of the equation, L_i is log of the odds ratio, which is not only linear in Xi but also linear in the parameters, X_i is vector of relevant independent variable.

If the disturbance term (Ui) is introduced, the logit model becomes:

$$Z_i = \beta_0 + \beta_1\chi_1 + \beta_2\chi_2 + \ldots\beta_n\chi_n + U_i \quad (6)$$

RESULTS AND DISCUSSION

It presents respondent's demographic profile, community's perspectives, and the role of irrigation use on production, employment, income, asset endowment, consumption and poverty.

Demographic profile of respondents

In this study, 145 irrigators and 168 non irrigators were compared. Table 1 indicates that the proportion of women irrigators was 19.3%, which implies that women's access to irrigation is by far below that of men. The report of Kinfe et al. (2012) also revealed that women's access to irrigation is limited in Northern Ethiopia. The minimum and maximum age limits are 18 and 82 respectively with mean age of 40. There seems no disparity by age towards accessing irrigation and there is a tendency for young farmers to engage in irrigation farming. Education is one of the most pertinent factors that affect human behavior. About 37% of respondents are illiterate; of which 37.8 and 36.3% respectively are irrigation users and non-users. This means that 37% of the respondents cannot read and write and there is no wide variation in the education attained between irrigation and non-irrigation households. The rest 23.5, 29.3 and 10.2% completed 1 to 4, 5 to 8 and 9 to 12 grades respectively (Table 1). The average household size was 6.6 persons, with 1 and 13 being the minimum and the maximum respectively. There is also no significant variation with respect to household size and number of dependents between irrigation users and non-users.

Community perspectives on the role of irrigation

The investigated community has perceived that SSI is a

Table 1. Socio-demographic profile of respondents.

Gender	User	Non user	Total		
Female	19.3	12.5	15.7		
Male	80.7	87.5	84.3		
Age				**Minimum**	**Maximum**
Mean	40.4	39.6	39.98	18	85
Standard deviation	11.82	11.87	11.3		
Education level				**0**	**12**
Illiterate	37.8	36.3	37.0		
1-4 grade	22.4	24.4	23.5		
5-8 grade	27.3	31.0	29.3		
9-12 grade	12.6	8.3	10.2		
Household size				1	13
Mean	6.7	6.57	6.6		
Standard deviation	2.55	2.28	2.4		
Dependency ratio				**0**	**10**
Mean	1.08	1.04	1.06		
Standard deviation	1.14	0.87	1.00		

pillar to improve rural livelihoods. According to focus group participants, almost all of the irrigation users in their specific localities have improved their livelihoods as a result of irrigation. Many of irrigation users have constructed corrugated iron sheet house, been able to educate their children, become food self-sufficient either through own production or purchasing from market, started local investment like petty trading; grain mill factory, buying vehicle (*Isuzu*) for transport facility etc. According to most focus group discussants, the proportion of irrigation users with investing in local business like rural shops, petty trades; did not exceed 25% of irrigation beneficiaries. The rest majority were unsuccessful due to lack of capital, limited potential and low bargaining power. This indicate that majority of the users are not gaining the intended benefit for one or another reasons. Furthermore, lack of efficient market and frequent fall of commodity price are mentioned as the major sources of failure. In addition, during the focus group discussions, we have investigated that there are several challenges in water use administration. There are no strong and functional water user associations (WUAs) in all of the investigated schemes. The WUAs have a weak coordination skill to solve scheme related problems like water theft and conflict between users.

The role of irrigation in production, employment and poverty

Irrigation may lead to poverty reduction via increased yields, increased cropping areas and higher value crops, by these means raises employment (directly of farm workers, indirectly of other workers if wages are bid up). Increased mean yields can mean increased food supplies, higher calorie intakes and better nutrition levels. This study investigated that there were significant differences in levels of production, employment, asset endowment, consumption, and income between irrigation users and non-users as follows:

Irrigation increased production

Comparative yields analysis by crop type could not be done because of lack of uniformity in the use of inputs. However, gross yield for major crops by access to irrigation was presented in Figure 2.

As expected, irrigation use has significantly contributed towards achieving household's goal of increased production and this result is similar to other reports (Getaneh, 2011). Data analysis of major cereals and horticultural crops showed that mean crop yield per household for *teff*, maize, green pepper, potato, tomato, red onion, cabbage and barely is highest for irrigation users than non-users. This evidence has ensured that irrigation use is a guarantee for increased food supply and ensured food security. Some crops like tomato, onion, pepper and cabbage are only grown by those households with access to irrigation. This is also an indication of the fact that irrigation use increases cropping diversification and intensity.

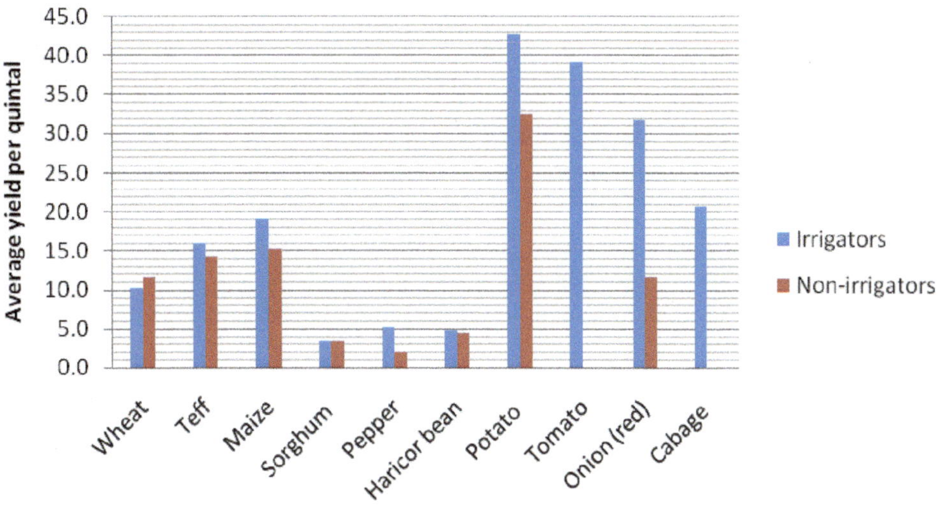

Figure 2. Average crop yields per quintal per household (1 quintal = 100 kg).

Table 2. Labor hour and cost by irrigation use.

Average labour hour	Irrigation use	Mean	Standard deviation	t/p
Plowing	User	76.90	84.38	8.464/0.000***
	Non user	21.71	25.61	
Weeding	User	90.79	116.18	7.085/0.000***
	Non user	26.51	38.11	
Harvesting	User	87.31	97.74	7.445/0.000***
	Non user	28.33	39.66	
Trashing	User	70.98	76.12	7.113/0.000***
	Non user	24.23	38.81	
Labor cost per ha in Ethiopian Birr (ETB)	User	535.94	800. 95	2.988/0.003***
	Non user	305.92	495.36	

***, Significant at less than 1% probability level, SD; standard deviation.

Irrigation enhanced employment opportunities

Conceptually, among the many benefits of irrigation, employment generation is crucial. The beneficiaries have shifted from once a year (rainy season) to two and three harvests and labor use efficiency were improved due to irrigation. Table 2 shows that mean hour invested on irrigated farm is significantly higher than the rain fed only farm for all activities from plowing to trashing. Similarly, the average labor cost (calculated only for hired labor) for irrigation user is more than double of the non-user households. This implies that, irrigation is a stimulus to increased employment opportunity. Most smallholder activities all draw from the same family labor sources, supplement for certain operations by neighbor help and casual wage labor. The development of the irrigation schemes has created job opportunities for the nearby farmers in addition to the irrigation users in the traditionally slack dry times.

Irrigation increased income

It is expected and revealed that irrigation would improve income earning (Getaneh, 2011; Hussain, 2004; Kinfe et al., 2012). Similarly, irrigation beneficiaries earned an annual mean income of 10161.5 Birr per household, which is 33.6% higher than that of non-users. Irrigation use has a positive impact on households earning from crop, and livestock, while the value of off farm income

Table 3. Income earned by households with and without irrigation.

Income source (ETB)	Irrigation use						t
	User			Non user			
	Mean	Standard deviation	% share	Mean	Standard deviation	% share	
Livestock	1451.6	2826.6	13.5	1070.2	2150.3	13.7	1.324
Crop	8138.5	6012.1	76.0	5520.9	3879.3	70.5	4.635***
Off farm	1125.2	2549.6	10.5	1234.7	2239.9	15.8	-0.0.4
Total	10161.5	5612.7	100	7606.0	4280.6	100	4.562***

***Significant at less than 1% probability level.

Table 4. Asset endowments by households with and without irrigation.

Assets owned	Irrigation use	Mean	Standard deviation	t
Total value of asset (ETB)	User	2060.16	6510.74	2.500**
	Non user	597.58	3450.67	
Total size of plots (ha)	User	1.50	1.00	3.84***
	Non user	1.12	0.76	
Total livestock (TLU)	User	5.45	3.80	2.008/**
	Non user	4.55	3.88	

***, ** Significant at less than 1 and 5% probability levels.

earning was higher for non-users. Close examination of the data exhibit that remunerative off farm income sources like cart and trade were the results of irrigated agriculture whereas inferior livelihood activities like fire wood and charcoal selling, and causal work were dominated by non irrigators. This finding is similar to the findings of Getaneh (2011) which states small-scale irrigation has a negative impact on non-farm incomes. Income share by category indicate that 76 and 70.5% of total incomes for users and non-users respectively come from crop, while the rest from livestock and off farm activities. Irrigators earned 47.4% higher than that of non-irrigators from crop alone and this difference is statistically significant (Table 3).

Irrigation improved asset endowments

Irrigation allows a greater area of land to be used for crops and asset ownership increases with access to irrigation (Hussain, 2004). This study paid attention to the basic production resources like land and livestock, as well as total value of household goods (farm tools and furniture's) estimated at purchase price. Accordingly, the value of asset owned by irrigators is three fold of non irrigators. Access to irrigation increases mean land ownership by 0.38 ha and it enhance livestock ownership by a factor of 0.91 tropical livestock unit (TLU) (Table 4).

Irrigation improved household consumption

In order to measure the impact of irrigation on household consumption, expenditure pattern was used as a proxy indicator for standard of living. This usually refers to the ability of the household to produce/purchase a basket of goods containing the minimum quantity of calories and non-food commodities. Accordingly, the average consumption expenditure per adult equivalent (AE) per annum for irrigators is more than twofold of non irrigators. Similarly, the value of home consumption, food and non food expenditures are significantly higher than that of non-users. For instance non irrigators consumption from own production is only about 51% of that of irrigation beneficiaries. This indicates that access to irrigation improves food security through home consumption by increasing the frequency of production. It also enhances the capacity to access food through purchase by 50.7%. Thus, there is a positive correlation between nutritional status and irrigation access. It has also a positive impact on non food consumption. The non food consumption value of non-users was 60.8% of that of irrigators (Table 5). Thus, this study could argue that irrigation access improves overall welfare of rural households through

Table 5. Expenditure pattern of households with and without irrigation.

Expenditure (ETB)	User		Non user		F	P
	Mean	Standard deviation	Mean	Standard deviation		
Food	3467.8	2965.2	1715.6	1813.6	40.934	0.000***
Non food	2540.6	4725.5	1546.5	2052.2	6.073	0.014**
VOC	5968.9	19828.1	3047.1	2660.7	3.57	0.060*

***, **, * Significant at less than 1, 5 and 10% probability levels; VOC, value of own consumption.

Table 6. Poverty status and indices by access to irrigation.

Irrigation use	Poverty status				X^2	P
	Non-poor		Poor			
	N	%	N	%		
User	130	89.7	15	10.3	**51.152**	**0.000*** **
Non user	88	52.3	80	47.6		
Total	218	69.6	95	30.4		

Irrigation use	Head count index (α = 0)	Poverty gap (α = 1)	Squared poverty gap (α = 2)
User	0.10	0.042	0.02
Non user	0.48	0.17	0.09

***Significant at less than 1% probability level.

improved food access, non food consumption and asset accumulation.

Irrigation contributed to poverty reduction

Local poverty line: There are many different concepts of poverty in various disciplines. It has been increasingly realized that poverty is a multidimensional concept, extending from low levels of incomes and expenditures to lack of education and poor health, and includes other social dimensions such as powerlessness, insecurity, vulnerability, isolation, social exclusion and gender disparities. This study made use of cost of basic needs to set poverty lines. The first activity in this approach is to identify a bundle of food and non food items usually consumed by the 20% lowest income quartile and estimating the cost of meeting this need (Ravallion, 1994). Accordingly, the food poverty line (FPL) for this study is 1016.49 ETB per AE per year, whereas the total non food expenditure is 310.64 birr per AE per year which covers clothing, medication, tax and social obligation costs. Adding all these expenditures from the lowest income group will make the total poverty line beyond which an individual is considered to be non-poor. Thus, the poverty line was 1016.49 birr per AE per year.

Poverty status and indices by access to irrigation: Table 6 shows from the 313 sample households, 30.4% of them are poor, which accounts for 47.6% of non-users and 10.3% of the users, which implies that poverty incidence is 37.3% higher in rain-fed only farm than irrigation. The rest 89.7% of the users and 52.3% of non-users respectively are non-poor. This confirms that irrigation development is a key for poverty reduction. The fact that 10.3% of irrigation beneficiaries being poor entails, on one hand access to irrigation is a necessary, but not a sufficient condition for poverty alleviation, and on the other hand, poverty may be adversely affected where irrigation is mismanaged leading poverty. In addition, one has to understand that poverty is a complex phenomenon. The study showed that 48 and 10% of the non user and user households respectively were living below the locally determined poverty line on the head count basis. The corresponding poverty gap by irrigation use was 0.042 and 0.17 for user and non user, respectively; whereas poverty severity index was 0.02 and 0.09 for users and non-users respectively (Table 6). Thus, poverty is more severe and widespread among non irrigators than irrigators.

Determinants of poverty: Binomial logit model was used to identify factors pushing in or pulling out households of poverty. As the major focus of this study aims to investigate the role of irrigation in poverty reduction; poverty is considered as the dependent variable of the model, while the variables are listed in

Table 7. Binomial logit model result for determinants of poverty.

Irrigation user	B	Standard error	Wald	Significant difference	Exp (B)
	-0.572	**0.289**	**3.900**	**0.048****	**0.565**
Age of head	0.045	0.014	9.845	0.002***	1.046
Household size	0.521	0.095	29.847	0.000***	1.683
Dependency ratio	0.267	0.140	3.636	0.057*	1.306
Farm size	-0.859	0.230	13.967	0.000***	0.424
Livestock holding (tlu)	-0.153	0.063	5.893	0.015**	0.858
Education of head	0.098	0.048	4.185	0.041**	1.103
Distance to market	-0.017	0.038	0.201	0.654	0.983
Constant	-3.993	0.812	24.157	0.000***	0.018
Pearson X^2	5.109***				
-2 Log likelihood	308.208				
Sample size	313				

***, **, * significant at less than 1, 5 and 10% respectively.

Table 7 including irrigation use are independent variables that determine the likelihood of being poor or not. Before running the model, the study used the variance inflation factor and contingency coefficients to check for multicollinearity among continuous and discrete variables respectively. According to the test result, multicollinearity was not a serious problem among the continuous variables. However, there is strong association between irrigation use and sex of household head. As a result, sex of head was removed from the model. The regression classification table revealed that binomial logistic model managed to predict 82.6% of the responses correctly. The model chi-square statistic for Hosmer and Lemeshow test also showed the chi-square value was found to be 5.025 and the overall model was found non-significant at 0.755 levels stating that the model adequately fits the data (Table 7).

Interpretation of significant variables

The results of binomial logit verify that most of the explanatory variables in the model have the signs that conform to our prior expectations, except education of head. Thus, irrigation use with the odds of being poor over non-poor was negatively correlated and significant. This means the probability of being poor decreases by a factor of 0.565 for those households with access to irrigation keeping other factors constant. This suggests that the probability of being poor decreases if one has access to irrigation. This finding is incongruent to the findings of Ayalneh and Korf (2009) and Getaneh (2011). Hussain (2004) also noted that irrigation contributes to poverty alleviation both directly and indirectly. It may lead to poverty reduction via increased yields, increased cropping areas and higher value crops and raising employment opportunities (FAO, 2003). Among demographic factors, age of household head was positively and significantly related to the probability of being poor; hence, old age is the cause of poverty. That means as age of the household head increases, this contributes to household poverty. The probable reason is that with age asset depletes for example land decreases upon inheritance to children. The results are consistent with the study of Gyekye and Akinboade (2001) and Sabir et al. (2006). But, it is against the findings of Ayalneh and Korf (2009); which stated that older households have greater likelihood of being non-poor. Household size positively affected the probability of a household to be poor; a unit increase in household size increased the probability of being poor by 1.683. This finding is consistent with that of Alemu et al. (2009) and Ayalneh and Korf (2009).

Similarly, dependency ratio was found to positively and significantly affect the probability of being poor by a factor of 1.3. This ratio allows one to measure the burden weighing on members of the labor force within the household. It is also in agreement with findings of Gyekye and Akinboade (2001), which stated that poverty is more likely to be associated with large households with a high dependency ratio. As expected, ownership of land and livestock showed strong negative effects on the probability of households to be poor. A unit increase in landholding and livestock holding increased the probability of a household being non-poor by 0.4 and 0.8 respectively. This finding is also similar to that of Alemu et al. (2009) and Ayalneh and Korf (2009). Contrary to expectation education of head was found to influence poverty positively and significantly at $P<0.05$. It seems illogical, but the possible reason is that the educational attainment of sample households was below the level that guarantee employment and it did not allow them to generate income as a result of their education. People who have obtained jobs in the urban areas are in general better educated (at least completed 10th grade level), which only few of the sample households achieved. Thus, unlike some findings of Ayalneh and Korf (2009), educational level of household heads was not found to have a negative effect on poverty.

CONCLUSION AND RECOMMENDATION

This study assessed the role of small scale irrigation on poverty based on 313 households of the rift valley river basin. The roles that SSI played were seen in terms of increasing production, income, assets, and employment opportunity, as well as poverty reduction. Both the descriptive and econometric analysis showed that irrigation use has a positive effect on farm production, income, asset endowment, and employment opportunity and poverty reduction.

Thus, it is pertinent to conclude from this study that irrigation development helps to increase household income and reduces the incidence of poverty at the household level. It can benefit the poor through raising yields and production and nonfarm employment. However, the economic performances of irrigation systems in the study areas were constrained due to imperfect market structure and financial shortages.

The following recommendations were given based on the findings of the study. Water users associations should be organized and empowered in order to improve the performance of SSI schemes; simultaneously, cooperatives should be encouraged and empowered in order to solve the marketing constraints of members. In this regard, agricultural extension should be improved and include market information and business training. The most crucial ones are linking the traders and the producers to work as partners. Institutional support towards capacitating, training, and coordinating rural cooperatives would play an inevitable role in enhancing the effect of irrigation on poverty reduction.

ACKNOWLEDGEMENTS

The authors would like to thank UNESCO-IHE [the University Water Sector Partnership- Demand Driven Action Research (DDAR) project] for financing this research, and Arba Minch and Hawasa Universities for providing the necessary facilities. Our special thanks will be for Oromiya Water Mines and Energy bureau, SNNPR Irrigation Development and Scheme Administration Agency.

We are also highly indebted to Dr. Flip Wester, Dr. Frank van Steenbergen, Mrs. Susan Graas, Mr. Yohannis Geleta and Gezahegn Wejebo for their all round support to realize this research.

REFERENCES

Alemu D, Bewket W, Zeleke G, Assefa Y, Trutmann P (2009). Extent and determinants of household poverty in rural Ethiopia: A study of six villages. Muse project, https://docs.google.com/document/preview?hgd=1&id=12KsqSBLg-h-ZZjqtTbaaKP8TwkgdfzOlJaLWthMYm9U&pli=1

Ayalneh B, Korf B (2009). Analysis of poverty and its covariates among smallholder farmers in the eastern Hararghe highlands of Ethiopia. Proceedings of the International Association of Agricultural Economists Conference. August 16-22, 2009, Beijing China.

Bekele S, Denekew A, Loulseged M, Loiskandl W, Ayana M, Alamirew T (2007). Water Resources and Irrigation Development in Ethiopia. Colombo, Sri Lanka: International Water Management Institute. P.78.

Berhanu GB (2001). Food Insecurity in Ethiopia: The Impact of Socio-political Forces. Development research Series Research Center on Development and International Relations (dir) Working Paper no. 102.

Bruce JW, Hoben A, Dessalegn R (1994). After the Derg: An Assessment of Rural Land Tenure Issues in Ethiopia. Land Tenure Center, University of Wisconsin-Madison.

FAO (2003). Preliminary review of the impact of irrigation on poverty with special emphasis on Asia Land and water development division Water resources, development and management service Food and Agriculture Organization of the United Nations Rome.

Farah KO, Nyariki DM, Noor AA, Ngugi RK, Musimba NK (2001). The socio-economic and ecological impacts of small-scale irrigation schemes on pastoralists and drylands in Northern Kenya. J. Soc. Sci. 7:267-274.

FDRE (2003). The new coalition for food security in Ethiopia food security programme M and E Plan. Volume I. http://www.food-security.nl/resource/new-coalition-food-security-ethiopia-food-security-programme-me-plan

FDRE (2010). Ethiopia's Agricultural Sector Policy and Investment Framework (PIF) 2010-2020. Federal Democratic Republic Government of Ethiopia, Ministry of Agriculture and Rural Development.

Foster J, Greer J, Thorbecke E (1984). A Class of Decomposable Poverty Measures. Econometrical 52(3):761-766.

Getaneh KA (2011). The impact of selected small-scale irrigation schemes on household income and the likelihood of poverty in the Lake Tana basin of Ethiopia. M.Sc Thesis, Cornell University.

Gujarati DN (2003). Basic Econometrics, fourth edition. McGrew-Hill, Inc. New York.

Gyekye AB, Akinboade OA (2001). Analysis of poverty in the northern province of South Africa: Implications for empowerment policy. Proceedings of the 75th Anniversary Conference of the Economic Society of South Africa at Glenburg Lodge, September 13, 2001, Johannesburg. http://www.iwmi.cgiar.org/propoor/files/ADB_Project/Research_Studys/poverty%20study%20-ethiopia.pdf

Hussain I (2004). Assessing Impacts of Irrigation on Poverty: Approaches, Methods, Case Studies and Lessons. International Water Management Institute (IWMI). http://www.iwmi.cgiar.org/propoor/files/ADB_Project/Research_Studys/poverty%20study%20-ethiopia.pdf

Kinfe A, Chilot Y, Rajan S (2012). Effect of small-scale irrigation on the income of rural farm households: the case of Laelay Maichew district, central Tigray, Ethiopia. J. Agric. Sci. 7:1

Mekuria T (2003). Small-scale irrigation for food security in sub-Saharan Africa. Report and recommendations of a CTA study visit Ethiopia, 20–29 January 2003CTA Working Document Number 8031.

MOFA (2007). Food security strategy for Ethiopia. Ministry of Foreign Affairs of (FDRE) Federal Democratic Republic Government of Ethiopia.

MoFED (2006). Building on progress: A plan for accelerated and sustained development to end poverty 2005/6–2009/10. Ministry of Finance and Economic Development Ethiopia: Addis Ababa, Ethiopia.

MoFED (2010). The Millennium Development Goals Report. Trends and prospects for meeting MDGs.http://www.un.org/millenniumgoals/11_MDG%20Report_EN.pdf .

MoFED (2012). Progress towards eradicating poverty. An Interim

Report on Poverty Analysis Study.
Oni SA, Maliwichi LL, Obadire OS (2011). Assessing the contribution of smallholder irrigation to household food security, in comparison to dryland farming in Vhembe district of Limpopo province, South Africa. Afr. J. Agric. Res. 6:2188-2197.
Ravallion M (1994). Poverty Comparisons, Harwood Academic Publishers, Chur, Switzerland.
Sabir HM, Hussain Z, Saboor A (2006). Determinants of small farmers poverty in the central punjab (Pakistan). J. Agric. Soc. Sci. 2:11-12.
Samuel G (2006). Intensification of Smallholder Agriculture in Ethiopia. Proceedings of the Options and Scenarios Study prepared for the Future Agricultures Consortium Meeting. 20- March 22, 2006, Institute of Development Studies.
World Bank (2006). Ethiopia: Managing Water Resources to Maximize Sustainable Growth. Water Resources Assistance Strategy for Ethiopia. Washington DC.

Eradicating extreme poverty among the rural poor in Uganda through poultry and cattle improvement programmes - A Review

J. Ekou

Department of Animal Production and Management, Faculty of Agriculture and Animal Sciences, Busitema University, P. O. Box 236, Tororo, Uganda.

Uganda is still struggling with chronic poverty and malnutrition especially among women and children despite the targeted efforts towards eradicating poverty by transforming subsistence agriculture to intensive or commercial agriculture that have been in place for the last 13 years. The aim of this paper was therefore to analyze the Republic of Uganda's current strategy on eradication of extreme hunger and poverty, identify the major constraints and suggest possible strategic interventions using examples of poultry and cattle. The article has identified unemployment and persistent inflation as major constraints to modernization of agriculture in Uganda and has offered development of small holder poultry production, promotion of poultry crossbreeds, rehabilitation of community dip tanks and promotion of dairy crossbreeds as strategic intervention areas to alleviate extreme hunger and poverty in Uganda. In conclusion, the strategies used by the government of Uganda to fight poverty and hunger did not favour the rural poor but have supported the rich and thus failed to cause a notable impact. It is therefore recommended that adoption of the suggested intervention areas shall overcome the said bottlenecks and accelerate eradication of hunger and poverty.

Key words: Livestock improvement, poverty eradication, rural farmers.

INTRODUCTION

Over 85% of Uganda's population lives in rural areas where agriculture is the major contributor to their livelihoods. Livestock accounted for 16% of agricultural Gross Domestic Product (GDP) in 2003, fisheries (12%), cash crops (17%) and the rest came from food crops (RoU, 2004). Despite the fact that agriculture supports most livelihoods and is the main source of exports, contributing as much as 85% of export earnings, the overall share of agriculture in GDP has declined in recent years, from 50% in the early 1990s to 23% in 2008 (RoU, 2004; FAO, 2009). A slowdown in the growth of agricultural production besides declining agricultural prices and insecurity in northern and eastern Uganda have all contributed to the drag on agriculture (FAO, 2010). Consequently, many communities in Uganda struggle with chronic malnutrition, especially among children. The prevalence of stunting growth among children less than five years of age is nearly 40% across the country, and is higher in Karamoja and the southwest, where it exceeds 50% (Kikafunda et al., 1998; FAO,

2010; WFP, 2011). Malnutrition accounts for 40%of all child deaths in Uganda (Bridge et al., 2006), and the prevalence of childhood anemia exceeds 70% (WFP, 2011). According to von Grebmer et al. (2011), Uganda has a Global Hunger Index (GHI) score of 16.7, placing it 42nd out of 81 countries ranked in 2011 and a hunger situation considered as being *serious*. The Millennium Development Goal (MDG) which seeks to cut the number of underweight children in half before 2015 thus remains elusive. However, the agricultural sector presents a great opportunity for poverty eradication since it employs over 80% of the labour force (MAAIF, 2000).

The Plan for Modernization of Agriculture (PMA) has been part of the government of Uganda's broad strategy on poverty eradication contained in the Poverty Eradication Action Plan (MAAIF and MFPED, 2004). The mission of the PMA has been to eradicate poverty by transforming subsistence agriculture to commercial agriculture. Under PMA, the government of Uganda formulated the National Agricultural Advisory Services (NAADS) programme (RoU, 2001) to transform natural resource based communities out of poverty to better livelihoods. The programme was meant to increase farmer access to information, knowledge and technology through effective, efficient, sustainable and decentralized extension service delivery with increased private sector involvement. However, poverty levels in Uganda have remained high. Uganda's national poverty levels stood at 24% in 2010 with Northern region registering 46.2%, the highest in the country (UBOS, 2010). The impact of poverty reduction strategies has been minimal (Kaduru, 2011). The aim of this article was therefore to analyse the Republic of Uganda's current strategy on eradication of extreme hunger and poverty, identify the limitations and suggest possible strategic interventions using examples of poultry and cattle.

OVERVIEW OF UGANDA'S NAADS PROGRAMME

The NAADS programme has gone through a number of policy changes. The strategic frame work that guided NAADS Phase I was a decentralized, farmer owned public/private sector serviced extension system contributing to the realization of the agricultural sector objectives (RoU, 2001). Advisory services were provided by service providers termed Agricultural Advisory Service Providers (AASPs) contracted by the sub-county on behalf of sub-county farmer forum. In contracting AASPs, the following guiding principles were observed: Farmer participation, competitive and transparent selection process, previous performance, qualification and integrity. Sources of AASPs have been contracted service providers, researchers and private sector providers under public/private sector partnerships. These were supported by subject matter specialists (SMSs), model farmers and community based facilitators.

In 2010, the National Development Plan was formulated (NDP, 2010).

Thus, NAADS was redesigned under phase II to align NAADS with National Development Plan and Strategic Investment Plan (MAAIF, 2010; NDP, 2010; MFPED, 2010) and to enhance the effectiveness of Agricultural Advisory Services delivered within the frame work of NAADS vision, mission and principles.The strategic elements of NAADS phase II were to create option for financing and delivery of extension/advisory and technical services appropriate for different farmer categories, gradually reduce the share of public financing of public advisory/extension service cost to the extent that by the end of 25 years of NAADS it will account for less than 50% through PPP, utilize professional and certified service providers competitively recruited and on performance based contracts, empower subsistence and other farmer categories to access extension/advisory services and relevant information for informed decision making and to develop public and private sector capacity professional capability and service systems. Under NAADS Phase II (MAAIF, 2010), the Agricultural technology and agribusiness advisory services (ATAAS) project was aligned to NDP and the DSIP. ATAAS was designed to support implementation of programmes of the National Agricultural Research Organisation (NARO) and NAADS. The project is to support key activities along research, extension, farmer-market value chain continuum through five components.

The mandate of ATAAS is to promote productivity and Agricultural Production through close collaboration between research by NARO and extension by NAADS under the DSIP of MAAIF. The ATAAS Project began in July 2011 (MAAIF, 2010). The first component comprises technology promotion and farmer access to information. This targets advisory service provision to accelerate farmer progression towards commercialization while ensuring food security. Component two is concerned with agribusiness development for supporting market access so to shift from direct intervention by NAADS in supporting farmer enterprises to facilitate the performance for value chain and functioning or emergence of businesses that can provide production support services that contribute to value chain development and input and output market access among others. The third component comprises institutional development and programme management. This handles roles, responsibilities and institutional relationships between NAADS and all institutions (public and private) especially local government to accommodate new changes to the NAADS mandate. Component four involves planning, monitoring and evaluation to enhance the use of participatory monitoring and evaluation (PM&E) information in management decision making and to develop human organizational financial and network capacity for PM&E (MAAIF, 2010). Studies conducted have however demonstrated that NAADS interventions have not had a notable impact (Benin et al., 2007; EPRC, 2011).

MAJOR CONSTRAINTS TO MODERNISATION OF AGRICULTURE IN UGANDA

Like most developing countries, Uganda's economy is dominated by two large problems: unemployment and inflation. These form the major constraints for the planning of the development of the livestock sector. This section will discuss these constraints and how they could have impacted on the modernisation of the livestock sector in Uganda.

Unemployment

According to UBOS (2010), 66% of all working persons in Uganda are employed in agriculture, 82% of the workers live in rural areas where agriculture is the main source of livelihood, 70% of the labour force does not have any formal education while the unemployment rate stands at 4.2%. If the increase in production is to be brought about by modernisation and intensification of animal production, and technically it could be, this will only occur in a capitalist form that is to say by substituting capital for labour by relatively elaborate techniques which will not permit the use of the rural youth and the unemployed (Tacher, 1992). Introduction of modern large-scale industrial livestock units would lead to problems for smallholder producers who cannot compete for available feed resources and markets (Ogle and Phuc, 1997) and may not have the skills for the more sophisticated management which is required (Preston, 1995). Uganda's strategy of distributing broiler and layer chicks to rural youth groups could have failed due to limited management skills. This usually results in a reduction of rural employment opportunities and easily turns the problem of rural unemployment into an even greater problem of urban unemployment.

Besides, often these techniques cause a transfer from the peasant class towards the well-off classes of the population, classes more open to modern technology (Tacher, 1992). For example an increase in milk production can mainly be brought by a leap in technology, which often can only be done by people with sufficient means who have received some level of education. This type of person is found among those sectors of the population already in the favoured position. In Uganda over two thirds of the working population work in the rural areas and 70% of the labour force does not have any formal education (RoU, 2004; UBOS, 2010) and often provide production which is insufficient at the national level. Modernisation might then permit an increase in production but might never help in the resolution of the crucial problem of unemployment and might even risk accentuating it. The disadvantaged people will be faced with impoverishment, the gap between the few rich and the major poor will become larger and larger, and social evils will become a heavy burden (Trach, 2009). Already, data indicates that Uganda's Gini coefficient has been rising over the years, and currently stands at 0.426, (UBOS, 2010) indicating a worrying trend towards increased income inequality. Intensification could moreover have two other major disadvantages. It might be situated in the areas around towns, which might increase regional disequilibrium and it might not favour the fight to improve the living standards of the rural poor, which could ensure the continuation of the actual vicious cycle of poverty: the crisis of the rural areas leads to shrinking of aggregate demand, which in turn is the source of unemployment (Tacher, 1992).

Persistent inflation

Uganda has witnessed regular surges in inflation over the past years. In October 2011, for example, Uganda recorded the second highest level of inflation (30.5%) in the East African region (Kabundi, 2012). This rise in price has been an issue of concern for policymakers and the general public. Constraints on agricultural production together with high demand both domestically and from neighbouring countries push domestic food price and hence creates a rise in overall price level, given high age share of food price in Consumer Price Index (CPI) (RoU, 2004; FAO, 2009; Kabundi, 2012). As explained by Tacher (1992), politicians are extremely sensitive to variations of food prices. In developing countries such as Uganda, nominal salaries are very low and low food prices create a salary which finds itself increased accordingly, in real terms.

However, an increase in prices always leads to claims for salary increases. Trying to stifle this inflation generates tensions, particularly in towns, and so governments seek to keep prices as low as possible. This has frequently been done without success by fiscal means. There is evidence of considerable rise in real money growth, attaining a maximum of 36% in November of 2010, prior to the rise in inflation (Kabundi, 2012). As suggested by Nachega (2001), monetary aggregate portrays an equilibrium relationship with Inflation. It means that expansionary policy that drives up money supply is inflationary over the long-run. As a result of lack of success, the government then tries to keep the consumption price at a low level by other means such as liberal import policies. However, the liberal import policy stifles the development of local production (IMF, 2011). That is even truer with exports from developed countries, with which the local product must compete, and which are often subsidised in a more or less disguised way. This pressure on prices is thus implicitly a transfer from livestock rearing to the rest of the economy (Tacher, 1992).

WHAT SHOULD BE THE STRATEGIC INTERVENTIONS FOR THE LIVESTOCK SECTOR?

Poultry

The total poultry population in Uganda was projected to be about 32.6 million birds for 2006/2007 year from 23.5 million in 2002. Of this, about 80% is free-range indigenous type of breeds while the commercial types mainly composed of exotics are about 20% (Byarugaba, 2007). Similarly, the economic and nutrient contribution of the indigenous free-range poultry has been estimated to be over 80% of the per capita consumption of poultry meat and eggs (Byarugaba, 2007). This is therefore an excellent intervention area. The approaches below are thus suggested as strategic interventions areas in the poultry enterprise.

Smallholder poultry (village chicken) development

The capacity for broodiness has been bred out of commercial-strain layer and broiler hens so as to maximize meat and egg production. They are therefore incapable of natural reproduction, and their value in a village environment is thus quite limited. Besides, they also have limited ability to withstand the harsh environment (FAO, 2010). The development of smallholder poultry production systems in Uganda, particularly village chicken production is thus crucial in meeting the nutritional, income, employment and gender needs of rural people (Kusina and Kusina, 1999). The hens become broody, so can reproduce without the need for artificial incubation and brooding; They are agile and can run fast, fly and roost in trees, so can escape predators; They have been shown to be more resistant to bacterial and protozoan diseases and to parasitic infestations than commercial broilers and layers. Their meat and eggs are generally preferred to those from commercial birds, not only by rural communities but also by urban dwellers because of their taste, leanness and suitability for special dishes (Ssewanyana et al., 2001; FAO, 2010). The major disease constraints affecting production include Newcastle disease, fowl typhoid, Salmonelloses among others. Fortunately, there are effective vaccines against these diseases. This is where the government should heavily invest in addition to distribution of commercial broiler or layer chicks.

Development of crossbred chicken

Indigenous hens often lay only 40 to 60 eggs per year while commercial layers developed from imported parent stock have the capacity to lay more than 300 eggs per year. The growth rate of indigenous genotype of chickens is also generally much slower than that of commercial chicken. Indigenous cocks often weigh not more than 1.0 kg at 20 weeks while broilers under typical confinement may reach 2.0 kg live weight at five weeks of age (FAO, 2010). There is therefore need to integrate the good features of local chicken that make them adaptable to the local conditions and those that are responsible for high productivity among exotic chicken (Kyule, 1994). To improve productivity of the indigenous chicken, the Malawi government introduced the Smallholder Poultry Improvement Programme (SPIP) in the 1950s with the main objective of increasing egg and meat production of indigenous chicken through crossbreeding with the Black Australorp (Safalaoh, 2001).

The future of cross bred chicken is promising in Uganda. At the National Semi-Arid Resources Research Institute (NaSARRI) chicken breeding project, the growth capacity of local chicks was compared with chicks from crosses of the local bird with the Bovans Brown and raised at the farms from day old by using ordinary feeding regimen at the farm. The results showed that 50% of the Bovans gene had higher body weights compared with local chicks and chicks with 25% Bovans gene. Furthermore, at 120 days of age, the difference was at maximum when expressed as a proportion of body weight of local chicks. However, they also recommended studies on the reproductive capacities of the crossbreeds (Sorensen and Ssewanyana, 2003).

Ssewanyana et al. (1998) further studied on-farm performance of the crossbreeds and found that; mean body weight of F_1 chickens (50% BBxLH) increased by 34%, egg number per clutch by 94%, egg weight by 29%, egg circumference by 10% and egg length by 2% over that of the local chickens. These results indicated that the technology of improving indigenous chickens through crossbreeding with exotic cocks was technically and economically viable even under on - farm conditions. Therefore, poultry production based on a genetic improvement of the local breeds is possible in using appropriate breeding programme that includes traits of importance for reproduction and survival under the smallholder environment, which also enhances the welfare of hens that can incubate and brood chicks as required by the rural farmers (FAO, 2010).

Cattle

Rehabilitation of community dip tanks

Tick-borne diseases are a major constraint to the improvement of livestock production in the developing world particularly the sub-Saharan Africa (Norval et al., 1992; Bell-Sakyil et al., 2004). The cost of controlling ticks and tick-borne diseases was reported to constitute about 85.6% (pastoral) and 73.8% (ranches) of total disease control costs (Ocaido et al., 2009). The major tick-borne diseases in Uganda are anaplasmosis, babesiosis, cowdriosis and East Coast fever (ECF). Together, these constitute the single most important constraint to the livestock production in Uganda. Originally, government of Uganda under the Ministry of Agriculure, Animal Industry and Fisheries (MAAIF) had established community dip tanks in strategic locations at all districts in the Country. However, following decentralisation, these communal dips have either collapsed or are neglected. The presence of many non functional dip tanks in the districts has been attributed to failure by the districts to rehabilitate and maintain the communal dips (King and Mukasa-Mugerwa, 2002; RoU, 2009). Dipping of livestock in dip tanks as a practice to kill off ticks and nuisance flies is no longer undertaken in most districts of Uganda. Farmers have since then failed to dip their livestock resulting in high prevalence of ticks and tick-borne diseases. There is therefore need to rehabilitate and construct community dip tanks as a key intervention in the control of ticks and tick-borne diseases.

Promotion of dairy crossbreeds

Currently the average live weight by a cow in most herds of Uganda is 180 to 350 kg and this requires about 5 years to be attained even if pastures are available (Jain and Muladno, 2009). However, they carry great advantages of relative resistance to tropical diseases, and ability to survive in high temperature zones and these should be conserved. By choosing exotic breeds from industrialized countries had been intensively selected for increased productivity and crossbreeding with local breeds, rapid improvement of local breeds can be realized. Crossbreeding, which uses complementary breed differences, avoids antagonistic genetic relationships and utilizes heterosis, is recommended for genetic improvement of farm animals (Smith, 1964; Moau, 1966; Dickerson, 1969, 1972). Such an approach would focus on crossbreeding of the indigenous breeds of cattle (Zebu) and Ankole longhorn (Sanga cattle) with large fast growing dairy breeds of Friesian, Jersey, Guernsey, Ayshire. Half crosses would be the most suitable genotype in most parts of Uganda. Previous studies have shown that at various heat intensities above 27°C, half Friesian-Zebu cattle produce more milk when compared to the three-quarter cross during the stage of maximum lactation inspite of the higher genetic potential of the latter (Igono and Aliu, 1982).

CONCLUSION

The strategies used by the government of Uganda to fight poverty and hunger, did not favour the rural poor since they were not considered in the conceptualization, formulation and implementation of these strategies. Instead, they have supported the rich and thus failed to cause a notable impact. It is therefore recommended that adoption of the suggested intervention areas will overcome the said bottlenecks and accelerate eradication of hunger and poverty in the country.

REFERENCES

Bell-Sakyil L, Koneye EBM, Dogbeyo O, Walker AR (2004). Incidence and prevalence of tick-borne haemoparasites in domestic ruminants in Ghana. Vet. Parasitol. 124:25-42.

Benin S, Nkonya E, Okecho G, Pender J, Nahdy S, Mugarura S, Kato E, Kayobyo G (2007). Assessing the Impact of the National Agricultural Advisory Services (NAADS) in the Uganda Rural Livelihoods. IFPRI Discussion Paper 00724. Available at: http://www.ifpri.org/publication/assessing-impact-national-agricultural-advisoryservices-. (Accessed on 10/5/2013).

Byarugaba D (2007). The structure and importance of the commercial and village based poultry systems in Uganda. Kampala, FAO. (Consultancy report). Available at:apf-uganda.ning.com/forum/attachment/*download?id*. (Accessed on 12/3/3013).

Dickerson G (1969). Experimental approaches in utilizing breed resources, Anim. Breed. 37:191-202.

Dickerson G (1972). Inbreeding and heterosis in animals, Proc. Anim. Breed. Genet. Symp. in honour of J. L. Lush, Champaign, IL. pp. 54-77.

Economic Policy Research Centre-EPRC (2011). An insight into the performance of the National Agricultural Advisory Services programme: A briefing paper presented to The Sessional Committee on Ministry of Agriculture, Animal Industry and Fisheries (MAAIF), Republic of Uganda. Available at: www.eprc.or.ug/pdf_files/briefingpaper_naads.pdf (Accessed on 10/5/2013).

FAO (2010) Uganda Nutrition Profile 2010. Nutrition and Customer Protection Division. Food and Agriculture Organization of the United States. Available at: ftp://ftp.fao.org/ag/agn/nutrition/ncp/uga.pdf. (Accessed on 12/04/2011).

FAO (2009). More people than ever are victims of hunger"; Press release, FAO, Rome; *www.fao.org/* fi leadmin/user_upload/newsroom/docs/pdf. (Accessed on 11/4/2011).

FAO (2010). Chicken genetic resources used in smallholder production systems and opportunities for their development, FAO Smallholder Poultry Production Paper No. 5. Rome.

Jain AK, Muladno M (2009). Selection criteria and breeding objectives in improvement of productivity of cattle and buffaloes In Selection and Breeding of Cattle in Asia: Strategies and Criteria for Improved Breeding. IAEA-TECDOC-1620.

Igono MO, Aliu YO (1982). Environmental Profile and Milk Production of Friesian-Zebu Crosses in Nigerian Guinea Savannalnt. J. Biometeor. 26(2):115-120.

IMF (2011) Uganda," IMF Country Report 11/308.

Kabundi A (2012). Dynamics of Inflation in Uganda, Working Paper Series N° 152, African Development Bank, Tunis, Tunisia.

Kaduru R (2011). Strategies for eradicating poverty in LDCs: Findings from research by the Chronic Poverty Research Centre. Proceedings of the Fourth United Nations Conference on the Least Developed Countries. Available at: www.sustainabledevelopment.un.org/futurewewant.html (Accessed on 20/5/2013).

Kikafunda JK, Walker AF, Collet D (1998). Risk factors for early childhood malnutrition in Uganda: A Cross Sectional Study. Pediatrics 102(4):45-53.

King A, Mukasa-Mugerwa E (2002). Livestock Marketing in South Sudan with particular reference to cattle trade between South Sudan and Uganda. Available at: www.eldis.org/vfile/upload/1/document/0708/DOC11410.pdf (Accessed on 10/5/2013).

Kusina JF, Kusina NT (1999). Feasibility study of agricultural and household activities as they relate to livestock production in Guruve

District of Mashonaland Province with emphasis on poultry production. Report for the Household Agricultural Support Programme (HASP), Zimbabwe.

Kyule SK (1994). Improvement of Poultry through management and breeding. Proceedings of the refresher course on appropriate technology options for small livestock husbandry. Nairobi, Kenya.

MAAIF (2010). NAADS Implementation guidelines: Farmer selection and support for progression from subsistence to commercialization; NAADS Secretariat, MAAIF, Republic of Uganda. Available at: www.drtug.org/.../NAADS%20PhaseII%20implementation%20...(Accessed on 10/5/2013).

MAAIF (2000). Plan for Modernisation of Agriculture: Eradicating Poverty in Uganda: Government Strategy and *Operational Framework*. Kampala: Ministry of Agriculture, Animal Industry and Fisheries (MAAIF), and Uganda Ministry of Finance, Planning and Economic Development.

MFPED - Ministry of Finance, Planning and Economic Development (2010). Support to Uganda's Agricultural Sector Development Strategy and Investment Plan (DSIP): Funding proposal For Global Agriculture and Food Security Program (GAFSP). Kampala, Uganda. Available at: http://www.gafspfund.org/gafsp/sites/gafspfund.org/files/Documents/Uganda_3_of_6_Proposal.pdf (Accessed on 06/5/2013).

MAAIF and MFPED (2004). Poverty Eradication Action Plan 2004/5-2007. Kampala: Government of Uganda.

Moau R (1966). Specialized sire and dam lines: Economic evaluation of cross-breds, Anim. Prod. 8:193-202.

Nachega JC (2001). Financial Liberalization, Money Demand, and Inflation in Uganda, IMF Working Paper 01/118. Available at: www.imf.org/external/pubs/ft/wp/2001/wp01118.pdf (Accessed on 20/3/2010).

NDP (2010). The National Development Plan 2010/11 - 2014/15. Kampala: Government of Uganda.

Norval RAI, Perry BD, Young AS (1992). The Epidemiology of Theileriosis in Africa. Academic Press Ltd., London.

Ocaido M, Muwazi RT, Opuda JA (2009). Economic impact of ticks and tick-borne diseases on cattle production systems around Lake Mburo National Park in South Western Uganda. Trop. Anim. Health Prod. 41(5):731-739.

Ogle B, Phuc BHN (1997). Sustainable intensive livestock-based systems in Vietnam. IRD Current. 14:16-22.

Preston TR (1995). Strategy for sustainable use of natural renewable resources: constrains and opportunities. Tropical Feeds and Feeding Systems. FAO. Available at: www.fao.org/ag/aga/AGAP/FRG/ECONF95/PDF/TRP1.PDF (Accessed on 24/5/2013)

RoU (2001). National Agricultural Advisory Services Act. Kampala, Uganda.

RoU (2004). Increasing incomes through exports: a plan for zonal agricultural production, agro processing and marketing, Republic of Uganda. Available at: www.ugandaexportsonline.com/strategies/zoning_plan.pdf (Accessed on 26/2/2012).

RoU (2009). Value for money Audit Report on the prevention and control of livestock diseases by the Department of livestock health and entomology in the Ministry of Agriculture Animal industry and Fisheries. Office of the Auditor General, Republic of Uganda. Available at: www. afrosai-e.org.za/.../Uganda (Accessed on 10/5/2013).

Safalaoh ACL (2001). Village chicken upgrading programme in Malawi. World's Poultry Sci. J. 57:179-188.

Smith C (1964). The use of specialized sire and dam lines in selection for meat production. Anim. Prod. 6:337-344.

Sorensen P, Ssewanyana E (2003). Progress in SAARI Chicken Breeding Project: Analyses of Growth Capacity. Proceedings of the Livestock Systems Research Programme (LSRP) Annual Scientific Workshop, Kampala, Uganda.

Ssewannyana E, Oluka J and Masaba J (1998).On-Farm Evaluation and Multiplication of SAARI Crossbred Chickens for increased Meat and Egg Production. Paper presented at the Inception Workshop of the Livestock Research Systems Research Programme (LSRP) Mukono, Uganda: DFI.

Ssewanyana E, Onyait A, Ogwal J, Mukasa B, Nsamba P, Masaba, J (2001). Characterization of rural chicken production in Uganda. Paper presented at the NARO Scientific Conference 5–10 December 2001. Kampala.

Tacher G (1992). Postscript: Economic problems and the future methods of livestock production In Pagot, J (Ed). Animal production in the Tropics and Subtropics. The Mackmillan Press Ltd, London and Basingstoke.

Trach NX (2009). Agricultural Modernization in Conjunction with Rural Development in Vietnam. KKU Sci. J. 37:90-95.

UBOS (2010). Uganda National Household Survey 2009/2010. Republic of Uganda. Available at: www.ubos.org/UNHS0910/unhs200910.pdf (Accessed 21/5/2013).

von Grebmer K, Torero M, Olofinbiyi T, Fritschel H, Wiesmann D, Yohannes Y, Schofield L, von Oppeln O (2011). 2011 Global Hunger Index-The challenge of hunger: Taming price spikes and excessive food price volatility. Washington, DC: International Food Policy Research Institute. Available at: www.ifpri.org/sites/default/files/publications/ghi11.pdf (Accessed on 20/5/2013).

WFP (2011). Uganda Overview. Available at: www.wfp.org/countries/Uganda/Overview (Accessed on 12/05/2013).

Economic assessment of yam production in Kabba-Bunu Local Government Area of Kogi State, Nigeria

Ibitoye, Stephen Jimoh[1] and Onimisi J. A.[2]

[1]Department of Agricultural Economics and Extension, Kogi State University, Anyigba, Nigeria.
[2]Department of Integrated Science, Kogi State College of Education, Ankpa, Nigeria.

This study examined the economic assessment of yam production in Kabba- Bunu Local Government Area of Kogi State, Nigeria. Data used for the study were obtained using structured questionnaire. The questionnaire was administered to 150 randomly selected yam farmers in the area. Descriptive statistics, multiple regression and gross margin (GM) analysis were used to analyze the data. The regression result showed that, farm income (2.778), age (1.820) and education level (2.334) have significant effects on yam output in the area. The GM analysis also revealed that, yam production is profitable in the study area with an average profit of ₦ 121,200 ha^{-1}. It was therefore recommended that, farm inputs be made available to farmers at subsidized prices as a way of improving income from yam production.

Key words: Yield, profit, farmers, variables and socioeconomic.

INTRODUCTION

Yam *(Dioscorea* spp.) is an annual tuber and monocotyledonous crop. The plant Genus comprises of over 600 species with only 10 species producing edible tuber. Six of these edible species are cultivated in Africa and only 3 of them are available in Nigeria. In Nigeria, the primary species cultivated are the white yam *(Dioscorea rotundata*), yellow yam *(Dioscorea cayensis*) and water yam (*Dioscorea alata*), (Amusa, 2000).

FAO (2002) reported that Nigeria accounted for about 71% (26 000 000 tons) of the total world production of yam harvested from 2,760 ha. Yam production in Nigeria has more than tripled over the past 45 years from 8.7 000 000 tons in 1961 to 31.3 million tons 2006. This increase in output is attributed more to the large area planted to yam than to increased productivity (Izekor and Olumese, 2010). Though the area cultivated to yam production is still being increased, production growth rate declined tremendously from average of 27.5% between 1986 and 1990 to 3.5% in the period between 1991 and 1999 (FAO, 2002). However, between 2001 and 2006 production growth rate increased by 31.3%. Record of yield showed similar trend during the same period. Average yield per hectare dropped from 14.9% between 1986 and 1990 to 2.5% in the period between 1991 and 1999. However, the period between 2001 and 2006 recorded 23.4% increase in the average yield (Izekor and Olumese, 2010).

Yam production trend in Kogi State has been observed to be fluctuating for the past 15 years and has not kept pace with other major yam producing states in the country. The production index was estimated at 1.174 000 000 metric tons in 2000. Yam production output in the State dropped to 1.00331 000 000 metric tons in 2003, there was significant rise to 1.26428,000,000

Table 1. Area cultivated and production of yam in Kogi State between 1994 and 2010.

Year	Area cultivated ('000 ha)	Production ('000 metric tons)	Yield (mt/ha)
1994	92.15	912.96	9.91
1995	87.50	911.18	10.41
1996	89.25	929.40	10.41
1997	92.09	1093.40	11.87
1998	88.823	11038.74	11.69
1999	110.75	1393.03	12.58
1999	110.75	1393.03	12.58
2000	100.03	1174.00	11.74
2001	94.00	1089.70	11.59
2002	92.20	1015.41	11.01
2003	89.46	1003.31	11.23
2004	91.28	1100.00	12.05
2005	101.89	1153.54	11.32
2006	120.43	1264.28	10.50
2007	100.06	1226.35	12.26
2008	104.56	1286.96	12.31
2009	109.37	1361.60	12.45
2010	114.62	1480.11	12.91

Source; Kogi ADP crop area and yield survey, 2011.

metric tons in 2006 with the cultivated area of 120,400 ha. In 2008, the total area cultivated for the state reduced to 104,560 ha and the corresponding production output was 1286.96 metric tons (Table 1).

The production figure for 2008 marked the beginning of increase yam production in the state as the production of yam increased to 1.36160 000 000 metric tons in 2009 with cultivated area of 114620 ha.

On the basis of quantity of root and tuber crops produced in Nigeria, yam ranks second to cassava. Yam is the perfect stable food appreciated in its state and cultural role. It is a major source of energy in diet of Nigeria people. Yam can be eaten when boiled, roasted, baked or fried. It can also be processed into crude flour by drying thin slices in the sun and then pound or ground into flour. Yam can further be processed into instant flakes producing a food similar to instant potato and can also be made into fried chip. Most of starch industries also make use of yam as one of their important raw materials. It provides job opportunities and income to both the producers and the marketers. Yam peels serve as feed for livestock and as a good component of farm yard manure. It is used as laboratory crop for scientific investigations.

As food crop, the place of yam in the diet of Nigerians cannot be overemphasized. It contribute more than 200 dietary calories daily, for more than 150 million people in West Africa as well as serving as an important source of income (Babaleye, 2003). According to Okenwe, Orewa and Emokaro (2008), yam contains a high value of rotein (2.4%) and substantial amount of vitamins and minerals than some other common tuber crops. It is also comparable to any starchy root crops in energy and the fleshy tuber is one of the main sources of carbohydrates in the diet of most Nigerians. Yam also plays vital roles in traditional culture, rituals and religion as well as local commerce of African people (Izekor and Olumese, 2010). Yam is reported to be part of the religious heritage of several Nigerian tribes and often play key role in religious ceremony (Amusa, 2000). Due to the importance attached to yam, many communities in Nigeria celebrate the new yam festival annually.

In Nigeria, some of the constraints to yam production are unavailability of planting materials, soil degradation, poor handling and storability, pest and disease and other environmental factors (Ibitoye and Attah, 2012). Seed yam for cultivation has continued to be a problem for the farmers. The cost of producing yam is also observed to be higher compared with other tubers in the country. This is largely due to the high cost of seed yam. On the average, about 25% of the annual yam harvest is used as seed yam (Kushwaha and Polycarp, 2001). This situation has caused yam cultivation to suffer a severe setback due to high cost of production. It is in light of these problems that, the study assessed the economic performance of yam production in Kabba-Bunu Local Government Area of Kogi State, Nigeria.

MATERIALS AND METHODS

The study area is Kabba-Bunu Local Government Area of Kogi State, Nigeria. Kabba-Bunu local Government Area is one of the 21 Local Government Areas in Kogi State. It is located in the western senatorial district of Kogi State. The Local Government was created in 1991. It is bounded in the North by Lokoja Local Government and by Ijumu Local Government to the South, Yagba- East and Mopamuron Local Government share boundary with the Local Government to the west and to the East by Okehi Local Government Area. According to the National Population Census (2006), Kabba-Bunu Local Government Area has a population of 145,446 people which is made up of 74,289 males and 71,157 females. It has land area of about 2,706 km^2.

The local government usually experience 2 district seasons, the wet and dry seasons. The wet season usually spans from the middle of March to October while the dry season cover the period between November and early March. The vegetation of the area comprise of derived savannah and rain forest in some areas. There are vast available lands for farming. Agriculture is the most important economic activities in the Local Government as majority of the population derive their livelihood from it. Agricultural practice in the area is still at subsistence level, which invariably makes the farmers vulnerable to poverty. The soil is viable for growing crops such as yam, maize, cassava, sorghum, cashew, cocoa, oil palm and coffee.

A total of 5 communities were purposively selected from the 2 districts of Kabba-Bunu Local Government Area for this study. Odolu and Okedayo were selected in Kabba district while Edumo, Iluke and Apaa were selected in Bunu district. They were purposively selected because of their high levels of involvement in yam production in the area. Twenty five respondents were randomly selected from each of the 5 communities to have a grand total of 150 respondents for the administration of the questionnaire. Well structured questionnaires were used for the collection of primary data. The questionnaire elicits information on the socioeconomic characteristics of the farmers, problems militating against yam production in the area and other related information on the inputs and output of yam production. Descriptive and inferential statics such as age (%), mean, gross margin (GM) analysis and multiple regression were used to analyze the data.

Model specification

The regression model was specified as follows:

$$Y = \alpha + b_1x_1 + b_2x_2 + b_3x_3 + b_4x_4 + b_5x_5 + b_6x_6 + b_7x_7$$

Where, Y = Yam output (Tons), α= Constant (Intercepts), X_1 = Farm size (hectare), X_2 = Farm Income (naira), X_3 = Age (years), X_4 = Farming experience (Years), X_5 = Sex (male = 1, female = 0), X_6 = Family size (number), X_7 = Educational level (Years), B_1.....b7= Coefficients of independent variables, ei = Stochastic error term, GM analysis was used to determine the cost and returns in yam production in the studied area. The model used is specified thus:

$$GM = TR - TVC$$

Where, GM = Gross margin, TR = Total revenue, TVC = Total variable cost.

Gross ratio of the farm was also calculated. Gross ratio is a profitability ratio that measures the overall success of the farm. The lower the ratio, the higher the return per naira invested (Ekunwe et al., 2008):

$$GR = \frac{TFE}{G1}$$

Where, GR = Gross ratio, TFE = Total farm expenses and GI = Gross income (total revenue).

Return on capital invested (ROI) was also calculated. ROI uses accounting information as revealed by the financial statement to measure the profitability of an investment (Ekunwe et al., 2010). According to Izekor and Olumese (2010), the ROI measure the returns per naira invested. Any investments in which the ROI is greater than 1, indicates a potentially profitable venture and if less than 1, it shows a potentially unprofitable venture. ROI is the GM divided by total variable cost:

$$R1 = \frac{GM}{TVC}$$

Where, R1 = Return on capital invested, GM = Gross margin, and TVC = Total variable cost

RESULTS AND DISCUSSION

The results of socioeconomic variables of the respondents are presented in Table 2. The socioeconomic variables considered includes: Age, family size, sex, farming experience, farm size, educational status, and farm income. The study shows that, majority of the respondents (90%) are still within the productive age bracket of 21 to 60 years. The mean age of the respondent was 46 years. Odinwa et al. (2011) observed similar age bracket among yam farmers in Northern area of River State. The result generally reveals that, majority of the respondents are still energetic to carry on with yam production.

Family size of the respondent shows that majority of them (68%) belong to the family size of 6 to 10 members. The mean family size was found to be 7 members per family. The mean family size recorded for the study is lower than 13 members per family recorded by Pius and Odjurwuedernie (2006) for the Northern part of Nigeria.

Gender distribution of the respondents revealed that, 78% of the farmers are males while the remaining 20% are females. The result of farming experience also showed that, all of them had above 5 years experience in yam production. About 42% of the respondents had no formal education. About 33% others had primary education while about 25% of the remaining respondents attained either secondary or tertiary education. It is then obvious that, the educational standard of the respondents are generally low. Formal education enables the farmers to obtain useful information from media and other sources. Formal education aids farmers to accept new technologies.

The analysis of farm size showed that, 82% of the respondents had between 1 and 5 ha of farmland. The result of farm income of respondents showed that, about 52% of the farmers had less than ₦ 100,000.00 as annual farm income. About 22% had between ₦100,000 and ₦200,000 as annual farm income. The remaining 26% had above ₦ 200,000 as annual farm income. Going by the small farm size of the respondents in this study, couples with their low levels of farm income, it can be conducted that, most yam farmers in the study area are still operating at the subsistence level. This is in

Table 2. Distribution of respondents according to socioeconomic variables.

Socioeconomic variable	Frequency (No)	Age (%)
Age (year)		
Less than 21	0	0
21 – 40	45	30
41 – 60	91	60
Above 60	14	10
Total	150	100
Family size (number)		
1-5	18	12
6-10	101	68
Above 10	31	20
Total	150	100
Sex		
Male	117	78
Female	33	22
Total	150	100
Farming experience (year)		
Less than 6	0	0
6 – 15	88	59
Above 15	62	41
Total	150	100
Farm size (hectare)		
Less than 1	54	36
1 – 5	69	46
Above 5	27	18
Total	150	100
Educational status		
No formal education	63	42
Primary education	49	33
Secondary and above	38	25
Total	150	100
Farm income (naira)		
Less than ₦100,000	78	52
₦100,000 – ₦ 200,000	33	22
Above ₦200,000	39	26
Total	150	100

agreement with the opinion of Izekor and Olumese (2010) that over 90% of the country food supply comes from smallholder farmers.

The effect of socioeconomic variables of respondents on yam production is presented in Table 3. Some of the socioeconomic variables that were regressed on yam output (tones) includes: farm size (x_1), farm income (x_2), age (x_3), farming experience (x_4), sex (x_5), family size (x_6) and educational levels (x_7). The regression result of the estimated double log equation showed that, the coefficient of multiple determinants (R^2) is 0.79 which implies that, 79% variability in the output of yam was explained by the variables in the model while the remaining 21% could be attributed to error and omitted variable. The f-.value of 2.258 is significant at 1% level which confirms the significance of the entire model.

Farm income is positively related to yam output and significant at 1% level. This implies that, an increase in the income level of farmers will translate into increase in yam output. This result validates the findings of Ibitoye et al. (2012), who reported a positive and significant relationship between farmer's income and rice output. The educational level was found to be negatively related to output of yam production and significant at 1% level. This implies that, an increase in the number of years spent in school will lead to reduction in yam output. This may be attributed to the fact that, most of the respondents with higher qualification were not full time farmers but have other major occupations from which they earn their income. The regression result further showed that, age (X_3) is negatively related to yam output and only significant at 10% level. This implies that, as the farmer is ageing their productivity on the farm will decline. The significance of farm income, educational level and age of farmers is in conformity with earlier findings by Ibitoye et al. (2012), Pius and Odjurwuedernie (2006) and Ekunwe et al. (2008). Other variables like farm size (x_1), farming experience (x_4), sex (x_5) and family size (x_6) were found to be insignificant and therefore have no serious impact on yam production in the area.

Result of cost and return analysis in Table 3 suggests that, an average of 4,000 kg of yam tuber was realized from a hectare of yam farm. About ₦58,800 was spent on hiring labour and this constituted about 21% of total variable costs. The amount spent on procuring yam sett was 200,000 which is about 72% of the total cost of production. The cost of yam sett is still a major concern in yam production. The GM calculated for yam production per hectare of farmland was ₦121,200. This implies that, every one naira invested on yam production in the area generate a revenue of ₦1.43. This shows that, yam production in the study area is profitable.

The cost and return analysis of yam production per hectare in the study area is presented in Table 4:

(i) Gross margin (GM) Analysis of yam production:
Total revenue (TR) – Total variable cost (TVC)
GM = ₦ 400,000 – ₦ 278,800 = ₦ 121,200.
(ii) Return on investment (RI) = ₦121,200 / ₦278,800 = 0.43
(iii) Gross ratio (GR)= ₦278,800// ₦ 400,000 = 0.70

The ROI was 0.43 which implies that, every one naira invested in yam production generated a profit of ₦0.43. The gross ratio was also found to be 0.70 which is less than 1. This further confirmed that, yam production in the

Table 3. Regression results of the effect of socioeconomic variables on yam output.

Variable	Estimated coefficient	t- Statistics	Levels of significance
Farm size (x_1)	21.171	0.483	Not significant
Farm income (x_2)	0.012	2.778	1% level
Age (x_3)	-27.127	1.820	10% level
Farming experience (x_4)	10.233	0.855	Not significant
Sex (x_5)	-122.463	-0.323	Not significant
Family size (x_6)	-47.973	-1.374	Not significant
Educational levels (x_7)	-38.808	2.334	1% level
Constant	1257.523	2.984	
R^2= 0.79	f-value=2.258		

Table 4. Cost and returns analysis of yam production per hectare in the study area.

Items/operation	Unit of measurement	Unit cost (₦)	Total quantity	Total value (₦)
Labor cost				
Land clearing	Man day	800	18	14,400
Land cultivation	Man day	800	18	14,400
Planting	Man day	600	10	6,000
Fertilizer Application	Man day	600	5	3,000
Weeding	Man day	600	15	9,000
Staking	Man day	600	5	3,000
Harvesting	Man day	600	15	9000
Total				58,800
Other farm tools				
Yam sett	Sett	40	5,000	200,000
Fertilizer	50 kg bag	2000	4	8,000
Simple farm tools	Lump sum	-	-	4,000
Transportation	Lump sum	-	-	4,000
Miscellaneous expenses	Lump sum	-	-	4,687
Total farm input cost				220,000
Total variable cost (A + B)				278,800
Yam output				
Revenue				
Yam tuber	Kg	100	4,000	400,000

study area is profitable. This result agrees with the findings of Odinwa et al. (2011) who in their studies found yam mini sett production to be profitable.

CONCLUSION AND RECOMMENDATIONS

Yam is a valuable source of carbohydrate for human consumption. It can be processed into various staple, intermediate and end product forms which are used for direct consumption by both human and animals. It is used as basic ingredient for snacks or made into flour and for making instant chips. Judging by the value of yam in the society coupled with the fact that yam production is a profitable venture in the area, yam production will continue to play a prominent role in the area.

In order to ensure a better profitability level and a rapid improvement in yam production, it is recommended that:

(i) Agricultural mechanization should be encouraged as it would reduce labor cost. This can be achieved through the provision of tractors to farmers groups at subsidized prices and establishment of tractor hiring centre's at affordable prices.

(ii) Agro-chemicals especially fertilizers should also be provided by government to farmers at subsidized rate. This will also help to reduce the cost of farm inputs and increase productivity.

(iii) The cost of planting materials (yam sets) constitutes major part of variable costs of yam production. Government should therefore commercialize yam mini sett technique and make it available and affordable for rural farmers in the state.

REFERENCES

Amusa NA (2000). Screening cassava and yam cultivars for resistance to anthracnose using toxic metabolic of collectorichim species. Mycopathologia 150:137-142.

Babaleye T (2003). Improving yam production technology in West Africa. Supplement Issues/Edition. P. 463.

Ekunwe PA, Orewa SI, Emokaro CO (2008). Resource use efficientcyin yam production in Delta and Kogi States of Nigeria. Asian J. Agric. Res. 2:61-69.

Food and Agricultural Organization (FAO), (2002). Food and Agricultural Organization Year Book. P. 56.

Ibitoye SJ, Attah S (2012). An assessment of yam mini-sett utilization and profit level in Kogi State, Nigeria. Int. J. Applies Res. Technol. (IJART). Esxxon Publishers USA 1(5):8-14.

Ibitoye SJ, Orebiyi JS, Shaibu M (2012). Economic effect of inorganic pesticide use on fadama II rice farming in Ibaji LGA of kogi state, Nigeria. Int. J. Agric. Rural Develop. 15(2):1063-1070.

Izekor OB, Olumese MI (2010). Determinants of yam production and profitability in Edo State, Nigeria. Afr. J. General. Agric. 6(4):30-35.

Kushwaha S, Polycarp IM (2001). Economics of small scale yam production in Qua'an Pam LGA of Plateau Satae, Nigeria. Proceeding of the 34th Annual Conference of Agricultural Society of Nigeria. Held at AbubakarTafawaBalewa University, Bauchi. October 15-19, 2001. pp. 69-74.

National Population Commision, (NPC) (2006). National Population census figures, NPC Bulletin.

Odinwa AB, Ali N, Abali I, Ahiskwo AA, Odinwa .AN (2011). Economic viability of yam.mini-sett production and the problems affecting mini-sett enterprise in Northern River State, Nigeria. Mediterranean J. Social Sci. 2(4):67-62.

Pius CI, Odjurwuedernie EI (2006). Determinants of yam production and economic efficiency among small holders farmers in Southern Nigeria. J. Central Eur. Agric. 7(2):333-342.

Understanding the performance of food production in sub-Saharan Africa and its implications for food security

M. Demeke[1], F. Di Marcantonio[1] and C. Morales-Opazo[1,2]

[1]Agricultural Development Economics Division, Food and Agriculture Organization of United Nations, Rome 00153, Italy.
[2] European Commission, Joint Research Centre, Institute for Prospective Technological Studies, Edificio Expo, c/Inca Garcilaso 3, 41092 Seville, Spain.

The performance of the food production sector in Sub-Saharan Africa (SSA) is critical for a number of reasons. Domestic production is the principal avenue to ensuring access to affordable food in poor countries with limited capacity to import food. The multiplier effect of sustainable food production on the economy is considerable as it has direct linkage with other activities such as transportation, marketing, tourism and local trade. In countries where the growth of food production outpaces demand, social and political harmony as well as macroeconomic stability can be maintained, paving the way for sustained economic growth. This paper examines changes in food production performance among 30 SSA countries over the period of 1968 to 2008. The results support previous findings that not many countries have managed to achieve a food production growth rate in excess of 3% per annum. Annual food production performance averaged 3% or more in 60% of the sample countries following the policy reforms. Nevertheless, rates of output growth varied from one period to the other and the recent improved performances were achieved not only through unsustainable expansion of land under cultivation, but also failed to satisfy the rapidly growing food demand. With domestic supply lagging behind, most countries have experienced unaffordable food import bills. Addressing political instability and building institutions that foster partnership between governments, farmers, traders and other operators along the food value chain to address market failures and inefficiencies in input, output, credit and risk managementis critical to ensure food availability, accessibility and stability in SSA.

Key words: Sub-Saharan Africa, food production performance, food security.

INTRODUCTION

Over the past thirty or forty years, there has been a growing concern over poverty and food insecurity in Sub-Saharan Africa (SSA). Various attempts and commitments have been made to address the problem both at global and continental levels. At the global level, the most prominent commitments include the 1996, 2003 and 2009 World Food Summit which pledges to achieve food security; the 2000 Millennium Development Goals (MDGs) declaration of the UN which specifically sets the objective of halving the proportion of the world's poor and hungry people by the year 2015; and the 2009 L'Aquila Food Security Initiative (G8, 2009) that announced a goal of mobilizing US$20 billion over a period of three years for increasing G8 assistance to Agriculture and Food Security (GAFS). Priority is often given to SSA countries in the global initiatives. At the continental level, there have been numerous commitments from as far back as the 1980s in the Lagos Plan of Action for the Economic Development of Africa (UNECA, 1979) and in the early 1990s in the African Economic Community initiative (AEC, 1991). One of the more recent ones include the Maputo Declaration on Food Security (African Union, 2003); the committed member countries to allocate at least 10% of national budgetary resources to agriculture and rural development policy implementation within five years. Nevertheless, the reform institutionalized under the name Structural Adjustment Programs (SAPs) is by far the most significant and far-reaching policy initiative (Rono, 2002; FAO, 1999; Zawalinska, 2004; Munthali, 2004; Maxwell, 1999; Lele, 1990).

On the basis of this framework, many SSA countries implemented, to varying extents, reforms which included, among others, macroeconomic stabilization, trade liberalization and reduction in the public sector. It was expected from the outset that through an implementation of a set of macroeconomic and microeconomic policy reform measures, SSA countries would see enhanced food production as well as sustainable growth and development.

The aim of this paper is to contribute to the understanding of food production performance of SSA countries and its implications for food security. This analysis is carried out by paying particular attention to good and poor performing countries and assessing differences between growth trends across three different periods: before the SAPs or pre-reform period (1968 to 1983), the introduction of SAPs or transition period (1984 to 1993), and after SAPs implementation or post-reform period (1994 to 2008). Thus, the study performs an indirect assessment of the impact of these reforms in different countries, with also documenting available research evidences on key challenges in addressing the long standing concerns over sustainable increase in food production and food security objectives. To analyze the performance of different SSA countries, the paper uses the Food Production Index (PIN) of the FAO statistics (FAOSTAT). The food PIN measures the value of the final food output in 'international dollars', which are the same in all countries, implying that the weight given to each commodity is the same across different countries[1].The analysis covers the performance of 30[2] SSA countries over four decades (1968 to 2008).

The least-squares growth rate is used to measure food production performance. The least-squares growth rate, *r*, is estimated by fitting a linear regression trend line to the logarithmic annual values of production in the relevant period (OECD, 2005; Kakwani, 1997). Least-squares growth rates are used whenever there is a sufficiently long time series to permit a reliable calculation. No growth rates are calculated if more than half the observations in a period are missing. The regression equation takes the form:

$$\ln X_t = a + b_t.$$

This expression is equivalent to the logarithmic transformation of the compound growth equation,

$$X_t = X_o (1 + r)^{*}t,$$

Where: X is the variable, t is time, and $a = \ln X_o$ and $b = \ln (1 + r)$ are parameters to be estimated. The calculated growth rate is an average rate that is representative of the available observations over the entire period. It does not necessarily match the actual growth rate between any two periods[3]. On the basis of their food production performance, the countries have been grouped into three: better performers [with a food PIN average annual growth rate (AAGR) greater than 3%], medium performers (with a Food PIN AAGR between 2 and 3%) and poor performers (with a food PIN AAGR for 1968 to 2008 of less than 2%).

FOOD PRODUCTION PERFORMANCE IN SSA: BETTER, MEDIUM AND POOR PERFORMERS

As a region, SSA relies heavily on agriculture. The sector accounts, on average, for close to 20% of total gross domestic product and about 60% of the region's total labour force – although many countries in the region depend on agriculture to a much greater extent than indicated by these regional averages (FAO, 2008; World Bank, 2009). Traditionally, the agricultural sector has been the overwhelming driving force for Africa's economic growth and development (World Bank, 2008; IFPRI, 2011; AFDB, 2010; ECOWAS, 2009; Kydd et al., 2007). The output and employment multiplier effects of food production on the economy are considerable as it has direct linkage with other activities such as transportation, marketing, warehousing, food processing, tourism and local commerce. Figure 1 presents the challenge facing SSA countries. The net value of production doubled more than in the period of 1968 to 2000, increasing from about I$35,000 million to about I$100,000 million. However, the net per capita production decreased by about 10% during this period indicating that food production did not keep pace with population

[1]The food production index number (PIN) includes commodities that are considered edible and that contain nutrients. As such, cocoa is included in the food PIN but excludes coffee and tea, although edible, along with inedible commodities, because they have little to no nutritive value.

[2] We considered countries with significant agricultural sector and with complete or nearly complete food production index (PIN) database in the FAOSTAT of FAO.

[3]The World Bank website: http://data.worldbank.org/about/data-overview/methodologies

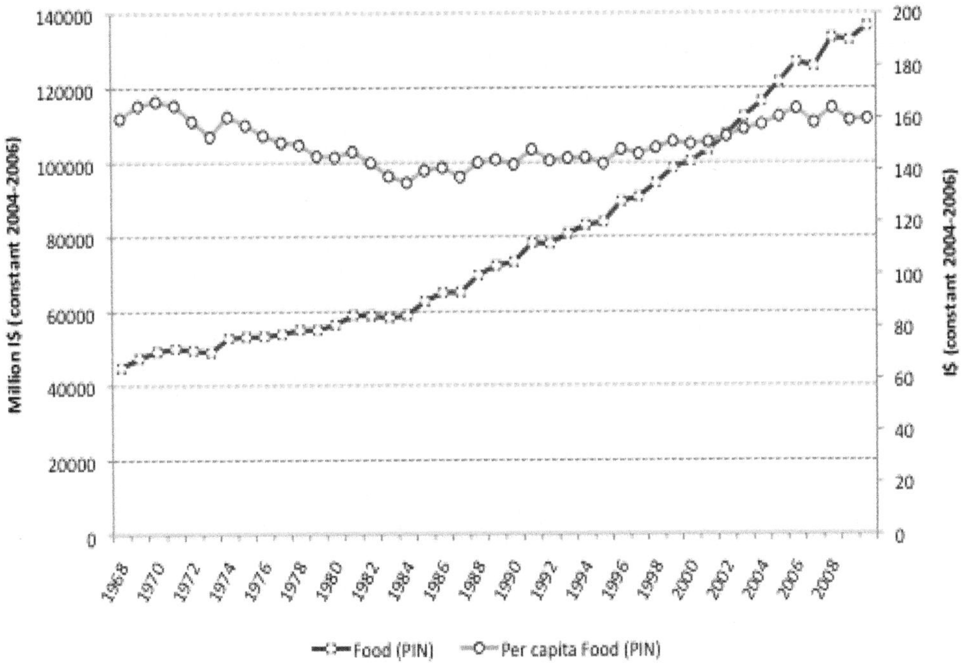

Figure 1. Production indices (PIN) - trends in net production and net per capita PIN base. Source: Authors' elaborations based on FAOSTAT data.

growth in the region. However, this general trend conceals many differences between countries and periods. Table 1 in the annex compares the growth of food production in the pre-reform, transition and post-reform period, and also presents a ranking of the good and poor performing countries.

As indicated earlier, this ranking of 30 countries, which is obtained by looking at food production performance for the whole period (1968 to 2008), gave rise to three groups referred to as better, medium and poor performers. Over the last 40 years, only nine of the 30 countries (30%) (group one) were able to achieve a food production average performance of 3% or more per annum. The majority of these better performing countries (six of them) are from West Africa (Benin, Burkina Faso, Nigeria, Cote d'Ivoire, Ghana and Niger). The other three countries, namely Sudan, Malawi and Kenya, are from eastern and southern Africa. Nevertheless, growth rates varied between countries and across the different periods. In Malawi, for instance, negative growth was registered during the transition period but very high growth in the post-reform period (Chilowa, 1998; Harrigan, 2003, 1997). Growth during the pre-reform period was poor for group one as a whole (2.2%) with negative growth rate in Ghana -possibly due to macroeconomic instability (Weissman, 1990)- and almost zero growth in Nigeria (Moser et al., 1997). Benin had one of the highest growth rates, especially during the transition and post-reform periods and this could be attributed to the dramatic increase in rice production following the 1994 currency devaluation which made imported rice expensive and raised domestic prices for farmers (Noameshi et al., 2007). In addition, the increase of manioc production, in response to increasing commercialization and cross-border exports, has contributed to the country's agricultural growth (Kherallah et al., 2001). In countries such as Ghana and Nigeria, good performance in the transition and post-reform period was partly related to a sharp increase in cassava production, resulting from widespread adoption of high-yielding varieties and improved pest management practices (Ugwu, 1996, Nweke, 2004; Camara, 2000). This has contributed in making Nigeria the world's top producer of cassava.

For the second group of nine countries (medium performers), the growth rate of food production averaged 2.6% per annum over the period of 1968 to 2008. The group, which includes Central African Republic, Mali, Cameroon, Guinea, Chad, Togo and Congo from central and western Africa, and Tanzania and Zambia from eastern and southern Africa, had an average growth rate of food production which fell well short of population growth rate in the pre-reform period and barely caught up with it in the transition period. Within the group, the performance of Congo was particularly poor in the pre-reform and transition period, probably because of the Marxist policies and political/macroeconomic instabilities during this period (Clark, 2002). Tanzania also performed poorly in the transition period probably due to disruptive policies such as villagization program, removal of fertilizer subsidies and bad weather (Skarstein, 2005; Wobst, 2001). On the other hand, annual growth rate of food

Table 1. Country performance ranking based on food (PIN) average annual growth rate (AAGR) (1968 to 2008) - net production value (constant 2004 to 2006 1000 I$).

Country	Annual average growth rate food PIN			
	Pre-reform	Transition	Post-reform	Whole period
	1968-1983	1984-1993	1994-2008	1994-2008
Better performers				
Benin	1.8	3.8	4.6	4.1
Burkina Faso	1.5	6.1	3.8	4.1
Nigeria	0.1	8.0	3.5	4.0
Côte d'Ivoire	4.9	3.3	2.3	3.8
Ghana	-1.4	5.7	4.4	3.5
Niger	2.6	4.7	6.4	3.3
Sudan	3.7	2.8	3.0	3.2
Malawi	3.2	-0.1	6.9	3.1
Kenya	3.6	4.4	3.5	3.1
Mean	2.2	4.3	4.3	3.6
Medium performers				
Central African Republic	2.8	3.1	3.0	2.8
Mali	2.7	3.1	4.4	2.8
U.R. of Tanzania	3.8	1.1	4.4	2.8
Cameroon	2.2	2.3	3.4	2.6
Guinea	1.3	3.4	2.9	2.6
Chad	0.9	3.9	3.3	2.6
Togo	0.9	3.2	2.7	2.5
Zambia	2.5	3.8	2.7	2.3
Congo	1.1	0.5	3.0	2.1
Mean	2.0	2.7	3.3	2.6
Poor performers				
Rwanda	3.8	1.1	6.4	2.0
Ethiopia	1.5	1.0	4.9	2.0
Uganda	0.5	4.0	2.7	1.8
Mauritania	0.7	2.3	2.3	1.6
Senegal	0.2	2.4	2.2	1.5
Madagascar	1.5	1.5	1.6	1.5
Liberia	2.9	-3.6	3.5	1.2
Mozambique	-0.7	-1.0	2.6	1.2
Sierra Leone	1.2	0.2	3.8	1.1
Somalia	2.3	-2.4	1.2	1.1
Democratic Republic of Congo	2.0	3.1	-1.2	1.0
Burundi	0.7	2.8	1.1	1.0
Mean	1.4	1.0	2.6	1.4

Source: Authors' calculations based on FAOSTAT database, Note*: All selected African countries.

production among poor performing countries averaged only 1.4% in 1968 to 1983 and 1.0% in 1984 to 93. There has been a recovery in the post-reform period of 1994 to 2008 (2.6%) but the growth was mainly driven by strong performance in Rwanda, Ethiopia, Sierra Leone and Liberia.

Following the introduction of the policy reforms, food production performance has improved in most countries in all the three groups. Annual growth rates in group one countries averaged 4.3% during the transition and post-reform period, compared to 2.2% during the pre-reform period. In group two countries, average annual growth rates increased to 2.7% in the transition period and 3.3% in the post-reform period, compared to 2.0% in the pre-reform period. Growth rates declined to 1% in the transition period but increased to 2.6% in the post-reform period in group three countries. Overall, growth rates averaged or exceeded 3% in 18 of the 30 countries (60%) between 1994 and 2008.

IMPACT OF FOOD PRODUCTION PERFORMANCE ON FOOD SECURITY

The impact of food production performance on food security is assessed in terms of per capita food supply, number of undernourished people and food self-reliance. Although, a direct correlation between food production performance and food security has not been proven, it is observed that domestic production plays a critical role in food security, particularly for regions like sub-Saharan Africa where it represents the main source of food consumption (FAO, 2012; O'Connell, 2008; Boon, 2007). Our results confirm that better food production performance is correlated with food security as measured by per capital food supply and FAO's estimations of undernourished people (Table 2 annex). Among the three groups, better performing countries experienced a longer period (1968 to 2007) of sustained increase in per capita food supply, increasing from 2,077 kcal/capita/day in 1968 to 1983, to 2,118 in 1984 to 1993 and 2,337 in 1994 to 2007 (Table 2 Annex). In the case of medium performers, average per capita food supply declined in the transition period and slightly improved in the post-reform period. The situation rather worsened in the case of poor performing countries, which lessen the level of food supply from 2,145 in 1968 to 1983 to 2,042 kcal/capita/day in 1994 to 2007. The difference between the three groups is more striking in terms of undernourishment. The prevalence of undernourishment in the total population averaged 14% in 2006 to 2008 among better performers, compared with 30% of medium and poor perfuming countries (Table 2). However, there are again marked variations between countries within each group.

Kenya and Malawi have the highest prevalence rate of undernourishment among the better performers while Niger, Cote d'Ivoire and Burkina Faso have the lowest rate (8% or less). The dominance of maize in daily diets and low rates of dietary diversity among low-income groups may explain the high rate of undernourishment in Kenya and Malawi (Smale et al., 2011). Levels of undernourishment are particularly high in Burundi, Ethiopia and Mozambique (among poor performers) as well as Zambia, Central African Republic and Chad (among medium performers). Over the years, there has been a steady decline in the proportion of undernourished people in most SSA countries over the past 18 years (2008 as compared to 1990) (FAO, 2011a), but the decline is more significant in better performing countries (-40%) than in medium (-23%) or poor performing countries (-8.5%) (Table 2). It is also evident that here are countries that are still at risk in better as well as medium and poor performing countries. For some countries such as Kenya (group one), Zambia (group two), and Uganda, Liberia and Burundi (group three), the proportion of undernourished people has changed very little or even increased (Table 2). Many SSA countries are not food self-sufficient and are expected to have export earnings that allow them to meet their food import needs (Fafchamps, 1992; Kasfir, 1986). This study compared the food trade balance and share of food import in total merchandise export to determine if SSA countries are exposed to food security risks emanating from trade. The results show that the food trade balance of good performing countries is better than medium or poor performing countries but the trend over time is one of a growing deficit in nearly all cases. For instance, the average food trade balance of the first group, which was positive for six of the nine countries during the pre-reform period (1968 to 1983), turned negative for all countries except Cote d'Ivoire and Ghana in 1994 to 2008. It should be noted that the positive food trade balance in Cote d'Ivoire and Ghana is mainly due to the fact that the two countries are the first and second biggest producers and exporters of cocoa in the world (ODI, 2007). For the first group as a whole, food trade balance changed from surplus in the pre-reform and transition period to negative in the post-reform period (Table 3a).

The food trade balance in group three countries as a whole worsened considerably over the years: the food trade deficit increased sharply in the transition and post-reform periods, averaging 1.65 billion USD per annum. No group three country registered a positive food trade balance in 1994 to 2008. The deficit also increased considerably in group two countries in the post-reform period. In general, the number of net food exporting countries declined from 15 in 1968 to 1983 to only 4 in 1994 to 2008 (Table 3a). The negative food trade balance on its own may not signify any danger to self-reliance in SSA countries. It is thus important to check if a country has sufficient export earnings to meet its food import needs. Indeed, with few exceptions, SSA countries have very limited importing capacity. Even among group one countries, food imports accounted for about 28% of the total value of export earnings during the period of 1968 to 2008, compared to a threshold share of 8.8% or less which is considered as food self-reliant or food secure at national level (Breisinger et al., 2010). For the poor performing countries, the share increased from 25% in the pre-reform period to 74% in the post-reform period, averaging 49% for the entire period. Food imports are relatively more affordable for the second group of countries, mainly because they have significant foreign exchange earnings from export of oil and other minerals [for example, Central Africa Republic (diamonds), Cameroon (petroleum), Chad (petroleum) Guinea (bauxite), Zambia (copper) and Congo (petroleum)]. For

Table 2. Performance in terms of food supply and undernourishment.

Country	Average food supply (kcal/capita/day)				Prevalence of undernourishment in total population	Proportion of undernourished in total population
	Pre-reform	Transition	Post-reform	Whole period	(%)	Change so far
	1968-1983	1984-1993	1994-2007	1968-2007*	2006-2008	1990-2008
Better performers						
Benin	1929	2127	2398	2143	12	-41
Burkina Faso	1723	2260	2574	2155	8	-40
Nigeria	1808	2097	2588	2154	11	-61
Côte d'Ivoire	2690	2539	2459	2572	6	-2
Ghana	1964	2074	2633	2226	14	-83
Niger	1979	1987	2136	2036	5	-55
Sudan	1956	1952	2155	2025	16	-44
Malawi	2322	1960	2052	2137	22	-37
Kenya	2319	2070	2042	2160	33	-1
Mean	2077	2118	2337	2179	14	-40.4
Medium performers						
Central African Republic	2335	1878	1900	2069	40	-8
Mali	1650	2094	2379	2016	12	-56
U.R. of Tanzania	2007	2174	1964	2034	34	15
Cameroon	2199	2024	2142	2135	22	-33
Guinea	2283	2393	2460	2373	16	-18
Chad	1773	1638	1940	1798	39	-36
Togo	1969	1866	2053	1973	30	-31
Zambia	2314	2037	1908	2103	44	23
Congo	1992	2037	2267	2099		-68
Mean	2058	2016	2113	2067	30	-23.6
Poor performers						
Rwanda	2180	1973	1898	2029	32	-28
Ethiopia	1644	1579	1794	1680	41	-40
Uganda	2296	2202	2254	2258	22	15
Mauritania	2091	2555	2746	2436	8	-34
Senegal	2230	2206	2190	2210	19	-14
Madagascar	2536	2304	2089	2322	25	24
Liberia	2402	2361	2140	2300	32	7
Mozambique	1844	1761	1960	1864	38	-36
Sierra Leone	2086	1950	2050	2040	35	-22
Somalia						
Democratic Republic of Congo	2227	2171	1649	2011	13	
Burundi	2056	1873	1689	1882	62	43
Mean	2145	2085	2042	2094	30	-8.5

Source: Authors' calculations based on FAOSTAT database, Note*: Data for 2008 not available.

Table 3a. Food trade balance (US$).

	Trade balance (1000$)			
	Pre-reform	**Transition**	**Post-reform**	**Whole period**
Country	**1968-1983**	**1984-1993**	**1994-2008**	**1968-2008**
Better performers				
Benin	-9140	-68966	-143027	-72715
Burkina Faso	-14428	-57204	-104203	-57706
Nigeria	-493362	-491904	-1444114	-840843
Côte d'Ivoire	340521	860851	1621715	936160
Ghana	376707	280625	335236	338100
Niger	13358	-38972	-78001	-32830
Sudan	32330	-27196	-191886	-64219
Malawi	19649	-23950	-18576	-4969
Kenya	18401	-13060	-144584	-48901
Mean	31560	46692	-18605	16898
Medium performers				
Central African Republic	-10432	-13403	-10087	-11030
Mali	26896	18126	-64715	-8760
U.R. of Tanzania	7474	-25004	-150318	-58176
Cameroon	117333	53850	3308	60133
Guinea	-8263	-75225	-153763	-77827
Chad	27048	19656	2088	16113
Togo	3168	-29923	-26493	-15754
Zambia	-51739	-38047	-36647	-42878
Congo	-19711	-71296	-137883	-75526
Mean	10197	-17919	-63834	-23745
Poor performers				
Rwanda	-10609	-41481	-63831	-37610
Ethiopia	10857	-169683	-195155	-108547
Uganda	-15016	-15891	-153927	-66051
Mauritania	-27990	-72881	-171876	-91580
Senegal	-34469	-174660	-533891	-251377
Madagascar	44039	61294	-12443	27583
Liberia	-39736	-65931	-95452	-66509
Mozambique	31748	-150614	-213827	-102575
Sierra Leone	-25483	-62621	-122084	-69883
Somalia	18115	-28758	-68211	-24900
Democratic Republic of Congo	-89502	-187735	-316656	-196567
Burundi	-11499	-20409	-35305	-22382
Mean	-12462	-77447	-165222	-84200

Source: Authors' calculations based on FAOSTAT database., Note*: Data for 2008 not available.

most of the non-mineral exporting countries, the share of food in total merchandise export is much higher.

Overall, only three countries, namely, Nigeria (better performer), Zambia and Congo (medium performers) registered a share below the critical threshold of 8.8% from the entire sample countries. SSA has become a net importer of food and of agricultural products, despite the region's vast agricultural potential and improved performance in recent years. For instance, Nigeria has a considerable agricultural potential, its food production performance averaged 3.5% in 1994 to 2008, and the country became the largest producer of cassava. At the same time, Nigeria's import of wheat increased from 675,282 tons in 1994 to 3,079,637 tons in 2008. Rice import increased from 350,000 to 971,815 tons over the same period. By 2010, wheat and rice import further

Table 3b. Share of food import on total merchandise exp.

Country	Pre-reform 1968-1983	Transition 1984-1993	Post-reform 1994-2008	Whole period 1968-2008
Better performers				
Benin	73.6	33.9	56.6	57.7
Burkina Faso	87.3	67.9	45.0	67.1
Nigeria	9.8	7.2	6.9	8.1
Côte d'Ivoire	13.6	12.4	9.1	11.7
Ghana	11.4	13.1	19.3	14.7
Niger	19.8	30.2	42.5	30.6
Sudan	24.4	41.0	23.9	28.2
Malawi	9.7	17.9	17.8	14.7
Kenya	9.5	13.8	20.0	14.4
Mean	28.8	26.4	26.8	27.5
Medium performers (AAGR for 1968-2008 greater than 2% and less than 3%)				
Central African Republic	20.7	18.2	16.2	18.4
Mali	37.1	31.4	17.8	28.7
U.R. of Tanzania	16.9	20.6	31.9	23.3
Cameroon	9.4	12.8	12.0	11.2
Guinea	14.2	14.1	22.5	17.2
Chad	20.8	17.4	11.5	16.6
Togo	17.3	27.3	18.7	20.3
Zambia	5.6	5.4	8.9	6.8
Congo	13.6	8.4	6.2	9.6
Mean	17.3	17.3	16.2	16.9
Poor performers (AAGR for 1968-2008 less than 2%)				
Rwanda	21.7	43.5	70.9	45.0
Ethiopia	11.6	65.8	48.6	38.3
Uganda	6.8	10.8	27.5	15.3
Mauritania	30.5	27.5	37.1	32.2
Senegal	41.5	40.4	53.6	45.6
Madagascar	18.5	16.6	28.1	21.5
Liberia	11.8	19.4	55.2	29.5
Mozambique	30.6	165.0	53.1	71.6
Sierra Leone	33.1	56.8	330.9	147.8
Somalia	60.2	99.3	84.9	78.8
Democratic Republic of Congo	13.2	25.2	34.3	23.8
Burundi	22.1	22.7	64.3	37.7
Mean	25.1	49.4	74.0	48.9

Source: Authors' calculations based on FAOSTAT database.

increased to nearly 4 and 2 million tons, respectively. Wheat import in Ethiopia (with one of the fastest production growth in recent years) increased from about 0.55 million tons in 1994 to 1.74 million tons in 2009[4]. It is also estimated that a significant proportion of sugar and sugar products, vegetable oils and milk products consumed in the countries of sub-Saharan Africa comes from abroad (OECD, 2008). Despite the recent improved performance, Africa's agricultural GDP per capita is the lowest in the world (one-fourth of world's average) (Rakotoarisoa et al., 2011), resulting in stagnat or declining per capita food production as shown earlier.

Various studies (Omamo et al., 2006; Diao et al., 2008; FAO, 2011a, 2010; UNCTAD, 2009) have documented the persistent rise in net food imports in Africa, and we will examine some of the key structural bottlenecks

[4] FAOSTAT: http://faostat.fao.org/site/535/DesktopDefault.aspx?PageID=535#ancor

subsequently.

SOURCES OF GROWTH AND THE KEY CHALLENGES TO SUSTAINABLE PRODUCTION INCREASE

Some experts agree that population growth, rising incomes, and urbanization will continue to drive demand growth for some foodstuffs, especially vegetable oils and livestock, with a higher derived demand for feed and industrial products (FAO, 2009; USDA, 2007; WHO, 2008). To cope with the rapidly growing demand, food production would need to increase significantly (Von Braun, 2008). The traditional response is expanding area under cultivation but a more sustainable option is yield improvement. The World Bank's report suggests that without improving technologies and raising yield levels, the "land rush" is unlikely to slow (Deininger et al., 2010). SSA countries have relied more on land expansion than on intensification to increase food production.

Sources of growth - area under cultivation and yield levels

Expansion of area under cultivation has remained the main strategy of increasing production in SSA countries. On average, area under cultivation expanded by 2.2% per annum over the last 40 years for better performers (Table 4a). The rate of expansion was greater than 3% per annum in the case of Sudan and Niger (Mosley, 2011). The pattern is similar in the medium performing countries, with an average expansion rate of 1.7% per annum. By contrast, growth in area under cultivation was slow (1.1%) among poor performers. Area growth rates were negative in Ethiopia [pre-reform and transition), Mauritania (pre-reform and post-reform), Liberia (transition), Somalia (transition and post-reform)]. While there is a substantial difference in the average growth rate of area harvested between the three groups, this rate drops to no significant difference when yield growth is considered. Yield growth averaged 0.6% in group one (better performing) countries, compared to 0.3% in group two and -0.3% in group three countries over the period of 1968 to 2008 (Table 4b). Within the first group, Benin experienced a negative yield growth rate over the 40 year period, while Malawi managed a relatively better performance (1.8%) and this is largely due to a significant yield improvement in the post-reform period (6.1% per annum). The yield performance of Chad (1.8%) and Mali (1.6%) is better than the other countries within medium performers, while Togo is the only country with negative growth rate (-2.9) among the group. On the other hand, within the poor performing countries, yield growth was negative for five out of the 12 countries and almost zero for further four countries (Table 4b). It is important to note that the performance of yield in Africa compares very unfavourably with the Green Revolution of Asia.[5]

African farmers have traditionally depended on shifting cultivation in response to the challenges of population growth and declining soil fertility. As shown earlier, opening new land is still a common strategy to increase production in nearly all the sample countries. The advantages of extensification are clear: new land means additional output at a lower cost than purchasing fertilizers and other inputs to increase yield on already cultivated lands. However, extensive agriculture is unsustainable for most countries (Dorward et al., 2004; Reardon et al., 2002). Average farm sizes are small (less than 1 ha for over 50% of the farms) and declining in sub-Saharan Africa, due to population pressures and an exhaustion of the arable land frontier, especially in the productive highland regions (Jayne et al., 2003). Extensive agriculture has also got major environmental drawbacks: extensification into permanent pasture, forest and watershed lands may lead to loss of biological diversity and land degradation (UNEP, 2012). On the other hand, increasing productivity on existing land enhances the economic value of food and agricultural production through forward and backward linkages in the form of input and output marketing, transport, export and processing increases. Increasing productivity also avoids greenhouse gas emissions and the large-scale disruption of existing ecosystems due to bringing new land into production (Edgerton, 2009). As discussed as follows, improving yield on a sustainable basis has eluded SSA countries.

Challenges to sustainable increase in food production

Sustainable intensification is producing more output from the same area of land while reducing the negative environmental impacts. It is commonly achieved through the use of high yield crop varieties along with fertilizers (both organic and inorganic) and management practices which conserve and improve soil and water productivity (FAO, 2011b). A number of challenges have made it very difficult for SSA countries to achieve a sustainable increase in food production.

Limited utilization of inputs and irrigation

There are no reliable data on the use of improved seeds but available evidences indicate adoption rates are very low in SSA. For instance, the adoption of improved open-pollinated varieties and hybrids of maize is estimated at 44% of maize area in Eastern and Southern Africa in 2006 to 2007, excluding South Africa. Some 56% of

[5]Between 1965 and 1982, average rice, maize and wheat yields increased by 2.54, 3.48 and 4.07% per year, respectively in Asia countries that adopted Green Revolution technologies. Cultivated area expanded by only 0.7, 1.09 and 1.3%, respectively, over the same period (Thapa and Gaiha, 2011).

Table 4a. Performance in terms of area harvested (average annual growth rate).

Country	Annual average growth rate			
	Pre-reform	Transition	Post-reform	Whole period
	1968-1983	1984-1993	1994-2008	1968-2008
Better performers (AAGR for 1968-2008 greater than 3%)				
Benin	0.8	2.5	2.6	2.5
Burkina Faso	0.6	3.7	2.3	1.9
Nigeria	-5.0	8.1	1.4	2.6
Côte d'Ivoire	1.8	3.3	-0.7	0.9
Ghana	-0.1	0.6	2.5	2.2
Niger	3.3	8.4	2.5	3.9
Sudan	5.2	1.1	0.5	3.2
Malawi	0.8	1.9	2.6	1.5
Kenya	0.6	4.0	0.5	1.0
Mean	0.9	3.7	1.6	2.2
Medium performers (AAGR for 1968-2008 greater than 2% and less than 3%)				
Central African Republic	1.2	0.3	2.3	0.3
Mali	1.5	7.7	3.1	2.8
U.R. of Tanzania	2.8	0.6	5.3	2.3
Cameroon	1.1	2.4	3.8	1.2
Chad	-1.5	3.5	3.8	2.4
Guinea	1.6	3.5	4.0	3.0
Togo	1.9	3.5	0.7	2.9
Zambia	-4.6	4.2	0.7	-0.4
Congo	-0.2	-1.2	1.9	0.5
Mean	0.4	2.7	2.8	1.7
Poor performers (AAGR for 1968-2008 less than 2%)				
Rwanda	3.6	2.5	5.2	2.2
Ethiopia	-2.0	-1.1	3.0	1.3
Uganda	-0.2	3.1	1.9	1.3
Mauritania	-3.8	1.4	-1.2	1.3
Senegal	-0.6	-0.7	1.1	0.5
Madagascar	1.7	0.8	0.9	1.0
Liberia	2.1	-7.8	5.3	-0.3
Mozambique	2.4	0.4	0.4	1.8
Sierra Leone	1.6	2.1	8.1	1.6
Somalia	2.4	-6.2	-0.6	0.6
Democratic Republic of Congo	2.2	3.4	-0.8	1.3
Burundi	1.1	1.3	0.6	0.8
Mean	0.9	-0.1	2.0	1.1

Source: Authors' calculations based on FAOSTAT database.

smallholders have no access to improved varieties (Smale et al., 2011; Langyintuo et al., 2008). The situation is relatively better in West Africa where the adoption rate of improved seeds was estimated at 60% in 2005 (Smale et al., 2011). As already indicated, most of the better performing countries are from West Africa. Several factors have hampered the emergence of an efficient seed market in Africa: i) inadequate certification, licensing and enforcement capacity, ii) lack of knowledge on varietal characteristics and performance, iii) lack of credibility and adulteration of seed, and iv) limited access to credit facilities. These problems are clear indications of market failure, market inefficiency and institutional weaknesses. Adoption is often constrained by lack of finance. Lack of partnership and collaboration between public and private seed companies is a major gap in SSA

Table 4b. Performance in terms of yield (average annual growth rate).

Country	Annual average growth rate			
	Pre-reform	Transition	Post-reform	Whole period
	1968-1983	1984-1993	1994-2008	1968-2008
Better performers				
Benin	0.7	-3.0	1.9	-1.6
Burkina Faso	n.a	n.a	n.a	n.a
Nigeria	0.5	2.2	0.3	0.6
Côte d'Ivoire	1.7	-1.0	1.3	1.1
Ghana	0.0	5.2	0.6	1.5
Niger	1.7	1.0	2.5	0.0
Sudan	0.5	0.4	-1.1	0.3
Malawi	0.4	0.7	6.1	1.8
Kenya	1.2	3.5	1.0	1.3
Mean	0.7	1.0	1.4	0.6
Medium performers				
Central African Republic	1.0	0.3	0.1	0.5
Mali	-0.9	3.5	3.8	1.6
U.R. of Tanzania	3.2	0.2	0.5	0.5
Cameroon	1.0	0.2	-1.0	0.3
Chad	0.6	1.8	1.1	1.8
Guinea	0.1	-0.1	-0.1	0.0
Togo	-0.7	-3.6	0.0	-2.9
Zambia	0.5	0.3	-0.2	0.1
Congo	1.5	1.1	0.3	1.0
Mean	0.7	0.4	0.5	0.3
Poor performers				
Rwanda	-0.1	-4.1	1.8	-1.0
Ethiopia	1.8	0.2	0.9	0.6
Uganda	0.2	0.8	0.9	0.5
Mauritania	-0.6	-0.6	-3.4	-0.9
Senegal	0.3	-2.8	-0.2	-1.0
Madagascar	0.1	0.6	0.4	0.2
Liberia	1.0	-0.1	0.1	0.3
Mozambique	-0.2	-0.6	0.7	0.3
Sierra Leone	0.3	-3.7	-0.4	-2.6
Somalia	-0.9	-1.6	-1.4	-0.9
Democratic Republic of Congo	0.2	0.2	-0.2	0.2
Burundi	0.0	1.5	1.0	0.6
Mean	0.2	-0.9	0.0	-0.3

Source: Authors' calculations based on FAOSTAT database. , n.a = not available or reliable.

(Erenstein et al., 2011; Odame and Muange, 2011; Scoones and Thompson, 2011). As far as fertilizer is concerned, FAOSTAT data shows that consumption is slightly higher among the first group: on average, better performing countries used 8 kg of fertilizer (plant nutrients) per ha, compared to 6 and 3 kg/ha among medium and poor performing countries, respectively (Table 5a). Despite the inherently low soil fertility, the quantity of fertilizer consumption per unit of farmland in SSA is the lowest in the world and well below the level that sustains rapid yield increases. Fertilizer application rates in SSA as a whole is only 10 kg of nutrients per hectare (ha) of arable land, compared with 86 kg/ha in South Asia, 118 kg/ha in Latin America, 198 kg/ha in an average middle-income country, and 288 kg/ha in a high-income country (Hernandez and Torero, 2011).

Apart from shortage of complementary inputs such as high yield varieties and irrigation, both supply and

Table 5a. Fertilizer consumption nutrient per ha of arable land.

Country	Kilograms fertilizer nutrient per hectare of arable land			
	Pre-reform	Transition	Post-reform	Whole period
	1968-1983	1984-1993	1994-2008	1968-2008
Better performers				
Benin	2	5	8	5
Burkina Faso	1	4	7	4
Nigeria	2	9	6	5
Côte d'Ivoire	9	9	23	14
Ghana	6	3	6	5
Niger	0	0	0	0
Sudan	4	5	4	4
Malawi	9	21	23	17
Kenya	13	20	28	20
Mean	5	8	12	8
Medium performers				
Central African Republic	1	1	0	0
Mali	4	9	9	7
U.R. of Tanzania	3	5	4	4
Cameroon	3	4	6	4
Guinea	2	1	2	2
Chad	1	1	1	1
Togo	1	4	5	3
Zambia	23	33	26	27
Congo	2	2	4	3
Mean	4	7	6	6
Poor performers				
Rwanda	0	1	2	1
Ethiopia	2	6	12	6
Uganda	1	0	1	1
Mauritania	3	8	3	4
Senegal	5	4	7	6
Madagascar	3	2	3	3
Liberia	8	2	0	4
Mozambique	5	1	3	3
Sierra Leone	3	3	1	2
Somalia	2	2	0	1
Democratic Republic of Congo	1	1	0	1
Burundi	1	3	2	2
Mean	3	3	3	3

Source: Authors' calculations based on FAOSTAT database.

demand constraints have made fertilizer expensive in Sub-Saharan Africa. The cost of importing fertilizer is high because of the small volume that many countries import and inadequate port facilities. Transport and logistics costs in Africa are often very high, making fertilizer in Sub-Saharan Africa at least double more expensive than in Asia and the US (Smale et al., 2011; Morris et al., 2007). Market failures and inefficiencies affecting seed markets (discussed earlier) have also constrained the emergence of a viable fertilizer markets in SSA. Turning to irrigation system, it is clear that SSA countries have made little effort to expand area under irrigation. The share of arable land under irrigation showed no change over the past 40 years and stayed at an average of 2.3% among group one or better performing countries (Table 5b). The only country with

Table 5b. Share of total area equipped for irrigation in total arable land.

Country	Period average (%)			
	Pre-reform	Transition	Post-reform	Whole period
	1968-1983	1984-1993	1994-2008	1968-2008
Better performers				
Benin	0.4	0.6	0.5	0.5
Burkina Faso	0.3	0.5	0.6	0.5
Nigeria	0.7	0.7	0.8	0.8
Côte d'Ivoire	1.8	2.5	2.6	2.2
Ghana	1.0	1.0	0.8	0.9
Niger	0.2	0.5	0.5	0.4
Sudan	13.9	13.9	10.9	12.8
Malawi	0.6	0.9	1.7	1.1
Kenya	1.0	1.1	1.7	1.3
Mean	2.2	2.4	2.2	2.3
Medium performers				
Central African Republic	0.0	0.0	0.1	0.0
Mali	3.2	3.2	4.1	3.6
U.R. of Tanzania	1.0	1.6	1.8	1.4
Cameroon	0.2	0.4	0.4	0.3
Guinea	8.8	11.4	6.9	8.7
Chad	0.3	0.5	0.7	0.5
Togo	0.1	0.3	0.3	0.2
Zambia	0.7	1.5	5.6	2.7
Congo	0.2	0.2	0.4	0.3
Mean	1.6	2.1	2.3	2.0
Poor performers				
Rwanda	0.6	0.5	0.8	0.7
Ethiopia	1.2	1.6	2.6	1.8
Uganda	0.1	0.2	0.2	0.2
Mauritania	16.2	13.6	10.2	13.4
Senegal	2.2	2.7	3.4	2.8
Madagascar	21.3	34.9	37.1	30.4
Liberia	0.5	0.7	0.8	0.6
Mozambique	1.6	3.0	2.8	2.4
Sierra Leone	3.2	5.8	4.2	4.2
Somalia	11.8	19.7	18.2	16.1
Democratic Republic of Congo	0.0	0.1	0.2	0.1
Burundi	1.5	1.7	2.2	1.8
Mean	5.0	7.0	6.9	6.2

Source: Authors' calculations based on FAOSTAT database.

significant level of irrigated agriculture is Sudan and the trend over time is a decline instead of an expansion. In Zambia, area equipped for irrigation increased from 46,000 in 1996 to 156,000 ha in 2008. The level of irrigation is relatively better among poor performing (Group 3) countries with an average of 6.2% of arable land under irrigation over the period of 1968 to 2008. The proportion also increased over time, from 5% in 1968 to 1983 to 6.9 to 7.0% in the transition and post-reform period (Table 5b). However, the extent of irrigation in the group is influenced mainly by three countries, Madagascar, Mauritania and Somalia. The latter two countries are arid and rely on irrigation for much of their agricultural production, while Madagascar has a traditional irrigation-based rice cultivation system. Given the poor performance of the three countries, it appears that irrigation systems have not been effectively utilized probably due to lack of complementary inputs and effective management practices. Lack of technical expertise, inconsistent and poor government policies and programs, weak research capacity, and underdeveloped markets, among others, seem to have hampered the emergence of productive irrigation agriculture in SSA (ICID, 2010).

Inadequate policy support to intensification

Between the mid-1980s and early 1990s, several African countries adopted the structural adjustment programs (SAPs) that included liberalization of output and input prices, devaluation of local currency, removal of subsidies and dismantling of parastatals. Most SSA countries moved to market-determined exchange rates and open trade regimes. Net taxation of agriculture decreased, which, together with competition in the market place, created a more positive environment for agricultural investment (Anderson and Masters, 2009; Jayne et al., 2002). However, the fact that the production increase was obtained largely through area expansion indicating that the reform has not provided sufficient incentive to intensify production even in better performing countries. Besides, currency devaluations and subsidy removals tended to temporarily reduce fertilizer consumption in nearly all regions of SSA (Kelly, 2006). The cost of inputs such as fertilizer rose sharply, making it unaffordable for many smallholders. As a result, some countries were forced to re-introduce subsidy programs. For instance, Malawi (from group one) implemented a large scale input subsidy program known as 'starter pack' in the 1998/1999 and 1999/2000 agricultural seasons. The program was scaled down to 'targeted input program' in 2000/2001 but expanded as large scale Agricultural Input Subsidy Program in 2005/2006 (Dorward, 2009). Malawi's exceptional performance in the post-reform period (1994 to 2008) could be related to the fertilizer and hybrid seed subsidy program of the government that has been in force since the late 1990s (Buffie and Atolia, 2009).

In Nigeria (group one), the Developing Agricultural Inputs Markets in Nigeria (DAIMINA) project introduced vouchers in 2004 to support agro-dealer development and improve producer access to and use of inputs. Other countries, including Burkina Faso, Ghana, and Kenya from group one as well as Mali, Tanzania and Zambia from group two, and Rwanda and Senegal from group three have implemented input subsidy programs since 2007/2008 (Druilhe and Barreiro-Hurlé, 2012). Inadequate rural financial services following the economic reform and the poor performance of state-owned agricultural development banks have also constrained farmers' access to input loans and discouraged intensification. Private commercial banks have shown limited interest in expanding their operations in rural areas, following the financial liberalization programs (Gonzalez-Vega, 2003). The failure of special credit lines to agriculture has left gaps in financial services in many countries (World Bank, 2008). In Nigeria, for instance, agricultural financing has a long history and various attempts to supply loans to farmers was met with limited success (Mafimisebi et al., 2010; Ugwu and Kanu, 2012).[6] State-owned agricultural development banks were allowed to operate in some countries but they performed poorly, although, there have been some notable exceptions. The reformed Banque Nationale de Développement Agricole of Mali is currently operating as a second-tier institution offering refinancing facilities and savings products (Making Finance Work for Africa Secretariat, 2012). Liberalization of output markets was expected to raise producer prices and improve the incentive to use inputs. However, the performance of food staple markets in SSA is often hampered by poor infrastructure, limited capacity of grain traders, inadequate support services, and weak institutions, thus giving rise to high transaction costs and price volatility.[7] In Ethiopia, for instance, maize prices collapsed from 150 Birr/quintal to 20 to 30 Birr/quintal in 2001 to 2002. Ethiopian farmers could not repay their production loans and a major crisis occurred. Farmers generally consider such price collapse a principal factor discouraging new technology introduction (Sanders and Shapiro, 2006). In Tanzania, for instance, producer prices have shown considerable seasonal variability after market liberalization, falling to very low levels immediately after harvest (when most farmers sell their produce) and rising to very high levels just before the next main harvest. Price uncertainty has not encouraged investment in inputs (Skarstein, 2005). Unpredictable government operations in grain markets, lack of quality standards with respect to moisture content, and threat of grain confiscation, among others, have discouraged investment in market stabilizing activities such as grain storage (Jayne et al., 2010)[8]. A major problem in SSA is the absence of risk management tools to deal with price and

[6]Agricultural loans were given at concessionary interest rates and beginning in 1972 commercial and merchant Banks were mandated to extend a prescribed minimum percentage of their loan portfolio to agriculture. However, such measures were found inconsistent with financial-sector reform and the policy was abolished in 1996. Cooperatives, friends, and family members dominate the sources of farm credit among small farmers in Nigeria, and the total amount obtained from these sources is often very limited compared to the amount that formal financial institutions would have offered (Phillip et al., 2009).

[7]In Africa, high transport costs due to poor roads, high fuel prices, administrative procedures which cause delays, etc. have resulted in high marketing costs which lower grain prices for producers and raise prices for consumers. In East Africa, for instance, prices [per ton-kilometer (tkm)] on the Mombassa – Kampala (linking Uganda with Kenya) are more than two times higher than in Brazil and four times higher than in Pakistan (Teravaninthorn and Raballand, 2009).

[8] Jyne et al. (2010) Patterns and Trends in Food Staple Markets in Eastern and Southern Africa: Towards the Identification of Priority Investments and Strategies for Developing Markets and Promoting Smallholder Productivity Growth, MSU International Development Working Paper No. 104.

Table 6. Average number of conflict-related deaths.

Country	Number of people			
	Pre-reform	Transition	Post-reform	Period
	1968-1983	1984-1993	1994-2008	1968-2008
Better performers				
Benin	0	0	0	0
Burkina Faso	0	100	0	100
Nigeria	60484	0	124	60608
Côte d'Ivoire	0	0	1265	1265
Ghana	74	0	0	74
Niger	0	172	1284	1457
Sudan	23380	35389	21030	79799
Malawi	0	0	0	0
Kenya	318	0	0	318
Mean	9362	3962	2634	15958
Medium performers				
Central African Republic	0	0	546	546
Mali	0	150	247	397
U.R. of Tanzania	0	0	0	0
Cameroon	0	500	0	500
Guinea	19643	12402	5257	37302
Chad	0	0	1174	1174
Togo	0	262	0	262
Zambia	0	0	0	0
Congo	0	660	9945	10605
Mean	2183	1553	1908	5643
Poor performers				
Rwanda	0	10000	14454	24454
Ethiopia	224065	126882	9802	360749
Uganda	63613	50880	8137	122629
Mauritania	2615	0	0	2615
Senegal	0	384	889	1273
Madagascar	128	0	0	128
Liberia	27	15298	3469	18794
Mozambique	41253	123751	0	165005
Sierra Leone	0	1400	12812	14212
Somalia	1828	60761	12339	74928
Democratic Republic of Congo	919	0	151618	152537
Burundi	0	1984	9563	11547
Mean	27871	32612	18590	79073

Source: Encyclopedia of the Nations, http://www.nationsencyclopedia.com/WorldStats/WDI-poverty-conflict-fragility-deaths.html

production risks. Price and market stabilization schemes or commodity exchange systems with futures and options for price risk management are largely non-existent in SSA (Demeke et al., 2012).

High incidence of external shocks

Apart from deficient policies and inadequate economic incentives, external shocks in the form of conflict and uncertain rains have affected investment in farm inputs and technology. The majority of the 12 countries within the poor performers are known to have gone through some armed conflict and severe political instability during the period under consideration, namely: Rwanda, Ethiopia, Uganda, Mauritania, Senegal, Madagascar, Liberia, Mozambique, Sierra Leone, Somalia, Democratic Republic of the Congo and Burundi (Tables 6 and 7). Strong production performance in Rwanda, Ethiopia, Sierra Leone and Liberia in the post-reform period is also

Table 7. Average annual rainfall (mm).

Country	Average rainfall in mm			
	Pre-reform	Transition	Post-reform	Whole period
	1968-1983	1984-1993	1994-2000*	1968-2008
Better performers				
Benin	994	1002	1051	1017
Burkina Faso	745	715	771	747
Nigeria	1114	1097	1172	1131
Côte d'Ivoire	1309	1280	1326	1308
Ghana	1143	1137	1146	1143
Niger	151	145	166	155
Sudan	404	396	438	414
Malawi	1098	1034	1091	1080
Kenya	619	603	669	633
Mean	842	823	870	848
Medium performers				
Central African Republic	1335	1293	1350	1330
Mali	293	276	313	296
United Republic of Tanzania	1016	995	1014	1010
Cameroon	1581	1539	1598	1577
Guinea	1646	1542	1690	1636
Chad	325	298	351	328
Togo	1150	1139	1172	1156
Zambia	992	926	993	976
Congo	1603	1582	1605	1599
Mean	1105	1065	1121	1101
Poor performers				
Rwanda	1139	1036	1091	1096
Ethiopia	782	725	765	762
Uganda	1171	1120	1168	1157
Mauritania	79	77	92	83
Senegal	649	643	712	671
Madagascar	1418	1479	1481	1456
Liberia	2399	2264	2481	2396
Mozambique	983	938	1016	984
Sierra Leone	2460	2287	2403	2397
Somalia	235	238	274	250
Democratic Republic of the Congo	1510	1499	1480	1496
Burundi	1248	1215	1205	1224
Mean	1173	1127	1181	1164

Sources: Country aggregated rainfall time-series dataset was created by Hideki Kanamaru, NRC, FAO from CRU TS 3.1 of University of East Anglia Climate Research Unit (CRU). [Phil Jones, Ian Harris]. CRU Time Series (TS) high resolution gridded datasets 3.1, [Internet]. NCAS British Atmospheric Data Centre, 2008. Available from http://badc.nerc.ac.uk/view/badc.nerc.ac.uk__ATOM__dataent_1256223773328276, accessed April, 2011.

associated with the end of serious civil wars (Collier et al., 2002). Persistent conflict has resulted in poor or negative food production performance in Somalia, Democratic Republic of the Congo and Burundi. Among the first group of countries, Sudan is the most affected by violent conflict but it has managed to sustain its better performance because production is concentrated in irrigated areas where the problem of conflict is limited (Keen, 1994, 1998, 2000). Similarly, the recent conflict in Cote d'Ivoire can be considered as a major contributing factor to the slow growth in the post-reform period, which lowered the growth of food production to 2.3% compared to 3.3% recorded during the transition period (Table 1). Comparably, Nigeria's poor growth during the pre-reform period could also be associated with the violent – the Biafra War – of 1967 to1970 (Richards, 2006).

The impact of conflict on food production and food security has been documented by other studies. Messer et al. (1998), for instance, estimated that food production in 13 war-torn SSA countries during 1970 production in 13 war-torn food security has been mpared to peace years. FAO study also estimated that conflict induced losses of agricultural output totalled $121 billion in real terms (or an average of $4.3 billion annually) during the period of 1970 to 1997 (FAO, 2000). Climate variability is another major risk constraining the adoption of improved technologies and inputs in many parts of Africa (Barret, 2002). Globally, Africa faced the highest frequency during the period of 1960 to 2006 with a total of 382 reported drought events, compared to 165 in Asia, the region the next highest frequency (Gautam, 2006). It is reported that about 60% of SSA is exposed to drought and 30% so extremely. Part of the Sahel as well as Eastern and Southern Africa are among the most affected (Benson and Clay, 1998). African countries are also affected by floods that cause loss of life, damage to property, and promote the spread of diseases such as malaria, dengue fever and cholera. Rainfalls accompanying tropical cyclones often result in flood disasters in Mozambique. Many parts of Ethiopia, Kenya and Somalia are also vulnerable to flood. Madagascar and Mozambique are among countries most often affected by cyclones (ICSU Region Office for Africa, 2007). With little or no access to insurance or other production risk management tools, hazards related to weather, pests and diseases have impeded technology adoption, resulting in poverty traps in SSA.

CONCLUSION

The performance of the food production sector in Sub-Saharan Africa (SSA) is critical for a number of reasons. First domestic production is the principal avenue to ensuring access to affordable food in poor countries with limited capacity to import food. The multiplier effect of sustainable food production on the economy is considerable as it has direct linkage with other activities such as transportation, marketing, warehousing, food processing, tourism and local commerce. In countries where the growth of food production outpaces demand, social and political harmony as well as macroeconomic stability can be maintained, paving the way for sustained economic growth. This paper examines changes in food production performance among 30 SSA countries over the period of 1968 to 2008. The countries were grouped into three: the comparatively better, medium and poor performers – based on their food production performance. Three different periods, representing the pre-reform, transition and post-reform, were also identified to pay particular attention to the impact of policy reforms. The results support previous findings that not many countries have managed to achieve a food production average growth rate in excess of 3% per annum. Indeed, only 9 or 30% of the sample countries achieved such growth rates over the period of 1968 to 2004.

Annual food production performance averaged 3% or more in 60% of the sample countries following the policy reforms. Nevertheless, rates of output growth varied from one period to the other and the recent improved performances were achieved largely through unsustainable expansion of land under cultivation. At an average application rate of 10 kg of nutrients per ha of arable land, fertilizer use levels in SSA are only 5% of the level in an average middle income country. Not surprisingly, Africa's agricultural GDP per capita is only one-fourth of world's average and per capita food production is stagnant or declining. Better food production performance is associated with better availability of food supply and lower rates of undernourishment. The prevalence of undernourishment in the total population averaged 14% in 2005 to 2008 among better performers, compared with 30% among medium and poor perfuming countries. However, most countries have failed to achieve food self-reliance. With domestic supply lagging behind rapidly expanding demand, most countries have experienced a substantial increase in their food import spending in recent years (1994 to 2008) and have faced serious food security concerns following the hikes in international food price and high levels of volatility that began in 2007/2008. The deficit in food trade balance has sharply increased in nearly all cases, and with few exceptions, SSA countries have very limited capacity of financing their food import bills. Even among better performing counties, food imports accounted for about 27% of the total value export earnings during the period of 1994 to 2008. The share averaged 74% among poor performing countries over the same period. Imported food items are also expensive in the local markets owing to high cost of freight, port charges, domestic transport and marketing margins. Locally produced staples are cheaper, but a more sustainable and higher growth of domestic production is constrained by lack of public support programs and absence of effective measures to address market failures and inefficiencies in input, output, credit, land and risk management markets. Weak institutions, which are often aggravated by conflict and political instability, have hampered investment in the food value chain and made it very difficult for farmers to access inputs and new technologies.

Building institutions that foster partnership between governments and farmers, traders, processors and other stakeholders along the value chain and facilitate the emergence of a stable and competitive markets is critical to ensure food availability, affordability/ accessibility and stability in SSA. Without a significant increase in budgetary allocations to build institutional capacity, develop market infrastructure, expand irrigation schemes, ensure sustainable natural resource utilization, transform agricultural research and development, and build capacity for climate change mitigation and adaption, as rightly advocated by the Comprehensive Africa Agriculture Development Program (CAADP) of the African Union, the food security situation is likely to worsen further for many SSA countries. The views expressed are purely those of the writers and may not in any circumstances be regarded as stating an official position of FAO and the European Commission.

REFERENCES

AEC (1991) 'Treaty of African Economic Community'.

AFDB OECD, UNECA (2010). African Economic Outlook – Public Resource Mobilisation and Aid, OECD Development Centre and ADB, Paris.

African Union (2003). "Maputo Declarations", Assembly of the African Union, Second Ordinary Session. Addis Abbaba, P. 10.

Anderson K, Masters W (2009). Distortions to Agricultural Incentives in Africa, Washington DC: World Bank.

Benson C, Clay E (1998). The Impact of Drought on Sub-Saharan African Economies: A Preliminary Examination. Washington, DC: World Bank. Technical 401:1998.

Boon EK (2007). Food Security in Africa: Challenges and Prospect, Report for Regional Sustainable Development Review: Africa.

Breisinger C, van Rheenen T, Ringler C, Pratt A, Minot N, Aragon C, Yu B, Ecker O, Zhu T (2010). Food Security and Economic Development in the Middle East and North Africa Current State and Future Perspectives. IFPRI Discuss. P. 00985.

Buffie EF, Atolia M (2009). Agricultural input subsidies in Malawi: Good, bad or hard to tell? FAO Commodity and Trade Policy Research Working Paper No. 28, August.

Camara Y (2000). Profitability of cassava production systems in West Africa: A comparative analysis (Cote d'Ivoire, Ghana and Nigeria), CAB. P. 200.

Chilowa W (1998). The Impact of Agricultural Liberalization on Food Security in Malawi. Paper presented at a seminar on Economic Liberalization: A review of Policies, Chancellor College, Zomba, Malawi.

Clark J (2002). 'The neo-colonial context of the democratic experiment of Congo-Brazzaville'. Afr. Affairs 112:446.

Deininger K, Byerlee D, Lindsay J, Norton A, Selod H, Stickler M (2010). Rising Global Interest in Farmland. Can It Yield Sustainable and Equitable Benefits? Washington, D.C: World Bank.

Diao X, Fan S, Headey D, Johnson M, Nin Pratt A, Yu B (2008). Accelerating Africa's food production in response to rising food prices: Impacts and requisite actions. IFPRI Discuss. P. 825.

Dorward A, Shenggen F, Kydd J, Lofgren H, Morrison J (2004). 'Institutions and Policies for Pro-poor Agricultural Growth'. Dev. Pol. Rev. 22:611-622.

Dorward A (2009). Rethinking agricultural input subsidy programmes in a changing world, Paper prepared for the Trade and Markets Division, FAO, April.

Druilhe Z, Barreiro-Hurlé J (2012). Fertilizer subsidies in sub-Saharan Africa, ESA Working paper Num 12:04, FAO, July.

ECOWAS (2009). Ecowas Agricultural Policy Compressive African Agriculture- Report Ghana.

Edgerton MD (2009). Increasing Crop Productivity to Meet Global Needs for Feed, Food and Fuel'. Plant Physiol. 149:7-13.

Erenstein O, Kassie GT, Mwangi W (2011). Challenges and opportunities for maize seed sector development in eastern Africa, Paper presented at the conference: Increasing Agricultural Productivity and Enhancing Food Security in Africa: New Challenges and Opportunities, 1-3 November 2011, Africa Hall, UNECA, Addis Ababa, Ethiopia.

Fafchamps M (1992). 'Cash Crop Production, Food Price Volatility, and Rural Market Integration in the Third World'. Am. J. Agric. Econ. 74:90-92.

FAO (1999). The Impact of the Structural Adjustment Programmes on Family Farms. Working Paper Investment Center. Rome, Italy

FAO (2009). The State of the Food Insecurity in the world 2009: Economic crises - impacts and lessons learned. Rome, Italy.

FAO (2010). The State of the Food Insecurity in the world 2010: Addressing food insecurity in protracted crises. Rome, Italy.

FAO (2011a). The State of the Food Insecurity in the world 2011: How does international price volatility affect domestic economies and food security? Rome, Italy.

FAO (2011b). Save and Grow: A Policymaker's Guide to Sustainable Intensification of Smallholder Crop Production. Rome, Italy.

FAO (2012). The State of the Food Insecurity in the world 2012: Economic growth is necessary but not sufficient to accelerate reduction of hunger and malnutrition. Rome, Italy.

FAO (2000). The state of food insecurity in the world. Rome, Italy.

G8 (2009). 'L'Aquila' Joint Statement on Global Food Security.L'Aquila Food Security Initiative (AFSI) declaration.

Gautam M (2006). Managing Drought in Sub-Saharan Africa: Policy Perspectives, Invited paper prepared for an Invited Panel Session on Drought: Economic Consequences and Policies for Mitigation, at the IAAE Conference, Gold Coast, Queensland, Australia, August 12-18.

Gonzalez-Vega C (2003). Deepening Rural Financial Markets: Macroeconomic, Policy and Political Dimensions, paper prepared for the conference "Paving the Way Forward: An International Conference on Best Practices in Rural Finance". Washington, D.C., 2-4 June.

Harrigan J (1997). 'Modeling the impact of World Bank policy-based lending: the case of Malawi's agricultural sector'. J. Dev. Stud. 33:6.

Harrigan J (2003). 'U-Turns and Full Circles: Two Decades of Agricultural Reform in Malawi 1981–2000'- World Dev. 31:847-863

Hernandez MA, Torero M (2011). Fertilizer Market Situation: Market Structure, Consumption and Trade Patterns, and Pricing Behavior, IFPRI Discuss. P. 01058.

ICID (2010). Irrigation development challenges for the least developed countries in Africa, International Commission on Irrigation and Drainage (ICID).

ICSU Region Office for Africa (2007). Natural and Human-induced Hazards and Disasters in sub-Saharan Africa, International Council for Science (ICSU).

IFPRI (2011). Agricultural Productivity and Policies in Sub-Saharan Africa. IFPRI Discuss. P. 01150.

Jayne TS, Yamano T, Weber M, Tschirley D, Benfica R, Chapoto A. and Zulu B (2003). 'Smallholder Income and Land Distribution in Africa: Implications for Poverty Reduction Strategies'. Food Policy, 28:253-275.

Jayne TS, Govereh J, Mwanaumo A, Nyoro J, Chapoto A (2002). 'False Promise or False Premise? The Experience of Food and Input Market Reform in Eastern and Southern Africa'. World Development, 30:1967-1985.

Kakwani N (1997). On Measuring Growth and Inequality Components of Poverty with Application to Thailand', The University of New South Wale. School Econ. Discussion Paper.

Kasfir N (2008). 'Are African Peasants Self-Sufficient?', Dev. Change 17:335-357.

Keen D (1994). The Benefits of Famine: A political economy of famine and relief in south western Sudan, 1983-1989. Princeton, NJ (USA): Princeton Univ. Press.

Keen D (1998). Aid and Violence, with Special Reference to Sierra Leone'. Disasters 22:318-327.

Keen D (2000). Managing Armed Conflicts in the 21st Century. 'War and Peace: What's the difference?', International Peacekeeping 7:1-22.

Kelly V (2006). Factors affecting demand for fertilizer in Sub-Saharan

Africa, Agriculture and Rural Development Department, World Bank, Washington D.C.

Kherallah M, Minot N, Kachule R, Soule BG, Berry P (2001). Impact of agricultural market reforms on smallholder farmers in Benin and Malawi, IFPRI Collaborative Research Project: http://www.ifpri.cgiar.org/sites/default/files/pubs/divs/mtid/dr/200102/v1.pdf.

Kydd J, Dorward A, Morrison J, Cadisch G (2007). Agricultural development and pro-poor economic growth in sub-Saharan Africa: potential and policy'. Oxford Dev. Stud. 32:37-57.

Lele U (1990). Structural adjustment, agricultural development and the poor: Some lessons from the Malawian experience'. Word Dev. 18:1207-1219.

Mafimisebi T, Oguntade A, Mafimisebi O (2010). Re-engineering Agriculture For enhanced Performance through Financing. J. Econ. Finan. Admin. Sci. 15:35-49.

Making Finance Work for Africa Secretariat (2012). Policy brief on agricultural finance in Africa, ADB, Tunisia.

Maxwell D (1999). The Political Economy of Urban Food Security in Sub-Saharan Africa'. Word Dev. 27:1939-1953.

Messer E, Cohen MJ, D'Costa J (1998). Food from peace: breaking the links between conflict and hunger. In: 2020 Vision for Food, Agriculture, and the Environment. Discussion, IFRI, Washington DC. P. 24.

Morris M, Valerie A, Kelly, Ron J. Kopicki, and Derek Byerlee (2007). Fertilizer Use in African Agriculture: Lessons Learned and Good Practice Guidelines. Washington DC: World Bank.

Moser G, Rogers S, Reinold van Til (1997). Nigeria: Experience with Structural Adjustment, IMF Occasional P. 148, Washington: IMF

Munthali T (2004). The Impact of Structural Adjustment Policies (SAPs) on Manufacturing Growth in Malawi, Macroeconomics 0410002, EconWPA

Nweke F (2004). 'New Challenges in the Cassava Transformation in Nigeria and Ghana', EPTD Discussion paper No 118, IFPRI, Washington, DC.

O'Connell S (2008). 'The Political Economy of Economic Growth in Africa, 1960-2000: An Overview' in B. Ndulu, A. O'Connell, J.P. Azam, R. Bates, A. Fosu, J. Gunning and D. Njinkeu (eds.) The Political Economy of Economic Growth in Africa, 1960-200. Vol. 2. Country Case Studies. Cambridge: Cambridge University Press.

Odame H, Muang E (2011). 'Can Agro-dealers Deliver the Green Revolution in Kenya?'. IDS Bull. 42:78-89.

OECD (2005). 'Data and Metadata Reporting and Presentation Handbook', Paris.

OECD (2008). African Economic Outlook 2008. Paris, France.

Omamo S, Diao X, Wood S, Chamberlin J, You L, Benin S, Wood-Sichra U, Tatwangire A (2006). 'Strategic Priorities for Agricultural Development in Eastern and Central Africa', Research Report no. 150, IFPRI, Washington, DC.

Phillip D, Nkonya E, Pender J, Oni OA (2009). Constraints to Increasing Agricultural Productivity in Nigeria: A Review, IFPRI/NSSP.

Rakotoarisoa MA, Lafrate M, Paschali M (2011). Why has Africa Become a Net Food Importer: Explaining Africa agricultural and food trade deficits, Trade and Markets Division, Rome: FAO.

Reardon T, Barret Ch, Valerie K (2002). 'Policy Reform and Sustainable Agricultural Intensification in Africa'. Dev. Policy Rev. 17:375-395.

Richards P (2006). The History and Future of African Rice: Food Security and Survivalian in West African War Zone', African Spectrum, 41:77-93. (Not cited in the article)

Rono JK (2002). The impact of structural Adjustment Programs on Kenyan Society. J. Soc. Dev. Afr. 17:81-98.

Sanders JH, Barry S (2006). 'Policies and market development to accelerate technological change in the semiarid zones: A focus on Sub-Saharan Africa' in Payne et al (ed.), Dryland Agriculture, 2nd Edition, Agronomy Monograph No. 23. American Society of Agronomy, Inc. Crop Science Society of America, Inc. Soil Science Society of America, Inc. 677 South Segoe Raod Madison, Wisconsin 53711:879-900

Scoones I, Thompson J (2011). The Politics of Seed in Africa's Green Revolution: Alternative Narratives and Competing Pathways', IDS Bull. 42:1-23.

Skarstein R (2005). 'Economic liberalization and smallholder productivity in Tanzania: From Promised success to real failure, 1985-1998. J. Agrar. Change 5:334-362.

Skarstein R (2005). 'Economic Liberalization and Smallholder in Tanzania. From Promised Success to Real Failure, 1985-1988'. J. Agrarian Change 5:334-362.

Smale M, Byerlee D, Jayne T (2011). 'Maize Revolutions in Sub-Saharan Africa', Policy Research Working Paper, 5659, The World Bank Development Research Group Agriculture and Rural Development Team.

Teravaninthorn S, Raballand G (2009). 'Transport Prices and Costs in Africa: A Review of the Main International Corridors', African Infrastructure Country Diagnostic, Working P. 14, The World Bank.

Thapa G, Gaiha R (2011). Smallholder farming in Asia and the Pacific: Challenges and Opportunties', Conference on New Directions for Smallholder Agriculture, 24-25 January, Rome, IFAD HQ.

Ugwu B (1996). Increasing cassava production in Nigeria and prospects for sustain the trend'. Outlook Agric. 25:179-185.

Ugwu DS, Kanu IO (2012). Effects of agricultural reforms on the agricultural sector in Nigeria', J. Afr. Stud. Dev. 4:51-59.

UNCTAD (2009). Food security and agricultural development in times of high commodity prices, Geneve.

UNECA (1979). Lagos Plan Action for the Economic Development of Africa 1980-2000', Addis Abbaba.

UNEP Year Book (2012). Emerging Issues in our Global Environment.

USDA (2007)- 'The Impact of Agricultural Research in Tropical Africa', Study P. 21.

Weissman S (1990). 'Structural adjustment in Africa: Insights from the experiences of Ghana and Senegal'. World Dev. 18:1621-1634.

WHO (2008). 'Global and regional food consumption patterns and trends'. Working P. 3.

Wobst P (2001). 'Structural Adjustment and Intersectoral Shifts in Tanzania: A Computable General Equilibrium'. International Food Policy Research Institute, Working, Washington, DC, P. 109.

World Bank (2008). 'World Bank Assistance to Agriculture in Sub-Saharan Africa', Washington DC: World Bank.

World Bank (2009). 'African Agriculture—What Do We Not Know? What Do We Need to Know?', Washington DC: World Bank.

Zawalinska K (2004). 'What has been an economic impact of Structural Adjustment Programs on households in Transition countries?', Essay no. 1 for P. 2 Institutions and Development Submitted in part-fulfilment of the requirements for the MPhil in Development Studies at the University of Cambridge.

25

Tenure security and soil conservation investment decisions: Empirical evidence from East Gojam, Ethiopia

Akalu Teshome

Amahar Agricultural Research Institute, Ethiopia P. O. Box 527, Bahir Dar, Ethiopia.

The purpose of this study is to investigate the mutual relation between tenure security and soil conservation investment and to examine the influence of other socio-economic and institutional factors on soil conservation investment and tenure insecurity. A formal survey is conducted in two districts of East Gojam Zone of Amhara region. The Zone and the districts are selected because of their long time experience with soil conservation development activities and land re-distribution. A two-stage random sampling procedure is used to obtain sample households. Because the structural model represents a simultaneous binary choice system, the investment and insecurity equations are estimated using a two-stage probit method. The results show that tenure insecurity is an important variable that affects the probability of investing in soil conservation technologies. However, the reverse relation is insignificant. Farmers' soil conservation investment decisions are positively and significantly related to slope, age, education level and public investment, whereas, tenure insecurity and distance from the main road have a negative significant influences on soil conservation investments. The analysis of tenure insecurity reveals that expectation of redistribution and farm size has a negative influence on tenure security, whereas education level has a reverse effect.

Key words: Tenure security, soil conservation investment, simultaneous binary choice system, two-stage probit, Ethiopia.

INTRODUCTION

Land degradation is one of the major environmental problems in developing countries. Soil erosion by water is the principal cause of land degradation, and a major constraint to agricultural development in developing countries (de Graaff, 1996). Globally, the problem of soil erosion is widely recognized and millions of dollars are spent every year on soil conservation projects. But in spite of money being spent and great effort being made, the problem is gradually getting worse. An important factor in land degradation and farmers' investment in soil conservation in developing countries is the change in the socio-economic environment of farm households resulting from policy reform measures taken at higher levels (Heerink et al., 2001). As population increases and land becomes scarce, land demand by the growing number of land claimants may be met by non-market mechanisms such as state land redistribution, informal land contracts and customary inheritance. The persistence of such mechanisms and absence of an established legal rights land system has resulted in increasing tenure insecurity and continued land fragmentation (Amare, 1998). The absence of tenure security is highly linked to poor land use which in turn leads to environmental degradation (Otsuka and Place, 2001; Wannasai and Shrestha, 2008).

Ethiopia is one of countries that is heavily dependent on peasant agriculture and is affected by extensive degradation of agricultural land. Coupled with the poor performance of the agricultural sector, high population growth, land scarcity, technological stagnation, misguided policies and deficient institutional structure hinder sustainable utilization of agricultural land (Shiferaw, 1998).

Reducing resource degradation, increasing agricultural productivity, combating poverty, and achieving food security are major challenges of the nation. The poor agricultural practice and the country's intrinsic fragile biophysical conditions have resulted in large areas becoming severely degraded. Land degradation is most severe in highlands (over 1500 m altitude), which account for more than 43% of the country, 95% of the cultivated area, 75% of the livestock and host about 88% of the population. Hurni (1988) estimates that the annual rate of soil loss on crop land is on average 42 t/ha per year. If soil erosion continues with this rate, by the year 2010 some 60000 km^2 of agricultural land will have disappeared.

The Amhara National Regional State (ANRS) is one of the nine regional states of Ethiopia. The Region is endowed with huge potential of land and water for agriculture, but these are now under the threat of land degradation due to soilerosion. A recent study by Gete (2003) revealed that Western Amhara (Gojam) which was once known as bread basket of Ethiopia is now at severe risk due to soil degradation. The cause of soil erosion in ANRS is a combination of natural factors such as topography, erratic and erosive rainfall patterns and human actions including destruction of vegetation cover through deforestation, overgrazing, and inappropriate agricultural practices that are not in harmony with the environmental conditions. In this regard, dense population, primitive farming practices combined with intensive rains and rugged topography intensified land degradation (Betru, 2003).

To alleviate these problems, a massive conventional soil conservation program has been launched since 1975 (Shiferaw and Holden, 1998). In spite of the effort in introduction, the adoption rate has been minimal (Shiferaw and Holden, 1998, 1999).

Investment in soil and water conservation practices are influenced and constrained by socio-economical and institutional factors (de Graaff, 1993; Shiferaw et al., 2009). Soil conservation investment may be undertaken when sufficient returns are expected for a considerable period of time in comparison with the situation when such investments are not made. This is possible with a secure land tenure system.

Since the beginning of the twentieth century, Ethiopia has implemented different types of interventions in the area of land tenure. Currently, land is the state property and redistribution is the sole mechanism through which land transfer to accommodate new demands. The majority of the smallholders in Ethiopia (76%) are not sure whether their current land will belong to them in five years time (Ethiopia Economic Association, 2002 cited in Dessalegn, 2004). Benin and Pender (2001) in their study of the incidence of land redistribution in the Amhara region of Ethiopia revealed that every community has experienced at least one redistribution since 1974, and nearly half had a land redistribution since 1991, mainly in the recent redistribution since 1997. And also about four-fifths of the communities expect redistribution in the future. The stronghold of the state over rural land and subsequent action of land allocation through redistribution has given rise to tenure insecurity by rural farmers (Dessalegn, 2004). Cognizant of this problem, the government of Ethiopia has introduced land certification very recently to increase tenure security and farmers' propensity to investment.

Lack of secure rights on land decreases farmers' incentives to invest in land improvement (Besley, 1995; Otsuka and Place, 2001; Mekonnen, 2009). Moreover soil conservation investment is constrained and influenced by credit facilities, extension service, infrastructure availability, household endowment and household and farm characteristics. This implies that there are a lot of institutional and socio-economical factors that might hinder farmers to invest their own conservation measures.

In view of this, it is important to investigate the factors influencing subsistence farmers' soil conservation decisions in the context of northwestern Ethiopia. The main objectives of the study are to investigate the mutual relation between tenure security and individual soil conservation investment and to examine the influence of other socio-economic and institutional factors on individual soil conservation investment and tenure insecurity

Analytical model

Following the above description of the relationship between property rights, uncertainty and utility maximization, we use a one-period household model to assess the impact of tenure security on land related investments as the conceptual basis for our empirical investigation. There are two alternatives hypotheses related to tenure security and investment. The first one is that more secure land rights will have a positive impact on investment. In this case tenure security is exogenous. The other hypothesis is that investment is undertaken to enhance tenure security, in this case tenure security is endogenous.

In our situation, the farmer willingness to invest may be affected by the perception of risk. A farmer decides whether to invest in soil conservation technologies by considering risk of losing land, due to redistribution sometimes in the future. When the farmer feels secure about the tenure system he may decide to invest in soil conservation technologies and his production may increase as the result of the investments. Meanwhile he may lose his investment some time in future due to redistribution. If a farmer feels insecure about tenure he may decide not to invest in soil conservation technologies and his production may decrease due to soil erosion. But he will not lose any investment when redistribution is undertaken. The farmer decides whether to invest or not by considering the above scenarios.

Assuming that farmers maximize expected utility, the decision whether to invest (I_i =1) or not (I_i = 0) is based on a comparison of expected utilities of investing or not investing soil conservation technology. Using the difference in expected utilities gives the following decision rule:

$$I_i = \begin{cases} 1, if, E[U_I^I - U_I^O | z_i, \gamma_i] > 0 \\ 0, if, E[U_I^I - U_I^O | z_i, \gamma_i] < 0 \end{cases} \quad (1)$$

Where E denotes expectation of a farmer which is conditional on household and farm characteristics (z_i) and perception of risk (γ_i). U_I^I denotes utility of investing soil conservation and U_I^O is utility of not investing soil conservation. The utility level of investing or not investing depends up on the expectation of income with the presence or absence of soil conservation technologies. Considering an individual farmer with utility function $U(\omega)$, where ω is income which depends on individual and farm characteristics (z_i) and variables affecting perception of risk (γ_i). Farmer's expected utility is assumed to be increasing in income $[\omega]$, as indicated in Equation (3)

$$\max E[U(\omega) | z_i, \gamma_i] \quad (2)$$

Subject to

$$\omega = \sum_{i=1} (Y_i A_i p_i - w_i I_i - C) \quad (3)$$

$$Y_i = Y(I_i, F_i, A_i, Z_{1i}) \quad (4)$$

$$I_i = I(T_i, Z_{2i}) \quad (5)$$

$$T_i = T(I_i, Z_{3i}) \quad (6)$$

This equation defines income $[\omega]$ as annual crop revenues minus the unit cost (w_i) of conservation investments (I_i) and other variable costs (C). Crop revenue is the product of crop price (p_i), yield (Y_i) per hectare and land area (A_i). Yield, in turn, is concavely increasing with the presence of soil conservation investments $(Y'(I_i) > 0)$ and also depends on fertilize use (F_i), land area (A_i) and other factors (Z_{1i}) such as soil fertility, pest and weather condition. Soil conservation investment (I_i) depends on tenure security $(I'(T_i) > 0)$. This implies that better land security leads to more land investment. Soil conservation investment (I_i) also depends on other factors (Z_{2i}) such as slope of the plot, farm size, distance of the homestead from the main road, age of the household, extension contact, public investment, number of oxen, and education level.

On the other hand, tenure security is endogenous; security of tenure can be enhanced through investment $(T'(I_i) > 0)$. Tenure security also depends on other factors (Z_{3i}) such as farm size, expectation of re-distribution, age and education level of the household.

Hypotheses

Investment hypotheses

From the theoretical framework, several hypotheses can be derived that serve in empirical examination. Investments are measured in this study with the presence of soil conservation technologies on farmers' fields (fanya juu terrace[1], fanya juu with plantation and perennial) or not. The farmers are asked whether they invest or not individual soil conservation measures on their own plots. Investments undertaken by mass mobilization are not considered. Therefore, our dependent variable represents the presence of investment or not on farmer fields and it is a function of social, institutional, physical, and economical and attitudinal factor.

Development of the model is influenced by a number of working hypotheses. Based on the literature reviewed it is hypothesized that farmers decision to invest in conservation measures is influenced by combined effects of social, economical and institutional factors. A number of variables are expected to influence investment in soil conservation measure explained as follows:

Slope of the plot (Slope): Slope is an indicator of the probability of erosion on the land (Laper and Pandy,

[1]A fanya juu terrace is made by digging a trench and throwing the soil uphill to form an embankment. In our case farmers plant a grass strip on the fanya juu.

1999). The steeper the slope, the more likely the land will erode. Hence, it is hypothesized that investment tends to be likely on steeper slope.

Tenure insecurity (T_i): Tenure insecurity measures the perceived risk of loss of land at some time in the future. Investment is undertaken when the household is assured that he will reap the benefit for a considerable time. The household that feels insecure will not invest in soil conservation measures. So it is hypothesized to negatively influence investment.

Farm size (Farms): To invest soil conservation measure, the farm size is the crucial matter. Farmers having a large farm invest more than the others (Shiferaw and Holden, 1998). So it is hypothesized to positively associate with investment decision.

Distance to the main road (Disth): Distance to the main road is hypothesized to be negatively related to the probability of investment of soil conservation measures, since households near to main road tend to have access to information and are more likely to be visited by extension agents (Laper and Pandy, 1999).

Family size (Shh): Larger families will be able to provide the labor that is required for soil conservation investment. So it is hypothesized to be positively related to soil conservation investment.

Level of Education (Edu): Level of education is assumed to increase a farmer's ability to obtain, process, and use information relevant to the investment of soil conservation decision (Laper and Pandy, 1999). Education is therefore expected to increase the probability of investment of soil conservation.

Number of oxen (Ox): Number of oxen is hypothesized to be positively related to the probability of investment. This is because oxen are indicator of wealth and it is used in digging while soil bund is constructed.

Age of the household (Age): The age of household is hypothesized to be negatively related to the probability of investment. This is because old farmers are more suspicious about new technologies than young (Shiferaw and Holden, 1998).

Extension contact (Extc): Farmers who have frequent contact with extension agent are positively influenced to invest (Makokha et al., 1999). So it is hypothesized to be positively related to soil conservation investment.

Public investment (Pubcon): Farmers who have public investment are expected to have positive attitude towards soil conservation. This is because they perceive the benefit of the measures. Here public investment is a soil conservation investment practice which is constructed on farmers' plots by mass mobilization of the community. From Equation (5) the model of investment is specified as follows:

$$\text{Investment} = f(\text{Slope, Tenure Insecurity, Farms, Disth, Shh, Edu, Ox, Age, Extc, Pubcon}) \quad (6)$$

Tenure insecurity hypotheses

Tenure insecurity is measured as the perceived risk of loss of land some time in future. The farmers are asked about their expectation of handling their lands at different time interval (1 year, 5 years, 10 years and throughout their life time). The response will fall in one category: Insecure land holding or secure land holding. Therefore, our dependent variable represents the feeling of tenure insecurity and it depends on a lot of factors. A number of variables are expected to influence tenure security, explained as follows.

Farm size of the household (Farms): During re-distribution, farm size was one of the yardsticks to lose land. In line of this, the household who has a large farm size fear the risk of losing his/her land. So it is hypothesized to be positively related to tenure insecurity.

Investment (I_i): Land tenure security is influenced by investment. Tenure security can be enhanced through investment. Land related investment is undertaken to enhance security of land holding (Brasselle et al., 2001). This implies that investment will be undertaken by insecure households in order to increase their security. So it is hypothesized to be positively related to tenure insecurity.

Expectation of re-distribution (Expredis): The household may expect re-distribution due to the government land re-distribution policy. So it is hypothesized to be positively related to tenure insecurity.

Education level of the household (Edu): Level of education is assumed to increase farmers' ability to obtain information about the tenure system. So it is hypothesized to positively relate to tenure insecurity.

Number of oxen (Ox): Number of oxen is the proxy variable for wealth and power. Wealthy households may f ear losing their land due to the past redistribution criteria. So number of oxen is hypothesized to be positively related to tenure insecurity.

From equation 3.6 the model of tenure insecurity is specified as follows:

$$\text{Tenure insecurity} = f(\text{Age, Edu, Farms , Ox, Expredis, Investment}) \quad (7)$$

METHODOLOGY

Study areas, sampling procedure and data collection

The study is undertaken in two major districts (Gozamen and Awabel) of East Gojam zone, Amhara region. The Zone and the Woredas are selected purposely because of their long time experience of soil conservation development activities and their land re-distribution experiences. Rigorous SWC activities were implemented in the study areas in 1999 by the District Agriculture Office with financial support from the Swedish International Development Agency (SIDA) as part of its on-farm research program in Amhara Region.

A two stage sampling procedures is used to select farmers for the study. Kebeles (Dijil Watershed and Gudalema Watershed) are selected using a random sampling procedure. Following the selection of Kebeles, 60 farmers are randomly selected from each Kebeles (Watersheds). Data are collected from primary and secondary sources. Secondary sources include published and unpublished information about soil conservation activities, agricultural production, farming system and other socio-economic information. This information is collected from the zonal and Woreda level office of Agriculture. Primary data are collected from sample farmers using a structural questionnaire. Moreover, group discussions are undertaken with opinion leaders of respective districts.

Empirical model and estimation

Investment and tenure insecurity equation and their estimation

In this study, we empirically investigate the relations laid out in the theoretical model by a system of binary choice equations. As discussed earlier, the influence of tenure insecurity on soil conservation investment is direct. Alternatively, some factors may simultaneously affect both tenure insecurity and investment. With a simultaneous equations model two or more endogenous variables are determined jointly within the model. Both are also depend on set of exogenous variables. Simultaneity induces correlation between error terms of each equation in the system. Ordinary least squares (OLS) can not be used to estimate this model, because the relationship specified by equations violates the OLS assumption of zero covariance between the disturbance term and the independent variables. Estimation of such model through OLS will lead to biased and inconsistent estimates of the coefficients (Verbeek, 2002). As a result, the main estimating technique is two stage least squares (2SLS) for continuous variable and two stage probit estimation in the case of binary choice (Maddala, 1983).

I_i^* and T_i^* are endogenous (latent) variables and β and γ are the set of parameters and the simultaneous equations model is written the following form:

$$I_i^* = \beta_{11}T_i^* + \gamma_1 Z_{1i} + u_1 \tag{8}$$

$$T_i^* = \beta_{21}I_i^* + \gamma_2 Z_{2i} + u_2 \tag{9}$$

Where, $I_i = 1$ if $I_i^*>0$; $I_i = 0$ otherwise; $T_i = 1$ if $T_i^*>0$; $T_i = 0$ otherwise.

For this study, our simultaneous probit equations model is:

$$I_i = f(Z_1, X, T_i) \tag{10}$$

$$T_i = f(Z_2, X, I_i) \tag{11}$$

Where investment (I_i) and tenure insecurity (T_i) are binary [0,1] indicator variables for a given household. The Z and X are vectors of observed exogenous variables representing household and farm specific characteristics and institutional setting. And f represents the non-liner transformation of I_i^* and T_i^*

To investigate the relationship between investment and tenure insecurity, we use a simultaneous probit equation model which consists of two simultaneous binary choice equations. The estimation procedure comprises the following steps: First, the reduced form of tenure insecurity (exogenous variable) is estimated and then its predicted value obtained. Second, the predicted value of insecurity $(\hat{T}_i^*)$ is used as a regressor in the investment equation. The process is repeated for insecurity equation using predicted value of investment $(\hat{I}_i^*)$.

Before running the model all the hypothesized explanatory variables were checked for the existence of multi-collinearity problem. In this study, variance inflation factor (VIF) and contingency coefficients were used to test multi-collinearity problem for continuous and dummy variables, respectively.

RESULTS AND DISCUSSION

Descriptive analysis of soil conservation investment

About 44% of the households in the study invest in their own conservation technology (Table 1). Family size, age and farm size variables are assumed to influence the decision to invest in the soil conservation technology. The average family size, farm size and age of the households' head in the study area are 5.6, 1.47 and 38.6, respectively. But in this study no significant differences in these variables between investing and non-investing households are found based on univariate t-test.

The number of oxen is hypothesized to influence the decision of soil conservation investment. This is because the number of oxen indicates the wealth status of the household. The average number of oxen per household is 1.88. A t-test indicates that investing households on average have significantly more oxen (2.11) than non-investing households (1.69). Education level is also assumed to influence decision of soil conservation investment. The majority of the households who invest in soil conservation technologies (69.2%) are literate. The chi- square test shows a systematic association between the level of education and soil conservation investments. Frequency of extension contact is also assumed to influence the decision of soil conservation investment. About 82.7% of investing households and 53% of non-investing households has high level of extension contacts. The chi-square analysis shows a systematic association between soil conservation investment and extension contact. Furthermore, the presence of public soil conservation investments on the plot of household is hypothesized to influence in decision of investments. This

Table 1. Characteristics of investing and non-investing households in the study area.

Characteristics		Investing HH (N=52)		Non-Investing HH(N=66)		t- statistic
		Mean	**S.D**	**Mean**	**S.D**	
Age(years)		40.59	12.01	37.06	11.95	0.11
Family size		5.94	1.77	5.48	2.00	0.19
Farm size (ha.)		1.54	0.79	1.40	0.71	0.37
Distance from main road (km)		0.95	1.09	1.32	1.0	0.05**
Number oxen		2.11	0.83	1.69	1.15	0.02**
		% of investing HH		**% of non-investing HH**		**χ^2 statistic**
Slope	Yes	86.4		50.0		0.000**
	No	15.4		50.0		
Tenure insecurity	Yes	30.8		56.1		0.006**
	No	69.2		43.9		
Level of Education	Literate	61.5		39.7		0.001**
	Illiterate	38.5		62.1		
Extension Contact	High	82.7		53.0		0.001**
	Low	17.3		47.0		
Public investment	Yes	59.6		34.8		0.007**
	No	40.4		65.2		

*Significance at the 10% level; **significance at the 5% level.

is because the households who have public investments may perceive the benefit of the technologies. About 59.6% of investing households and 34.5% of non-Investing households has public soil conservation investments.

The chi-square analysis reveals a systematic association between the presence of public investments and individual soil conservation decision. In addition insecurity of tenure is hypothesized to influence soil conservation investment. This is because investment may be undertaken when the household is assured that he will reap the benefit for a considerable time. The household who feels insecurity may not invested soil conservation measures. About 56.1% of non-investing households and 30.8% of investing households feel tenure insecurity. The chi-square analysis also reveals a systematic association between tenure insecurity and soil conservation investment.

Descriptive analysis of tenure security

Around 55% of the sample households feel secure about their landholdings (Table 2). Farm size is hypothesized to influence the feeling of tenure insecurity. The households who have a larger farm size may feel insecurity of tenure because they fear some plots of their land may be taken away through redistribution. Insecure households on the average have a larger farm size (1.7 ha) than secure households (1.29 ha). The t- test shows that this difference is significant. Moreover, the number of oxen is also assumed to influence the feeling of tenure security. This is because the number of oxen indicates the wealth status of the household. Insecure households on average have significantly more oxen (1.7) than secure households (1.29).

Expectation of redistribution is assumed to influence the feeling of tenure insecurity. This is due to the prevailing land tenure policy. About 92.2% of insecure households and 75% of secure households expect re-distribution in the future. This indicates that expectation of land redistribution does not totally lead to tenure insecurity. This is because farmers expect that there will be land redistribution for landless youth from large size holders or land from dead people or from grazing areas. Chi-square analysis reveals a systematic association between tenure insecurity and expectation of re-distribution in the future. Level of education is also hypothesized to influence the feeling of tenure insecurity. However, Chi-square analysis shows no systematic association between education level and tenure insecurity.

Table 2. Descriptive statistics results of tenure security and insecure households in the study area.

Characteristics		Insecure household (N=53)		Secure household (N=65)		t- statistic
		Mean	S.D	Mean	S.D	
Age (years)		39.4	12.5	37.9	11.71	0.5
Farm size (ha.)		1.7	0 .8	1.29	0.62	0.003**
Number oxen		2.0	1.1	1.7	0 .95	0.065*
		% Land insecure HH		% Land secure HH		χ^2 statistic
Investment	Yes	24.5		55.4		0.006**
	No	75.5		44.6		
Expectation of redistribution	Yes	92.2		75.4		0.018**
	No	7.8		24.6		
Level of Education	literate	58.5		46.2		0.182
	illiterate	41.5		53.8		

*Significance at the 10% level; **significance at the 5% level.

Table 3. Two- stage probit estimation results of investment of soil conservation.

Variable	Coefficient	Standard error
Age	0.033**	0.013
Level of education	0.678**	0.299
Family size	-0.037	0.096
Distance	-0.318**	0.131
Farm size	0.267	0.262
Extension contact	0.321	0.327
Number of oxen	0.188	0.160
Public conservation	0.815**	0.303
Slope	1.048**	0.333
Insecurity(predicted value)	-0.856*	0.518
Constant	-3.171	0.776
Regression diagnostic		
Chi-square	45.79	
Probability >Chi-square	0.000	
Pseudo R-square	0 .290	
Count R^2	0.756	
Base line for count R^2	0.56	
Number of observations	115	

*Significance at the 10% level; **significance at the 5% level.

Soil conservation investment model

Using the variables described in Equation (8) is estimated using two- stage probit method. Two-stage probit estimation results (Table 3) reveal that the investment decision of soil conservation technologies is influenced by different factors at different levels of significance.

Most of the regressors used in this model have signs in line with our prior expectations. The results show that farmers' soil conservation investment decisions are positively and significantly related to slope, age, education level and public investment. Similarly, tenure insecurity and distance from the main road have negative significant influence on soil conservation investment. Some variables like farm size, extension contact and number of oxen have positive signs but are not significant.

Consistent to our expectation, the level of the slope of plot is positively related to the decision of the soil conservation investment and statistically significant. This implies that farmers who operate on fields with steeper slope are more likely to invest in soil conservation technologies than the others. This may be explained by the positive relationships between slope and severity of soil erosion. This result is consistent with the findings of Shiferaw and Holden (1998) and Gebremedhin and Swinton (2003) in Ethiopia, and Lapar and Pandey (1999) in the Philippines. Therefore, the level of the slope of the plot is an important factor for the decision of soil conservation investment.

As expected, tenure insecurity has a significant negative influence on soil conservation investments. This suggests that tenure insecure households are less likely to invest in soil conservation technologies. It is argued that in Ethiopia, land is state property and farmers have only use and lease rights and redistribution of land is a common phenomenon.

For instance, in the Amhara region, Benin and Pender (2002) revealed that nearly half of the communities have experienced land redistribution since 1991 and about four-fifths of the communities expect redistribution in the future. This expectation of redistribution may erode tenure security and hence farmers may not undertake land improving investment because they may not be able benefit fully from the returns on their investments. This result is consistent with findings by Besley (1995) in Ghana, Hays et al. (1997) in Gambia, Gavian and Fafchamps (1996) in Nigeria, Ostuka et al. (2003) in Ghana, Winters et al. (2004) in Ecuador and Fraser (2004) in Southwest British Columbia. Moreover, a study conducted by Geberemedehin and Swinton (2003) in the Tigray region of North Ethiopia is in line with our result. Thus, tenure insecurity has a negative influence on soil conservation decision. Conversely, Benin and Pender (2001) in their study in Amhara found that land redistribution and expectation of land redistribution have a statistically insignificant effect on the influence of land investment.

Similarly, Holden and Yohannes (2002) in Southern Ethiopia revealed that tenure insecurity has no negative effect on long term investment like planting of perennials. These differences could be explained by the differences of socio-economic and land re-distribution experiences between Amhara and Southern regions, but the different results for the same region may be due to methodological difference.

As hypothesized, the presence of public soil conservation investments on a plot is positively related with individual soil conservation investments and statistically significant. This means that households who have public investments on their plots are more likely to invest in individual soil conservation technologies than others. This is because they perceive the benefits of soil conservation technologies. This result is consistent with the finding of Gebremedhen and Swinton (2003) in Ethiopia. They found that farmers, who have nearby public investment, are encouraged to invest in private soil conservation measures. Therefore, public soil conservation is stimulating individual soil conservation investment.

Consistent with our expectation, the coefficient of distance from the main road is negative and significant. It implies that farmers whose homesteads are far from the main road have a lower probability of investing in soil conservation technologies. This can be due to the fact that households near the main road tend to have access to information and are more likely to be visited by extension agents. Moreover, the transaction cost of searching for technical knowledge and information is lower for farmers living close to the road. This result is in line with the findings of Gebremedhen and Swinton (2003) in Ethiopia and Lapar and Pandy (1999) in the Cebu districts of Philippines. It can be concluded that distance from the main road is a crucial factor for the individual soil conservation decision.

Investment is found to be positively and significantly influenced by education status of the households. This suggests that literate farmers are more likely to invest in soil conservation measures than illiterate farmers. It is argued that literate farmers have the ability to obtain, process and use information related to soil conservation technologies and they are also taking more rational decisions. This result is consistent with the findings of Pender and Kerr (1999) in Aurepalle district of India and Lapar and Pandy (1999) in the Cebu district of Philippines. Thus, level of education has a positive influence on the decision of soil conservation investment.

Contrary to the hypothesis, the household age is found to have a significant and positive effect on the decision of soil conservation investment. This result implies that older farmers are more likely to invest soil conservation technologies. An explanation could be that farmers cognizant the problem of soil erosion through their life experience and hence they may take decisions of soil conservation investments. The overall model goodness of fit represents by model count R-square is satisfactory. Using the model we predict that 44 households would investing and that 71 would not investing.

In reality 50 households did invest and 65 did not. When we evaluate the predictions it is found that 33 of these 44 predictions of investing are correct and 11 not. Of the 71 predictions of non-investing 54 are correct and 17 not. So in total there are 33+54=87 correct predictions and 11+17=28 wrong predictions. Overall the predictive power of the model is 87/115=0.756. Moreover, the prediction of a model with only an intercept and no explanatory variables is 65 of the 115 observations. This gives us a baseline for predictions. So, the explanatory variables in the model give us an additional 22 correct predictions.

Table 4. Two stage probit result of determinants of tenure insecurity.

Variable	Coefficient	Standard error
Age	-0.003	0.011
Level of education	-0.511*	0.287
Farm size	0.491**	0.191
Number of oxen	0.063	0.135
Expectation of redistribution	0.609*	0.135
Investment (predicted Value)	0.262	0.367
Constant	-1.222	0.566
Regression diagnostic		
Chi-square	19.01	
Probability>Chi-square	0.0042	
Pseudo R-square	0.1203	
Count R^2	0 .678	
Base line for count R^2	0.55	

*Significance at the 10% level; **significance at the 5% level.

Tenure insecurity model

The results of two stage probit estimation for tenure insecurity are presented in Table 4. Most of the explanatory variables used in this model have signs similar with our prior hypothesis. The results indicate that tenure insecurity is positively and significantly related to farm size and expectation of redistribution. Age and number of oxen variables have negative and positive signs, respectively, but are not significant.

As expected, farm size is positively and significantly related to tenure insecurity. This result suggests that large farm size holders are more likely to feel tenure insecure. It is argued that owners of more than the average landholding may fear loss of some plots of land through redistribution. Moreover, the large farm size holders may feel tenure insecurity due to the past land redistribution policy as well as a great land holding inequality in the community. This result is also in line with the finding of Holden and Yohannes (2002) in some study sites of Southern Ethiopia.

Contrary to the hypothesis, the variable education is significant with a negative sign. This implies that literate farmers are less likely to feel tenure insecure. This may be explained by the fact that educated farmers may have alternative employment and the result they may give less attention to farm activities. Hence they may not feel tenure insecurity. However, Holden and Yohannes (2002) in their study in southern Ethiopia found that educated households (above grade 6) feel more tenure insecure. They argued that educated farmers may have better information about recent redistribution history of Amhara region.

As hypothesized, expectation of redistribution is positively and significantly related to the feeling of tenure insecurity. This result suggests that farmers who expect re-distribution are more likely to feel tenure insecurity.

It is argued that farmers may expect redistribution due to the government land policy as well as the past redistribution experience in the region and these perceptions may be a real source of tenure insecurity[2]. Therefore, expectation of redistribution due to land policy is the main source of tenure insecurity.

Age, whether investing soil conservation or not, and number of oxen do not have a significant effect on the feeling of tenure insecurity. Particularly, the finding of investment is not in line with the new finding of Brasselle et al. (2002) in Burkina Faso. They suggest that investment may be undertaken to enhance tenure security rather than as a consequence of more secure rights. The reason that our results are not in line with their findings may be that during the previous redistribution, investments did not guarantee tenure security and most farmers has lost what they invested and denied of their rights to compensation and payments for their investment. Investments may influence tenure security in flexible indigenous and customary land tenure system.

The model goodness of fit represents by model count R-square is acceptable. Using the model we predict that 40 households would feel insecurity and that 75 would secure. In reality 51 households felt insecurity and 64 secure. When we evaluate the predictions it is found that 27 of these 40 predictions of insecure are correct and 13 not. Of the 75 predictions of secure 51 are correct and 24 not. So in total there are 27+51=78 correct predictions and 13+24=37 wrong predictions. Overall the predictive power of the model is 78/115=0.678. Moreover, the

[2]Some authors use tenure insecurity and expectation of redistribution interchangeably. Here the two terms are different. Some farmers may expect re-distribution but they do not feel tenure insecurity due to their farm and personal characteristics. For instance small farm holders may expect re-distribution in the future but may not feel insecurity due to the size of their holding. This is because they are sure that they are not evicted from their small size

prediction of a model with only an intercept and no explanatory variables is 64 of the 115 observations. This gives us a baseline for predictions. So, the explanatory variables in the model give us an additional 14 correct predictions.

Conclusions

The result of the analysis indicates that tenure insecurity is an important factor that affects the probability of investing in soil conservation technologies. However, the reverse relation is insignificant. This shows that tenure insecurity has a negative impact up on the propensity to invest in soil conservation. This is because uncertainty in use rights leads to insecurity and reduced investment in land. Without clear and enforceable use rights, everyone is afraid others will reap the benefits of one's own investment. Under conditions of tenure insecurity, resource use and investment decisions regarding land cannot be made with the long term. Planning horizons will be short term, and oriented to maximizing immediate profits. In Amhara Region, tenure insecurity is the result of the past redistribution and government land policy and hence farmers do not undertake long term soil conservation investment. Therefore, the land policy should provide long-term and lasting tenure security to the peasant. Secure and stable rights to the land may possibly help in creating positive incentive to undertake long term investments by land users, stimulating the rural economy. Moreover, secure rights may increase the planning horizon of farmers. The new initiatives undertaken by the regional government to address the problem of tenure insecurity through a user right document is a promising start. But registration of land use rights without prior legal clarification of land rights may not increase tenure security[3].

The analysis of soil conservation investment equation reveals that age, education, distance from the main road, public conservation investment, slope and tenure insecurity are the main socio- economic and institutional factors that influence individual soil conservation decision. The study shows that the presence of public investments has a substantial positive impact on private soil conservation decision. This is because farmers perceive the benefit of soil conservation technologies. Thus, continuing and expanding public soil conservation measures that serve as demonstration sites with collaboration of research, extension and farmers is of paramount importance. In this study we find a negative relation between distance from the main road and decision to invest soil conservation. This results suggest that policy makers to give emphasis on expanding road facilities. Expansion of road network has facilitated access to market and as the result the improvement in communication has enabled farmers to keep better informed about outlets. This has provided farmers with strong incentives to seek ways of increasing production by better conservation technologies. The analysis also shows that literate farmers have higher probability of investing soil conservation technology compared to illiterate farmers.

The analysis of insecurity model reveals that expectation of redistribution and farm size have a positive influence on tenure insecurity, whereas education level has a reverse effect. Almost all farmers expect future redistribution and this may erode tenure insecurity. To reverse this situation, the regional government should consider a policy that may end up periodic land redistribution and there should be an awareness campaign to inform all the stakeholders about it to immune them from this problem. Additionally, improving access to land rather than redistribution through other means such as development of land rental markets and encouraging longer lease may be an alternative strategy in situation where formal land transactions are not possible. However, investment has no a significant influence on tenure insecurity.

To sum up, the implication of these findings is that tenure security should be the top priority agenda for the regional government in order to increase farmers' propensity to invest in soil conservation technologies and thereby to reduce soil erosion in particular, and to combat resource degradation in general. Finally, the study underlines the need to carry out future research to investigate the impact of land certification on tenure security.

[3]User rights documentation is being undertaken within the context of existing legislations. And these legislations are the ones that are responsible for promoting tenure insecurity.

REFERENCES

Amare T (1998). The effect of tenure security and land fragmentation on productivity and investment in smallholder agriculture: Evidance from Tiyo woreda, Arsi, Ethiopia.

Benin S, Pender J (2001). Impact of land redistribution on land management and productivity in the Ethiopian Highlands. Land Degrad. Dev. 12:555-568.

Besley T (1995). Property rights and Investment Incentives: Theory and Evidence from Ghana. J. Polit. Econ. 103(5):903-937.

Betru N (2003). Soil and water conservation program in the Amhara National Regional State. In: Tilahun A., (ed.).Natural Resource Degradation and Environmental Concerns in Amhara Regional State, Ethiopia: Impact on Food Security. Bahir Dar, Ethiopia.

Brasselle A, Gaspart F, Platteau J (2001). Land tenure security and investment incentives: Puzzling evidence from Burkina Faso. J. Dev. Econ. 67:373-418.

De Graaff J (1993). Soil conservation and sustainable land use: An economic Approach. Royal Tropical Institute. The Netherlands.

De Graaff J (1996). The Price of Soil Erosion: An economic evaluation of Soil conservation and watershead development Wageningen. The Netherlands.

Dessalegn R (2004). Searching for tenure security? The land system and new policy initatives in Ethiopia. Forum for social studies, Addis Ababa, Ethiopia.

Gavian S, Fafchamps M (1996). Land tenure and allocative efficiency in Nigeria. Am. J. Agric. Econ. 78:469-471.

Gebremedhin B, Swinton S (2003). Investment in soil conservation in northern Ethiopia: the role of land tenure security and public

programs.Agric. Econ. 29:69-84.
Gete Z (2003). Resource use and poverty in the Ethiopian highlands. In: Tilahun A., (ed.). Natural Resource Degradation and Environmental Concerns in Amhara Regional State, Ethiopia: Impact on Food Security. Bahir Dar, Ethiopia.
Hays J, Roth M, Zepeda L (1997). Tenure Security, Investment and productivity in Gambian Agriculture: A Generalized Probit Analysis. Am. J. Agric. Econ. 79:369-382.
Heerink N, Kuyvenhoven A, Wijk M (2001). Economic Policy Reforms and Sustainable Land use in Developing Countries: Issues and Approaches. In: Heerink, N., Keulen, H., Kuiper, M., (eds.). Economic Policy and sustainable Land Use: recent advances in quantitative analysis for developing countries. Physical Verlag, New York.
Holden S, Yohannes H (2002). Land Redistribution, Tenure Insecurity, and Intensity of production: A Study of Farm Households in Southern Ethiopia. Land Econ 78(4):573-590.
Hurni H (1988). Degradation and conservation of the resource in the Ethiopian highlands. Mt. Res. Dev. 8(2/3):123-130.
Lapar ML, Pandey S (1999). Adoption of soil conservation: The case of the Philippines uplands. Agric. Econ. 21:241-256.
Maddala GS (1983). Limited-dependent and qualitative variables in econometrics. Cambridge University Press.
Makokha M, Odera H, Maritam II.K, Okalebo JR, Iruria DM (1999). Farmers' perception and Adoption of soil management technologies in Western Kenya. Afr. Crop Sci. J. 17(4):549-558.
Mekonnen A (2009). Tenure security, resource endowments, and tree growing: Evidence from the Amhara region. Land Econ. 85(2):292-307.
Ostuka K, Place F (2001). Land tenure and Natural Resource Management: A comparative study of agrarian communities in Asia and Africa. IFPRI, Washington DC.
Otsuka K, Quisumbing A, Payongayong E, Aldoo JB (2003). Land tenure and the management of land and trees: the case of customary land tenure areas of Ghana. Environ. Dev. Econ. 8:77-104.
Pender J, Kerr J (1998). Determinants of farmers' indigenous soil and water conservation investment in semi-arid India. Agric. Econ. 19:113-125.
Shiferaw B (1998). Peasant Agriculture and Sustainable Land Use in Ethiopia: Econmic Analyses of Constraints and Incentives for Soil Conservation. Ph.d. Thesis. Agricultural University of Norway.
Shiferaw B, Holden S (1998). Resource degradation and adoption of land Conservation technologies in the Ethiopian High lands: A case study in Andit Tid, North Shewa. Agric. Econ. 18:233-247.
Shiferaw B, Holden S (1999). Soil Erosion and Smallholders' conservation Decisions in the Highlands of Ethiopia. World Dev. 27(4):739-752.
Shiferaw B, Okello J, Reddy RV (2009). Adoption and adaptation of natural resource management innovations in smallholder agriculture: reflections on key lessons and best practices. Environ. Dev. Sustain. 11:601-619.
Verbeek M (2000). A guide to Modern Econometrics. Chichester, John Wiley & Sons, Ltd.
Wannasai N, Shrestha RP (2008). Role of land tenure security and farm household characteristics on land use change in the Prasae Watershed, Thailand. Land Use Policy. 25:214-224
Winters P, Crissman C, Espinosa P (2004). Inducing the adoption of conservation technologies: Lessons from the Ecuadorian Andes. Environ. Dev. Econ. 9:695-719.

Oil pollution and technical efficiency of food crop farmers in the Niger Delta region of Nigeria

F. O. Idumah[1] and F. Y. Okunmadewa[2]

[1]Forestry Research Institute of Nigeria, Moist Forest Research Station, Benin City, Edo State. Nigeria.
[2]World Bank, Abuja, Nigeria.

A study was carried out to investigate into factors responsible for technical inefficiency of food crop farmers in the oil polluted and non-polluted areas of Niger Delta. Data were collected from 270 (140 for oil polluted and 130 in unpolluted area) farmers selected through a multistage random sampling technique. A stochastic frontier function that incorporated inefficiency effect was estimated using the Maximum Likelihood Estimation (MLE) technique. The MLE of the stochastic production function revealed mean technical, efficiency of 78% in polluted area while the corresponding values in unpolluted area were 88%. The most efficient farmer had the technical efficiency (TE) of 0.93 and least efficient farmer of 4.48. Farmers with efficiency index between 4.48 and 0.65 constituted 31% while 68.2% of the farmers had efficiency index between 0.70 and 0.95. The predicted technical efficiency varied widely across farms between 28 and 86% for farmers in polluted area while between 38 and 96% for the farmers in unpolluted area. The results show that farmers generally in the study area are not technically efficient, although the farmers in the unpolluted area are relatively more efficient than farmers in the polluted area. The implications are that the policies that would reduce oil pollution and encourage farmers to utilize their resources optimally should be put in place. Hence, in order to halt the continual degradation of the Niger Delta environment there is need for the enactment and enforcement of stringent environmental laws to protect the area.

Key words: Nigeria, Niger Delta, stochastic function, oil pollution, technical efficiency, food crops farmers.

INTRODUCTION

Food remains a major requirement for man's survival and the need to produce enough food to feed the teeming population continues to be a major focus in the developing countries. Efforts to produce enough food in countries like Nigeria are however being frustrated by a number of natural, human and economic factors. Food production in the Niger Delta zone which incidentally is the oil producing area of the country is hampered by a number of environmental problems and prominent among them is oil pollution occasioned by the oil exploration involving several million barrels of crude oil have been going on in that area. Hundreds of cases of oil spills reported (Eronmosele, 1998; Egwaikhide and Aregbeyan, 1999). It is also reported that an on average about 86% of the total gas production from 1970 to 1996 was flared. The effects of oil spillage and gas flaring have been a source of major concern. Indeed, gas flaring has been identified as the major cause of respiratory infection among the Niger Delta people including the farmers as well as the cause of reduced growth potentials of farm crops. Oil pollution has been identified among the factors causing land degradation which results in the reduction of the soil's ability to contribute to crop production and a change to the land that makes it less useful for human beings.

Chindah and Braide (2000) in a study on the effects of oil spill on crop production in the Niger Delta reported that oil spill caused great damage to the plant community due to high retention time of oil occasioned by limited flow. They observed that oil pollution affects the physiochemical properties of the soil such as temperature, structure, nutrient status and pH which results to wilting and die back of some plants. Benson and Odinwa (2010) found that cassava planted in oil polluted soil recorded low yield. Land degradation also reduces productivity thereby contributing to the low efficiency of the farmers. Inoni et al. (2006) observed that oil spill reduced crop yield, land productivity and greatly depressed farm income. They found out that a 10% increase in oil spill reduced crop yield by 1.3% while farm income declined by 5%. Orubu et al. (2004) discovered that oil pollution contributes to the depletion of the active labour force as well as the farm size which affect the efficiency and productivity of the farmers. Efficiency is a very important factor of productivity growth, especially in developing agricultural economies where resources are meager and opportunities for developing and adopting better technologies are dwindling (Ali and Chaudhry, 1990). In such economies inefficiency studies help to indicate the potential possibility to raise productivity by improving efficiency without necessarily developing new technologies or increasing the resource base (Bifarin et al., 2010). The concept of efficiency is concerned with the relative performance of the processes used in transforming given inputs into outputs. Economic theory identifies at least three types of efficiency. These are technical, allocative and economic efficiencies.

Allocative efficiency refers to the choice of an optimum combination of inputs consistent with the relative factor prices. Technical efficiency shows the ability of firms to employ the 'best practice' in an industry, so that no more than the necessary amount of a given sets of inputs is used in producing the best level of output. Economic efficiency is the product of technical and allocative efficiencies. Efficiency is a very important factor of productivity growth, especially in developing agricultural economies where resources are meager and opportunities for developing and adopting better technologies are dwindling (Ali and Chaudhry, 1990). It is often assumed that factors affecting farm households' technical efficiency (TE henceforth) are due to demographic and socio-economic characteristics. However, Pascual (2001) noted that input quality (and not just quantity) is important when deriving TE measures. Coelli (1995) recognized that environmental factors such as soil quality may also influence technical efficiency measures. This study is concerned with the assessment of the effect of oil pollution on farmers' efficiency. The outcome of the analysis is relevant for policy making in the Niger Delta. It will help to assess the role of environmental (soil) quality and relevant demographic and socio-economic factors affecting the agricultural performance of food crops farmers in the region.

METHODOLOGY

Data

Data used for this study were collected from 270 food crops farmers (140 from oil polluted area and 130 farmers from non- oil polluted area) in 31 villages in Rivers and Delta States of the Niger Delta Region of Nigeria through multi-stage sampling procedures. The data covered socio-demographic characteristics of the farmers, types of crop grown, labour used, membership of association, sources of fund for farming, land ownership status, incidence of oil pollution, prices of output and wages.

Theoretical framework

Several techniques have been developed for the measurement of efficiency of production. These techniques can be broadly categorized into two approaches: parametric and non parametric. Under the parametric technique we have deterministic parametric frontier (Afriat, 1972) and stochastic parametric frontier (Aigner et al., 1977). The parametric stochastic frontier production approach (Aigner et al., 1977); Meeusen and van den Broeck (1977) deals with stochastic noise and permits statistical test of hypotheses pertaining to production structure and the degree of inefficiency. As in Bravo-Ureta and Evenson (1994) and Bravo–Ureta and Rieger (1991), the parametric technique cost decomposition procedure is used to estimate technical, allocative and economic efficiencies. Following Sharma et al. (1999), the firm's technology is represented by a stochastic production frontier as follows:

$$Yi = f(Xi; \beta) + \varepsilon i \quad (1)$$

Where Yi denotes output of the ith firm, Xi is a vector of functions of actual input quantities used by the ith firm; β is in vector of parameters to be estimated and εi is the composite error term (Aigner et al., 1977; Meeusen and Van den Broeck, 1977) defined as:

$$ei = vi - ui \quad (2)$$

Where vis are assumed to be independently and identically distributed N (0.02) random errors, independent of the uis; and the uis are non-negative random variables, associated with technical inefficiency in production which are assumed to be independently and identically distributed and truncation (at zero) of the normal distribution with mean μ and variance σ_u^2 $|N(\mu u;\sigma_v^2)|$

The maximum likelihood estimation (MLE) of Equation 2 provides estimation for β and variance parameter $\sigma^2 = \sigma_u^2 + \sigma_v^2$, and $v = \sigma_u^2 / \sigma_v^2$. Subtracting vi from both sides of Equation 1 yield:

$$\Psi i = Yi - vi = f(Xi\beta) - ui \quad (3)$$

Where Ÿi is the observed output of the ith firm adjusted for the stochastic noise captured by vi.

Empirical model specification

Theoretically, a production frontier defines the maximum output attainable for a given level of inputs. Therefore, in order to estimate an efficient frontier, farm level data on input and output quantities are required. However, it is often the case that input quantity data are not available. Data are often available, however on farm output revenue and input expenditures. Therefore, a common approach is to use revenue and expenditure data as proxies for input and output quantities for example, Aly et al. (1987), Grabrowski et al. (1990) and Neff et al. (1991). In traditional agriculture, multiple outputs and inputs are common features and for the purpose of efficiency, analysis output is aggregated into one category and inputs are aggregated into seven categories namely: farm size, fertilizer, labour, capital, land that is, rental value of land, other variable inputs. The stochastic frontier production function used in this study is a linearized version of Cobb-Douglas production function. The stochastic frontier production function in Equation 4 and the inefficiency model in Equation 5 were simultaneously estimated as proposed by Battese et al. (1996).

Specification of technical efficiency model

$$\ln Y = \beta o + \beta_1 \ln X_{1ij} + \beta_2 \ln X_{2ij} + \beta_3 \ln X_3 ij + \beta_4 \ln X_4 ij + \beta_5 \ln X_5 ij + \varepsilon i \quad (4)$$

Where subscripts ij refer to the ith observation on the jth farmer; ln = denotes logarithm to base e; Y = represents the farm output in grain equivalent (Kg); X_1 = total farm size under cultivation (in hectares); X_2 = family labour used in production (mandays); X_3 = is hired labour used in production (in man-days); X_4 = is material inputs of seeds and other planting stocks (in kgs and cuttings); X_5 = quantity of fertilizer used (in kgs); εi = error term (vi - ui).

It is assumed that the technical efficiency effects are independently distributed and varies and uij arises by truncation (at zero) of the normal distribution with mean μ and variance σ^2; where uij is defined by equation.

Inefficiency model

$$Uij = \alpha o + \alpha_1 \ln Z_1 ij + \alpha_2 \ln Z_2 ij + \alpha_3 \ln Z_3 ij + \alpha_4 \ln Z_4 ij + \alpha_1 \ln D_1 1ij + \alpha_2 \ln D_2 1ij + \alpha_3 \ln D_3 1ij + \alpha_4 \ln D_4 1ij \quad (5)$$

Where uij represents the technical inefficiency of the ith farmer; Z1 denotes age; Z2 represents sex;

Z3 represents family size; Z5 represents years of schooling; D1 denotes dummy variable for membership of association; where one denotes membership of association and zero is otherwise. D3 denotes dummy variable for ownership of farmland; where one denotes who own their farmland zero is otherwise. D4 denotes dummy variable for source of fund for farming; where one represents those who depend on personal saving for their farming activities and zero is otherwise. D5 denotes dummy variable for pollution; where one denotes farmland where there is oil pollution and zero is otherwise.

The β and α coefficient are unknown parameters to be estimated together with the variance parameters. The parameters of the stochastic production function are estimated by the method of maximum likelihood, using FRONTIER 4.1* program (Coelli, 1994). The maximum likelihood estimation (MLE) procedure is used because it is asymptotically efficient; consistent and asymptotically normally distributed.

Description of variables

Farm output

Output is the total quantity of crop mix in each farm converted to their grain equivalent in kilograms.

Farm size (XI)

This is expressed in hectares. On the expected sign of the coefficient, there seems to be no consensus of opinion (Oredipe, 1998). Hence, the sign of the coefficient of the variable cannot be predicted a-priori.

Farmily labour (X2)

Because family labour is not paid for in the study area, large family labour may not reflect considerable increasing output nor be matched with increase in resource pool. Inefficiency may set in if there is excess labour on the farm. The coefficient of this variable is therefore expected to be negative.

Hired labour (X3)

Labour intensive technologies will require additional or specialized skill, which can be secured through hired labour. Hired labour constitutes a major constraint to attainment of optimal productivity level and is expected to be positively related to technical efficiency level.

Planting stock (X4)

The quantity and quality of planting stocks use in farming have considerable influence on the ultimate yield from the farm. Thus, it is expected that good quality planting stock will positively affect farm output.

Fertilizer (X5)

It is generally believed that fertilizer application improves the fertility of the soil and secures greater yield from the farm. This however depends on several factor like the quantity applied and the timing of application. The coefficient of the variable is expected to be positive to output.

RESULTS AND DISCUSSION

The socio-economic characteristics of the respondents are presented in Tables 1 and 2. They seem to exhibit similar pattern. This is quite understandable as they are people with the same cultural, historic and geographical background. The average age of the farmers is 43.3 years. The highest percentage of farmers (71.9%) is within the age bracket of 31 and 50 years. This shows that most farmers from the study areas are still young. On the gender aspect, male farmers are more than female farmers. The percentage of female farmers is 30.7%. This indicates that women involvement in farming in the study area is low. The average family size is 5.18. This large family size implies availability of family labour to the farmers. The literacy level of most farmers is relatively moderate with about 23% having no formal education while 18.1% had primary education. Over 53% of the

Table 1. Socio-economic characteristics of respondents in the study areas.

Demographic variables	Characteristics	Polluted		Unpolluted	
		Frequency	Percentage	Frequency	Percentage
Gender	Male	84	61.8	65	48.5
	Female	52	38.2	69	51.5
Age	21-30	19	14.0	12	9.0
	31-40	34	25.0	36	26.8
	41-50	41	30.1	45	33.6
	51-60	25	18.4	29	21.6
	61 and above	17	12.5	12	9.0
Marital status	Single	19	14	14	10.4
	Married	99	72.8	103	76.9
	Divorced	01	0.7	02	1.5
	Widow/widower	17	12.5	15	11.2
Household size	1-5	74	54.4	83	61.9
	6-10	62	45.6	51	38.1
Educational qualification	No formal education	26	19.1	31	23.1
	Primary education	18	13.2	25	18.7
	Secondary	29	21.3	30	22.4
	Tertiary	57	41.9	34	25.4
	Vocational	6	4.4	14	10.4
Years of farming	0-5	13	9.6	17	12.7
	6-10	30	22.1	23	17.2
	11-15	31	22.7	27	20.1
	16-20	15	11.0	19	14.2
	21-25	13	9.6	15	11.2
	26 and above	34	25	33	24.6
Land tenure	Family land	44	32.4	61	45.5
	Communal land	15	11.0	12	9.0
	Rented land	53	39.0	45	33.6
	Purchased land	24	17.6	16	11.9
Farming system	Mixed cropping	107	78.7	103	76.9
	Agroforestry	29	21.3	31	23.1
Farm size	0-2.0	102	75.0	103	76.9
	2.1-3.0	23	16.9	24	17.9
	3.1-4.0	2	1.5	2	1.5
	4.1 and above	9	6.6	5	3.7

Source: 2002.

farmers have post- primary education. The marital status of farmers shows that 13.7% of the farmers are singles while over 80% are married. Membership of co-operative societies is not very common among the farmers. Among the respondents only 22.6% belong to co-operative societies. This shows that majority of the farmers are not exploring the benefits accruable from co-operatives societies. The farming experience of farmers shows that most of the farmers have been in the farming business for an average of 16 years. Resulting from the vagaries of farming operation due to unfavourable environmental condition in the study area, 57% of the farmers engage in other jobs like fishing, trading etc, to supplement income from farming activities. The farmland ownership structure shows that most respondents (64.1%) farm on communal and leased lands.

Table 2. Summary statistics of socio- economic characteristics of respondents.

Variables	Oil polluted soil environment				Un- polluted soil environment			
	Sample mean	Minimum value	Maximum value	Standard deviation	Sample mean	Minimum value	Maximum value	Standard deviation
Age	42.4	20	59	10.7	42.95	20	45	11.3
Family size	5.06	1	9	2.13	5.34	1	9	1.9
Years in schooling	9.9	0	19	7.07	11.66	0	19	6.6
Years in farming	16.6	2	32	8.5	15.99	5	27	8.5
Farm size	1.5	0.2	6.07	0.87	1.59	.2	5.89	1.03
Family labour (man days)	82.8	10	215	44.5	83.82	11	200	42.3
Hired labour (man days)	2.95	0	15	2.7	3.55	0	15	3.37
Quantity of fertiliser used (kg)/ha	66.5	0	600	111.57	53.5	0	666.7	113.86
Total output (kg)/farmer	836.5				1546.7			
Average gross revenue (n)	28,834	5,000	300000	33,066	33361.51	5,000	200000	30290.0
Total cost (n)	7516	1200	28200	5064.4	8022.38	1600	24600	6511

Source: Field data (2002).

All the farmers in the area practice mixed cropping with over 50% planting between 4 to 7 different crops on the same plot. About 51.8% of the farmers attested to the pollution of their farm with petro-chemical products while 48.2% reported that there was presence of oil pollution in their farmlands. In summary, the socioeconomic characteristics of the farming households in the study areas seemed to exhibit similar pattern. This is quite understandable as they are people with the same socio-cultural background and within the same geographical setting. For example, while the average farm size in polluted area is 1.5 ha, that of the unpolluted area is 1.59 ha. Also, the average number of mandays used by households in polluted area is 82.8 and those in unpolluted area are 83.8. Meanwhile, farmers in the polluted area appeared to use more of inorganic fertilizer (66.5 kg/ha) than those in unpolluted area (53.5 kg/ha). There is however a marked difference in the average output between farmers in the unpolluted area (1546.7 kg/farmer) and the polluted area (836.5 kg/farmer). A plausible reason is most likely the effects of pollution.

Estimates of the parameters of the inefficiency factors

The estimated parameters and the related statistical tests results obtained from the analysis are presented in the Table 3. All the coefficients in the model have the expected signs and many are statistically significant at 10% or less.

Determinants of technical inefficiency

The coefficient of farm size was significant in the 5% that is, in polluted and non-polluted areas. Family labour was significant at the 10% in both polluted and non-polluted areas. Hired labour was not significant as it was observed that majority of the farmers did not engaged hired labourprobably due to high cost. The coefficient of planting materials, which include seeds, was not significant. Fertilizer was significant at 10% level in both cases. The coefficients of family size years schooling, crop diversification and membership of Farmers Association had negative sign in both polluted and unpolluted areas while family size was significant in both situations; years of schooling was significant in unpolluted area. The significance of these coefficients combined with their negative signs implies that these variables help to reduce inefficiency in the farmers. In other words, crop diversification for example, reduces farmers technical inefficiency (Amaza, 2000) while

Table 3. Maximum likelihood estimates of the parameters of the stochastic frontier production function (technical efficiency model).

Variables	Polluted area		Unpolluted area	
	Coefficient	Standard error	Coefficient	Standard error
Production factor constant	2.8845	0.3239***	2.9395	0.03656***
Farm size (X1)	0.2596	0.0095**	0.2853	0.1257**
Family labour (X2)	0.6544	0.0719***	0.7312	0.0779***
Hired labour (X3)	-0.0624	0.0712	-0.0586	0.0759
Fertilizer (X4)	0.0442	0.0260*	0.0417	0.0263
Planting stocks (X5)	0.0336	0.1384	0.0191	0.1532
Inefficiency factor				
Constant	0.5793	0.9477	0.8368	0.0102***
Sex (Z1)	0.8838	0.6188	1.040	0.5926*
Year of farming (Z2)	-1.500	0.4238**	-1.9574	0.8874**
Family size (Z3)	-0.7574	0.5244	-0.8728	0.7165
Year in schooling (Z4)	-0.2787	0.1367**	-0.3248	0.1492
Crop diversification (Z5)	-7.0753	0.3028**	-1.312	0.4525**
Membership of association (D1)	-0.3749	0.1705**	-0.3913	0.2064*
ownership of land (D2)	0.7839	0.3932**	0.9054	0.5519
Source of fund (D3)	0.8086	0.3559**	0.9591	0.5329*
Pollution (D4)	0.2205	0.0166**	-	-
Diagnostics statistics				
Livelihood ratio	41.73		39.92	
Sigma square (σ^2)	0.1209	0.0331***	0.1593	0.0552**
Gamma (γ)	0.6807	0.0856***	0.7707	0.0784***

Source: Computed from field data, ***1% level of significance, **5% level of significance, *10% level of significance.

membership of Farmers Association affords the farmers the opportunity to share information on new farming practices by interacting with other farmers thereby reducing their inefficiency. These findings are consistent with earlier findings by Bravo-Ureta and Evenson (1994), Ajibefun and Aderinola (2004) and Nwaru (2004). The coefficient of pollution (0.2205) had positive sign to technical inefficiency. In other words, it contributes to technical inefficiency among the farmers. This finding is however contrary to that of Hadri and Whittaker (1999) who assessed the effect of soil pollution on crop technical efficiency and found a positive relationship between technical efficiency and use of contaminants in a sample of farms in South West England.

Pascual (2001) also found out that soil quality affects technical efficiency in Mexico and attributed this to household response to ecological constraints who try to substitute lower soil quality for higher managerial ability. In this study, the effects of pollution on food production can be seen in the output of farmers. Whereas, the total output per farmer in the polluted area was 836.7 kg; that of the unpolluted area was 1546.7 kg per hectare for cassava. The coefficient of source of fund had positive and significant at the 5% level. The significance of this coefficient indicates that where the farmers source for fund for farming affects their efficiency. A situation where farmers depend largely on their personal saving as is the case with majority of the farmers in the area will adversely affect their efficiency.

The diagnostic statistics of the technical efficiency factors

The estimated sigma-squared (σ^2) in Table 3 for both polluted and unpolluted areas are large (0.12 and 0.15) and significantly different from zero at the 5% level. This indicates a good fit and the correctness of the specified distributional assumption of the composite error-term. In addition, the magnitude of the variance ratio defined as $\gamma = \delta u^2/(\delta u^2 + \delta v^2)$ is estimated to be as high as 68% for polluted area and 77% suggesting that systematic influences that are unexplained by the production functions are the dominant sources of errors. It also confirms the presence of one-sided error component in the model, thus rendering the use of the ordinary least squares (OLS) estimating technique inadequate in representing the data. This means that over 65% of the

Table 4. Frequency of technical efficiency in the study area.

Efficiency (%)	Technical efficiency	
	Oil polluted	Unpolluted
10-50	6	3
50-55	0	2
55-60	3	1
60-65	3	0
65-70	9	3
70-75	20	4
75-80	24	10
80-85	65	16
85-90	10	24
90-95	0	67
95-100	0	0
	140	**130**
Minimum value	28	38
Maximum value	86	96
Mean value	77.6	88

variations in output among the farms in both polluted and unpolluted areas are due to difference in technical efficiency. In other word the inefficiency effects indicated by the variance parameter are significant in determining the level and variability of output of farmers in the study area. The livelihood functions are estimated to be 41.73 and 39.92 for polluted and unpolluted areas, respectively. These values represent the values that maximize the joint densities in the estimated model.

Distribution of technical efficiency

The technical indices of farmers are derived from the analysis of the stochastic production frontier function in Equation 4. The technical efficiency of the sampled farmers in both polluted and unpolluted areas is less than 100 indicating that all the farmers are producing below the maximum efficiency frontier as shown in Table 4. A range of technical efficiency is observed across the sampled farmers. The best farmer in the polluted areas has a technical efficiency of 86% while the least efficient farmer has 28% whereas in the unpolluted area the most efficient has a technical efficiency of 96% and least efficient farmers has 38%. The mean technical efficiency is 77.6% for the polluted area and 88.5% for the unpolluted area. This implies that on the average the respondents were able to obtain a little over 77.6% of optimal output in the polluted area and 88.5% in the unpolluted area. Testing for significance difference reveals that the computed z- statistics is statistically significant at 1% level showing that farmers in the unpolluted area are more efficient than those in the polluted area. The hypothesis that states that there is no difference in the technical efficiency of farmers in the two areas is thereby rejected. A plausible reason for this could be the effects of oil pollution given the fact that farmers in the area operate under the same technical condition.

The distribution of technical efficiency group reveals that the highest proportion (46.4%) of the farmers in the polluted area falls between the efficiency ranges of 0.80 to 0.85 while the highest proportion (23.7%) falls between the efficiency ranges of 0.85 to 0.90 in the unpolluted area. The distribution of the technical efficiency shows efficiency at 77.6 and 88.5% for farmers in polluted and unpolluted area respectively implying that in the short-run there is scope for increasing technical efficiency in food crop production in the study area especially those in the polluted area. That is, if the problem of oil pollution is taken care of and if farmers would adopt the technology and production techniques currently used by the most efficient farmers.

Conclusion

Expanding population and economic development have generated a growing demand for various land based products leading to unnecessary pressure on soil, water resources and plants with the attendant consequences of deteriorating land resources, declining productivity and reduced income. This study has been able to quantitatively establish the fact that oil pollution in the area is having negative impacts on the food crop farmers resulting in reduced income from farm activities. In considering the results obtained from the analysis of technical efficiency effects of stochastic frontier production function, it is important to note that the production frontier involved are determined by models and within the sample values. This implies that there may be techniques of production practiced by some of the farmers in the sample, which yielded much higher output for the same level of inputs. Governments at both the Federal and State levels should ensure increase fund allocation to agriculture in the region as well as the provision of and distribution of farm inputs like fertilizers, chemical, capital, etc. so as to boost food production in that area. Government should also ensure that stringent environmental laws to protect the area are enacted and enforced.

REFERENCES

Aigner D, Lovell CAK, Schmidl P (1977). Formulation and Estimation of Stochastic Frontier Production Function Model. J. Econ. 6:21-27.

Ajibefun IA, Aderinola EA (2004). Determinants of Technical Efficiency and Policy implication of Traditional Agricultural Production: Empirical Study of Nigerian Food Crops Farmers. Final Report Presentation at Biannual Research Workshop on African Economic Research Consortium, Nairobi, Kenya. pp. 2-20.

Ali M, Chaudhry MA (1990). Interregional Farm Efficiency in Pakistan's Punjabi. A frontier production function study. J. Agric. Econ. 41:62-73.

Aly HY, Belbase K, Grabrowski R, Kraft S (1987). The Technical Efficiency of Illinois Grain Farms: An Application of a Ray-Homothetic Production Function. Southern J. Agric. Econ. 19:69-78.

Amaza PS (2000). Resource Use Efficiency in Food Crops Production in Gombe State Of Nigeria. Unpublished Ph.D Thesis. Department of Agricultural Economics. University of Ibadan.

Battese GE, Malik SJ, Gill MA (1996). An Investigation of Technical Inefficiencies of Production of Wheat Farmers in Four Districts of Pakistan. J. Agric. Econ. 47(1)37-49.

Benson DN, Odinwa AB (2010). Impact of crude oil exploration and production on the yield and Income of Cassava(Manihot spp) in Khana Local Government Area, Rivers State, Nigeria. Africa J. Agric. Res. Dev. 3(3):59-64.

Bifarin JO, Alimi T, Baruwa OI, Ajewole OC (2010). Determinant of technical, allocative and Economic efficiencies in the plantain (Musa spp.) production industry, Ondo State, Nigeria. In (Eds) Dubois T , Staver C, Coyne D, Hauser S *Proc. IC on Banana & Plantain in Africa* Kenya.

Bravo-Ureta BE, Evenson RE (1994) Efficiency in agricultural production: the case of peasant farmers in Eastern Paraguay. Agric. Econ. 10:27-37.

Bravo-Ureta BE, Rieger C (1991) Diary farm efficiency measurement using Stochastic frontier and neo-classical duality. Am. J. Agric. Econ. 73:421-428.

Chindah AC, Braide SA (2000) The impact of oil spils on the Ecology and Economy of the Niger Delta. In Proceedings of the workshop on Sustainable Remediation Development Technology held at the Institute of Pollution Studies, Rivers State University of Science and Technology, Port Harcourt.

Coelli TJ (1994) A Guide to Frontier 4-1: A Computer Programme for Stochastic Frontier Production Function and Cost Function Estimation" Unpublished Paper Department of Econometrics. University of New England Armdale P. 32.

Coelli TJ (1995) Recent developments in frontier modeling and efficiency measurement. Aus. J. Agric. Econ. 39(3):219-245

Egwaikhide FO, Aregbeyan O (1999). Oil Production and Externalities in the Niger Delta. Is Fiscal Solution feasible?. Proceeding of the 1999 Annual Conference of the Nigerian Econmic Society. 101-115.

Grabrowski R, Kraft S, Pasurka C, Aly HY (1990). A Ray- Homothetic Frontier and Efficiency: Grain Farms in Southern Illinois. Eur. Rev. Agric. Econ. 17:435-448.

Hadri K, Whittaker J (1999). Efficiency, environmental contaminants and farm size: Testing for links using stochastics production frontiers. J. App. Econ. 2(2):337-356.

Inoni OE, Omotor DG, Adun FN (2006). Effect of Oil Spillage on Crop Yield and Farm Income in Delta State, Nigeria. J. Central Eur. Agric. 7(1):41-48.

Meeusen W, van den Broeck J (1977). Efficiency Estimation from Cobb - Douglas Production Function with composed Error. Int. Econ. Rev. 18:435-444.

Neff DC, Grarcia P, Hornbaker RH (1991). Efficiency Measures Using the Ray- Homothetic function: A Multiperiod Analysis. Southern J. Agric. Econ. 23:113-121.

Nwaru JC (2004). Rural Credit Markets and Arable Crop Production In Imo State of Nigeria. Ph.D Dissertation. Michael Okpara University of Agriculture, Umudike, Nigeria.

Oredipe AA (1998). "An Analysis of Efficiency and Production Crops Among Farmers Adopting Improved Maize Technology in Ogun State, Nigeria. Unpublished Ph.D Thesis.University of Ibadan.

Orubu CO, Odusola A, Ehugreime W (2004). The Nigeria Oil Industry: Environment Strategies and Need for Community Involvement. J. Hum. Ecol. 3:204-224.

Pascual U (2001). Soil Degradation and Technical Efficiency in Shifting cultivation: The Case of Yucatan(Mexico). Discussion Paper Series No 16 University of Manchester. October 29th, P. 25.

Sharma KR, Leung P, Zaleski HM (1999). Technical, allocative, and economic efficiencies in swine production in Hawaii: A comparison of parametric and non-parametric approaches. Agric. Econ. Elservier 20:23-35.

Determinants of smallholder sweet potato farmers' participation in different market options: The case of Vihiga County, Kenya

Benjamin K. Mutai[1], Elvi N. Agunda[2], Augustus S. Muluvi[3], Lawrence K. Kibet[3] and Mary C. Maina[1]

[1]Department of Agricultural Economics/Agribusiness Management at Egerton University, Njoro Campus, Kenya.
[2]Department of Agricultural Economics/Agribusiness Management, Egerton University, Kenya.
[3]Department of Economics, Egerton University, Kenya.

Market participation has a potential to increase farmers' rural incomes and employment opportunities especially if farmers concentrate on production and marketing of local crops requiring low inputs such as sweet potatoes. The purpose of this research was to investigate the factors that determine farmers' shift in market participation from village to regional market in Vihiga County. Cross-sectional data was collected and a multinomial logit model was used for the analysis. Participation in local town market rather than village market was influenced by credit access, total income, transport mode to market, access to extension services, age, value addition done and the quantity of sweet potatoes supplied, while; transport mode, land size, quantity of sweet potatoes and gender determined participation for the regional option. It is recommended that the local and national government should: Increase its support in the establishment of sweet potato market; improve the rural road networks to cut down transport costs, and increase support to farmer groups or associations to increase farmers' market participation.

Key words: Determinants, smallholder sweet potato farmers, participation, market options.

INTRODUCTION

Poverty in Africa has been found to be predominantly a rural phenomenon. About 75% of the world's poor are believed to work and live in rural areas, and it is estimated that, by the year 2020, 60% of the poor will still be rural (Olwande and Mathenge, 2010). According to Omiti et al. (2009), agriculture supports the livelihoods of about 80% of the rural population in Kenya (about 85% of them being small-scale farmers). Only 22% of land in Kenya is arable though another 40% has potential for irrigated agriculture. The agricultural sector employs 70% of the national labor force through forward and backward industrial linkages, thus providing food and incomes to individuals and households (Omiti et al., 2009). Small-scale agriculture in Kenya is characterized by landholdings of less than 5 acres and no more than 20 ruminant animals (mainly cattle, sheep and goats) and a few chickens per farming household (Omiti et al., 2009). Crop-livestock production systems on small scale farms often entail very little use of purchased inputs and limited application of modern technology with majority of farmers producing for subsistence. Meeting the challenge of reducing poverty and improving rural incomes in Kenya will require some form of transformation out of the semi-subsistence production systems that currently characterize much of rural Africa to a more commercialized agriculture (Komarek, 2010). Omiti et al. (2009) in their study were of the opinion that commercial orientation of smallholder agriculture leads to a gradual decline in real food prices due to increased competition and lower costs in food marketing and processing.

Vihiga is one of the poorest and densely populated Counties in Kenya with an average household land size of 0.4 ha (FAO, 2007). According to Karanja (2006) the main food crops grown in the area are maize, beans, sorghum, finger millet, Irish potatoes, sweet potatoes, cassava and vegetables, while the main industrial ones are coffee, tea and sugar cane. The causes of poverty could be attributed to limited land, high poverty levels, and limited off-farm incomes. According to MOPND (2005), about 62% of the population (in what was by then) Vihiga District lives in absolute poverty and about 60% of the population is poor. Maize is the staple food for the residents of Vihiga thus its insufficiency is synonymous to food insecurity. Over the decade (1997 to 2006), the County's demand for maize outpaced the production level, worsening the already bad food situation (Nyangweso, 2007). This could be attributed to diminishing land sizes because of the increase in population, high costs of inputs for maize production thus making it uneconomical for production.

With the growing food crisis and high prices of mainstream food crops such as maize, there is a growing recognition of the importance of local crops such as sweet potatoes in supporting livelihoods for the poor. Rono et al. (2007) in their study in the North Rift Valley region of Kenya, which has almost similar agro-climatic conditions as Vihiga, found that 75% of the farmers get sweet potatoes from their own farms, 22% from the market and 1% get from their neighbors.

This crop has the potential to diversify the farming systems, spread risks, contribute to food security, and provide income opportunities for the most vulnerable and women in particular. If sweet potato commercialization efforts are to be put in place then a majority of the farmers would be better off. According to CPPMU (2010), the area under sweet potato production in Kenya in 2009 increased by 24%, production in tones by 16% while the unit price per 100 kg bag in Kenya shillings in various markets increased by 43%.

Despite this increase in production and prices, not all farmers participate in markets. According to Omiti et al. (2009), 52% of rural farmers, sweet potato included, participate in markets where they sell only less than 50% of their produce to the various market options. The factors that make the sweet potato farmers not to increase their participation especially in the regional market options are not clear and hence the need to investigate them through establishing the market options available and further determine what socio-economic characteristics determine participation in the different market options.

MATERIALS AND METHODS

Econometric model

To analyse the choice of market option, the multinomial logit model, whereby the dependent variable is the choice of market option while the independent variables are the explanatory variables predicted to have an influence on the choice of the market option, was used. According to Greene (2002), the model has a single decision among two or more alternatives. Unordered choice models can be motivated by a random utility model. For the ith farmer faced with j choices, suppose that the utility of choice j is:

$$U_{ij} = Z'_{ij}\beta + \varepsilon_{ij} \quad (1)$$

If farmer makes choice j in particular, then we assume that U_{ij} is the maximum among the j utilities. Hence the standard model will be driven by the probability that choice j is made which is,

Probability $(U_{ij} > U_{ik})$ for all other $k \neq j$ (2)

Assuming that Yi represents the choice taken, then with J disturbances being distributed identically and independently, the multinomial logit model will be represented as follows:

$$\Pr ob(Yi = j) = \frac{e^{\beta j x i}}{\sum_{k=0}^{J} e^{\beta k x i}}, \text{ where } j = 0, 1.........J \quad (3)$$

Equation 3 represents a multinomial logit model, that can provide a set of probabilities for the J+1 choices for the decision taker with characteristics x_i. This means that we can compute J log-odds ratio as in equation 4 below. From the point of view of estimation, it is useful that the odds ratio Pj/Pk does not depend on other choices which follow from the independence of disturbances in the original model (Greene, 2002).

$$In\left(\frac{Pij}{Pik}\right) = x_i(\beta_j - \beta_k) = x_i\beta_j \text{ if } k = 0 \quad (4)$$

Based on Equation 4, according to Greene (2002) and Mugisha et al. (2004) and the fact that farmers participate some times in more than one market option, their participation in different market options are categorized into alternatives, using those who participated in the village market option as the base alternative because it was common. The other alternatives include selling to neighbors (immediate neighbor with buyer and seller sharing a common boundary), local town and regional market options. Therefore, the multinomial logit regression model estimated 3 Equations (5, 6 and 7) simultaneously (Studenmund, 1992). The 3 equations are specified as:

$$ln\left(\frac{p_i}{p_o}\right) = Z_i = \alpha_o + \alpha_1 X_1 + \alpha_2 X_2 + \ldots\ldots\ldots \alpha_{11} X_{11} + \varepsilon_i \quad (5)$$

$$ln\left(\frac{p_e}{p_o}\right) = Z_e = \beta_o + \beta_1 X_1 + \beta_2 X_2 + \ldots\ldots\ldots \beta_{11} X_{11} + \varepsilon_e \quad (6)$$

$$ln\left(\frac{p_u}{p_o}\right) = Z_u = \gamma_o + \gamma_1 X_1 + \gamma_2 X_2 + \ldots\ldots\ldots \gamma_{11} X_{11} + \varepsilon_u \quad (7)$$

Table 1. Variables used in multinomial logit model.

Variable	Description	Unit of measurement	Expected signs
Totinc[a]	Monthly total income	Kenya shillings	(+)
Sacks[a]	Sacks of sweet potatoes taken to market option	100 Kg bags	(+)
Vadon^	Value addition done to sweet potatoes	Dummy (1 = sorting, washing and packing, 0 = otherwise)	(+)
Acext^	Access to sweet potato extension services	Dummy (0 = no, 1 = yes)	(+)
Accrdts^	Access to credit	Dummy (0 = no, 1 = yes)	(+)
Educ[a]	Education level	Years	(+)
Age[a]	Age	Years	(+)
Totassbas[a]	Total asset base	Kenya Shillings	(-)
Trusop^	Transport used to option(s)	Dummy (1 = better, 0 = otherwise)	(+)
Totland[a]	Total land owned	Acres	(+)
Gend^	Gender of farmer	(1 = male, 0 = female)	(+)
Hhsize[a]	Household size	No of males/females	(-)

[a] = Natural logarithm, ^ = dummy variables.

Where *po* = probability that a household chose to participate in the village market option (base alternative); *Pi, Pe* and *Pu* = the probability that a household chose to participate in the local town, regional and neighbor market options respectively; *Zi, Ze* and *Zu* = household participates in the local town, regional and neighbor market options, respectively (1 if household participates, 0 otherwise); *αo, βo* and *γo* = intercepts showing probability of participating in the local town, regional and neighbor market option respectively, if all other explanatory variables are kept constant at zero; *αi, βi* and *γi* = parameters of the *i*th set of local town, regional and neighbor market options respectively that are estimated, and *Xi* = explanatory variables of the market option participation *εi, εo,* and *εu* are the error terms.

Data collection

The data used in the analysis was collected between the months of May and June, 2011 in 3 divisions, Vihiga, Luanda and Sabatia in Vihiga County which were selected based on the population density. Random samples of 38, 48 and 34 sweet potato farmers both participating in markets and otherwise were selected from Luanda, Sabatia and Vihiga respectively, resulting to a total sample size of 120. Data was collected using interview schedules collecting information on farm and farmer characteristics, transaction costs and market related factors. Table 1 presents definitions for the variables used in the multinomial logit model.

RESULTS

Table 2 presents the descriptive statistics of the sweet potato farmers found to participate in the different market options available in Vihiga County. Sweet potato farmers who mainly sell their produce to the regional markets have the largest average land acreage of 4.03, followed by 3.03, 2.1 and 2 acres for local town, village and neighbor market options, respectively.

Land size was found to significantly vary across participants of the different markets options. Farmers owning larger farms engaged more in a wider market ranging from participant in neighborhood market with land size of 2 acres to that in the regional market of 4.4 acres. Income from sweet-potatoes and value addition were positively related to market participation thus explaining the monetary incentives that make the larger market other than the village market attractive to the farmers.

Table 3 presents a cross-tabulation of the market options and socio-economic characteristics of the farmers in discrete values such as gender, access to credit and extension. With regards to gender, 50% of female and 50% of males participated in the market. Majority (57.1%) are participating in the village market with females being 34.3% while males are 22.9%.

The socio-economic characteristics determining the various market options which farmers participate in are given in Table 4. According to the results from the multinomial logit model, 60.9% of the variation in the categorical dependent variable was explained by the model. The village market option is chosen as the base market option since it is common across all the 3 options and every other choice is then compared to the base.

DISCUSSION

Analysis of the data show that, total land ownership was significantly different at 10% level hence influencing choice of the farmer's market option. Land was a critical production asset having a direct bearing on production of a marketable surplus (Machethe et al., 2008). Means of the average distance to the farmers' market option was also significantly different with the farthest market option being regional with an average distance of 257, followed by 27.21 and 4.7 km for local town and village. The distances represent the geographical coverage of the market with the regional market being the largest and the village market being the smallest.

Table 2. Descriptive statistics of the participant farmers in the different market options.

Variable	Overall mean (n = 70)	Main market options				
		Means				
		Neigh (n = 1)	Village (n = 40)	Local town (n = 24)	Regional (n = 5)	F-Test
Age	45.23 (12.92)	48	44.22(13.52)	47.75(12.91)	40.60(7.99)	0.604
Household size	5.16 (2.50)	5	5(2.48)	5.54(2.73)	4.6(1.82)	0.316
Total land owned	2.58 (1.87)	2	2.1(1.32)	3.03(2.14)	4.4(3.05)	3.239*
Education	10.24 (3.03)	8	9.73(3.19)	10.83(2.93)	12(0)	1.462
Months per season	3.99 (0.53)	4	4.03(0.62)	3.96(0.20)	3.805(0.84)	0.296
Distance	30.4 (76.5)	0	4.7(5.2)	27.21(39.07)	257(144.5)	52.26*
Sweet potato income	2074(1868)	750	1477(1377)	2764(2090)	3805(2448)	4.705*
Cost of value addition	283.28(479.9)	25	127(148)	526(714)	420(435)	4.221*

*Significance level of 10%.

Table 3. Descriptive statistics of the participant farmers in discrete variables.

Variable		Market options					
		Farmers' percentage frequencies					
		All (n=70)	Neighbor (n=1)	Village (n=40)	Local town (n=24)	Regional (n=5)	Chi-square
Gender	Female	50	.0	34.3	12.9	2.9	0.231
	Male	50	1.4	22.9	21.4	4.3	
	Total	100	1.4	57.1	34.3	7.1	
Extension services	No	58.6	1.4	41.4	11.4	4.3	0.017*
	Yes	41.4	.0	15.7	22.9	2.9	
	Total	100	1.4	57.1	34.3	7.1	
Credit	No	77.1	1.4	45.7	25.7	4.3	0.711
	Yes	22.9	.0	11.4	8.6	2.9	
	Total	100	1.4	57.1	34.3	7.1	

*Significance level of 10%.

Mean income from sweet potato across the market options was statistically different at 10% level. This income influences the choice of market option because it increases the farmers' cash resources and hence could result in investment of sweet potato marketing. The highest average income, Kshs. 3805, was received by farmers selling to the regional market probably because of the good sweet potato prices. The mean cost of value addition is Kshs. 283.28 which is inclusive of water fees, labor for sorting the sweet potatoes and packaging materials for majority of the farmers. The cost was statistically different across the groups with those preferring local town markets incurring the highest average of Kshs. 526, followed by regional (Kshs.420) and Kshs. 127 for village.

Access to credit gives the farmer more cash resources hence it has an effect on his/her sweet potato marketing activities. According to the results, access to credit was significant at 10% level and negatively influences local town market participation. As the credit status of the sweet potato farmer changes from not accessing it to accessing, the probability of participating in the local town market than village reduces by 34.3% implying that the farmer will sell fewer sweet potatoes in the local town market as compared to the village market. This is inconsistent with the priori positive sign (Asfaw et al., 2010) likely because the farmer may increase participation of other farm and off-farm activities which are perceived to be more lucrative such as cash crops and dairy products. The margins from these alternatives are likely larger than sweet potatoes thus enabling the farmer to meet repayment of the credit and have some savings for his needs. Additionally, the study population was made up of smallholder farmers who have so diverse

Table 4. Model Estimates of the determinants of market option participation by farmers.

Variable	Coefficient	Z	p > |z|	Marginal effects
Local town market option				
Household size	0.99 (0.89)	1.12	0.262	0.224
Credit access	-1.94 (1.14)	-1.70	0.088*	-0.343
Total asset base	-0.48 (0.47)	-1.02	0.310	-0.108
Total income	0.99 (0.57)	1.73	0.084*	0.222
Transport mode to option	1.28 (0.49)	2.62	0.009*	0.287
Access to extension	2.04 (0.92)	2.23	0.026*	0.446
Total land owned	0.75 (0.81)	0.93	0.352	0.168
Education	-1.41 (1.07)	-1.31	0.191	-0.316
Sacks taken to option	1.05 (0.63)	1.68	0.094*	0.235
Value addition done	0.98 (0.49)	2.02	0.043*	0.220
Age	-3.24 (1.49)	-2.17	0.030*	-0.727
Gender	0.98 (0.88)	1.11	0.266	0.218
Regional market option				
Household size	1.33 (3.44)	0.39	0.699	3.49e-06
Access to credit	-0.92 (4.26)	-0.22	0.830	1.28e-06
Total asset base	-5.02 (3.27)	-1.53	0.125	-0.00002
Total income	2.37 (2.83)	0.84	0.401	7.16e-06
Transport mode to option	6.52 (3.25)	2.00	0.045*	0.00002
Access to extension	-9.18 (7.11)	-1.29	0.196	-0.00022
Total land owned	9.40 (5.45)	1.73	0.084*	0.000032
Education	-5.51 (5.44)	-1.01	0.310	-0.000017
Sacks taken to option	5.87 (3.03)	1.94	0.053*	0.0000194
Value addition done	-3.35 (2.84)	-1.18	0.238	-0.0000129
Age	4.42 (8.08)	0.55	0.584	0.0000194
Gender	8.50 (4.71)	1.80	0.071*	0.0002025

MAINMKT = Village market outlet is the base outcome

Number of observations = 68

LR chi 2 (20) = 91.05

Log likelihood = -29.179805

Prob> X^2 = 0.0000

Pseudo R^2 = 0.6094

*Significance levels at 10%; Figures in parentheses () are standard errors.

household needs to be met from the same limited resources, such that their behavior may not be identical to that of a firm. This implies that other studies of different populations may be needed to understand better this unique behavior.

The results indicate that the number of sacks taken to the market positively influences local town market participation. As the number of sacks the farmer takes to market increases by 1 bag, the probability of participating in local town than village increases by 23.5%. The increase in sacks taken to the market will make the farmer sell his produce in the local market which has a larger population of buyers hence increasing his chances of selling most or all of his produce as compared to the village which has less.

Total income which is a summation of both farm and off-farm sources positively influences local town market participation. An increase in the monthly income by 1% increases the probability of participating in the local town market than village by 22.2%. The increase in cash resources will make the farmer invest more in sweet potato production and marketing activities resulting to more surplus driving him to sell to local town which is a larger market compared to village.

On transport mode used to reach the market, the results show that it positively influences local town market participation. As the transport mode becomes better in terms of quick access to the market, affordability and convenience (collecting produce from home and carrying many bags at a time), the probability of participating in local town than village increases by 28.7%. Good transport reduces transportation costs for the farmer and hence makes it easy and cheaper for him/her to access local town market which has better market conditions in terms of big population of buyers and sellers and better prices compared to the village market option (Machethe et al., 2008).

Access to extension positively influences local town market participation such that, a change in a farmer's status from no access to extension to access increases the probability of local town market participation than village by 44.6%. This implies that extension will make the farmer participate more in the local town than the village market option. Extension services enable the farmer to improve his production methods hence leading to more output which in turn increases his/her marketed surplus hence market participation (Lapar et al., 2002), especially in larger markets such as local town.

Value addition done by the farmer does positively influence participation in local town market. If the farmer changes his value addition activities to include sorting, washing and packaging from otherwise, the probability of participating in local town market than village increases by 22%. Because of presence of many sellers in the local town compared to village, farmers have to ensure that their products are appealing to the buyers and hence have competitive advantage over other sellers.

Age of the farmer negatively influences local town market participation. This implies that an increase in the age by 1 year decreases the probability of participating in the local town than village market by 72.7%. This is inconsistent with the expected positive priori sign according to Machethe et al. (2008) such that, as the farmer gets old he will choose to participate less in the local town market than the village market option because of the loss in energy to sell in distant markets.

Results for the socio-economic factors determining participation in the regional markets, show that transport used to reach the regional market positively influences participation. An improvement in transport increases probability of participation in the regional market than the village market. Good transport acts as an incentive for farmers to sell to distant markets, which usually have better market conditions, because it reduces the cost of transportation and hence increases the farmers' profit margins.

Total land owned positively influences participation in the regional option. An increase in land owned increases the probability of participating in regional as compared to the village market. This is consistent with Machethe et al. (2008) who find that larger land sizes raise the probability of market participation for sellers since land is a critical production asset having a direct bearing on production of a marketable surplus, *ceteris paribus*. This implies that those with large tracts of land are likely to participate more in markets especially larger ones such as regional.

The sacks taken to the market positively influences regional market participation. An increase in the number of sacks increases the probability of farmers participating more in the regional market than village. As the number of sacks increases, farmers are likely to participate more in the regional than village market option because more sacks implies more economies of scale and hence it is better to sell to far away markets where prices are good.

Gender positively influences participation in the regional option and is significant at 10% level. As the gender of the participant changes from female to male, the probability of participating in the regional option being higher for male, implying that male farmers are likely to participate more in the regional market option than the village. Despite sweet potato production and marketing being mainly done by women (Nungo et al., 2007), men usually influence participation in distant market options such as regional.

In conclusion it can therefore be stated that the shift in participation of smallholder sweet-potato farmers from village market to a more income generating regional market is determined by; increased household income, land size, transport mode, gender of farmer, output of sweet-potato, extension services and value addition made.

Policy implications

Sweet potato production and marketing activities are mainly done by women in Vihiga County. To reduce poverty, the Government through agricultural officers, Non-governmental organizations and other stakeholders should first identify which member of the household has control over the crop in question then offer interventions which will take into consideration the gender of the person. For instance, if the interventions include programs such as giving financial aid to the farmers, they ought to take into consideration that most of these female farmers do not have collateral since title deeds and other property in most cases are in the name of their husbands who are less likely to allow the titles to be used as collateral.

The most common market option for majority of the farmers is the village market because of its close proximity that makes farmers to incur lower transportation costs. Consequently, the prices and income in this market are low because of the excess supply. Formal or informal institutional arrangements such as farmer groups or organizations should be encouraged through which farmers can collectively access distant markets which have good prices. The arrangements will facilitate: use of common transport; exchange of marketing information while strengthening negotiation; bargaining position of farmers, and; also make contracting and enforcement of contracts easier.

The size of the land owned was found to be low while the population in the area is high hence putting pressure on the land resources. To increase land productivity, the government needs to provide farmers with high yielding and disease resistant sweet potato varieties. The government also needs to avail extension officers in the area so as to advice farmers on new farming techniques. Since transport costs reduce market participation, the government should increase investment in the rural road transport network so as to ease movement of goods and reduce the transportation costs.

REFERENCES

Asfaw S, Amare M, Lipper L, Davis B, Simtowe F (2010). Smallholder market participation and rural poverty; evidence from tanzania and ethiopia. www.csae.ox.ac.uk/conferences/2011-EDiA/papers/306-Asfaw (accessed on 30th June, 2011).

CPPMU (2010). Economic review of agriculture. Central Planning and Project Monitoring Unit, (CPPMU), Ministry of Agriculture, Nairobi.

FAO (2007). Food Security Profiles. Food and Agriculture Organization, Nairobi, http://www.fao.org (Accessed on 17th July 2010).

Greene WH (2002). Econometric Analysis. 5th Ed. Pearson Education Inc., New Jersey.

Karanja F (2006). Cropwat model analysis of crop water use in six districts in Kenya. Department of Meteorology, University of Nairobi, Nairobi.

Komarek A (2010). The determinants of banana market commercialization in Western Uganda. Afr. J. Agric. Res. 5(9):775-784.

Lapar ML, Holloway G, Ehui S (2002). Policy options promoting market participation of smallholder livestock producers: A case study of the Philippines. Socio-economic and Policy Research Working Paper 47. ILRI, Nairobi.

Machethe C, Jagwe J, Ouma E (2008). Transaction costs and smallholder farmers' participation in banana markets in the Great Lakes Region. Department of Agricultural Economics, Extension and Rural Development, University of Pretoria.

MOPND (2005). Vihiga District Strategic Plan 2005-2010: for implementation of the national population policy for sustainable development. National Coordination Agency for Population and Development, Ministry of Planning and National Development (MOPND), Nairobi.

Mugisha J, Mbowa S, Kasibante F (2004). Determinants of use of Information and Communication Technologies by Agribusiness Firms in Uganda. East Afr. J. Rural Dev. 20:0377-7103.

Nungo RA, Ndolo PJ, Kapinga R, Agili S (2007). Development and promotion of sweet potato products in Western Kenya. Proceedings of the 13th ISTRC symposium. pp. 790-794.

Nyangweso O (2007). Household food security in Vihiga District, Kenya; Determinants of dietary diversity. African Crop Science Proceedings. 8:1383-1389.

Olwande J, Mathenge M (2010). Market participation among rural poor households in Kenya. Tegemeo Institute, Egerton University, Nairobi.

Omiti JM, Otieno DJ, Nyanamba TO, McCullough E (2009). Factors influencing the intensity of market participation by smallholder farmers; A case study of rural and peri-urban areas of Kenya. *AFJAE,* 3(1):71-82; http://www.kippra.or.ke (Accessed on 28th March, 2011).

Rono SC, Wanyama JM, Mose LO, Mutoko CM, Chesoro J, Lusweti CM, Wanjecheke E, Wachige B (2007). Assessing Household Sweet Potato-based Diet among Rural Communities in North Rift Region of Kenya. http://www.kari.org.

Impact of the adoption of semi-mechanized technologies of shea processing on rural women's income in Northern Benin (West Africa)

Jacob Afouda Yabi[1], Patrice Igue Adegbola[2], Dansi Silvere Tovignan[1], Morenike Cendrine Ahouandjinou[1] and Adeyemi Suleiman Adekambi[2]

[1]Department of Agricultural Economics, Faculty of Agronomy, University of Parakou, BP: 123 Parakou, Republic of Benin.
[2]Programme for Agricultural Policy Analysis (PAPA) of the National Institute of Agricultural Researches of Benin (NIARB), Porto Novo, Republic of Benin.

In Benin, the shea sector is one of the most promoted sectors in government's attempt to diversify sources of farm income. The objective of this study is to evaluate the impact of the adoption of the semi-mechanized shea processing on the income of rural women in northern Benin. In total, a random sample of 200 women processing shea was investigated. A probit model was firstly run to analyze the factors determining the adoption of the semi-mechanization. Then, a multiple regression model was used to assess the impact of adoption on women's income. The results highlighted that, the adoption of semi-mechanization was determined by availability of electricity, availability of a market, contact with extension services and shea processing as main activity for woman. Moreover, the adoption of semi-mechanization has induced a positive and significant increase of rural women's income up to 103 914 Francs CFA. As a result, it is important that agricultural policy reinforce the promotion of modern agricultural processing technologies to improve the added value and to reduce poverty in rural areas. As well, electricity and market access, women's level of education and contact with extension need to be improved.

Key words: Adoption, impact, shea processing, semi-mechanization, income, Benin.

INTRODUCTION

In Benin, cotton is the main cash crop which contributes over 75% to the export earnings (DSCRP, 2007). The crisis in the sector has made the Government to promote new export sectors such as shea processing. Traditionally, the nuts mainly transformed in shea butter are manually crushed in mortar and then ground to stone. This laborious work is very difficult and requires a high amount of man-power. However, following the diagnostic survey of the processing of food products in Benin, the mechanization of certain stages of shea processing (semi-mechanization) has been identified as priority need that can reduce drudgery in the sector (Kruit and Godjo, 1998). As well, Ahouansou and Singbo (2005) showed that, in shea processing, grinding and milling operations are most restrictive. In addition to these operations, women's exposure to heat and smoke for hours during the roasting process and low yield (10 kg of roasted product per and per person on average) are constraints weighing down on shea processing in Benin (Yabi et al., 2009).

For improving the processing conditions, several tools are developed by manufacturers among which stand the Benin Cooperative of Agricultural Equipment (COBEMAB[1]). Through the Agricultural Program and Food Technology (PTAA[2]), the National Agricultural research System (INRAB[3]) supports this initiative, and from 2002 to 2003, an adaptation test of motorized equipment was conducted by the Program in the areas of N'Dali, Banikoara, Djougou, and Natitingou. The results showed that, the complex equipment made up a mill of BCAE and a corn mill motor is effective in enabling women processing shea to save up to 75% of their initial working time (Singbo and Ahouansou, 2005).

Nowadays in Benin, equipment such as the grinder or crusher, the shea grinder, the complex made up grinder and shea mill, and the churn are introduced and made available for women processing shea. Despite their performance in terms of saving time, it is relevant to check out whether the introduction of these technologies for semi-mechanized shea processing has increased the income of rural women, contributing to reduce rural poverty. Moreover, since the introduction of these technologies in shea parks of northern Benin, evaluation studies of their economic impact on adopting women have not been conducted yet. Thus, this study aims at assessing the impact of the semi-mechanization of shea processing on the income of rural women in shea parks in northern Benin.

Theoretical background

Talking about impact assessment, the major problem is to isolate the effects of a project, a program or an innovation on the target group. To deal with this problem, several frameworks or approaches are available. The common ones are the "before-after" approach, the "with-without" approach, the so-called "naive" approach, the experimental approach, and the non-experimental approach (Yabi, 2008). The 'before-after' approach compares the performance of key variables after the introduction of technology with the one before its introduction. But this comparison of situations "before" and "after" the introduction of a given innovation does not isolate the effects related to exogenous factors (inflation, rainfall, natural disasters, economic and agricultural policies for instance) that may arise during the adoption process and that could determine the adoption rate as well as the impact on individual. Moreover, this approach imposes a baseline study before the introduction.

Otherwise, data of the conditions "before" are not available.

In comparison with the "before and after" approach, the "with-without" one seems clearer an easier (Scherr and Muller, 1991). It divides the potential target group into two subgroups. One subgroup has received the technology (beneficiary or treatment group) and the other which has not (non beneficiary or control group). In this approach, one compares the two subgroups. The problem while using this approach is to find out respondents that are similar enough so that only the adoption or not of the technology sets out the difference between them.

The "naïve" approach consists in taking a random sample of adopters and non-adopters of a given technology and using the simple difference in mean scores observed in both groups as the impact estimation. While this estimation method is commonly used in the literature (Adekambi, 2005), the estimator is potentially biased (Heckman, 1990; Diagne, 2003) and does not take into account the socio-economic characteristics of operators. The experimental approach consists in setting up a group of people having the same rights and agreeing to participate in the experiment. The selected people are randomly divided within two subgroups: one group of those who receive the intervention (treatment group) and the second group of those who do not receive (control group). Participants in this experiment are randomly selected and any difference with non-participants is only due to treatment. This experimental approach is considered to be the most reliable (unbiased estimates) and gives results easier to interpret (Cochrane and Rubin, 1973; Bassi, 1984). However, this type of evaluation is difficult to apply in practice because of ethical problems arising in the case of social phenomena (Diagne, 2003). In addition people adopting are likely to be willing to participate in the experiment and one might come up with zero non-adopters.

Economists widely use non-experimental approach based on economic theory and econometric analysis to guide and minimize potential errors in the impact estimation (Diagne, 2003). Non-experimental approaches are used when it is not possible to select a control group. In this framework, it is possible to compare program participants with non-participants by using statistical methods to control the observed differences between the two groups that could influence the impact indicator regardless to the program participation. Indeed, it is possible, using regression analysis, to obtain a "control" of age, income, gender and other characteristics of the participants. This assessment approach is relatively inexpensive and easy to apply, but the results interpretation is not straightforward and the results

[1] Coopérative Béninoise de Matériel Agricole du Bénin
[2] Programme Technologie Agricole et Alimentaire
[3] Institut National de Recherches Agricoles du Bénin

themselves may be less reliable (Diagne, 2003), if the researcher was not aware of econometrics methods.

In this study, we consider three main technologies aiming at semi-mechanized shea processing and introduced in northern Benin: the grinder, the shea mill, and the entire complex (grinder + mill). Since the technologies under consideration are already introduced, we apply an ex-post evaluation. Regarding the strengths and weaknesses of the different impact assessment frameworks, we used a non-experimental approach. This approach is based upon principles of the "with-without" approached improved with econometrics methods. Finally, following the problem statement of this study, we mainly focus on income as impact estimator.

MATERIALS AND METHODS

Study area and database

The study area includes the 3 shea parks (Parakou, Bembèrèkè, and Kandi) in northern Benin situated between 09° 20' and 12° 30' North latitude and 0°45' and 3°20' East longitude. Thus, it is located in a Savannah area with a Sudano-Guinean climate characterized by a wet season (mid April to late October to early November). The annual rainfall varies between 900 and 1200 mm. This climate is very suitable to the shea production, making the study environment an area of high production and consumption of shea butter in Benin.

In the study area, 20 villages well known in shea production and processing were selected with the support of agricultural extension officers. The respondents were women performing activities related to shea processing. The sampling was conducted randomly (based upon a list of women processing shea per village) and included an initial number of 200 women producers of shea butter (10 per village). Regarding the assessment approach adopted, the sampled women were set into two subgroups: one group using at least one of the three main technologies of semi-mechanization and the second control group using the traditional method for shea processing.

Primary quantitative and qualitative data such as women's characteristics, technologies adopted and the production of shea butter were collected through a survey method by using an individual questionnaire. As well, the triangulation of data was done through semi-structured interviews, focus group discussions and interviews with key informants.

Modeling the impact of adoption on income

We aim at determining the effect of technological change on an outcome indicator defined by y (income). Let us call y_1 the income level for an individual i if he uses the new technology and y_o the level if he does not use the new technology. Then, we consider Wi as is a binary variable taking the value 1 when the technology is adopted and 0, if not. The causal effect of technology adoption for that individual i is the difference between y_1 and y_o:

$$\Delta y = y_1 - y_0 \qquad (1)$$

The main problem while estimating Δy is that, for a given individual, the income is observed either following the adoption or before the adoption, but never both at once. Then, let us quote by P the probability of technology adoption. Having this, Δy is due to a change ΔP. Thus we might consider that:

$$\Delta y / \Delta P = \beta \qquad (2)$$

By moving to an infinitesimal change (d instead of Δ), we come up with:

$$dy / dP = \beta \qquad (3)$$

From equation [3], y can be expressed as a function of P and other factors Z. It comes out that:

$$y = F(P, Z, e) \qquad (4)$$

Where, Z stands for exogenous factors other than P and e for the error term assumed to have a normal distribution with mean 0 and constant standard deviation. Since P is by hypothesis determined by some characteristics C specific to the individual adopting the technology, we have:

$$P = G(C, v) \qquad (5)$$

Where, C stands for exogenous factors (education level, gender, contact with a project officer for instance) determining the probability of adoption, and v for the error terms assumed to be normally distributed with average 0 and constant standard deviation.

With such specification, we come up with two models: the adoption model (Equation 4) and the impact model (Equation 5). Wooldridge (2002) suggests estimating initially, the adoption model. Then, in a second step, the model impact by integrating an estimated probability $\hat{G}$ such as:

$$y = F(\hat{G}, Z, e) \qquad (6)$$

In case the estimated coefficient of $\hat{G}$ is significant, Rosenbaum and Rubin (1983) demonstrated that, we could identify an average causal effect of technological change within a population. Their idea is to pinpoint the difference between the average level of the indicator of beneficiaries and non beneficiaries of the technology. This gives the average treatment effect (ATE) which is defined by the difference between the estimated income $\widehat{y1}$ when the estimated probability is 1 and the estimated income $\widehat{y0}$ when the same probability is 0. Mathematically, we have:

$$ATE = [(\widehat{y1}/\hat{G} = 1) - (\widehat{y0}/\hat{G} = 0)] \qquad (7)$$

This indicator which measures the impact of the technology on an individual selected at random from the population is also equal to the average impact of technology on the entire population (Heckman, 1997; Wooldridge, 2002). Empirically, the logit or probit models was based upon the Maximum Likelihood method are widely used for estimating the adoption model (Equation 5). According to Heckman (1997), both logit and probit models do often generate similar results, and choosing one of them depends on the skill in data analysis. Thus, we used the probit model to estimate Equation 5. Thus, we get the following empirical equations:

$$G_i = \alpha_0 + \alpha_1 ELECTI_i + \alpha_2 EXPE_i + \alpha_3 CONTVUL_i + \alpha_4 MARCHE_i + \alpha_5 ACTPBEUR_i + \alpha_6 EDUCFT_i + v_i \qquad (8)$$

and

$$y_i = \beta_0 + \beta_1 \hat{G}_i + \beta_2 LEFTH_i + \beta_3 AUTREV_i + \beta_4 EXPE_i + \beta_5 ACTPBEUR_i + \beta_6 LDMARCHE_i + \beta_7 EDUCFT_i + \beta_8 ABLEFTHJi + \beta_9 ABAUTREVi + \beta_{10} ABLDMARCHEi + e_i \qquad (9)$$

Table 1. Descriptive statistics of women's characteristics.

Quantitative	Mean	Standard deviation
Experience in shea processing (years)	15.10	10.63
Qualitative	**Yes (%)**	**No (%)**
Adoption of semi-mechanized technology (G)	55.7	44.3
Availability of electricity in the village	45.75	56.25
Contact with extension	23.4	76.6
Availability of market in the village	52.08	47.92
Shea butter is the main activity of the respondent	73.96	26.04
The respondent received a formal education	4.69	95.31

Source: Authors' calculations.

Where, G stands for the distribution function of technology adoption and $\hat{G}_i$ for its value estimated from Equation 8. Similarly, y_i is the net annual income coming from shea processing of the ith woman. This income is obtained by subtracting from the total revenues the total costs related to processing. The coefficients α and β were estimated by using the Maximum Likelihoods (ML) and ordinary least squares (OLS) methods respectively.

Exogenous variables and hypotheses to be tested

The selection of the prospective exogenous or explanatory variables was grounded in the literature and in our field observations:

a) Availability of electricity in a village (ELECT): Most of the new processing technologies use electric power. Therefore, the availability of electricity in the village is a prerequisite for adoption of these new technologies.
b) Years of experience in shea processing (EXPE): Following Huffman (1977) and Kokoye et al. (2013), learning from experience reduces allocative errors. Thus, we hypothesize that women having long experience of the traditional way of shea processing will be less likely to adopt new technologies.
c) Contact with extension (CONTVUL): According to Maddison (2006), people enjoying free extension advice are likely to adapt. In this line, we expect a positive correlation between the contact with extension and the adoption of new processing technologies.
d) Availability of market in the village (MARKET): The shea processing is a market oriented activity in northern Benin. Thus, the market access is an incentive for women processing shea. Stating this, we hypothesize that the market availability is positively correlated with the adoption of new technologies.
e) Shea processing as main activity of the respondent (ACTPBEUR): We assume that women invest more time and money in their main activity. Therefore, we expect a positive correlation between having shea processing as main activity and adoption of new processing technologies.
f) The respondent's formal education status (EDUCF): Educated farmers are more likely to respond to environmental changes by adapting (Maddison, 2006). They might also be likely to understand easily the advantage of new processing technologies (time saving and health benefits for instance). Therefore, the education level is expected to have a positive effect on the adoption of technologies.
g) Quantity of family labor available for processing (LEFTHJ): Here we make the assumption that when the family labor is available women are less willing to invest in new technologies. Then the quantity of family labor available for processing is expected to have a negative impact on the adoption processing of new technologies.
h) Income from other activities of women (AUTREV): The availability of income sources can make women able to afford investments related to new technologies use. Thus, we expect a positive correlation between income from other activities and adoption of new processing technologies.

Interaction between adoption and amount of family labor available (ABLEFTHJ), interaction between adoption and income from other activities (ABAUTREV) and interaction between adoption and distance from the village to the nearest market (ABLDMARCHE) were defined to measure the impact of the interaction between the adoption of semi-mechanized shea processing and amount of family labor available (LEFTHJ), income from other activities (AUTREV), and distance from the village to the nearest market (LDMARCHE) respectively. They enable to isolate composite impacts on the income of women adopting the new technologies. These variables were obtained by multiplying the estimated probability $\hat{G}_i$ by the three variables LEFTHJ, AUTREV and LDMARCHE respectively. EXPE, ACTPBEUR, and EDUCFT are assumed to determine indirectly women's income. Therefore, they are introduced in both adoption and impact models.

RESULTS

Women's characteristics

Descriptive statistics of women's characteristics are presented in Table 1. These results indicate that, 56% of the respondents adopt new shea processing technologies. Women have a consistent experience in shea processing (median 15 years), but few (23%) of them have contact with extension services. The level of education very low (5%).

Estimation of the adoption probability

The results of the adoption model indicate that, the model is highly significant at 1% (Table 2). The coefficients of all explanatory variables, except the years of experience, positively influence the adoption of the semi-mechanized shea processing. Electricity availability, contact with extension, and market availability are significant at 1%

Table 2. Results of econometric estimation of the adoption model.

Variable	Coefficients	Standard error	p-value
Availability of electricity in the village	1.250 ***	0.2082	0.000
Experience in shea processing	-0.124	0.119	0.299
Contact with extension	0.858***	0.276	0.002
Availability of market in the village	0.771 ***	0.212	0.000
Shea processing as main activity	0.622**	0.248	0.012
Respondent received a formal education	0.090	0.511	0.860
Constant	-1.072 ***	0.366	0.003
Dependent variable		Adoption probability (G)	
Pseudo R^2		0.233	
Log pseudo-likelihood = -101.086		Wald χ^2 (6) = 58.54***	
Number of observations = 192		Prob = 0.000	

Note: *** and **: Significant at 1 and 5%, respectively. Source: Authors' estimations.

while having shea processing as main activity is significant at 5%. The formal education status is not significant in the model. Subsequently, the determinants of the adoption of at least one type of technology are the availability of electricity or market in the locality, the contact with extension services, and the shea processing as main economic activity.

As expected, although the coefficient of the years of experience in processing is not significant, the more women are experienced, the more they are more likely to use traditional methods for shea processing. The level of formal education should have a significant positive impact on adoption, but the small proportion of women who received this education (5%) made its impact positive but not significant.

Impact of adoption on the income of women

To estimate the impact of adoption on women's income, other variables were used in addition to the estimated probability of adoption as described in the empirical model. The descriptive statistics of these variables are presented in Table 3. On average, women interviewed did earn 173,218 Francs CFA as income from shea processing. Women adopting processing technologies have higher incomes than they fellow women who did not adopt. This difference in incomes that could come from the difference in technology between the two groups will be tested by the impact model.

Women who are not adopting travel a longer distance (0.899 km) than the ones adopting (0.478 km) in order to access market. This result could be explained by the fact that, most of women adopting processing technologies belong to village-groups and deal their products in group to traders who use to collect the butter in the village. The amount of available man-power within the household, the income from other activities, the experience in processing and the formal education status are more or less the same between women adopting and they fellow who did not adopt. Finally, a high proportion of women adopting the processing technology (90%) have shea processing as their main economic activity.

Table 4 presents the econometric results of the impact model. The Fisher F statistic indicates that, the model is highly significant at 1%. The over-identification test of Hansen used to test the independence between the instruments and the error term is not significant (p = 0.110). The significance of the interaction terms was tested by the Wald test. The value of this test is 74.29, and is significant at 1%. Thus, the null hypothesis which states that, all interaction terms are zero cannot be accepted.

The coefficient of the adoption probability $\hat{G}$ estimated from the adoption model is positive and significant at 1%. Therefore, there is a positive correlation between income and adoption. Put into a simplistic way, adopting the new shea processing technologies ensure an increase of the annual income. Furthermore, the average impact of the adoption of the semi mechanization of shea processing on the annual income of women is 103,914.1 Francs CFA/year.

In addition to the adoption, the family labor available, the shea processing as main activity and the income from other activities have positive and significant coefficients at 1 and 5%, respectively on the women income generated by shea processing. Indeed, women processing shea and having many children, so more workers, have higher incomes. The shea processing as main activity is the key to improving income. This result is not surprising because women having shea butter processing as main activity spend most of their time in doing the activity. Finally, revenue from other activities did promote capital accumulation for a better investment in shea processing. As a result, these revenues have a positive impact on income.

Considering the interactions between adoption and the three target variables, it comes out that, revenues from

Table 3. Descriptive statistics of explanatory variables used in the impact model.

Variables	Adopting	Not adopting	Total
Quantitative: Means (Standard deviation)			
Annual income from shea processing	225 175 (30 002)	121 261 (52 100)	173 218 (60121)
Estimated probability of adoption	0.96 (0.14)	0	0 0.57 (0.14)
Amount of family labor available	4.903 (0.758)	4.972 (0.828)	4.93 (1.12)
Income from other activities	117218 (94008)	101421 (54653)	109 319 (108 740)
Experience in processing	2.473 (0.830)	2.339 (0.864)	2.41 (1.20)
Distance from the village to nearest market	0.478 (0.708)	0.899 (0.950)	0.690 (1.18)
Interaction between adoption and amount of family labor available	-0.030 (0.758)	0	-0.030 (0.758)
Interaction between adoption and income from other activities	6993.69 (94008.3)	0	6993.69 (94008.3)
Interaction between adoption and distance from the village to the nearest market	-0.186 (0.708)	0	-0.186 (0.708)
Qualitative: Relative frequencies (%)			
Shea processing as main activity	90.4	57.6	74
Formal education	94.2	96.4	95.3

Source: Authors' calculations.

Table 4. Results of the impact model.

Variable	Coefficients	Standard Error	Probability
Constant	-331096.3 **	151548.8	0.029
Estimated probability of adoption	103,914.1***	33,517.14	0.002
Amount of family labor available	121118.3***	26406.15	0.000
Revenues from other activities	0.862**	0.388	0.026
Experience in shea processing	-13481.9	10116.4	0.183
Shea processing as main activity	87,007.48***	24020.91	0.000
Distance between village and nearest market	-29,876.21	22,102.64	0.176
Formal education status	33563.45	46969.87	0.475
Interaction			
Between adoption and amount of family labor available	-115777.3**	52033.01	0.026
Between adoption and revenue from other activities	4.520474 ***	0.585	0.000
Between adoption and distance from the village to the nearest market	84192.76**	43031.93	0.050
Average treatment effect (ATE)	103914.1***	37031.38	0.006
Dependent variable	Revenue of shea processing		
Wald test (H0: all interaction terms = 0)	χ^2 (3) = 74.29 (p = 0.000)		
Fisher F statistic	F =53.14; df1 = 11; df2 = 180 and p = 0.000		
Number of observations	192		

Note: *** and **: Significant at 1 and 5%, respectively. Source: Authors' estimations.

other activities and the distance from the nearest market have a positive impact on income generated by the processing activity. However, the interaction between adoption and amount of labor has a negative impact on the income of women adopting new technologies. As we hypothesize, this is simply due to the fact that women having higher labor available are already less willing to adopt new technologies of processing.

DISCUSSION

Three factors are identified as determinants of the adoption of semi-mechanized processing of shea nuts. It is about the availability of electricity or market in the locality, the contact with extension services and the shea processing as main activity for women. These finding confirm the theoretical hypotheses of the study. As well, several authors highlighted the key role of variables such as extension services in the adoption of new technologies (Bravo-Ureta et al. 2005; Glèlè et al. 2008).

By adopting the technologies of shea processing, the results pinpointed that women could improve their incomes. This supports the general hypothesis stating that adoption of agricultural innovations lead to improvement of incomes. Glèlè et al. (2008) found out that, adopters of improved cassava varieties derive more income per hectare than non-adopters and they earn on average 140,358 Francs CFA per hectare against 46,984 Francs CFA per hectare for their fellow non-adopters. Similar studies were also conducted in other African countries and the results are consistent with those found out in Benin.

According to Benin et al. (2011), the technologies adopted by farmers in Uganda, through the National Program for agricultural extension services, have a positive impact on incomes. The direct and indirect impacts of adopters are estimated between 37 and 95% higher in comparison with non-adopters. Benin et al. (2011) also reported that, this positive impact increases over time. For instance, between 2004 and 2007, it increased from 27% to 55% per capita. In the same vein, Kato et al. (2011) showed that in the Nile Basin in Ethiopia, technologies for soil and water conservation have a positive impact on agricultural output. Because the performance of agriculture determines the level of income, these technologies have a positive impact on income.

According to Hassan and Thurlow (2010), the water management strategy adopted by the South African producers had as impact, an improved domestic production by 0.4% annually from 1994 to 2007; implicitly improving producers' income over the same period. Bellemare (2010) found out in Madagascar that, the adoption of technologies promoted by extension agents improves productivity and results indicated that, the elasticity of performance in relation with the number of visits is between 1.3 and 1.7. In Kenya, the adoption of technologies (such as systems of rotation with green fertilizers) of the International Center for the Improvement of Maize and Wheat Improvement Center (ICIMWIC) allowed producers to improve their yields (De Groote et al., 2010). But, as several authors emphasized, the adoption of new technology can have a positive impact if the additional costs are offset by the additional production. As proof of this, LUCOP-TAN (2010) found out that the adoption of new technologies for the development of the valleys in the region of Tahoua in Niger had no impact on producers' income. Indeed, the costs associated with the development of the valleys are very high and the additional production value obtained does not cover the aforesaid costs.

Subsequently, the impact of adopting a new agricultural technology supports the economic theory of the producer's balance. The adoption becomes economically profitable from when the marginal revenue generated equal the marginal costs. In this case, the economic impact of the technology adoption is positive.

Conclusion

This study analyzed the impact of the adoption of semi-mechanized shea processing on income of rural women in northern Benin. The findings highlighted that, the adoption of shea processing technologies increases women's incomes. As policy implication, it is important to strengthen the promotion of new technologies in order to boost the agricultural development in general and the shea processing in particular. In this line, special emphasis should be put on modern technologies of processing agricultural products to create more value added, implying more wealth.

Moreover, electricity and market access, women's level of education and contact with extension need to be improved. This study focuses only on income. Future researches could address other variables such as education level of children and health.

Conflict of Interests

The author(s) have not declared any conflict of interests.

ACKNOWLEDGEMENT

The authors thank the National Institute of Agricultural Researches of Benin (NIARB) for funding this study through competitive grants.

REFERENCES

Adekambi SA (2005). Impact de l'adoption des variétés améliorées de riz sur la scolarisation et la santé des enfants au Bénin: Cas du Département des collines. Rapport d'étude technique. PAPA/INRAB, Porto-Novo, Bénin. P. 100.

Ahouansou R, Singbo A (2005). Etude technique des équipements de transformation du karité en beurre. Rapport d'étude technique. PTAA-PAPA/INRAB. Porto-Novo, Bénin. P. 13.

Bassi L (1984). Estimating the effects of Training Program with Nonrandom Selection; Rev. Econ. Stat. 66:36-43.

Bellemare MF (2010). Agricultural extension and imperfect supervision

in contract farming: evidence from Madagascar. Agric. Econ. 41:507-517.
Benin S, Nkonya E, Okecho G, Kato E, Randriamamonjy J, Lubadde G, Kyotalimye M (2011). Returns to Spending on Agricultural Extension: The Case of the National Agricultural Advisory Services (NAADS) Program of Uganda. Agric. Econ. 42(2):249-267.

Bravo-Ureta BE, Solis D, Cocchi H, Quiroga RE (2005). The impact of soil conservation and output diversification on farm income on Central American hillside farming. Agric. Econ. 35:271-275.
Cochrane W, Rubin PK (1973). Controlling bias in observational studies. Sankyha 35:417-446.
de Groote HD, Vanlauwe B, Rutto E, Odhiambo GD, Kanampiu F, Khan ZR (2010). Economic analysis of different options in integrated pest and soil fertility management in maize systems of Western Kenya, Agric. Econ. 41:471-482.

Diagne A (2003). Evaluation de l'impact: synthèse des développements méthodologiques récents, ADRAO/ Conakry. P. 15.
DSCRP (2007). Document de Stratégie de Croissance pour la Réduction de la Pauvreté. Ministère du Développement. Cotonou. République du Bénin. P. 117.
Glèlè EKA, Adekambi AS, Agli CK, Tamegnon BA, Adegbola PY (2008). Impact socio-économique de l'adoption des variétés améliorées de manioc au Bénin. P. 51.
Hassan R, Thurlow J (2010). Macro–micro feedback links of water management in South Africa: CGE analyses of selected policy regimes. Agric. Econ. 42:235-247.

Heckman J (1990). Varieties of Selection Bias. Am. Econ. Rev. 80:313-318.
Heckman J (1997). Instrumental variables: A study of the implicit assuptions underlying one widely used Estimator for Program Evaluations. J. Hum. Resour. 32:441-462.

Huffman WE (1977). Allocative Efficiency: The Role of Human Capital. Quart. J. Econ. 91:59-79.
Kato E, Ringler C, Yesuf M, Bryan E (2011). Soil and water conservation technologies: A buffer against production risk in the face of climate change? Insights from the Nile Basin in Ethiopia. Agric. Econ. 42(5):593-604.

Kokoye SEH, Tovignan DS, Yabi JA, Yegbemey RN (2013). Econometric modeling of Farm household land allocation in the Municipality of Banikoara in northern Benin. Land use Policy 34:72-79.

Kruit F, Godjo T (1998). Enquête diagnostique sur la transformation des produits vivriers au Bénin. Rapport d'étude technique. Ministère de l'Agriculture, de l'Elevage et de la Pêche (MAEP), Cotonou, République du Bénin. P. 108.
Maddison D (2006). The perception and adaptation to climate change in Africa. CEEPA Discussion Paper No 10. Centre for Environmental Economics and Policy in Africa, University of Pretoria, South Africa.
Rosenbaum PR, Rubin DB (1983). The Central role of Propensity Score in Observational Studies for Causal Effect. Biometrika 70(1):41-55.

Scherr SJ, Muller EU (1991). Technology impact evaluation in agroforestery projects. Agroforest. Syst. 13:235-257.

Singbo A, Ahouansou R (2005). Etude de faisabilité de l'implantation d'un atelier mécanisé de production de beurre de karité au Bénin (Afrique de l'Ouest). Bulletin de la Recherche Agronomique du Bénin. Numéro 50- Décembre 2005. pp. 38-46.
Wooldridge J (2002). Econometric analysis of cross-section and panel data. The MIT press, Cambridge, Massachusetts, USA. pp. 502-644.
Yabi AJ (2008). Impact of rural development projects on agricultural productivity in Adja and Nagot areas of Benin. Bulletin de la Recherche Agronomique du Bénin 62:24-33.
Yabi AJ, Ouinsavi C, Sokpon N (2009). Facteurs d'efficacité technico-économique des femmes transformatrices du karité en beurre au Nord-Bénin. Annales de l'Université de Lomé ; série Sciences Economiques et de Gestion. III:23-44.

Synergies between urban agriculture and urban household food security in Gweru City, Zimbabwe

Admire Jongwe

Economics and Markets Department, Ministry of Agriculture, Mechanization and Irrigation Development, Zimbabwe.

It is estimated that 25% of Zimbabwe's population lives in urban areas (CSO, 2002), 70% below the poverty line, and a million in the city of Gweru. The worsening macroeconomic situation in 2008 resulted in urban food insecurity. Households adopted different survival strategies, including the intensification of urban agriculture. In an effort to assess the extent to which urban household food insecurity is mitigated by UA activities, a comparative analysis between households practicing and not practicing UA was done in Gweru in 2009. Household size as well as household head sex, age and employment status were found to affect household UA practice. Results indicated that UA practicing households were food-secure than non-practicing households. Household size, UA participation, household income, household head sex, maize meal price affected household food expenditure. Household head sex, UA participation, household head age and informal activities carried out by household members significantly affected urban household food security. The study concluded that there are synergies that exist between UA and urban household food security.

Key words: Urban agriculture, household food security, Gweru.

INTRODUCTION

Urban food insecurity is a growing challenge emanating from rapid urbanization and rising poverty heightened by the HIV/AIDS epidemic. Rapid urbanization, declining rural productivity and poor marketing systems results in increased urban poverty and food insecurity. Urbanization increases resource competition, costs of supplying, distributing and accessing food, thus negatively impacting on urban household food security. The challenge of feeding cities lies in enhancing consumer access to food by ensuring increased local food production, processing and distribution as well as reversing dependence on distant production sites, thus enabling cities to become more autonomous in food production (Rabinowicz, 2002).

Macro-economic policies since the late 1990s have had a deleterious effect on wage-dependent workers, creating vulnerable urban people. Hovorka et al. (2009) highlighted that economic or political crisis drives UA, which provides safety net for the poor and for households seeking to augment dwindling incomes. It is estimated that one-fourth of Zimbabwe's population lives in urban centers, 70% living below the poverty line (STERP, 2009). The Zimbabwean annual real GDP has been declining an average of -5.9% since 2000 (STERP, 2009). Recession has culminated in non-wage unemployment rate soaring from 80% in 1995 to an estimated 94% in 2008. Food shortages in 2007 and 2008 resulted in inflation reaching 231 million percent (CSO, 2008).

UA is defined as mostly crop and livestock rearing on private, leased, or rented land in peri-urban areas, in backyards, on vacant public lands and in semi-public areas. The drivers of UA include: High food prices and rampant inflation; food shortages; growing joblessness; erosion of purchasing power of wages and pensions, and cultural attachment or hobby. UA, is a major coping strategy for poverty and food insecurity alleviation, has been increasing in the SADC region (Harare Declaration, 2003). UA is not a relic of the past, and will not fade away nor brought to the city by rural immigrants who will lose their rural habits over time as the city grows (Hovorka et al., 2009).

Urban agriculturalists are composed mostly of disadvantaged groups such as orphans, women, rural immigrants without jobs, and the elderly. However, there has been an influx of the lower and middle-income earners, as well as richer people seeking a good investment for their capital by undertaking UA for physical and or psychological relaxation (Hovorka et al., 2009). In 2007, it was estimated that 25% of the maize produced in Zimbabwe was produced in and around the cities (AGRITEX, 2008).

UA contributes to local economic development through boosting urban poor asset base, increasing income, alleviating poverty, and including the urban poor and women into mainstream economic activities, thus reducing vulnerability and food insecurity (Mbida, 1995). UA is a direct and indirect occupation provider in cities; it is estimated that 200 million urban residents world over provide food for the market and that 800 million urban dwellers are actively engaged in UA (RUAF, 2009).

In urban areas, virtually everything consumed is purchased; the low purchasing powers are undermined by economic shocks leaving households at the mercy of food insecurity. It is estimated that poor household devotes 60 to 80% of their income on food purchases. Income to purchase food matters less if the food is not available. At the heist of food insecurity in 2007, food was available in Zimbabwe's rural areas, whilst food shortages were prevalent in urban areas (ZIMVAC, 2008).

Food security

The four pillars of food security are access, availability, safety and stability. FAO (2001) defined food security as a state in which all people at all times have physical, social and economic access to sufficient, safe and nutritious food to meet their dietary needs and food preferences for an active and health life. Food security includes food supply; physical, social and economic access; adequacy; utilization; safety; nutritious and cultural acceptability.

Problem statement

Since the 1970s, UA has been growing in developing nations, in terms of land usage and number of farmers taking part. In the midst of loss of industrial jobs, decreasing income, and harsh macro-economic conditions, urban households have found it difficult to continue with rural agricultural activities, and these have led to increased UA activities. Despite UA growth in urban areas, households have continued being vulnerable amidst growing urban household food insecurity and outbreaks of food riots.

METHODOLOGY

Primary data was used as a main source of inference, while secondary data was used to validate the primary data collected. There was systemic and purposive Stratified sampling of UA practicing households. First areas practicing UA the most were selected and then the urban agriculturalists were randomly selected in the fields. Data was collected through structured and semi-structured questionnaires. The respondents were randomly selected with a bias towards the UA practicing households. A total of 150 questionnaires were administered of which 69.3% were UA-practicing households. The questionnaire captured data on household characteristics, economic activity, asset endowment, UA production and performance, non-participation in UA, food basket and consumption. The data was entered into the SPSS and STATA for analysis by cross tabulation, mean differencing, gross margin analysis, food security indexation and regression modeling. The study was undertaken in 2009, soon after the country adopted the multi-currency regime.

Descriptive statistics

Were used to describe the differences between UA-practicing and non-participating households. Cross tabulations were used to determine the interaction of socio-economic characteristics with UA participation. Descriptive statistics were also used to explore linkages between urban household food security and UA participation.

Determination of household food security status

According to FAO (2011) the energy requirement is 2100 kcal per person per day. Food security index was calculated by dividing household energy consumed per month by total requirements; households failing below 70% were deemed food-insecure. These Index values were used to explore food security differences between UA-participating and non-UA-participating households.

Regression analysis

Logit model examined the factors affecting UA participation. The study assumed the following model:

$$Y_0 = \alpha + \beta_0 X_1 + \beta_1 X_2 + \beta_2 X_3 + \beta_3 X_4 + \beta_4 X_5 + \beta_5 X_6 + \beta_6 X_7 + \beta_7 X_8 + \beta_8 X_9 + E$$

Table 1. Household demography characteristics and urban agriculture participation.

Demography characteristic	UA-practicing households	Non-UA-households	Z value
Household size	4.62 (1.66)	3.61 (1.82)	0.62**
Average age	41.41 (14.14)	42.53 (14)	0.65**
	UA participating households (%)	**Non-UA participating households (%)**	**Chi value**
HH Sex			
Female	33.62	54.05	6.65
Male	66.38	45.95	
HH employment status			
Formal employed	62	32	
Informal employed	38	68	

Source: Survey data.

Where: α, $\beta_0,\ldots,\ \beta_8$ are coefficients, and E=error term; X_1=Household head (HH) sex; X_2=HH age years; X_3=HH educational status; X_4=Household size; X_5=HH employment status; X_6=Total household income; X_7=Rural homestead; X_8=Informal activities, and X_9=Residential status.

Two minimum least squares regression model were used to examine the factors affecting urban household food security:

$$Y_1;\ Y_2= \alpha+\beta_0X_1+\beta_1X_2+\beta_2X_3+\beta_3X_4+\beta_4\ X_5+\beta_5X_6+\beta_6X_7+\beta_7X_8+\beta_8X_9+\beta_9X_{10}+E$$

Where Y_1= household food expenditure; Y_2= household energy adequacy; α, β_0–β_{10} are coefficients; E=error term; X_1=HH sex; X_2=HH age; X_3=HH educational status; X_4= HH size; X_5=HH employment status; X_6=Total HH income; X_7=Rural homestead ownership; X_8=Informal activities; X_9=Residential status; X_{10}=UA participation, and X_{11}=Maize meal price.

RESULTS AND DISCUSSION

Demographic and endowment characteristics

Despite respondents hailing from the same socioeconomic environment, heterogeneous traits between UA farming and non-farming households were noted in household head age, sex, employment status; and household size. UA practicing households were headed by slightly younger heads and also households with more members. It can be postulated that the motive behind the UA participation is to feed the large household. Therefore, comparatively more households with more members are expected to venture into UA activities. Older household members may shun UA because of its laborious nature, and these households may not require UA as a safety net (Table 1).

An association exists between HH sex and UA participation. UA participation is dominated by male-headed households. Households headed by the formally employed stand a better chance of getting UA plots. Low returns accruing from UA resulted in the unemployed households shunning it. Hyperinflation that eroded the formally employed incomes resulted in households participating in UA as a way to augment incomes.

Factors affecting household participation in urban agriculture

In order to better understand household decisions and insights into household factors significantly influencing UA participation, a logistic model was run. The decision to farm and the level of effort spent on UA are affected by household factors such as educational status, household size, and household head sex and age. The R^2 value of 0.618 implies a degree of weak relationship between the independent variables and the dependent variable (Table 2).

Household size is positively related to UA participation. Increase in household size increases vulnerability and leads to households venturing into UA as a coping mechanism.

Household head sex is positively related to UA participation. The haphazard manner of plot allocation, and conflicts inherent in UA plots, favor male participation as compared to females.

Household age is negatively related to UA participation, contrary to expectations. The life cycle hypothesis postulates that older households, after accumulating wealth, will be better cushioned against vulnerability, as they have more assets than younger ones.

Household head educational status is negatively related to UA participation. Well educated households are less likely to venture into UA, as they would be better paid at their workplace and will not require any income augmenting.

Urban agriculture participation

Multi-question interviews with urban farmers revealed

Table 2. Urban agriculture practice logistic regression model.

Urban agric participation	Coefficient	Standard error	P>|z|	Exp(α)
Residential status [+]	-0.48	0.83	0.56	0.62
Household size [+]	1.38	0.52	0.01***	0.92
HH sex [+]	0.10	0.31	0.93**	0.77
HH age [+]	-0.08	0.05	0.09*	3.97
Educational status [-]	-0.70	0.33	0.03**	0.49
Employment status [-]	-0.10	0.37	0.79	0.92
Informal activities [-]	-0.26	0.24	0.29	1.10
Rural home ownership [-]	1.14	1.00	0.26	3.13
Income range [-]	-0.12	0.33	0.73	0.89
Remittance	-0.36	0.38	0.31	0.68
Constant	3.08	3.61	0.39	21.72

[ ], A priori expected signs; ***, **, *significance at 1, 5 and 10% respectively. Source: Survey data.

that push factors into UA include cultural, economic and food security incentives. Ranked in order of importance, survey respondents gave the following reasons for engaging in UA:

1. Production for home consumption (96.2%),
2. Food shortages (74%),
3. Income enhancement (61.5%),
4. Hobby or tradition (37.5%),
5. Supplementary employment (9.6%).

The top three motivating factors for UA engagement are economic. A household's perception of food insecurity risk will affect its farming effort because of the insurance value of own food production (Seeth et al., 1998). Food insecurity, or the perceived risk of it, pushed 96.2% of the respondents into UA for production of food for home consumption so as enhance household food supplies. The food shortages prevalent in 2008 caused 74% of the respondents to venture and or intensify UA activities to alleviate the food shortages. About 61.5% highlighted that UA was a form of income-enhancing activity. Vegetables produced would be sold in the markets as well as surplus grain and other products yielding direct income.

UA yields both direct income through sales and indirect income through reduction of expenditures on food. UA offers direct and indirect employment opportunities in Gweru, highlighted by 9.6% of the respondents. During the summer, UA acts as a form of short-term employment.

Farmers migrating from rural areas would want to continue with their farming practices, and hence would look for UA plots. There has been an influx of the rich, who views UA as a hobby. These accounted for 37.5% of the respondents who highlighted that UA practicing was taken as a hobby.

The major crop produced was maize, cultivated by 99.3% of the respondents; this validates the notion that engagement of UA is mostly for food security reasons. This was followed by sweet potatoes, beans, groundnuts, round nuts, vegetables and cowpeas, at 72.1, 62.5, 14.4, 1.9, 0.9 and 4.8%, respectively. The crops produced are mostly for food and nutrition security, though a surplus could be sold to generate income.

Urban household food security

Urban household food security is a contentious issue in the endeavor to reduce food insecurity. Of importance is the question: how significantly does UA produce contribute to the household food basket?

Time family food runs dry and coping or adaptive strategies adopted

During the heist of food shortages in 2008, households were affected by food run-outs. Table 5 shows that non-UA participating households were mostly affected by food run-outs during the hyperinflationary era.

More non-UA practising households had food running out at the start of the month, middle of the month and month end as compared to UA-participating households, whilst more UA participating households never ran out of food during the hyper-inflationary era (Table 3). When households were affected by food shortages, they adopted strategies to see them through the month.

In the aftermath of rampant food shortages that bedevilled the country in 2008, 50% of the households borrowed food, 43.3% opted for less preferred foodstuffs, 45.3% reduced their number of meals, and 46% reduced meal quantity and frequency during times when food ran out. Approximately 22.7% of households reported that they would sometimes spend the night without eating,

Table 3. Household food shortage and urban agriculture participation.

Time food ran out	UA participating households (%)	Non-UA participating households (%)	Critical value	Chi value
Start of the month	26	30	7.81	9.42
Middle of the month	28	28		
Month-end	24	26		
Never	22	15		

Source: Survey data.

Table 4. Coping strategies when food runs dry.

Strategy	Urban agriculture participating households (%)	Non-urban agriculture participating households (%)
Opting for less preferred food	38.5	54.4
Reducing quantity consumed	36.5	65.2
Reducing number of meals	38.5	41.3
Borrowing food	45.2	60.9
Selling assets	33.7	21.7
Spending night without eating	21.2	26.1

Source: Survey data.

Table 5. Expenditure by income category.

Parameter	Income category		
	Lower 20%	Middle 60%	Upper 20%
Food proportion	0.5	0.48	0.46
Per-capita food consumption	29.1	26.59	23.4
Total non-food consumption	70.64	108.76	147.46

Source: Survey data.

whilst 30% liquidated assets to purchase food. Table 4 summarizes the coping strategies that were adopted by respondents when they were faced with food shortages.

Expenditure approach

Expenditure responses showed the proportion of income devoted to food; this is affected by household wealth, employment status, residential status and sex. A priori the proportion of food expenditure and per-capita food consumption would decrease as income increases, whilst non-food expenditure would increase with rise in income. Table 5 shows expenditure profiles according to income category.

This is consistent with the Engelian relationship between income and the proportion of income allocated to food. As income increases, the percentage of the budget allocation to food falls whilst the converse is true. The respondents in the study area purchased a variety of food commodities: staples, luxuries and inferior commodities. Foodstuffs that were classified as essential included maize grain, maize meal, flour, rice, potatoes, beans and beef. The proportion of households purchasing beef and rice increased with income levels. For the inferior foods chunks and kapenta, purchase decreased with increase in income. As income increases, the consumption of less desirable commodities decreases, as shown by chunks and kapenta. Table 6 shows the consumption of foodstuffs according to wealth category.

Table 7 shows expenditures of foods and non-foods according to UA participation. The mean food expenditure for UA participating households was significantly lower than for non-UA participating households as well as proportion of food purchased. Households engaged in UA do not purchase grain and other products such as vegetables and pulses. The mean percentage food expenditure for non-UA-practicing households is greater that of UA-practicing households,

Table 6. Food purchases according to wealth category.

Food items	Lower 20% (%)	Middle 60% (%)	Upper 20% (%)	Chi value	Critical value
Beef	73	76	87	45.06	18.3
Rice	53	58	100	35.80	23.7
Chunks (soyacake)	63.3	63.3	30.0	27.79	26.3
Kapenta	76.7	67.8	53.3	33.00	32.7

Source: Survey data.

Table 7. Expenditure profiles.

Households	Total food expenditure	Percentage expenditure on food	Non-food expenditure
UA participating households	$93.52	47.11	111.43
Non-UA participating households	$95.91	50.13	109.84

Values with** are significant at 5%; Source: survey data.

Table 8. Factors affecting household food expenditure.

Parameter	Coefficient	R^2	F	P
		0.5959	5.20	0
		Std. Error	t	P>|t|
UA participation [-]	-0.04*	6.33	-1.2	0.08
Residential status [+]	-6.76	6.27	-1.08	0.28
Educational level [+]	4.04	1.99	2.03	0. 44
Employment status [+]	0.51	2.84	0.18	0.86
Household size [+]	5.62***	2.03	2.77	0.01
HH age [+]	0.31	0.27	1.14	0.26
Rural home ownership [-]	-0.65	6.25	-0.1	0.92
Income [+]	4.52**	2.14	2.12	0.04
HH sex [+]	-9.94*	5.96	-1.67	0.09
Informal activity [+]	5.67	8.28	0.68	0.49
Maize meal price [+]	1.99***	0.68	2.95	0.01
Constant	43.43	22.50	1.93	0.06

***, ** and * indicate statistical significance at the 1, 5 and 10% levels respectively, [ ] indicate *a priori* expectations.

mainly because UA-practicing households purchase less grain, maize meal and vegetables, resulting in low food basket costs.

The mean non-food expenditures for UA-participating household are significantly greater than those of the non-UA-participating households. UA-practicing households purchased more non-food items as compared to non-UA practicing households; mainly because money saved from not purchasing grain and maize meal was spent on non-foodstuffs.

Factors affecting household food expenditure

Household food expenditure is affected by both social and economic factors. UA participation, household size, total household income, household head sex and maize meal price are shown to significantly affect household food expenditure (Table 8). The R^2 value shows that 60% of the variation in household food expenditure is explained by the model.

UA participation is significant at the 1% level, and an increase in UA participation lowers food expenditure. UA produce-mainly maize grain-results in reduced maize meal purchase and inevitably household food expenditure.

Male-headed households are likely to be food-secure than their female counterparts. This is because female-headed households are normally unemployed, and as such they are deprived of the much-needed finances to purchase adequate food as compared to male-headed households.

Table 9. Household own food security assessment.

Food status	UA-practising households (%)	Non-UA-practising Households (%)	Critical value	Chi-value
Food insecure	2	7		
Sometimes food insecure	44	54	5.99	7.017
Food secure	54	39		

Source: Survey data.

As the household size increases, the household food requirement also increases, as well as household food expenditure. Smaller households have less food expenditure than large households.

Household educational status increases employment opportunities as well as the adoption of new technologies, which includes consumption of new foodstuffs and as such changes in preferences. Usually these new foodstuffs cost more, and as such this increases the food expenditure of the household.

Income increase by the household will also result in increased food expenditure. As income increases, household food expenditure also increases. This stems from the fact that maize meal is one of the major components of the food basket, and, with an increment in its price, the household food expenditure will increase.

Caloric and energy requirement

Before calculating the caloric food intake, households were asked to judge their own food security status. This assessment reviews what households egoistically[1] felt their food security status to be. At the time of the survey, 56.7% of the households felt that they were food-secure, whilst 40% and 3.3% felt that they were sometimes food insecure and food insecure, respectively.

Participation in UA gives sovereignty to households, hence the likelihood of been food security increases with participating in UA activities (Table 9).

The foods consumed by the households were converted into calories using the UNHCR 1996 caloric requirements per day to obtain the monthly household energy requirements. UA participating households consumed more energy than non-UA participating households. The caloric and energy requirements showed that 47% of the households were food-secure whilst 53% of the households were food-insecure, with 52.9% of the UA-participating households being food-secure, compared to 47% from the non-UA-participating households. This is consistent with earlier assertions about household own assessment. Household food security was enhanced by UA participation.

Factors affecting urban household food security

To determine the factors that affect household food security, a regression model of caloric index[2] was run. The R^2 value was 0.6019, whilst the P value shows that the equation is significant at the 1% level (Table 10). The estimated model shows that, household head sex and age, UA participation, household size, and informal activity significantly affected urban household food security.

There is a positive relationship between UA participation and food security, significant at 5%. This is expected as UA produce fosters food base resulting in increased availability of grain to the household and as such improves the food security of the households. UA produce also lowers food costs. By not purchasing maize meal, money can be channeled to other foodstuffs, improving the food base and food security status of the household.

Household head sex is significant at 5% and is positively related to the food security. Male headed households during the heist of food shortages fared better as they could use their muscular power in queues to get foodstuffs.

Household size is significant at 1%, and is negatively related to the food security. Higher household size results in reduction in per-capita food consumption increased household food dependency greatly compromising food consumption. This is as expected, since the larger the household the more vulnerable it is to food insecurity.

Household head age is positively related to the food security status of the households, and is significant at the 10% level, consistent with the life cycle hypothesis. The life cycle hypothesis assumes that assets are accrued as one advances in age, and that these assets can be liquidated during times of income constraint. During the time of the survey, the country was experiencing hyperinflationary conditions, and, as such, the workers' income was greatly eroded and households failed to

[1]Ego of the respondent, which can be ego positive or ego negative. Ego positive respondents tend to overestimate their food security status, whilst ego negative respondents would underestimate their food security status.

[2] The caloric index was calculated by the calories consumed divided by the calories required and this was converted into a percentage

Table 10. Factors affecting household food security.

Variable	R^2	F	P	P>|t|
	0.6019	3.7071	0.0005	
	Coefficient	Std. error	t	
Rural homestead ownership [+]	0.14	10.84	0.01	0.99
HH sex [+] (0 female, 1 male)	0.18**	9.71	1.8	0.08
HH education[+]	1.62	3.53	0.46	0.65
Income range	1.12	3.40	0.33	0.74
UA [+]	0. 30**	0.14	2.13	0.04
Household size [-]	-1.24***	3.58	-3.47	0.00
HH age [+]	1.10*	0.56	1.95	0.06
Informal activity [+]	0.32**	3.21	-0.99	0.33
Maize meal price [-]	2.06	0.70	2.95	0.01
HH employment[+]	-0.52	4.20	-0.12	0.90
Residential status [±]	11.16	10.91	1.02	0.3
Constant	6.69	39.59	-0.21	0.83

***, ** and * indicate statistical significance at the 1, 5 and 10% levels respectively; [ ], *a priori* expectations. Source: Survey Data.

secure adequate food. Households endowed with assets could liquidate these so as to purchase or even import foodstuffs.

There is a positive relationship between food security and informal activity participation. This is consistent with expectations as participation of informal activities increases income, which can be used to purchase foodstuffs. During hyperinflation, foodstuffs were very expensive, and, by participating in many informal activities.

Conclusions

Household size as well as household head sex, age, and employment status were found to affect household participation in (UA) in Gweru city, Zimbabwe. UA participation is dominated by bigger families, male-headed, formally employed and relatively young-headed households. The risk of food insecurity drove people into UA activities, as shown by pull factors into UA participation, such as meeting household food consumption, food shortage, income enhancement, hobby and supplementary employment. Major UA products were predominantly food crops, showing the importance of UA in food provision. Household size, UA participation, income, household head sex, and maize meal price affected household food expenditure. Households participating in UA were found to have lower food costs whilst the non food costs were higher. Household food security was affected by household head sex and age, UA participation, and informal activities carried out by the household members. UA has a positive impact on urban household food security as it provides food as well as income through money saved from buying food, hence raising the standard of living.

REFERENCES

AGRITEX (2008). Second round crop and livestock assessment report. Harare. Zimbabwe. Central Statistic Office (2008). Report on Prices in Zimbabwe, Harare, Zimbabwe.

FAO (2001). Nutrition Country Profile for Zimbabwe, ESNA, FAO. Accessed on line on the 16th of March 2009. ftp://ftp.fao.org/es/esn/nutrition/urban/stellenbosch.pdf.

Harare Declaration (2003). The Harare Declaration on Urban and Peri-Urban Agriculture in Eastern and Southern Africa. Ministerial Conference on Urban Agriculture Held on the 28th and 29th of August, 2003. Victoria Falls, Zimbabwe.

Hovorka A, De Zeeuw H, Njenga M (2009). Women Feeding Cities. Mainstreaming Gender in Agriculture and Food Security. Practical Action Publishing, UK.

Mbida B (1995). Urban Agriculture in Zimbabwe, Implications for Urban Management and Poverty. Avebury. Aldershot.

Rabinowicz J (2002). Urban Food Security and the Potential for Urban Agriculture. Accessed on line on the 12th of June 2008. www.g24.org/ukeje.pdf.

Resource Center for Urban Agriculture Foundation (RUAF) (2009). Importance of Urban Agriculture. Accessed On Line On The 12th February 2009. http://www.ruaf.org/node/512.

Seeth H, Chachnov S, Surinov A, Von Braun J (1998). Russian Poverty: Muddling Through Economic Transition with Garden Plots. World Dev. 26(9).

STERP (2009). Getting Zimbabwe Moving Again. Presented By The Honorable Minister Of Finance March 2009. Harare, Zimbabwe.

ZIMVAC (Zimbabwe Vulnerable Assessment Committee) (2008). Zimbabwe Food Shortages. Accessed Online on 20 May 2008. http://www.afrol.com/articles/14473.

Consumer demand system of agri-food in Tanzania

Joel Chongela, Vasudev Nandala and Suhasini Korabandi

Department of Agricultural Economics ANGRAU[†], †ANGRAU - Acharya N. G. Ranga Agricultural University, Hyderabad 500-030, Andhra Pradesh, India.

This paper focused on consumer demand system of agri-food in Tanzania. The study investigated on the budget share of agri-food, estimated own price and income elasticities of demand of agri-food consumed by the households in Tanzanian context. The Almost Ideal Demand System (AIDS) was employed to estimate the consumer demand system of agri-food. The empirical results revealed that households' food budget share was 60% on average per month. The own price elasticity of demand for aggregated agri-food was inelastic (0.86). Income elasticity of demand for aggregated agri-food was 0.96. Moreover, inelastic demand of own price elasticity of agri-food products informed the government intervention in terms of agri-food price stabilization policies and programmes. Furthermore, income elasticity of < 1 suggested that interventions of Tanzanian government in terms of 'income support policies' can be done by increasing the incomes of lower and middle class income earners in rural and urban areas as consumers' economic stimulus package to increase normal food consumption. In rural areas where majority of farmers dwell, this can be done by promoting the price of farm products through minimum support price so that farmers can earn much more incomes from their economic activities as a result they can use it for purchasing normal food from the business sector.

Key words: Almost ideal demand system (AIDS), agri-food, consumer demand system, income and own price elasticities.

INTRODUCTION

Consumer demand system is the aggregate desire for goods and services expressed by all consumers in the economy. The essential element to consumer demand consists of desire for consumption, with the second condition of ability to pay to such kind of consumption. The desire for consumption must contain the willingness to consume at the prices demanded for that consumption, this establishing the price schedule for goods and services. The willingness to consume diminishes as the price of such consumption increases, on the other hand, willingness to consume increases as price reduces. The consumer demand system of agri-food in Tanzania is characterized by low purchasing power, and most consumers have a strong preference for traditional foods that are suited to their diverse social and ethnic backgrounds (Makweba, 2009). Despite this, the demand for processed and quality fresh food is growing, especially due to the recent increase in the middle class across the country. Both formal and informal food markets are changing constantly, driven by lifestyle changes brought about by urbanisation, income growth and changing of family structures. However, the transformation of agri-food markets in Tanzania has been characterised by different extents of supermarketisation, especially in the urban areas.

Panagiotis et al. (2011) estimated censored linear almost

almost ideal demand system of food in Pakistan, they revealed that all own-price elasticities but one are found to be negative and all total food expenditure elasticities are found to be positive. However, they suggested that the estimated elasticities can be used for policy analysis for international organizations and the national government to monitor the food security situation and to develop sound and proper intervention policies to mitigate the negative shock impact on food consumption and under nourishment. Taljaard et al. (2006) analysed a meat consumer demand in South Africa using Linearized Almost Ideal Demand System and Rotterdam models, they revealed that the Linearized Almost Ideal Demand System proved to fit better than Rotterdam model in South Africa meat demand market because of its flexibility to any form of utility function. Katchova and Chern (2004) examined a non-nested comparison between Quadratic Expenditure System (QES) and almost ideal demand system (AIDS) based on empirical data of food demand structure in China. They found that AIDS is the suitable model than QES for estimation of consumer demand system because of its unrestricting to any form of utility function.

The present study is focused on investigation of food budget share consumed by the households, to estimate own price and income elasticities of demand of agri-food consumed by the households in Tanzanian context.

METHODOLOGY

Types of data

Secondary data on 'agri-food consumer demand system' were collected from household budget survey (HBS) conducted in 2007 by the National Bureau of Statistics (NBS) of the United Republic of Tanzania.

Sampling technique

Households were selected from the National Master Sample (NMS) which is nation-wide covering both rural and urban areas. The sample size of 4680 households were selected by systematic sampling method, from 10466 households involved in HBS of 2007 for 21 regions included in the HBS namely; Dar-es-salaam, Arusha, Dodoma, Iringa, Mbeya, Morogoro, Kilimanjaro, Mwanza, Tabora, Tanga, Kagera, Pwani, Kigoma, Lindi, Mtwara, Mara, Shinyanga, Singida, Ruvuma, Rukwa and Manyara.

Agri-food categories

The 158 food items consumed by the households were categorized into 18 groups. Namely; cereals (paddy, rice, green maize cob, maize grain, maize flour, millet grain, millet flour, sorghum grain, sorghum flour, wheat grain, wheat flour, barley and other cereals). Cereal products (bread, biscuits, buns, cakes, chapatti, macaroni, spaghetti, cooking oats, and other cereal products). Roots and starches (cassava fresh, cassava dry, cassava flour, round potatoes, sweet potatoes, yams, cocoyam, cooking bananas / plantains and other starches). Sugar and sweets (sugar, honey, syrup, jams, ice cream, chocolate and sweets). Pulses (dry peas, green peas shelled or in pods, dry beans, green beans shelled or in pods, lentils, red grams, green grams, and other pulses). Pulse products (bagia). Nuts (groundnuts in shell, groundnuts shelled, coconuts, cashew nuts, and other nuts). Vegetables (carrots, radishes, beets, turnips, garlic, onions, spinach, lettuce, cabbage, other leafy vegetables, tomatoes, bitter tomatoes, ladies fingers / okra, cauliflower, cucumber, pumpkins, brinjals/eggplant, fresh green pepper, other cultivated vegetables, other wild vegetables, dried vegetables and canned vegetables). Fruits (bananas, oranges, limes, mangoes, avocado, pawpaw, sugar canes, apples, pears, other wild fruits). Meat and poultry products (mutton, beef, mince sausages, pork, pork sausages, goat meat, bacon, other domesticated animals, wild animals, offal (liver, kidney), dried or salted meat, canned meat, and other meat products, chicken, eggs and other poultry.

Milk and dairy products (fresh milk, yoghurt, cream, cheese, canned milk and powder milk). Fish and shellfish (fresh fish, shell fish, fresh dried fish, dried or salted fish/ shellfish fillets, and canned fish / shellfish). Oil seeds and fats (cottonseed oil, groundnuts oils, sesame oil, sunflower oil, olive oil, butter, margarines cooking fat, and other cooking oil). Spices and other foodstuffs (red / black peppers, curry powder and other spices as well as salt, yeast, baking powder, and other foodstuffs). Raw materials for drinks (tea powder, coffee powder and cocoa powder). Soft drinks (coca-cola, fanta, pepsi, mirinda, juice, mineral water), tea, coffee, cocoa and other beverages. Alcoholic drinks (Kilimanjaro beer, Safari beer, Guinness beer, Castle beer, Ndovu beer, Serengeti beer, other canned / bottled beer, chibuku and other local brews). Tobacco-cigarettes (sportsman, sweet menthol, nyota, snuff, pipe tobacco and other cigarettes). The categorizations of the food items were done to rule out the substitutability and complementary effects in the consumer demand system model.

Model specification

The almost ideal demand system (AIDS)

The almost ideal demand system (AIDS) proposed by Deaton and Muellbauer (1980), was employed to study the 'consumer demand system' of agri-food in Tanzanian context. It can be depicted as:

$$\omega_h = \alpha + \Sigma\gamma In P_h + \beta In\left(\frac{M_h}{P_h^*}\right) \quad (1)$$

Where: ω_h = Food budget share of the household, α = food budget shares parameter when all prices and real expenditure are equal to one (intercept), γ = price parameter, P_h = prices of food commodities consumed at the household, β = expenditure share parameter, M_h = total expenditure share of the household on all commodities (real income), *In* = natural logarithm, P^*_h = price index. Pr ice index can be defined as:

$$In P_h^* = \Sigma\omega_h In P_h. \quad (2)$$

The α parameter is the average food budget share when all prices and real expenditure are equal to one. The β parameter measures the change in the *i* th budget share with respect to a change in real income, all else factors held constant, and indicate whether goods are necessities or luxuries. If $\beta i < 0$, ωi decreases when real income *(Mh)* increases so that good *i* is a necessity. Conversely, if $\beta i > 0$, ωi increases with real income *(Mh)* so that good *i* is a luxury. The γ parameter measures the change in the *i* th budget share for a unit change in price *(Ph)* with real income held constant. The private household was represented as a single consumer. The rational consumer always aims to maximize utility subject to budget

Table 1. Parameter estimates of food consumer demand.

Variable	Adj R^2	Parameter	Parameter estimates	Standard error	t -Value	Pr	Food budget mean
Food budget share		α	0.03	0.03	1.00	0.3179	0.60
Food price	0.39	γ	0.07*	0.003	23.45	<.0001	
Real income		β	-0.03*	0.0007	-37.43	<.0001	

* Implies significant at 5% level of significance.

constraint for given prices and initial endowments.

Estimation of price and income elasticities of demand

The own price and income elasticities of agri-foods consumer demand system were calculated as:

$$\varepsilon_{ii} = -1 + \frac{\gamma_i}{\omega_h} - \beta i \quad (3)$$

$$\eta_i = 1 + \frac{\beta_i}{\omega_h} \quad (4)$$

Where: ε_{ii} = Own price elasticity of demand, ω_h = household food budget share, γ_i = price parameter, β_i = expenditure share parameter, η_i = income elasticity of demand.

Parameterization of the almost ideal demand system (AIDS)

The demand elasticities for the linearized AIDS model were computed on the basis of estimated parameters of its demand function. A stochastic element was introduced in the demand function to obtain the following equation model:

$$\omega_h = \alpha + \sum \gamma In P_h + \beta In\left(\frac{M_h}{P_h^*}\right) + \mu_h \quad (5)$$

Where: μ_h = Stochastic error term.

RESULTS

Estimation of almost ideal demand system (AIDS)

The empirical results for estimation of 'almost ideal demand system' are presented in Table 1. The results revealed that 39% of proportion of variation in food budget share of the household is explained by proportion of variation in food prices and real income jointly (Table 1). The empirical results for parameter estimates of the 'almost ideal demand system' of agri-food are presented in Table 1. The empirical results found that if food prices spike by 1%, the food budget share would increase by 7% under ceteris paribus assumption. On the other hand, if real income increases by one percent, the food budget share is expected to shrink by 3% (Table 1). This has been influenced by limits to the extra money people spend on food when their incomes rise. Consequently, the proportion of total spending devoted to food declines as income increases (Samuelson and Nordhaus 2008).

Food budget share

The households' food budget share for disaggregated food products are presented in Figure 1. The results revealed that cereals are the leading food budget share (26.89%), followed by vegetables (10.05%), meat and poultry products (9.69%), roots and starch (8.41%), pulses (6.99%), sugar and sweets (6.52%), fats and oil seeds (5.39%), soft drinks (5.26%), cereal products (5.04%), alcoholic drinks (3.79%), fish and shellfish (2.67%), milk and dairy products (2.31%), cigarettes (1.19%), spices and other food stuffs (0.98%), raw materials for drinks (0.78%), nuts (0.38%), fruits (0.29%) and pulse products (0.10%) (Figure 1).

Estimation of own price and income elasticities of demand for aggregated food groups

The empirical results of own price and income elasticities of demand for aggregated food groups are presented in Table 2. The own price elasticity of demand is negative 0.86 whereas income elasticity of demand is 0.96.

Estimation of own price and income elasticities of demand for disaggregated food groups

The empirical results of own price and income elasticities of demand for disaggregated food groups are presented in Table 3.

DISCUSSION

Parameter estimates

The implication of the empirical results is that 39% of

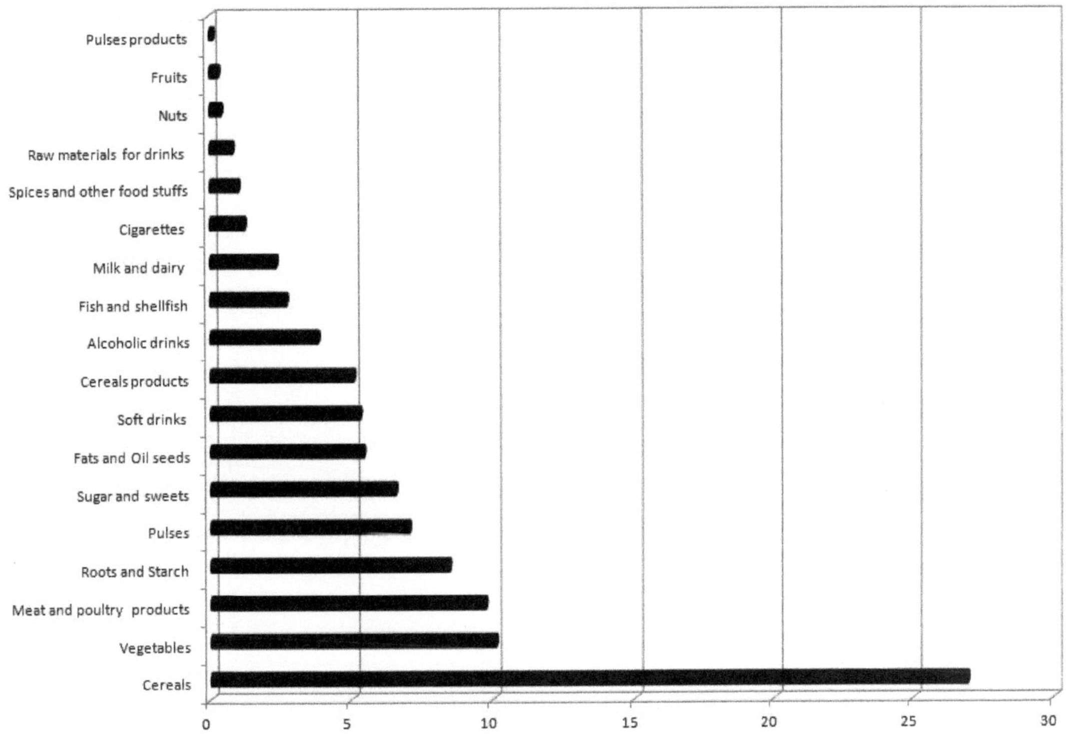

Figure 1. Tanzania Mainland: Budget share of food products for Households Budget Survey (HBS), 2007 (%).

Table 2. Own price and income elasticities of demand of aggregated food groups.

Own price elasticity	Income elasticity
- 0.86	0.96

proportion of variation in the household food budget share is driven by food prices and real income under *ceteris paribus* (Table 1). However, the adjusted R^2 is the goodness of fit of the econometric model measures the proportion of variation in the households' food budget share accounted for by the food prices and the real incomes jointly. Under *ceteris paribus*, empirical results revealed that there was a robust positive relationship between food prices and food budget share. If food prices spike by 1%, the food budget share is expected to increase by 7% because as food price spikes, consumers used to allocate much more money on food to withstand the robustness of food inflation. On the other hand, empirical findings revealed that if real income hikes by 1%, the food budget share is expected to decline by 3% because as income increases, consumers used to shift to luxurious goods consumption. Furthermore, the empirical results confirmed that food prices and real incomes are primary determinants of food budget share of the households at probability value of less than 0.0001; hence, food prices and real incomes are significant determinants of food budget share at 5% level of significance. If food prices and real incomes are minimal, the food budget share is expected to increase by 3% under ceteris assumptions (Table 1).

Food budget share

The empirical results found that households' food budget share is 60% on average for households' budget survey of 2007 (Table 1). The households' food budget share of 2007 declines by 5% from 65% of the 2000/2001 food budget share and shrink by 11% from 71% of the 1991/1992 food budget share. The implication of the findings is that household's budget devoted to food declines as the incomes of households rise. Similar findings have been reported by National Bureau of Statistics (2001) of the United Republic of Tanzania for households' budget surveys of 1991/1992 and 2000/2001. It reported that households' food budget shares for HBS 1991/1992 and 2000/2001 were 71 and 65% on average respectively which were consistent to incomes increase. However, households' budget devoted to food was expected to decline as incomes of households rose. The cereals are the leading food budget share (26.89%) because these are the main food staples consumed by the majority of food consumers in the country. The major consumed cereals include maize grain, green maize cob, maize flour, paddy, rice, sorghum grain, sorghum flour, wheat grain, wheat flour, millet grain,

Table 3. Tanzania mainland: Own price and income elasticities of demand of disaggregated food groups for households budget survey (HBS), 2007.

Food product	Own price elasticity (ε)	Income elasticity (η)
Cereals	-0.941	0.961
Cereal products	-0.786	0.946
Pulses	-0.956	0.987
Pulse products	-0.909	0.981
Fruits	-0.855	0.925
Vegetables	-0.971	0.982
Sugar and sweets	- 0.899	1.026
Roots and Starch	-0.828	0.988
Fats and oil seeds	- 0.802	0.967
Meat and poultry	- 0.990	1.070
Spices and other food stuffs	- 0.835	0.944
Milk and dairy products	- 0. 646	0.701
Tea, cocoa, coffee powders	- 0.524	0.844
Fish and shellfish	-0. 699	0.871
Soft drinks	- 0.823	1.028
Alcoholic drinks	-0.845	1.066
Cigarettes	- 0.985	1.031
Nuts	- 0.833	0.874

millet flour, barley and other cereals. The main consumed vegetables are tomatoes, other leafy vegetables, onions, pumpkin and cabbage. The major consumed meat and poultry products are beef, chicken, goat meat, pork meat, eggs, and other poultry (Figure 1).

The main consumed roots and starch are cassava flour, cooking banana / plantains, round potatoes, yam, cocoyam and sweet potatoes. The major consumed pulses are dry beans, broad beans, soya beans, peas, lentils and other pulses. The main consumed sugar and sweets are brown sugar, white sugar and sweets. The main consumed fats and oil seeds in the country are margarine, sunflower, groundnut, sesame, and other cooking oil. The major consumed soft drinks are mineral spring water, coca-cola, fanta, pepsi, tea without milk, tea with milk, fruit juice, mirinda and other soft drinks (Figure 1). Furthermore, the main consumed cereal products are bread, buns, chapatti and biscuits. The leading consumed alcoholic drinks are Safari beer, Kilimanjaro beer, Castle beer, Serengeti beer and local brew (chibuku and others). The major consumed fish and shellfish are dried sardines, fresh /chilled /frozen fish, dried / salted shellfish and dried fish. The main consumed milk and dairy products are fresh cow milk and yoghurt. The leading consumed cigarettes are sweet menthol, sportsman and nyota (Figure 1). The major consumed spices and other food stuffs are salt, other spices (masala), red and black pepper, other food stuffs, yeast and baking powder. The tea, coffee and cocoa powders are majorly consumed as raw materials for drinks. The groundnuts, cashew nuts, and coconuts are the main nuts consumed in the country. The major consumed fruits are pawpaw, orange, banana, mango and other wild fruits (Figure 1).

Estimation of own price elasticity of demand for aggregated food groups

In Table 2, empirical results revealed that the own price elasticity of demand is (-0.86), falls between negative 1 and zero ($-1 < \varepsilon p < 0$) which means that it is inelastic demand. The implication of inelastic demand is that food is a necessity good for life because as food price spikes the consumer expenditure share on food spikes as well, despite a decrease in consumers demand due to the fact that change in quantity demanded is smaller than change in price. The own price elasticity falls under demand curve which is negatively sloped (law of demand) implies that as price of food increases, consumer demand decreases due to increase in consumers expenditure share on food. The similar findings have been reported by Henderson and Quandt (2003) and Sadoulet and de Janvry (1995), they reported that as the price of a good increases, the consumer demand tend to shrink due to increase in consumers expenditure share on the particular good.

Estimation of income elasticity of demand for aggregated food groups

The empirical findings for income elasticity of demand for aggregated food groups are presented in Table 2. The empirical results found that income elasticity of demand for aggregated food is 0.96, implies that food is a necessity good for life. Also, this result implies that as income increases consumers used to consume normal goods by shifting from inferior goods. The similar results have been reported by Annabi et al. (2006) for functional forms and parameterization of computable general equilibrium (CGE) models, they pointed out that increase in incomes of the households reduced the demand for inferior goods due to change of consumer preferences from inferior to normal and luxury goods. Samuelson and Nordhaus (2008) reported on the similar results that there are however, limits to the extra money people will spend on food when their incomes rise. Consequently, the proportional of total spending devoted to food declines as income increases.

Estimation of own price elasticities of demand for disaggregated food groups

The empirical findings revealed that own price elasticities of demand for agri-foods were inelastic ($-1 < \varepsilon p < 0$) which implies that as the agri-food price spikes, consumers expenditure share increases despite a decrease in consumer demand due to the fact that the proportionate change in quantity demanded is less than proportionate change in prices consumers are able to pay. Moreover, the own price elasticities of the agri-food is negative because the corresponding demand curve is downward sloping (law of demand curve) (Table 3).

Estimation of income elasticities of demand for disaggregated food groups

The income elasticities of demand for most of disaggregated food products were < 1, implies that food are normal goods which used to have income elasticity of < 1 meaning that as incomes of consumers increases the expenditure share devoted to food increases less than increase in incomes because foods are the necessity goods for life (Table 3). However, income elasticities of demand for sugar and sweets, meat and poultry, soft drinks, alcoholic drinks and cigarettes were > 1 attributed to increase in consumer demand more than proportionally increase in income; hence expenditure share of these foods increases higher than incomes increase (Table 3). T he similar results have been reported by Panagiotis et al. (2011) for food consumer demand system in Pakistan, Taljaard et al. (2006) for meat consumer demand system in South Africa, Sarntisart and Warr (2005) for food consumer demand system in Thailand, Katchova and Chern (2004) for food consumer demand system in China, and Ananda et al. (2003) for food consumer demand pattern in Tanzania.

They reported that food own price elascitictie of demand tend to be inelastic because as price spikes willingness to consume tend to decline due to increase in consumers expenditure share whereas income elasticities for food tend to be < 1 because as income hikes the expenditure share devoted to food used to decline despite the fact that food is a necessity good for life as compared to luxury goods which used to have elasticities of > 1.

CONCLUSION AND POLICY IMPLICATIONS

The food consumption priority is for cereals, followed by vegetables, meat and poultry products, roots and starch followed by pulses, sugar and sweets. Moreover, cereal products, fats and oils seeds, and soft drinks are the next major products consumed in the country. However, pulses products consumption is meagre, which implies that value added in pulses is less. Therefore, the priority sectors may be encouraged to have perfect market competition so that fair and free competition exist, so as consumers can get the necessity products at reasonable price.

The inelastic demand of own price elasticity for food products informs the government intervention on food price stabilization policies and programmes which can be achieved through subsidizing the prices of staple food which have higher household budget shares such as maize, rice, sorghum, cassava, pulses, sweet potatoes consumed by the majority of lower and middle class income earners in rural and urban areas in the country. The high expenditure elasticities of milk and dairy, cereals, cereal products, pulses, fish, fruits, vegetables, roots and starches, fats, and oil seeds, suggest that income support programmes are likely to be good policy tools to promote consumption of these staple food items among urban and rural household consumers. Furthermore, income support policies can be done through increasing the income of lower and middle class income earners in rural and urban areas as consumers' economic stimulus package to enable them to increase consumption of normal foods by shifting from inferior foods.

The highest expenditure elasticity also suggests that the consumer demand for meat and poultry, sugar and sweets, soft drinks, alcoholic drinks, and cigarettes, is likely to expand as the economy develops.

ACKNOWLEDGMENTS

Sincere appreciation to Dr. Albina Chuwa, Director of National Bureau of Statistics United Republic of Tanzania for availing statistical data of great value for the study. Also, sincere thanks to Mwalimu Nyerere, African Union Scholarship Scheme (MNAUSS) under Africa-India Fellowship Programme, operated by African Union Commission (AUC) in Addis Ababa, Ethiopia for their valuable financial support for my PhD programme in incredible India. Great thanks to Department of Agricultural

Economics, Acharya N. G. Ranga Agricultural University (ANGRAU) for their valuable academic support for the entire period of study. Valuable thanks to the advisory committee, Professor Vasudev Nandala, Professor Suhasini Korabandi, Professor Sreenivasa Rao and Associate Professor Sri Bhave M. H. V for their great advice for the present manuscript.

REFERENCES

Ananda W, Nyange D, Tsujii H (2003). Food Demand Patterns in Tanzania: A Censored Regression Analysis of Microdata. Sri Lankan J. Agric. Econ. 5(1):9-27.

Annabi N, Cockburn J, Decaluwe B (2006). Functional Forms and Parameterization of Computable General Equilibrium (CGE) Models. J. Econ. Lit. 27(4):3-44.

Deaton A, Muellbauer J (1980). Almost Ideal Demand System. Am. Econ. Rev. 70:312-326.

Henderson JM. Quandt RE (2003). Microeconomic Theory: A Mathematical Approach. McGraw-Hill. New York. pp. 22-24.

Katchova LA, Chern WS (2004). Comparison of Quadratic Expenditure System and Almost Ideal Demand System Based on Empirical Data. Inter. J. Appl. Econ. 1(1):55-64.

Makweba J (2009).Supply Chain Management and Challenges Facing Food Industry Sector in Tanzania. J. Inter. Bus. Manage. 4(12):70-80.

National Bureau of Statistics (2001). Households' Budget Surveys of 1991/1992 and 2000/01. National Bureau of Statistics, Dar es salaam, Tanzania.

Panagiotis K, Steven TY, Cheng F (2011). Bayesian Estimation of a Censored Linear Almost Ideal Demand System: Food Demand in Pakistan. Am. J. Agric. Econ. 93 (5):1374-1390.

Sadoulet E, de Janvry A (1995). Quantitative Development Policy Analysis. The John Hopkins University Press. Baltimore and London. pp. 1- 438.

Samuelson PA, Nordhaus WD (2008). Economics. Tata McGraw- Hill Publishing Company Limited, New Delhi. pp. 446-447.

Sarntisart I, Warr PG (2005). Estimates of Consumer Demand System in Thailand. Thammasat Econ. J. 1 (4):139-187.

Taljaard PR, Schalkwyk HD, Alemu ZG (2006). Choosing between the Almost Ideal Demand System and Rotterdam models: A Meat Demand Analysis Case Study. Agrekon. 45(2):158-172.

Analysis of determinants of farm-level adaptation measures to climate change in Southern Africa

Charles Nhemachena[1], Rashid Hassan[2] and James Chakwizira[3]

[1]Human Sciences Research Council, 134 Pretorius Street, Pretoria 0002, South Africa.
[2]Centre for Environmental Economics and Policy in Africa, University of Pretoria, Department of Agricultural Economics and Rural Development, Pretoria 0002, South Africa.
[3]University of Venda, School of Environmental Sciences, P/Bag X5050, Thohoyandou, 0950, South Africa.

The study examines farmer adaptation strategies to climate change in Southern Africa based on a cross-section database of three countries (South Africa, Zambia and Zimbabwe). A multivariate discrete choice model was used to analyse the determinants of farm-level adaptation strategies. Results confirm that access to credit, free extension services, awareness of climate change are critical determinants of farm-level adaptation options. Policies aimed at easing identified key limits to farmers' capacity to adapt to climate change need to emphasize the critical role of: extension services; provision of improved climate, production and market information as well as the means to implement adaptation through affordable credit and insurance against climate risks (safety nets).

Key words: Climate change, farm-level adaptation, Southern Africa.

....

INTRODUCTION

Climate change models for southern Africa indicate that the region will face increased challenges due to projected changes in climate (IPCC, 2007; Hulme et al., 2005). Further evidence (e.g. IPCC, 2007; Tadross et al., 2005, 2009) predict reductions in rainfall and increased rainfall variability for most parts of southern Africa. In addition, the predictions point to a higher climate variability and increased frequency and intensity of extreme weather conditions in Africa (Klein et al., 2007). The implications for Southern Africa are that the region would generally get drier and experience more extreme weather conditions, particularly droughts and floods, although there would be spatial variations within the region with some countries experiencing wetter than average climate.

Local ecosystems provide the main source of livelihood for many of the world's poor. Most of the rural poor in sub-Saharan Africa rely for their livelihood and food security on highly climate-sensitive rain-fed subsistence or small-scale farming, pastoral herding and direct harvesting of natural services of ecosystems such as forests and wetlands (IPCC, 2007; Mitchell and Tanner, 2006). The expected long-term changes in rainfall patterns and shifting temperature zones are expected to have significant negative effects on agriculture, food and water security and economic growth in Africa

(Dinar et al., 2008; Nhemachena et al., 2010; Klein et al., 2007; Kurukulasuriya et al., 2006).

Further changes in climate are unavoidable even under stringent mitigation measures over the next few decades due to high concentrations of greenhouse gasses (higher than pre-industrial levels) and high residual levels of greenhouse gasses in the atmosphere (Klein et al., 2007; Houghton et al., 1996). Mitigation efforts to reduce the sources of or to enhance the sinks of greenhouse gasses will take time. Furthermore, effective mitigation requires collaboration and commitment from many countries (Klein et al., 2007). Adaptation is therefore critical and of concern in developing countries, particularly in Africa where vulnerability is high because ability to adapt is low. Adaptation helps reduce the impacts of climate change in the short to medium term, and is motivated from local priorities or regional risks, without requiring multi-country commitments (Hassan and Nhemachena, 2008; Nhemachena, 2009). The benefits of adaptation are realised in the short term and are felt at the local community level. Adaptation measures are therefore critical in the short to medium term, while in the long run mitigation, efforts are required to reduce risks and create sinks for further greenhouse gas emissions. It is therefore imperative to help identify ways of strengthening adaptation capacity of local communities, local, and national systems to enable them to cope with climate change and variability contributing to social and economic progress of local vulnerable communities.

A better understanding of current farmer climate change adaptation measures and their determinants is key in policy planning for future successful adaptation in the agricultural sector. This paper provides highlights on current farmer adaptation options and their determinants. The study suggests that a better knowledge of the current local adaptation measures that are already being used by farmers provide better ways of building support on farmers' local adaptation measures to enhance use and adoption of adaptation measures in the agricultural sector. Supporting the coping strategies of the local farmers has potential for facilitating widespread use and adoption of adaptation measures and to have great beneficial impacts in reducing the predicted negative effects of changes in climatic conditions on agricultural production. Support for local coping strategies require a better understanding of the local practices that will be important in helping designing focussed policies aimed at enhancing adaptation to climate change in agriculture.

To our knowledge, no studies published to date investigated the determinants of farm-level adaptation options in Southern Africa. Understanding the determinants of household choice of adaptation options may provide policy insights in identifying target variables for enhancing use of adaptation measures in agriculture. The adaptation study that used the same Global Environment Facility/World Bank (GEF/WB) data did not distinguish the determinants of each of the major adaptation options available to farmers, but instead aggregated adaptation options into two options of whether a farmer adapts or not. The decision of not adapting was then used in a sample selection Heckman model to analyse the determinants of not adapting to changes in climatic conditions (Maddison, 2007). Other studies that analysed adaptation using the same data set considered single adaptation options focusing mainly on climate related factors (Kurukulasuriya and Mendelsohn, 2007a, b; 2008) studies on modelling endogenous irrigation and crop choice, respectively and Seo and Mendelsohn (2007) (a micro-economic analysis of livestock choice).

This study adds to these analyses by distinguishing household and other socio-economic factors affecting propensity of use of each of the main adaptation measures available to farmers. In addition, this study uses a different approach to examine the determinants of use of various adaptation measures. The study by Maddison (2007) used a heckman sample selection model in trying to cater for sample selection bias and used no adaptation as the dependent variable. The approach considered factors that affect the decision not to adapt to changes in climatic conditions and did not consider determinants of the multiple adaptation options being used by farmers. This study uses a multivariate probit model to examine the determinants of various adaptation measures while allowing for the correlation across error terms due to unobservable explanatory variables.

The distinguishing feature of this study is that it uses a multivariate discrete choice econometric model to simultaneously examine the relationships between each adaptation option and a common set of explanatory variables. The advantage of using this approach as opposed to univariate (single-equation) technique is that it explicitly recognises and controls for potential correlation among adaptation options and therefore provides more accurate estimates of relationships between each adaptation option and its explanatory variables. The univariate technique on the other hand is prone to biases due to common factors in situations where there are unobserved and unmeasured common factors affecting the different adaptation options.

METHODOLOGY

Analytical framework

A multivariate probit econometric technique is used to analyse the determinants of adaptation measures (relationships between identified adaptation measures and the explanatory variables). The multivariate probit model simultaneously models the influence of the set of explanatory variables on each of the different adaptation measure while allowing the unobserved and unmeasured factors (error terms) to be freely correlated (Lin et al., 2005; Green, 2003; Golob and Regan, 2002). Complementarities (positive correlation) and substitutabilities (negative correlation) between different options may be the source of the correlations between error terms (Belderbos et al., 2004). Another source of positive correlation is the existence of unobservable household-specific factors that affect choice of several adaptation options but are not easily measurable such as indigenous knowledge. The correlations are taken into account in the multivariate probit model.

Another approach would be to use a univariate technique such as probit analysis for discrete choice dependent variables to model each of the adaptation measures individually as functions of the common set of explanatory variables. The shortfall of this approach is that it is prone to biases caused by ignoring common factors that might be unobserved and unmeasured and affect the different adaptation measures. In addition, independent estimation of individual discrete choice models fails to take into account the relationships between adoptions of different adaptation measures. Farmers might consider some combinations of adaptation measures as complementary and others as competing. By neglecting the common factors the univariate technique ignores potential correlations among the unobserved disturbances in adaptation measures, and this may significantly lead to statistical bias and inefficiency in the estimates (Lin et al., 2005; Belderbos et al., 2004; Golob and Regan, 2002).

A multinomial discrete choice model is another alternative to a complicated multivariate model with seven endogenous discrete choice variables. In the multinomial discrete choice model the choice set is made up of all combinations of adaptation measures or 2^7 = 128 available alternatives. With a problem of this size (128 alternatives and 19 explanatory variables) estimating a multinomial logit (MNL) model is possible. The shortfall of this technique is that interpretation of the influence of the explanatory variables on choices of each of the seven original separate adaptation measures is very difficult. The usefulness of a MNL is limited by the property of independence of irrelevant alternatives (IIA). In such situations estimation of multinomial probit (MNP) and "mixed" or random-coefficients MNL are more appropriate and simulation methods both Bayesian and non-Bayesian can be used to estimate parameters of large MNP and mixed logit models (Golob and Regan, 2002). The shortfall of this technique is that all multinomial replications of a multivariate choice system have problems in interpreting the influence of explanatory variables on the original separate adaptation measures.

This study uses a multivariate probit econometric technique to overcome the shortfalls of using the univariate and multinomial discrete choice techniques. Following Lin et al. (2005) the multivariate probit econometric approach used for this study is characterised by a set of n binary dependent variables y_i (with observation subscripts suppressed), such that:

$$y_i = 1 \text{ if } x'\beta_i + \varepsilon_i > 0,$$

$$= 0 \text{ if } x'\beta_i + \varepsilon_i \leq 0,\ i = 1,2,\ldots,n, \quad (1)$$

where x is a vector of explanatory variables, $\beta_1, \beta_2, \ldots, \beta_n$ are comformable parameter vectors, and random error terms $\varepsilon_1, \varepsilon_2, \ldots, \varepsilon_n$ are distributed as multivariate normal distribution with zero means, unitary variance and an $n \times n$ contemporaneous correlation matrix $R = [\rho_{ij}]$, with density $\phi(\varepsilon_1, \varepsilon_2, \ldots, \varepsilon_n; R)$. The likelihood contribution for an observation is the n-variate standard normal probability.

$$\Pr(y_1,\ldots,y_n \mid x = \int_{-\infty}^{(2y_1-1)x'\beta_1} \int_{-\infty}^{(2y_2-1)x'\beta_2} \ldots \times \int^{(2y_n-1)x'\beta_n} \phi(\varepsilon_1,\varepsilon_2,\ldots,\varepsilon_n; Z'RZ)\, d\varepsilon_n \ldots d\varepsilon_2 d\varepsilon_1, \quad (2)$$

Where $Z = \text{diag}\,[2y_1 - 1, \ldots, 2y_n - 1]$. The maximum likelihood estimation maximises the sample likelihood function, which is a product of probabilities (2) across sample observations. Computation of the maximum likelihood function using multivariate normal distribution requires multidimensional integration, and a number of simulation methods have been put forward to approximate such a function with the GHK simulator (Geweke et al., 1997; Hajvassilion et al., 1996) being widely used (Belderbos et al., 2004). This study follows the GHK simulator approach that uses Stata routine due to Cappellari and Jenkins (2003) to estimate the model[1].

The marginal effects of explanatory variables on the propensity to adopt each of the different adaptation measure are calculated as:

$$\partial P_i / \partial x_i = \phi(x'\beta)\beta_i,\ i = 1,2,\ldots,n \quad (3)$$

where P_i is the probability (or likelihood) of event i (that is increased use of each adaptation measure), $\phi(\cdot)$ is the standard univariate normal cumulative density distribution function, x and β are vectors of regressors and model parameters respectively (Hassan, 1996).

Econometric analysis with cross-sectional data is usually associated with problems of heteroscedasticity and multicollinearity and the effect of outliers in the variables. Multicollinearity among explanatory variables can lead to imprecise parameter estimates. To explore potential multicollinearity among the explanatory variables, we calculated the Variance Inflation Factor (VIF) for each of the explanatory variables. The VIFs ranges from 1.07 to 1.53 which does not reach convectional thresholds of 10 or higher used in regression diagnosis (Lin et al., 2005). In the analysis, multicollinearity does not appear to be a problem. To address the possibilities of heteroscedascticity in the model, we estimated a robust model that computes a robust variance estimator based on a variable list of equation-level scores and a covariance matrix (Stata 9 help robust).

Description of data

This study used cross-sectional data obtained from the Global Environment Facility/World Bank (GEF/WB)-CEEPA funded Climate Change and African Agriculture Project: Climate, Water and Agriculture: Impacts on and Adaptations of Agro-ecological Systems in Africa. The study involved eleven African countries: Burkina Faso; Cameroon; Egypt; Ethiopia; Ghana; Kenya; Niger; Senegal; South Africa; Zambia and Zimbabwe. For the purpose of this paper, only data from the Southern African region (South Africa, Zambia and Zimbabwe) were used for empirical analyses. (For more information on the survey method and the data collected see Dinar et al. (2008) After data cleaning, a total of 1719 surveys were usable for the Southern African region. This paper used part of the large dataset for the project that included farmer perceptions on climate change, adaptation strategies being used by farmers and perceived barriers to responding to perceived climate changes.

Temperature and precipitation data came from Africa Rainfall and Temperature Evaluation System (ARTES) (World Bank, 2003). This dataset created by the National Oceanic and Atmospheric Association's Climate Prediction Center is based on ground station measurements of precipitation.

Dependent and independent variables

The dependent variables for the model were seven dummy variables: using different varieties; planting different crops; crop

[1] "Hajivassilion and Ruud (1994) proved that under regularity conditions the simulated maximum likelihood estimator is consistent when both the number of draws and observation goes to infinity. Gourieroux and Monfort (1996); show that it has the same limiting distribution as the (infeasible) maximum likelihood of the number of observations as the number of draws approaches zero", Belderbos et al. 2004).

Table 1. Main farm-level adaptation strategies in Southern Africa (% of respondents).

Adaptation	Southern Africa	South Africa	Zambia	Zimbabwe
Different varieties	11	5	13	15
Different crops	4	4	6	3
Crop diversification	9	6	9	12
Different planting dates	17	7	5	38
Diversifying from farming to non-farming activity	8	5	11	7
Increased use of irrigation / groundwater / watering	9	18	5	6
Increased use of water conservation techniques	5	6	3	7
Number of observations	1719	236	829	654

diversification; different planting dates (given the high perceptions that the timing of rains is changing); diversifying from farming to non–farming activities; increased use of irrigation and increased use of water soil conservation techniques) equal to one if the household used the adaptation option and zero otherwise. Summary statistics of the identified main adaptation measures are presented in Table 1. The explanatory variables included in the model are based on review of adoption literature studies and our view of theoretical work; however this remains rather explorative given the lack of straight forward available theoretical predictions. The independent variables in this study represent some of the many factors that affect use of adaptation options at the farm-level. Although, there might be many factors affecting farmer use of adaptation options, this study identified seventeen independent variables listed in Table 2 to be most appropriate in explaining use of different adaptation options by farming households. In the empirical model, each explanatory variable is included in all seven equations to help test if the impacts of variables differ from one adaptation option to the other.

Descriptive statistics of the explanatory variables and their expected impacts of adaptation options are presented in Table 2 and a detailed description of the variables is presented in Appendix A. Appendix B presents a correlation matrix of the independent variables. Household socio-economic characteristics like farming experience; access to free extension services, credit; mixed crop and livestock farming systems; private property and noticing climate change are expected to have significant positive impact on use of adaptation measures at the farm-level.

RESULTS AND DISCUSSION

The study estimated a multivariate probit model and for comparison a univariate probit model for each of the seven adaptation options. Results from the multivariate probit model of determinants of adaptation measures are presented in Table 3. The results of the correlation coefficients of the error terms are significant (based on t-test statistic) for any pairs of equations indicating that they are correlated. The results on correlation coefficients of the error terms indicate that there are complementarities (positive correlation) between different adaptation options being used by farmers. The results supports the assumption of interdependence between the different adaptation options which may be due to complementarity in the different adaptation options and also from omitted household-specific and other factors that affect uptake of all the adaptation options. Another important point to note from the results is that there are substantial differences in the estimated coefficients across equations that support the appropriateness of differentiating between adaptation options.

The univariate probit models can be viewed as a restrictive version of the multivariate probit model with all off-diagonal error correlations set to zeros (that is, $\rho_{ij} = 0$ for $i > j$) (Lin et al., 2005; Belderbos et al., 2004). A likelihood ratio test based on the log-likelihood values of the multivariate and univariate models indicate significant joint correlations $\chi^2(21) = 57.867$; probability > χ^2 = 0.0000 justifying estimation of the multivariate probit that considers different adaptation options as opposed to separate univariate probit models and consequently the unsuitability of aggregating them into one adaptation or no adaptation variable as was the case by Maddison (2007).

Female-headed households are more likely to take up adaptation options. The possible reason for this observation is that, in most rural smallholder farming communities in the region, much of the agricultural work are done by women. Since women do much of the agricultural work and men are based in towns, women have more farming experience and information on various management practices and how to change them based on available information on climatic conditions and other factors such as markets and food needs of the households. The important policy message from this finding is that targeting women groups and associations in smallholder rural communities can have significant positive impacts in increasing uptake of adaptation measures by smallholder farmers.

Farmer experience increases the probability of uptake of all adaptation options. Highly experienced farmers are likely to have more information and knowledge on changes in climatic conditions, crop and livestock management practices. Experienced farmers are usually leading and progressive farmers is most rural communities and these can be targeted in promoting adaptation management to other farmers who do not have such experiences and are not yet adapting to changing climatic conditions. Making use of the local successful farmers as entry points in promoting adaptation among smallholder farmers can have significant positive impacts in increasing use of various adaptation options.

Table 2. Summary statistics of independent variables and their expected impacts on adaptation measures.

Variable	Mean	Standard deviation	Minimum	Maximum	Expected impact
Female-headed household	0.82	0.38	0.00	1.00	±
Age of household head	47.41	14.61	16.00	100.00	±
Household size	5.57	2.43	1.00	22.00	±
Farming experience (years)	16.31	12.88	1.00	80.00	+
Farm size	21.16	12.54	0.04	346.00	±
Free extension services	0.64	0.48	0.00	1.00	+
Mixed crop/livestock farms	0.22	0.41	0.00	1.00	+
Household has electricity	0.14	0.33	0.00	1.00	+
Access to credit	0.15	0.36	0.00	1.00	+
Subsistence	0.43	0.49	0.00	1.00	±
Mean annual temperature	21.79	2.57	16.08	26.79	+
Mean annual precipitation	69.47	13.47	20.44	97.88	+
Noticed climate change	0.65	0.48	0.00	1.00	+
Have animal power	0.30	0.46	0.00	1.00	±
Have heavy machines	0.37	0.28	0.00	1.00	+
Have tractor	0.07	0.26	0.00	1.00	+
Income per cap	451.63	131.34	0.00	2892.34	±
Private property	0.52	0.50	0.00	1.00	+

Noticing climate change increases the probability of uptake of adaptation measures. Farmers who are aware of changes in climatic conditions have higher chances of taking adaptive measures in respond to observed changes. It is important that, this is an important precondition for farmers to take responsive measures in adapting to changes in climatic conditions (Madison, 2007). Raising awareness of changes in climatic conditions among farmers would have greater impact in increasing adaptation to changes in climatic conditions. It is therefore important for governments, meteorological departments and ministries of agriculture to raise awareness of the changes in climatic conditions through all possible alternative communication pathways that are available to farmers such as extension services, farmer groups, input and output dealers, radio and televisions among others. This need to be accompanied by the various crop and livestock management practices, farmers can take to respond to the forecasted changes in climatic conditions such as varying planting dates, using irrigation, growing crop varieties suitable to the predicted climatic conditions.

Access to free extension services significantly increases probability of taking up adaptation options except moving from faming to non-farming. Extension services provide an important source of information on climate change as well as agricultural production and management practices. Farmers who have high extension contacts have better chances to be aware of changing climatic conditions and also of the various management practices that they can use to adapt to changes in climatic conditions. Improving access to extension services for farmers has potential to significantly increase farmer awareness of changing climatic conditions as well as adaptation measures in response to climatic changes.

Farmers with access to credit and markets have high chances of adapting to changing climatic conditions. Access to cheap credit increases financial resources of the farmers and their ability to meet transaction costs associated with the various adaptation options they might want to take. With more financial and other resources at their disposal, farmers are able to change their management practices in respond to changing climatic and other factors and are better able to make use of all the available information they might have on changing conditions both climatic and other socio-economic factors. For instance, with financial resources and access to markets farmers are able to buy new crop varieties, new irrigation technologies and other important inputs they may need to change their practices to suit the forecasted and prevailing climatic conditions.

Increasing mean annual temperature increases the probability of farmers to respond to changes in terms of changing management practices. Increasing warming is associated with decreases in water resources (surface and ground), high evapo-transpiration rates and this increases water scarcity and shortages for food production and other uses. In response to increasing temperatures, farmers tend to change their crop and livestock management practices to suit the changing temperature regimes. For instance farmers need to change to growing drought resistant crops; varying planting dates, so that critical crop growth stages do not coincide with peak temperature periods; diversifying crop and non-farming income options; use water and soil conservation techniques to conserve the little rain that is received as

Table 3. Results of multivariate probit analysis of determinants of adaptation measures.

Variable	Different crops	Different varieties	Crop diversification	Different planting dates	Increase irrigation	Increase water conservation	Farming to non farming
Log farmland	0.109**	0.021***	0.004*	0.017*	0.013**	0.304*	-0.102***
Free extension services	0.071***	0.152*	0.287**	0.106**	0.338***	0.476***	-0.370**
Farming experience (yrs)	0.009*	0.014*	0.011*	0.005	0.019**	0.012*	0.011
Total household workers	0.004	0.002	0.015***	0.003*	0.004**	0.014***	0.003
Mixed crop-livestock farm	0.306**	0.185**	0.095***	0.380**	0.018***	0.163*	-0.031
Income per cap	0.001	0.000	0.003	0.001	0.007*	0.000*	0.000*
Female headed household	0.047*	0.464**	0.024*	0.071*	0.058*	0.660**	0.266
Age of household head	0.006	0.001	-0.006	0.005	0.002	0.030**	-0.009
Household has electricity	0.414**	0.157*	0.314*	0.278***	0.321***	0.558***	0.223*
Subsistence	0.230*	0.102*	0.502***	0.488***	0.115	1.362***	0.148*
Log distance to selling market	0.039*	0.562**	0.305*	0.007	0.033***	0.135*	0.764***
Access to credit	0.180*	0.218*	0.288***	0.004*	0.435***	0.254**	0.157*
Noticed climate change	0.776***	0.929***	0.289*	0.005*	0.413**	0.726***	0.508***
Mean annual temperature	0.046*	0.309***	0.175***	0.081*	0.307***	0.093***	0.181***
Mean annual precipitation	0.012*	0.044***	0.001	-0.004	-0.008*	-0.022*	-0.004**
Have tractor	0.045	0.269*	0.086**	0.575*	0.134***	0.431***	0.827*
Have heavy machines	0.092*	0.291**	0.190*	0.167*	0.624***	0.269*	0.547**
Have animal power	0.171**	0.301*	0.558***	0.033*	0.452**	0.750***	-0.154
Private property	0.005	0.058	0.107*	0.219**	0.215*	0.042*	0.354**
Zambia	0.182*	0.371	0.047	0.791	0.829***	0.735*	-0.793***
Constant	-4.345***	-12.237***	-5.435***	1.644	-1.220	-6.208***	1.576
	Rho1	**Rho2**	**Rho3**	**Rho4**	**Rho5**	**Rho6**	
Rho2	0.167*						
Rho3	0.279***	0.051**					
Rho4	0.225*	0.003***	0.163*				
Rho5	0.054	0.039*	0.016***	0.249**			
Rho6	0.202*	0.027	0.315***	0.167*	0.190**		
Rho7	-0.012	0.557***	-0.222*	-0.156	-0.389*	0.247*	
Observations				846			
Log Likelihood				-1249.7669			
Wald $\chi^2(140)$				415.96			
Prob > χ^2				0.0000			

Likelihood ratio test of rho21 = rho31 = rho41 = rho51 = rho61 = rho71 = rho32 = rho42 = rho52 = rho62 = rho72 = rho43 = rho53 = rho63 = rho73 = rho54 = rho64 = rho74 = rho65 = rho75 = rho76 = 0: $\chi^2(21)=59.521$, Prob > χ^2 = 0.0000. *; **; *** Significant at 10; 5 and 1%, respectively.

well as using irrigation technologies to supplement rainwater and increase the crop growing period.

Increasing mean annual precipitation increase, the probability of farmers changing their management practices that include: growing crop varieties that suit the prevailing and forecasted precipitation. Less precipitation increases the probability of farmer to efficiently use water resources for food production and other uses as well as irrigation water and use water conservation techniques. Use of water conservation techniques increases with decreasing precipitation probably because farmers have learnt from drought experiences to conserve rainwater in times of good rains so that it is available for future use in dry periods. Increasing knowledge and empowering communities to use water conservation techniques such as water harvesting can significantly help farmers cope with changing rainfall and temperature regimes.

Private property increases uptake of adaptation measures. Farmers with secure tenure on their farm households have high propensity to invest in adaptation options compared to where tenure is insecure. The implication of this finding is that, it is important for governments to ensure that even in the communal systems that characterise most of the smallholder farming systems in the region, tenure arrangements are secure to facilitate investments in long-term adaptation options by farmers. Secure tenure gives farmers a feeling of ownership of the land and acts as a positive incentive in facilitating farmer investments on their farms that include investments in adaptation and good crop and livestock management practices. Conservation technologies have high chances of being taken where farmers feel secure on their ownership of the land and this can be very important in promoting use of soil and water conservation techniques as important adaptation options for farmers.

Mixed crop and livestock farmers are associated with positive and significant adaptation to changes in climatic conditions compare to specialised crop and or livestock farmers. The results imply that mixed farming systems are better able to cope with changes to climatic conditions through undertaking various changes in management practices. An important reason for this observation is that mixed farming systems are already diversified and they have a number of alternative crops and livestock options that can ensure that if one option fails the other will do well even if there are changes in climatic conditions. Diversification in farming systems is therefore important for farmers to adapt to changes in climatic conditions.

Subsistence farmers are more likely to take variations in planting dates, crop diversification, and use of water conservation techniques as their adaptation options. The important reason for this is that subsistence farmers usually produce one staple food crop, maize, sorghum or millet in most cases and it is easier for them to incorporate other crops in their current options than completely changing to different crops or using expensive irrigation technologies. Promoting cheap adaptation options among smallholder, farmers can positively and significantly increase subsistence farmers' adaptation to climate change.

Households with access to electricity, tractors, heavy machines and animal power are usually mechanised and have better chances of taking up other adaptation options. With access to technology farmers are able to vary their planting dates, switch to new crops, diversify their crop options and use more irrigation and water conservation techniques as well as diversifying into non-farming activities. Farmers with better technologies usually have access to market and they produce for sale which gives them better chances to change their management practices in respond to changing climatic and other conditions such as prices and market chances. Ensuring availability of cheap technologies for smallholder farmers can significantly increase their use of other adaptation options.

Country fixed effects were also included and the results for Zambia are shown. Including either South Africa or Zimbabwe resulted in each being dropped due to multicollinearity. The country effects from Zambia have significant effect on adaptation indicating the importance of national policies concerning adaptation to climate change.

Conclusion

This study was based on micro-level analysis of adaptation that focuses on tactical decisions farmers make in response to seasonal variations in climatic, economic and other factors. These tactical decisions are influenced by a number of socio-economic factors that include household characteristics, household resource endowments, access to information (seasonal and long-term climate changes and agricultural production) and availability of formal institutions (input and output markets) for smoothening consumption. Farm-level decision making occurs over a very short time period usually influenced by seasonal climatic variations, local agricultural cycle, and other socio-economic factors. Adaptation is important for farmers to achieve their farming objectives such as food and livelihood security and high incomes and significantly reduce potential negative impacts that are associated with changes in climatic and other socio-economic conditions (that include: climate variability, extreme weather conditions, volatile short-term changes in local and large scale markets).

This paper explored the determinants of household use of seven different adaptation measures (using different varieties, planting different crops, crop diversification, different planting dates (given the high perceptions that the timing of rains is changing), diversifying from farming to non–farming activities, increased use of irrigation, increased use of water and soil conservation techniques) using a multivariate probit model. The model allowed us to simultaneously model the determinants of all seven adaptation options, thus limiting potential problem of correlation between the error terms. The model help reflect that households simultaneously consider decisions to use various adaptation options. Correlation results between error terms of different equations were significant (positive) indicating that various adaptation options tend to be used by households as complementary, although this could also be due to unobserved household socio-economic and other factors.

Multivariate probit results confirm that access to credit, free extension services, farming experience, mixed crop and livestock farms, private property and perception of climate change are some of the important determinants of farm-level adaptation options. Use of different adaptation measures significantly increase for farming household with more access to these factors among others. Designing policies that aim to improve these factors for smallholder farming systems have great potential to improve farmer adaptation to changes in climate as a way of ensuring food and livelihoods and income objectives of the farmers are achieved among other goals. For example, more access to credit facilities, information (climatic and agronomic) as well as access to markets (input and output) can significantly increase farm-level adaptation. Government policies need to support research and development that prepares the appropriate technologies to help farmers adapt to changes in climatic conditions. Government responsibilities are usually through conscious policy measures to enhance the adaptive capacity of agricultural systems. Examples of these policy measures include crop development, improving climate information forecasting, or promoting and even subsidizing certain farm-level adaptations such as use of irrigation technologies.

Conflict of Interests

The author(s) have not declared any conflict of interests.

ACKNOWLEDGEMENTS

This paper was funded by GEF and the World Bank. It is part of a larger study on the effect of climate change on African agriculture coordinated by the Centre for Environmental Economics and Policy in Africa (CEEPA), Pretoria, South Africa. The authors are very grateful for review comments from Siwa Msangi, Kato, Mahmnud and Claudia Ringler. All the errors are solely for the authors.

REFERENCES

Belderbos R, Carree M, Diederen B, Lokshin B, Veugelers R (2004). Heterogeneity in R&D cooperation strategies. Int. J. Indust. Organ. 22:1237-1263.

Cappellari L, Jenkins SP (2003). Multivariate probit regression using simulated maximum likelihood. Stata J. 3:278-294.

Dinar A, Mendelsohn R, Hassan R, Benhin J (2008). Climate Change and Agriculture in Africa: Impacts Assessment and Adaptation Strategies. Earthscan. London.

Golob TF, Regan AC (2002). Trucking industry adoption of information technology: A multivariate discrete choice model. Transportation Research Part C, 2005, 10:228.

Green WH (2003). Econometric Analysis. 5th Edition, Prentice Hall, New Jersey.

Hajivassiliou V, McFadden D, Ruud P (1996). Simulation of multivariate normal rectangle probabilities: methods and programs. J. Econ. 72: 85-134.

Hassan RM (1996). Planting strategies of maize farmers in Kenya: a simultaneous equations analysis in the presence of discrete dependent variables. Agric. Econ. 15: 137-149.

Hassan R, Nhemachena C (2008). Adaptation to climate change in Africa: multinomial choice analysis of determinants of farm strategies, African J. Agric. Resour. Econ. 2(1):83-104.

Houghton J, Meira FL, Callander B, Harris N, Kattenberg A, Maskell K (Eds.) (1996). Climate change 1995: The science of climate change. Cambridge University Press, Cambridge.

Hulme M, Doherty R, Ngara T, New M (2005). Global warming and African climate change: A re-assessment. Climate Change in Africa. (ed Low, P.S), Cambridge University Press.

IPCC (Intergovernmental Panel on Climate Change) (2007). Fourth Assessment Report. Climate change 2007: synthesis report. Summary for policymakers. Geneva, Intergovernmental Panel on Climate Change. Cambridge University Press, Cambridge UK.

Klein RJT, Huq S, Denton F, Downing TE, Richels RG, Robinson JB, Toth FL (2007). Inter-relationships between adaptation and mitigation. Climate Change 2007: Impacts, adaptation and vulnerability. Contribution of Working Group II to the Fourth Assessment Report of the Intergovernmental Panel on Climate Change. In Parry ML, Canziani OF, Palutikof JP, van der Linden PJ, Hanson CE (Eds.), Cambridge: Cambridge University Press.

Kurukulasuriya P, Mendelsohn R (2007a). Endogenous irrigation: the impact of climate change on farmers in Africa. Policy Research Working Paper 4278. Development Research Group, Sustainable Rural and Urban Development Team. The World Bank, 1818 H Street NW, Washington, DC 20433.

Kurukulasuriya P, Mendelsohn R (2007b). Crop selection: adapting to climate change in Africa. Policy Research Working Paper 4307. Development Research Group, Sustainable Rural and Urban Development Team. The World Bank, 1818 H Street NW, Washington, DC 20433.

Kurukulasuriya P, Mendelsohn R (2008). Crop Switching as an Adaptation Strategy to Climate Change. Afr. J. Agric. Resour. Econ. 2(1):105-126.

Kurukulasuriya P, Mendelsohn R, Hassan R, Benhin J, Deressa T, Diop M, Eid HM, Fosu KY, Gbetibouo G, Jain S, Mahamadou A, Mano R, Kabubo-Mariara J, El-Marsafawy S, Moula E, Ouda S, Ouedraogo M, Séne I, Maddison D, Seo SN, Dinar A (2006). Will African agriculture survive climate change? The World Bank Econ. Rev. 20(3):367-388.

Lin CTJ, Jensen KL, Yen ST (2005). Awareness of foodborne pathogens among US consumers. Food Qual. Prefer. 16:401-412.

Maddison D (2007). The perception of and adaptation to climate change in Africa. Policy Research Working Paper 4308. Development Research Group, Sustainable Rural and Urban Development Team. The World Bank, 1818 H Street NW, Washington, DC P. 20433.

Mitchell T, Tanner T (2006). Adapting to climate change-challenges and opportunities for the development community. Institute of Development Studies and Tearfund, Teddington, UK.

Nhemachena C, Hassan R, Kurukulasuriya P (2010). Measuring the economic impact of climate change on African agricultural production systems. Clim. Change Econ. 1(1):33-35.

Nhemachena C (2009). Agriculture and Future Climate Dynamics in Africa: Impacts and Adaptation Options. PhD Thesis. Department of Agricultural Economics, Extension and Rural Development. University of Pretoria.

Seo N, Mendelsohn R (2007). Climate change adaptation in Africa: a microeconomic analysis of livestock choice. Policy Research Working

Paper 4277. Development Research Group, Sustainable Rural and Urban Development Team. The World Bank, 1818 H Street NW, Washington, DC P. 20433.

Tadross MA, Hewiston BC, Usman MT (2005). The interannual variability of the onset of the maize growing season over South Africa and Zimbabwe. J. Clim. 18:3356-3372.

Tadross MA,Pablo S, Alex L, Sepo H (2009). Growing-season rainfall and scenarios of future change in southeast Africa: implications for cultivating maize. Clim. Res. 40:147-161.

World Bank (2003). Africa Rainfall and Temperature Evaluation System. World Bank, Washington DC.

Appendix A. Description of independent variables.

Variable	Description of variable
Female-headed household	Dummy variable for household gender (female-headed households)
Age of household head	Age of household head
Household size	Size of the household
Farming experience (years)	Farming experience based on number of farming years
Farm size	Total household farm size
Free extension services	Dummy variable for households with access to free extension services (climate, crop and livestock farming)
Mixed crop/livestock farms	Dummy variable for mixed farming households with both crop and livestock
Household has electricity	Dummy variable for farming households with electricity
Access to credit	Dummy variable for households who have access to credit
Subsistence	Subsistence farming households
Mean annual temperature	Mean annual temperature
Mean annual precipitation	Mean annual precipitation
Noticed climate change	Dummy variable for households who noticed changes in climatic variables
Have animal power	Dummy variable for household possession of animal power
Have tractor	Dummy variable for household possession of a tractor
Have heavy machines	Dummy variable for household possession of heavy machines
Income per cap	Household income per capita
Private property	Dummy variable for tenure system

Appendix B. Correlation matrix of the independent variables.

Variable	Female-headed household	Age of household head	Household size	Farming experience	Farm size	Free extension services	Mixed crop/livestock farms	Irrigation	Access to credit	Subsistence	Mean annual temperature	Mean annual precipitation	Noticed climate change	Have animal power	Number of crops	Head non farm	Private property
Female-headed household	1.0000																
Age of household head	-0.0457	1.0000															
Household size	0.1345	0.1156	1.0000														
Farming experience	-0.0749	0.4996	0.0793	1.0000													
Farm size	0.0426	-0.0344	-0.0709	0.0434	1.0000												
Free extension services	0.0583	0.0739	0.0763	0.0153	-0.0066	1.0000											
Mixed crop/livestock	-0.0805	-0.0178	-0.2062	-0.0941	-0.0265	-0.0191	1.0000										
Irrigation	0.0701	0.0662	-0.0620	-0.0342	0.1528	0.2579	0.1584	1.0000									
Access to credit	0.0320	-0.0118	0.0474	-0.0581	-0.0184	0.1375	-0.0273	0.1457	1.0000								
Subsistence	-0.1166	0.0315	-0.0238	0.0624	-0.0630	-0.1486	0.0583	-0.2290	-0.1246	1.0000							
Mean annual temperature	-0.0471	-0.0620	0.0328	0.0714	-0.1535	-0.0873	-0.0607	-0.4143	-0.0463	0.3081	1.0000						
Mean annual precipitation	-0.0002	-0.0263	0.0657	0.0008	-0.1835	-0.1393	-0.1003	-0.2375	-0.0217	-0.0897	-0.0752	1.0000					
Noticed climate change	-0.0046	0.0087	0.0097	-0.0124	0.0071	0.1626	-0.0163	0.0520	0.0475	0.0699	-0.0264	-0.1723	1.0000				
Have animal power	0.0633	0.0294	0.1230	0.1118	0.0711	0.1718	-0.2015	0.0176	0.0793	-0.0600	0.0565	0.0038	0.0850	1.0000			
Number of crops	-0.0207	0.0791	0.1775	0.0632	-0.0436	0.1654	-0.0890	0.2255	0.2424	-0.2990	-0.0677	0.1268	0.0380	0.1344	1.0000		
Head non farm	0.0375	-0.2261	0.0571	-0.1731	-0.0283	-0.1094	0.0502	0.0171	-0.0171	-0.0091	-0.1582	0.1185	-0.0193	-0.0455	0.0324	1.0000	
Private property	0.0343	0.1346	-0.0091	0.0298	0.0559	0.0865	0.0141	0.0035	-0.0010	0.0246	0.1672	-0.2357	0.0317	-0.0032	-0.0470	-0.2281	1.0000

Permissions

All chapters in this book were first published in JDAE, by Academic Journals; hereby published with permission under the Creative Commons Attribution License or equivalent. Every chapter published in this book has been scrutinized by our experts. Their significance has been extensively debated. The topics covered herein carry significant findings which will fuel the growth of the discipline. They may even be implemented as practical applications or may be referred to as a beginning point for another development.

The contributors of this book come from diverse backgrounds, making this book a truly international effort. This book will bring forth new frontiers with its revolutionizing research information and detailed analysis of the nascent developments around the world.

We would like to thank all the contributing authors for lending their expertise to make the book truly unique. They have played a crucial role in the development of this book. Without their invaluable contributions this book wouldn't have been possible. They have made vital efforts to compile up to date information on the varied aspects of this subject to make this book a valuable addition to the collection of many professionals and students.

This book was conceptualized with the vision of imparting up-to-date information and advanced data in this field. To ensure the same, a matchless editorial board was set up. Every individual on the board went through rigorous rounds of assessment to prove their worth. After which they invested a large part of their time researching and compiling the most relevant data for our readers.

The editorial board has been involved in producing this book since its inception. They have spent rigorous hours researching and exploring the diverse topics which have resulted in the successful publishing of this book. They have passed on their knowledge of decades through this book. To expedite this challenging task, the publisher supported the team at every step. A small team of assistant editors was also appointed to further simplify the editing procedure and attain best results for the readers.

Apart from the editorial board, the designing team has also invested a significant amount of their time in understanding the subject and creating the most relevant covers. They scrutinized every image to scout for the most suitable representation of the subject and create an appropriate cover for the book.

The publishing team has been an ardent support to the editorial, designing and production team. Their endless efforts to recruit the best for this project, has resulted in the accomplishment of this book. They are a veteran in the field of academics and their pool of knowledge is as vast as their experience in printing. Their expertise and guidance has proved useful at every step. Their uncompromising quality standards have made this book an exceptional effort. Their encouragement from time to time has been an inspiration for everyone.

The publisher and the editorial board hope that this book will prove to be a valuable piece of knowledge for researchers, students, practitioners and scholars across the globe.

List of Contributors

S. T. Olowogbon
Occupational Health and Safety Division of Vertext Media Ltd., Ibadan, Oyo State, Nigeria

S. B. Fakayode
Department of Agricultural Economics, University of Ilorin, Ilorin, Kwara State, Nigeria

A. J. Jolaiya
Department of Agricultural Economics, University of Ilorin, Ilorin, Kwara State, Nigeria

A. Y. Adenrele
Occupational Health and Safety Division of Vertext Media Ltd., Ibadan, Oyo State, Nigeria

B. O. Asante
Crops Research Institute, Box 3785, Kumasi, Ghana

M. K. Osei
Crops Research Institute, Box 3785, Kumasi, Ghana

A. A. Dankyi
Crops Research Institute, Box 3785, Kumasi, Ghana

J. N. Berchie
Crops Research Institute, Box 3785, Kumasi, Ghana

M. B. Mochiah
Crops Research Institute, Box 3785, Kumasi, Ghana

J. N. L. Lamptey
Crops Research Institute, Box 3785, Kumasi, Ghana

J. Haleegoah
Crops Research Institute, Box 3785, Kumasi, Ghana

K. Osei
Crops Research Institute, Box 3785, Kumasi, Ghana

G. Bolfrey-Arku
Crops Research Institute, Box 3785, Kumasi, Ghana

Shiferaw T. Feleke
Agricultural and Resource Economics, University of Tennessee, 2621 Morgan Circle Drive, Knoxville, TN 37996, United States

Jane H. Starnes
Agricultural and Resource Economics, University of Tennessee, 2621 Morgan Circle Drive, Knoxville, TN 37996, United States

Harwood D. Schaffer
Agricultural and Resource Economics, University of Tennessee, 2621 Morgan Circle Drive, Knoxville, TN 37996, United States

Kelly J. Tiller
Agricultural and Resource Economics, University of Tennessee, 2621 Morgan Circle Drive, Knoxville, TN 37996, United States

L. Musemwa
Department of Agricultural Economics and Extension, University Of Fort Hare, P. Bag X1314, Alice 5700, RSA

L. Zhou
Risks and Vulnerability Centre, University of Fort Hare, P. Bag X1314, Alice 5700, RSA

S. Ndhleve
Department of Agricultural Economics and Extension, University Of Fort Hare, P. Bag X1314, Alice 5700, RSA

F. Aghdasi
Faculty of Science and Agriculture, University of Fort Hare, P. Bag X1314, Alice 5700, RSA

S. N. Mlote
Ministry of Livestock and Fisheries Development, P. O. Box 9152, Dar es Salaam, Tanzania

N. S. Y. Mdoe
Department of Agricultural Economics and Agribusiness, Sokoine University of Agriculture, P. O. Box 3007, Morogoro, Tanzania

A. C. Isinika
Institute of Continuing Education, Sokoine University of Agriculture, P. O. Box 3044, Morogoro, Tanzania

L. A. Mtenga
Department of Animal Science and Production, Sokoine University of Agriculture, P. O. Box 3004, Morogoro, Tanzania

Price Edwin
578 John Kimbrough Blvd, AGSV Building, Room 240, College Station, TX, USA, 77843

Kibriya Shahriar
578 John Kimbrough Blvd, AGSV Building, Room 240, College Station, TX, USA, 77843

Zhang Yu
578 John Kimbrough Blvd, AGSV Building, Room 240, College Station, TX, USA, 77843

Ramatu M. Al-Hassan
Department of Agricultural Economics and Agribusiness, University of Ghana, Legon, Ghana

Irene S. Egyir
Department of Agricultural Economics and Agribusiness, University of Ghana, Legon, Ghana

James Abakah
Department of Agricultural Economics and Agribusiness, University of Ghana, Legon, Ghana

Carel Ligeon
Auburn University, Montgomery, Alabama, USA

Curtis Jolly
Alabama Agricultural Experiment Station, Auburn University, Alabama, USA

Nelly Bencheva
Agricultural University in Plovdiv, Bulgaria

Stanko Delikostadinov
Institute of Plant Genetic Resources in Sadovo, Bulgaria

Naveen Puppala
New Mexico State University, Las Cruces, New Mexico

O. D. Adisa
Department of Agricultural Economics, Obafemi Awolowo University, Ile-Ife, Osun State, Nigeria

N. A. Sofoluwe
Department of Agricultural Economics, Obafemi Awolowo University, Ile-Ife, Osun State, Nigeria

C. M. Tasie
Department of Agricultural Science, Ignatius Ajuru University of Education, Rumuolumeni, Port-Harcourt, Rivers state, Nigeria

Amare Demissie
International Committee for the Development of Peoples (CISP), P. O. Box 888, Harar, Ethiopia

Belaineh Legesse
Agricultural Economics, Haramaya University, P. O. Box 138, Haramaya, Ethiopia

Prince Maxwell Etwire
CSIR-Savanna Agricultural Research Institute (CSIR-SARI), P. O. Box TL 52, Tamale, Ghana

Tahirou Abdoulaye
International Institute of Tropical Agriculture (IITA), Ibadan, Nigeria

Kwadwo Obeng-Antwi
CSIR-Crops Research Institute (CSIR-CRI), Kumasi, Ghana

Samuel S. J. Buah
CSIR-Savanna Agricultural Research Institute (CSIR-SARI), P. O. Box TL 52, Tamale, Ghana

Roger A. L. Kanton
CSIR-Savanna Agricultural Research Institute (CSIR-SARI), P. O. Box TL 52, Tamale, Ghana

Henry Asumadu
CSIR-Crops Research Institute (CSIR-CRI), Kumasi, Ghana

Mashark S. Abdulai
CSIR-Savanna Agricultural Research Institute (CSIR-SARI), P. O. Box TL 52, Tamale, Ghana

Alidu Haruna
CSIR-Savanna Agricultural Research Institute (CSIR-SARI), P. O. Box TL 52, Tamale, Ghana

John C. Etwire
Faculty of Mathematical Sciences, University for Development Studies, Navrongo, Ghana

J. Bonabana-Wabbi
Department of Agribusiness and Natural Resource Economics, Makerere University, P. O. Box 7062 Kampala, Uganda

S. Ayo
Department of Agribusiness and Natural Resource Economics, Makerere University, P. O. Box 7062 Kampala, Uganda

B. Mugonola
Department of Agribusiness and Rural Development, Gulu University. P. O. Box 166, Gulu, Uganda

D. B. Taylor
Department of Agricultural and Applied Economics, VirginiaTech, Blacksburg, VA, 24060, USA

J. Kirinya
Department of Agribusiness and Natural Resource Economics, Makerere University, P. O. Box 7062 Kampala, Uganda

M. Tenywa
Department of Agricultural Production, Makerere University, P.O. Box 7062 Kampala, Uganda

Mario Miguel Amin Garcia Herreros
Centro de Ciências Naturais e Tecnologia, Universidade do Estado do Pará, Belém, Brazil

Heriberto Wagner Amanajás Pena
Centro de Ciências Naturais e Tecnologia, Universidade do Estado do Pará, Belém, Brazil

Allan Klynger da Silva Lobato
Núcleo de Pesquisa Vegetal Básica e Aplicada, Universidade Federal Rural da Amazônia, Paragominas, Brazil

Raimundo Thiago Lima da Silva
Centro de Ciências Rurais, Universidade Federal de Santa Maria, Santa Maria, Brazil

Sunday B. Akpan
Department of Agricultural Economics and Resources Management, Akwa Ibom State University, Ikot Akpaden, Mkpat Enin, Akwa Ibom State, Nigeria

Inimfon V. Patrick
Department of Agricultural Economics and Resources Management, Akwa Ibom State University, Ikot Akpaden, Mkpat Enin, Akwa Ibom State, Nigeria

Daniel E. John
Department of Agricultural Economics and Extension, University of Uyo, Akwa Ibom State, Nigeria

Samuel J. Udoka
Department of Agricultural Economics and Resources Management, Akwa Ibom State University, Ikot Akpaden, Mkpat Enin, Akwa Ibom State, Nigeria

Mahdi Babaei
Young Researchers Club, Neyriz Branch, Islamic Azad University, Neyriz, Iran

Elahe Asemani
Department of Agriculture Economic, University of Zabol, Iran

Saeid Gazmeh
Department of Computer and Information Technology (IT), Science and Research Branch, Islamic Azad University, Fars, Iran

Soheyla Rezazadeh Shamkhal

Ebrahim Shirmohammadi

Sayed CHEHAIBI
High Institute of Agronomy, Chott Meriem 4042, Sousse, Tunisia

Wissem HAMDI
High Institute of Agronomy, Chott Meriem 4042, Sousse, Tunisia

Khaoula ABROUGUI
High Institute of Agronomy, Chott Meriem 4042, Sousse, Tunisia

Rohit Singla
Department of Agricultural Economics, McGill University, Montreal, QC, Canada

Phillip Johnson
Department of Agricultural and Applied Economics, Texas Tech University, Lubbock, TX, U.S.A

Sukant Misra
Department of Agricultural and Applied Economics, Texas Tech University, Lubbock, TX, U.S.A

Tewodros Tefera
School of Environment, Gender and Deveopment Studies, Hawassa University, Hawassa, Ethiopia

D. V. Subaro
Acharya RNGA Agricultural Universities, Hyderabad, India

James Masvongo
Department of Agricultural Economics and Extension, Faculty of Agriculture, University of Zimbabwe, Box MP 167, Mount Pleasant, Harare, Zimbabwe

Jackqeline Mutambara
Department of Agricultural Economics and Extension, Faculty of Agriculture, University of Zimbabwe, Box MP 167, Mount Pleasant, Harare, Zimbabwe

Augustine Zvinavashe
Department of Agricultural Economics and Extension, Faculty of Agriculture, University of Zimbabwe, Box MP 167, Mount Pleasant, Harare, Zimbabwe

Adugna Eneyew
Department of Agricultural Economics and Extension, Jimma University, P. O. Box 307, Ethiopia

Ermias Alemu
Department of Water Resource and Irrigation Engineering, Arba Minch University, Ethiopia

Mekonnen Ayana
Department of Water Resource and Irrigation Engineering, Arba Minch University, Ethiopia

Mihret Dananto
School of Bio-systems and Environmental Engineering, Hawasa University, Ethiopia

J. Ekou
Department of Animal Production and Management, Faculty of Agriculture and Animal Sciences, Busitema University, P. O. Box 236, Tororo, Uganda

Ibitoye, Stephen Jimoh
Department of Agricultural Economics and Extension, Kogi State University, Anyigba, Nigeria

J. A. Onimisi
Department of Integrated Science, Kogi State College of Education, Ankpa

M. Demeke
Agricultural Development Economics Division, Food and Agriculture Organization of United Nations, Rome 00153, Italy

F. Di Marcantonio
Agricultural Development Economics Division, Food and Agriculture Organization of United Nations, Rome 00153, Italy

C. Morales-Opazo
Agricultural Development Economics Division, Food and Agriculture Organization of United Nations, Rome 00153, Italy
European Commission, Joint Research Centre, Institute for Prospective Technological Studies, Edificio Expo, c/ Inca Garcilaso 3, 41092 Seville, Spain

Akalu Teshome
Amahar Agricultural Research Institute, Ethiopia P. O. Box 527, Bahir Dar, Ethiopia

F. O. Idumah
Forestry Research Institute of Nigeria, Moist Forest Research Station, Benin City, Edo State. Nigeria

F. Y. Okunmadewa
World Bank, Abuja, Nigeria

Benjamin K. Mutai
Department of Agricultural Economics/Agribusiness Management at Egerton University, Njoro Campus, Kenya

Elvi N. Agunda
Department of Agricultural Economics/Agribusiness Management, Egerton University, Kenya

Augustus S. Muluvi
Department of Economics, Egerton University, Kenya

Lawrence K. Kibet
Department of Economics, Egerton University, Kenya

Mary C. Maina
Department of Agricultural Economics/Agribusiness Management at Egerton University, Njoro Campus, Kenya

Jacob Afouda Yabi
Department of Agricultural Economics, Faculty of Agronomy, University of Parakou, BP: 123 Parakou, Republic of Benin

Patrice Igue Adegbola
Programme for Agricultural Policy Analysis (PAPA) of the National Institute of Agricultural Researches of Benin (NIARB), Porto Novo, Republic of Benin

Dansi Silvere Tovignan
Department of Agricultural Economics, Faculty of Agronomy, University of Parakou, BP: 123 Parakou, Republic of Benin

Morenike Cendrine Ahouandjinou
Department of Agricultural Economics, Faculty of Agronomy, University of Parakou, BP: 123 Parakou, Republic of Benin

Adeyemi Suleiman Adekambi
Programme for Agricultural Policy Analysis (PAPA) of the National Institute of Agricultural Researches of Benin (NIARB), Porto Novo, Republic of Benin

Admire Jongwe
Economics and Markets Department, Ministry of Agriculture, Mechanization and Irrigation Development, Zimbabwe

Joel Chongela
Department of Agricultural Economics ANGRAU†, †ANGRAU - Acharya N. G. Ranga Agricultural University, Hyderabad 500-030, Andhra Pradesh, India

Vasudev Nandala
Department of Agricultural Economics ANGRAU†, †ANGRAU - Acharya N. G. Ranga Agricultural University, Hyderabad 500-030, Andhra Pradesh, India

Suhasini Korabandi
Department of Agricultural Economics ANGRAU†, †ANGRAU - Acharya N. G. Ranga Agricultural University, Hyderabad 500-030, Andhra Pradesh, India

Charles Nhemachena
Human Sciences Research Council, 134 Pretorius Street, Pretoria 0002, South Africa

Rashid Hassan
Centre for Environmental Economics and Policy in Africa, University of Pretoria, Department of Agricultural Economics and Rural Development, Pretoria 0002, South Africa

James Chakwizira
University of Venda, School of Environmental Sciences, P/Bag X5050, Thohoyandou, 0950, South Africa